Introduction to Linear Algebra

Introduction to Linear Algebra

ROBERT F. V. ANDERSON
University of British Columbia

Holt, Rinehart and Winston
*New York • Chicago • San Francisco
Philadelphia • Montreal • Toronto
London • Sydney • Tokyo • Mexico City
Rio de Janeiro • Madrid*

CANADIAN CATALOGUING IN PUBLICATION DATA

Anderson, Robert F. V.
  Introduction to Linear Algebra

Includes index.
ISBN 0-03-921835-X.

1. Algebras, Linear. I. Title.

QA184.A52   1986   512′.5   C85-098447-5

PUBLISHER: Ron Munro
ACQUISITIONS EDITOR: Larry Hopperton
MANAGING EDITOR: Mary Lynn Mulroney
COPY EDITOR: David Friend
TYPESETTING AND ASSEMBLY: Compeer Typographic Services
DESIGN: Brant Cowie/Artplus Ltd.

Printed and bound in Canada by John Deyell Company

1 2 3 4 5 90 89 88 87 86

This book is dedicated to my wife, Joan.

Acknowledgements

The author is grateful to the several reviewers whose careful reading and perceptive comments resulted in many improvements to the drafts of this book. Their names and institutional affiliations are as follows:

Allan Candiotti — *Drew University*
Frank DeMeyer — *Colorado State*
Robert DeVaney — *Boston University*
Michael Doob — *University of Manitoba*
William R. Fuller — *Purdue University*
Ray Geremai — *Goncher College*
Martin Gutterman — *Tufts University*
Richard Koch — *University of Oregon*
Stanley O. Kochman — *University of Western Ontario*
Roger Marty — *Cleveland State*
Joseph Repka — *University of Toronto*
Norman Rice — *Queen's University*

The development of this book has also benefited from extensive class testing over the past five semesters at the University of British Columbia. Early drafts were used as texts by quite a number of the author's colleagues in their sections of our linear algebra course, and their many constructive comments are greatly appreciated. Particular thanks are due to Charles Lamb and Rimhak Ree for their substantial critical contributions to the revision process. Of course none of these people are in any way responsible for any errors and shortcomings of the final product.

I also want to thank our former department chairman at UBC, Benjamin Moyls, for encouraging me to embark on the writing of this book.

I've enjoyed the constant encouragement of the staff at Holt, Rinehart and Winston, initially in the person of Ron Munro, and later, during the detailed development process, by Larry Hopperton. The production process has been expertly managed by Mary Lynn Mulroney.

Typing of the first drafts of several chapters of the manuscript was facilitated, often under conditions of time pressure, by the Mathematics Department office at UBC. I'd like to thank Elizabeth Orne and Mary-Margaret Daisley for organizing this help.

Finally, I would like to express my appreciation to my wife, Joan, for her patience, encouragement, and support during the long months of intensive writing.

Introduction

Many important problems in mathematics and its applications involve *several unknown quantities*, or variables. These may be the coordinates of a point in space, the populations of a few species sharing an ecosystem, or the quantities of a hundred different products shipped from a manufacturing plant.

Naturally enough, such problems can be hard to solve; but for a large and important subclass of them known as "linear problems," effective techniques have been developed. These are the subject of this book.

The simplest linear problems are just sets of equations, such as you have solved in elementary algebra. We begin by presenting an orderly method for solving all problems of this kind. But then we adopt a tactic typical of mathematics: we *improve our notation*. Instead of writing out long strings of variables, we combine them into streamlined objects called "vectors," and we learn to manipulate them by using another new category of mathematical objects called "matrices."

The step-by-step development of the properties of vectors and matrices is easy and straightforward; yet it rapidly snowballs into a powerful mathematical instrument known as "matrix algebra."

We find that matrix algebra can be used to study geometric operations such as rotations and reflections; to classify surfaces in space; to analyse the stability of ecosystems; or to predict the future development of markets — and recall their past histories.

At the same time, matrix algebra, viewed abstractly, is one of the most beautiful and tidy of mathematical theories. It is nowadays essential to most branches of pure mathematics.

Finally, linear algebra is the foundation for many other important theories. In particular, it is the starting point for "linear programming," which is introduced in the last chapter of this book, and is used to solve such modern problems as the maximization of flows through pipeline networks, and the optimal scheduling of airlines.

The author hopes that this book will help you to enjoy smooth sailing through the varied and fascinating world of linear algebra.

Preface

Linear algebra has been a mainstay of the undergraduate curriculum for only a few decades. Improvements in its mathematical methodology are still accumulating, and the range of its applications continues to expand. The purpose of this book is to present a *state-of-the-art* introduction to the subject, with only high school algebra as prerequisite, which will be of maximum benefit to a broad cross-section of students.

A first step towards this goal is to develop the techniques of matrix algebra as quickly and efficiently as possible. The vector space R^n, and the concept of *linear transformation*, are introduced early. Then the active interpretation of matrices as implementing linear transformations can be emphasized, and the matrix product defined as a composite transformation. This strategy greatly facilitates the rapid development of the matrix algebra itself, its geometric interpretation, and its application to the modelling of composite and iterative processes.

Efficiency is also enhanced by the systematic use of *elementary matrices*, starting with the construction of the inverse matrix and the determinant, and continuing right through to the chapters on numerical methods and linear programming at the end.

A consensus is developing that thorough geometric interpretation of matrix algebra is an important aid to the understanding of the theory. For example, such familiar transformations of the plane as rotations and reflections can be used to illustrate the non-commutative character of matrix algebra. I have found that it makes good sense, in general, to concentrate on *two dimensional interpretations*, which our students can visualize, and work with, relatively well.

Coordinate vectors and *position vectors* are the only kinds of vectors needed in this book as a foundation for the core algebraic material and its geometric interpretation. This feature is designed to help those instructors who have to present a comprehensive account of the subject in a short course (as happens only too often with our crowded modern curricula!). Others, with a little more time at their disposal, may wish to cover the optional section 2.4. This provides an introduction to Euclidean vectors, and their use in the study of lines and planes in space — surely one of the most interesting and immediate applications of the theory of linear systems.

With microcomputers doing the number crunching nowadays, our concern in teaching linear algebra is, more than ever, explication of the concepts and encouragement of model-building and example-constructing skills.

A feature of this book is the abundance of *numerically simple, yet algebraically typical, examples* which provide the raw material for the development of such expertise. The examples are distributed throughout the narrative so as to complement each incremental advance in the theory. The exercise sections contain a wide variety of material carefully graded in difficulty, and will, I hope, further the same objectives. A special, worked, "Problem" at the end of each section of the narrative eases the transition to the exercises which follow.

The design of this book is such that *it can be read at more than one level*. The examples and exercises, and the low-keyed narrative, will support a rather informal approach; but the mathematical theory is complete. Full proofs are given for all important theorems, including some usually skipped or glossed over, such as the Principal Axes Theorem. This feature, in conjunction with the ample supply of more challenging and starred problems late in the exercise sections, makes the book perfectly usable in a math majors course.

In order to demonstrate the practical value of linear algebra to a wide audience of students from varied backgrounds, I have tried to select *plausible, everyday applications* which all can understand. They are intended to be sufficient to illustrate fully the stages of development of the subject, but not so numerous as to clutter the book and obscure its underlying structure. Applications requiring technical expertise in a special field have been avoided.

Here is a chapter-by-chapter summary of some of the features of this book:

In *Chapter 1*, row operations on matrices are used to solve linear systems. A "crossing line" with "corners" is drawn across a matrix to help describe the steps by which it is reduced. Section 1.4 is devoted to applications.

In *Chapter 2*, coordinate vectors are introduced and linear systems rewritten in matrix-vector notation. The transition to the "active" interpretation of matrices is made with the introduction of linear transformations in section 2.3. The last part of this section, on the concepts of kernel and range, can be postponed till later. As mentioned earlier, section 2.4, on lines and planes, is optional.

In *Chapter 3*, transformation methodology and elementary matrices allow a rapid development of matrix algebra. Iterative processes are discussed in section 3.2, and the inverse matrix is used to track these backwards in time in section 3.4.

Chapter 4 surveys geometric transformations, mainly in the plane, and their matrix representations. A selection from this material will suffice for most courses.

In *Chapter 5*, the method of elementary matrices makes possible a complete theory of the determinant which, it is hoped, is relatively neat and concise. Alternatively, emphasis can be placed on the practical methods of computing determinants which are developed alongside. Section 5.3 sketches the geometric interpretation of the determinant.

In *Chapter 6*, special attention is given to the intuitive justification of spanning sets and bases as a means of getting a grip on the important vector spaces which float mysteriously within R^n. Efficient practical methods for checking linear independence, and computing coordinates, are also featured. In my classes, I try to describe fully the important spaces Ker (A) and Range (A), and the link between their dimensions; although I seldom have time for all the details of dimension theory in section 6.3, or for much of the material on transformations and bases in section 6.4.

In *Chapter 7*, the use of orthonormal bases is motivated in terms of their geometric simplicity and computational advantages. They lead naturally to orthogonal matrices which, as the carriers of the congruence concept of Euclidean geometry, are fun to teach.

In *Chapter 8*, eigenvectors are introduced and the emphasis is, at first, on their utility in various matrix computations. In particular, the iterative processes introduced in chapter 3 are studied further by means of the new technique. Of course eigenvectors give a lot of insight into the structure of matrices, especially symmetric ones, and the application to quadratic surfaces which caps the chapter is always a source of pleasure.

All the core material for an introductory course is contained in Chapters 1 through 8. There may not be much time for the extra topics covered in Chapters 9, 10, and 11, but their inclusion is worthwhile for several reasons:

Chapter 9, on Jordan Form, is included partly because many math students need to refer to it at some stage of their studies. The treatment given here is less algebraically formal, and more constructive, than most. I hope this improves the accessibility of the subject. The extensive applications may also be convenient for reference purposes.

Chapter 10 is a short overview of numerical methods. One objective is to alert students to the possible pitfalls of the computer software which directs the number crunching. It is also important for them to know just which aspects of a computation use up the computer time and money.

Chapter 11 is an introduction to linear programming, and is intended to make students at least aware of the existence of this optimization technique which is of such vast importance in the modern world. The ideas are illustrated with small-scale, numerically simple examples so as to keep the computations under control. A concise treatment of duality theory is made possible by the algebraic methodology developed early in the book; this is fortunate since the topic is both mathematically rewarding and valuable for the applications.

For some years I have been teaching a standard, one-semester course to a mostly sophomore-level audience from a wide variety of programs, and have found that I can comfortably cover the following sections of the text:

STANDARD ONE-SEMESTER COURSE

Chapter 1	4 lectures
2.1–2.3	4
3	7
4 (Selected material)	2
5	3
6.1–6.3	6
7.1–7.2	3
8.1–8.3; 8.3–8.6	8
	37 lectures

I usually mention the simplest cases of complex eigenvalues; but omit discussion of other complex-number topics such as Hermitian matrices. I also skip the proofs of a few major theorems; say the uniqueness of dimension (section 6.3) and the Principal Axes Theorem (section 8.5).

The instructor's choice of applications will be one of the factors determining the number of lectures devoted to each chapter. For classes in which a minimal knowledge of calculus can be assumed, the application to systems of differential equations can be sketched in as little as $1\frac{1}{2}$ lectures.

Note that Chapter 11, on linear programming, can be taken up any time after Chapter 3. An interesting, non-standard course for business students might run as follows:

COURSE FOR COMMERCE STUDENTS

Chapter 1	4 lectures
2.1–2.3	4
3	7
11	9
5.1–5.2	3
6.1–6.2	4
8.1–8.3	6
	37 lectures

Thorough study of the entire book would take two semesters.

A degree of flexibility in the selection of material is indicated by the following chart:

LOGICAL DEPENDENCE CHART

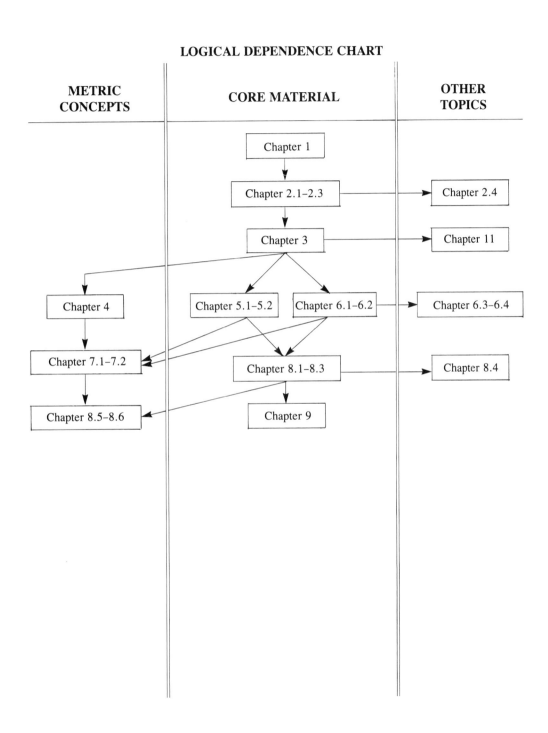

Table of Contents

Solution of Linear Systems

1.1 Introduction

We are concerned in the first instance with solving *systems of linear equations*, of which a simple example is:

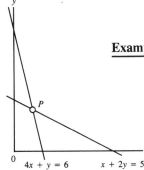

Figure 1.1

Example 1
$$x + 2y = 5$$
$$4x + y = 6 \ .$$

This system has a familiar geometric interpretation. Each equation represents a line in the xy plane, and a solution of the system represents a point P of intersection of the two lines. (See Figure 1.1.)

Linear equations occur frequently in practical problems. Suppose for example that grains of three types, A, B, and C, contain respectively one, four, and two units per bushel of a certain nutrient, and that a mixture of the grains containing a total of 270 units of the nutrient is needed. Then if x, y, and z represent the quantities in bushels of grains A, B, and C to be used, the nutrient requirement is expressed by the linear equation:

$$x + 4y + 2z = 270 \ .$$

A linear system will occur if, say, the grains contain respectively one, three, and one units per bushel of a second nutrient that is needed in the total amount of 190 units. Then the second requirement, written together with the first, gives the following system of two equations in three unknowns:

Example 2
$$x + 4y + 2z = 270$$
$$x + 3y + z = 190 \ .$$

Note that by introducing more types of grain, and more nutrients, we could obtain linear systems with more unknowns and more equations. The unknowns in a linear system are often quite numerous and need not have any geometric significance.

Linear systems are encountered in physics, chemistry, economics, other branches of mathematics, statistics, and just about any field of knowledge whose problems are susceptible to quantitative formulation. Numerous applications are presented throughout this book. One reason for the importance of linear systems is that, in general, linear problems can be solved by standard methods, and so one tries to formulate quantitative problems in such a way that the mathematical equations obtained are linear. Faced with equations which are *not* linear, we usually can't solve them, so we often have to "approximate" these equations with equations that *are* linear, and so can be solved!

In this chapter we will develop a procedure for solving an arbitrary system of linear equations. The procedure will be introduced through a series of examples, and once the method is clear, it will be easy to write down a theorem summarizing the method.

1.2 The Matrix Reduction Method

Our first examples of linear systems are simple enough to solve without any particular technique, but we will solve them methodically so as to introduce the procedure that will apply to every linear system, however complicated.

Example 1
$$x + 2y = 5$$
$$4x + y = 6 \ .$$

To solve, we first subtract four times the first equation from the second. Then the variable x is eliminated from the second equation, and the system becomes:

$$x + 2y = 5$$
$$E_2 - 4E_1 \qquad -7y = -14 \ .$$

The new form of the second equation is labelled $E_2 - 4E_1$ to indicate that it was obtained by the operation "equation two minus 4 times equation one."

Now we add $\frac{2}{7}$ times the second equation to the first, so as to eliminate the variable y from the first equation. Symbolically, the operation is $E_1 + \left(\frac{2}{7}\right)E_2$, and the system becomes:

$$E_1 + \left(\frac{2}{7}\right)E_2 \quad x \qquad = 1$$
$$-7y = -14 \ .$$

Finally, we multiply the second equation by $-\frac{1}{7}$, so as to change the coefficient of y in the second equation to 1. This operation is denoted $\left(-\frac{1}{7}\right)E_2$. The solution of the system is then:

$$x \qquad = 1$$
$$\left(-\frac{1}{7}\right)E_2 \qquad y = 2 \ .$$

The geometric interpretation of this result is that the two lines represented by the original equations intersect at the unique point $P(1,2)$.

We used two types of operations in solving the above system:

Type A: *Add* a multiple of one equation to another equation.
Type M; *Multiply* an equation by a nonzero constant.

Uses of Type A: first step, operation $E_2 - E_1$, multiple -4.
second step, operation $E_1 + \left(\frac{2}{7}\right)E_2$, multiple $\frac{2}{7}$.

Uses of Type M: third step, operation $\left(-\frac{1}{7}\right)E_2$, constant $-\frac{1}{7}$.

A third type of operation, not needed in example 1, is often useful:

Type E: *Exchange* two equations.

We will see that operations of types A, M, and E suffice to solve any linear system.

Example 2 $2x + 3y - z = 0$
$6x + 8y + 2z = 0$

First we eliminate x from the second equation:

$$2x + 3y - z = 0$$
$$E_2 - 3E_1 \qquad - y + 5z = 0 \quad .$$

Next we eliminate y from the first equation:

$$E_1 + 3E_2 \qquad 2x \qquad + 14z = 0$$
$$- y + 5z = 0 \quad .$$

Finally we change the coefficients of x and y to unity:

$$\tfrac{1}{2}E_1 \qquad x \qquad + 7z = 0$$
$$-E_2 \qquad y - 5z = 0 \quad .$$

To obtain a numerical solution, or *particular solution*, of the system, we choose any value for z, say $z = 2$, and then compute x and y from the solution formulas:

$$x = -7z$$
$$y = 5z \quad .$$

That is, a particular solution is: $(x, y, z) = (-14, 10, 2)$.
    Note that we obtain a different solution of the linear system for each choice of the unspecified number, or *parameter*, z.

From the examples so far, we can see that the variables x, y, z always stay in the same places throughout the solution process. So we can save effort by not writing them; instead we just write the array of coefficients at each step. Thus the linear system given in example 2 can be represented as:

$$\begin{bmatrix} 2 & 3 & -1 \\ 6 & 8 & 2 \end{bmatrix} .$$

Such a rectangular array of numbers is called a *matrix*, and in this case it is also called the *coefficient matrix of the linear system*.

The matrix has two *rows*, $R_1 = [2 \quad 3 \quad -1]$ and $R_2 = [6 \quad 8 \quad 2]$, corresponding to the two equations of the system; and also three *columns*,

$$\begin{bmatrix} 2 \\ 6 \end{bmatrix} , \begin{bmatrix} 3 \\ 8 \end{bmatrix} , \begin{bmatrix} -1 \\ 2 \end{bmatrix} ,$$ corresponding to the three variables of the

system. The numbers in the matrix are called *entries*.

Let us repeat the solution process, operating on the rows R_1 and R_2 of the matrix, just as we operated on the equations E_1 and E_2 of the system.

At the first step, use the entry, 2, in the upper left corner of the matrix to eliminate the other entries in the first column. That is, subtract 3 times the first row from the second row, or, symbolically, do the operation $R_2 - 3R_1$:

$$R_2 - 3R_1 \begin{bmatrix} 2 & 3 & -1 \\ 0 & -1 & 5 \end{bmatrix} .$$

Next, use the entry -1 in column two to eliminate the other entry in the column by means of the operation $R_1 + 3R_2$:

$$R_1 + 3R_2 \begin{bmatrix} 2 & 0 & 14 \\ 0 & -1 & 5 \end{bmatrix} .$$

Finally, change the first nonzero entry in each row to 1 by the operations $\frac{1}{2}R_1$, $-R_2$:

$$\begin{matrix} \frac{1}{2}R_1 \\ -R_2 \end{matrix} \begin{bmatrix} 1 & 0 & 7 \\ 0 & 1 & -5 \end{bmatrix} .$$

Having completed the calculation, we put the variables back in to obtain the same result as before:

$$x \quad - 7z = 0$$
$$y - 5z = 0 \quad .$$

The operations we are using to solve linear systems can therefore be represented as operations on the rows of a matrix, and are called the *elementary row operations*. These are:

Type A: *Add* a multiple of one row to another row.
Type M: *Multiply* a row by a nonzero constant.
Type E: *Exchange* two rows.

We now systematize the method of solution by introducing the construction of *corners* and the *crossing line* with the following example.

Example 3
$$2x + 4y - 3z - w = 0$$
$$4x + 8y - 6z - 2w = 0$$
$$6x + 12y - 6z + 6w = 0$$

Coefficient matrix:

$$\begin{bmatrix} 2 & 4 & -3 & -1 \\ 4 & 8 & -6 & -2 \\ 6 & 12 & -6 & 6 \end{bmatrix}$$

In the upper left corner of the coefficient matrix is a nonzero entry equal to 2. This entry is used to *eliminate* the other entries in the first column. So we emphasize this entry by putting a *corner* around it, and then do the elimination in column one. The first corner is also the beginning of the *crossing line*.

Next, we search the part of column two below the crossing line just formed for a nonzero entry to use to eliminate the other entries in the column. But both the possible entries are zero, and so no elimination is possible in column two. Instead, we extend the crossing line across column two as shown.

$$\begin{array}{c} \\ R_2 - 2R_1 \\ R_3 - 3R_1 \end{array} \begin{bmatrix} 2 & 4 & -3 & -1 \\ 0 & 0 & 0 & 0 \\ 0 & 0 & 3 & 9 \end{bmatrix}$$

Next, we look down the part of column three below the line until we come to a nonzero entry, 3. We bring it up to row two by exchanging rows two and three, and then put a corner around it.

$$\begin{bmatrix} 2 & 4 & -3 & -1 \\ 0 & 0 & 0 & 0 \\ 0 & 0 & 3 & 9 \end{bmatrix}$$

Then we use the new corner to eliminate the other nonzero entries in column three.

Now we search column four below the line. There is only a zero there, so we complete the crossing line by extending it across column four.

$$\begin{array}{c} \\ R_3 \\ R_2 \end{array} \begin{bmatrix} 2 & 4 & -3 & -1 \\ 0 & 0 & 3 & 9 \\ 0 & 0 & 0 & 0 \end{bmatrix}$$

Finally we change the corner entries to unity.

The final form of the system is then:

$$x + 2y \qquad + 4w = 0$$
$$z + 3w = 0$$
$$0 = 0 \quad .$$

$$R_1 + R_3 \begin{bmatrix} 2 & 4 & 0 & 8 \\ 0 & 0 & 3 & 9 \\ 0 & 0 & 0 & 0 \end{bmatrix}$$

We now define the variables corresponding to columns with corners as *corner variables*, and the other variables as *free variables*.

$$\begin{bmatrix} 2 & 4 & 0 & 8 \\ 0 & 0 & 3 & 9 \\ 0 & 0 & 0 & 0 \end{bmatrix}$$

We can then write down the *general solution in parametric form* by moving the free variables to the right side:

$$x = -2y - 4w$$
$$z = \qquad - 3w \quad .$$

$$\begin{array}{c} \frac{1}{2}R_1 \\ (\frac{1}{3})R_2 \end{array} \begin{bmatrix} 1 & 2 & 0 & 4 \\ 0 & 0 & 1 & 3 \\ 0 & 0 & 0 & 0 \end{bmatrix}$$

To obtain a particular solution, we assign arbitrary values to the free variables, or parameters, y and w, and then compute the corner variables, x and z, from the parametric formulas just obtained. For example, if we choose $y = 2$, $w = 10$, then $x = -4 - 10 = -14$, $z = -30$ and so the particular solution is $(x, y, z, w) = (-14, 2, -30, 10)$.

Let us analyze the form of the matrix

$$\begin{bmatrix} 1 & 2 & 0 & 4 \\ 0 & 0 & 1 & 3 \\ 0 & 0 & 0 & 0 \end{bmatrix}$$

that resulted from the row operations of example 3. This matrix has several properties that follow from the crossing line construction.

(i) *Zero rows.* Any rows consisting entirely of zeros (the *zero rows*) are below the other rows (the *nonzero rows*).

(ii) *Corner entries.* In any nonzero row, the first nonzero entry (the *corner entry*) is a 1.

(iii) *Corner progression.* The corner entry in a nonzero row is at least one column further to the right than the corner entry in any preceding row.

(iv) *Eliminated entries.* All entries above corner entries are zeros.

Any matrix having these four properties is called a *reduced matrix*.

For any given matrix A, the crossing line construction prescribes a definite sequence of elementary row operations that transform A into a specific reduced matrix R. This matrix, R, is called the *reduced form* of A. We also say that A reduces to R.

Example 4
$$x_1 + 2x_2 + 8x_3 = 0$$
$$2x_1 + 4x_2 + 18x_3 = 0$$
$$x_1 + 4x_2 + 20x_3 = 0$$

Solving the system amounts to reducing the coefficient matrix. The calculation is more efficiently presented when successive steps are written across the page:

$$A = \begin{bmatrix} 1 & 2 & 8 \\ 2 & 4 & 18 \\ 1 & 4 & 20 \end{bmatrix} \begin{matrix} \\ R_2-2R_1 \\ R_3- R_1 \end{matrix} \begin{bmatrix} 1 & 2 & 8 \\ 0 & 0 & 2 \\ 0 & 2 & 12 \end{bmatrix} \begin{matrix} \\ R_3 \\ R_2 \end{matrix} \begin{bmatrix} 1 & 2 & 8 \\ 0 & 2 & 12 \\ 0 & 0 & 2 \end{bmatrix} \cdots$$

$$\cdots \begin{matrix} R_1-R_2 \\ \\ \end{matrix} \begin{bmatrix} 1 & 0 & -4 \\ 0 & 2 & 12 \\ 0 & 0 & 2 \end{bmatrix} \begin{matrix} R_1+2R_3 \\ R_2-6R_3 \\ \end{matrix} \begin{bmatrix} 1 & 0 & 0 \\ 0 & 2 & 0 \\ 0 & 0 & 2 \end{bmatrix}$$

$$\cdots \begin{matrix} \\ \frac{1}{2}R_2 \\ \frac{1}{2}R_3 \end{matrix} \begin{bmatrix} 1 & 0 & 0 \\ 0 & 1 & 0 \\ 0 & 0 & 1 \end{bmatrix} = R .$$

All columns of the reduced form R have corners, so all variables are corner variables, and the general solution of the system is:

$$x_1 \quad\;\; = 0$$
$$x_2 \quad = 0$$
$$x_3 = 0 \;,$$

or $(x_1, x_2, x_3) = (0, 0, 0)$. Thus the solution of this linear system is unique. And since all the variables equal zero, this solution is called the *zero solution*.

The procedure for solving linear systems introduced in this section, and developed further in the next section, is called *Gauss-Jordan elimination*.

The matrix R used in the Gauss-Jordan procedure goes under several different names, such as "row reduced echelon matrix." The frequency of reference to R throughout the book led us to adopt the concise phrase "reduced matrix" for R.

■ **Problem**   Determine whether the matrix $\begin{bmatrix} 1 & 3 & 4 & 5 \\ 0 & 0 & 1 & 7 \end{bmatrix}$ is in re-

duced form.

■ **Solution**   Let us suppose the matrix is in reduced form and try to locate the crossing line. In fact the first nonzero entry in each
row must be a corner, and so the placement of the crossing
line must be as follows:

$$\left[\begin{array}{cc|cc} 1 & 3 & 4 & 5 \\ \hline 0 & 0 & 1 & 7 \end{array}\right] .$$

The corner progression and the 1 in each corner are correct. But we can
now see that the entry above the corner in column 3 is not 0 but 4. So property iv of reduced matrices, concerning eliminated entries, is not satisfied,
and the given matrix is not a reduced matrix.

■ EXERCISES 1.2

For each of the following linear systems, write down the coefficient matrix
A, compute its reduced form R, and deduce the general solution of the system.

1. (a) $2x + 3y = 0$
$ 8x + 6y = 0$

   (b) $x - y = 0$
$ 4x + 3y = 0$

   (c) $4x_1 - 8x_2 = 0$
$ -2x_1 + 4x_2 = 0$

   (d) $3x_1 + 3x_2 = 0$
$ 2x_1 + 2x_2 = 0$

2. (a) $2x + 6y - 9z = 0$
$ 6x + 7y + 6z = 0$

   (b) $5x + 2y + 14z = 0$
$ -10x + 11y + 2z = 0$

   (c) $2x_1 + 4x_2 + 3x_3 = 0$
$ -4x_1 - 8x_2 - 5x_3 = 0$

   (d) $x_1 - 3x_2 + 2x_3 = 0$
$ 3x_1 - 9x_2 + 6x_3 = 0$

   (e) $4x_1 + x_2 - 13x_3 + 16x_4 = 0$
$ 8x_1 + 3x_2 + 16x_3 + 11x_4 = 0$

   (f) $3x_1 + 4x_2 + 4x_3 + 6x_4 + x_5 = 0$
$ 3x_1 + 5x_2 + 4x_3 + 7x_4 + x_5 = 0$

3. (a) $x + 2y + 8z = 0$
$ 2x + 6y + 26z = 0$
$ x + 4y + 20z = 0$

   (b) $x + 3y - 2z = 0$
$ 2x + 5y - 3z = 0$
$ -4x - 9y + 5z = 0$

   (c) $2x_1 + 6x_2 - 3x_3 = 0$
$ 4x_1 + 12x_2 - 7x_3 = 0$
$ -2x_1 - 6x_2 + 7x_3 = 0$

   (d) $3x_1 + x_2 + 4x_3 = 0$
$ 9x_1 + 5x_2 + 16x_3 = 0$
$ 21x_1 + 11x_2 + 36x_3 = 0$

(e) $x_1 - 2x_2 - 4x_3 = 0$ (f) $2x_1 - 2x_2 - x_3 = 0$
       $4x_1 - 8x_2 - 15x_3 = 0$ $2x_1 - 2x_2 + 8x_3 = 0$
       $3x_1 - 6x_2 - 9x_3 = 0$ $2x_1 + x_2 + 2x_3 = 0$

4. (a) $3x + 5y = 0$ (b) $5x + 2y = 0$
       $6x + 7y = 0$ $10x + 4y = 0$
       $9x + 12y = 0$ $-5x - 2y = 0$

    (c) $x_1 + 2x_2 + 3x_3 = 0$
       $3x_1 - 2x_2 + 2x_3 = 0$
       $2x_1 - 4x_2 - x_3 = 0$
       $7x_1 - 2x_2 + 7x_3 = 0$

5. (a) $x + 2y - 2z + w = 0$ (b) $x + y - z + w = 0$
       $2x + 2y - 2z + w = 0$ $2x + 3y + z + 2w = 0$
       $2x - 2y + z + w = 0$ $4x + 5y - z + 5w = 0$

    (c) $x - y + z + 2w = 0$ (d) $x_1 + 3x_2 + x_3 + 2x_4 = 0$
       $2x - 2y + 3z + w = 0$ $2x_1 + 6x_2 + 2x_3 + 3x_4 = 0$
       $-4x + 4y - 5z - 5w = 0$ $3x_1 + 9x_2 + 4x_3 + 5x_4 = 0$

6. (a) $x - y + 2z + 2w = 0$ (b) $2x + z + 4w = 0$
       $x - y + 2z + 5w = 0$ $2x + 4y + w = 0$
       $2x - 2y + 5z + 7w = 0$ $4x + 3z + 2w = 0$
       $3x - 3y + 7z + 9w = 0$ $6x + 4y + 4z + w = 0$

    (c) $3x_1 + 2x_2 + x_3 + 3x_4 = 0$
       $6x_1 + 4x_2 + 3x_3 + 5x_4 = 0$
       $9x_1 + 6x_2 + 4x_3 + 8x_4 = 0$
       $3x_1 + 2x_2 + 2x_3 + 2x_4 = 0$

    (d) $x_1 - x_2 + 4x_3 + x_4 + 3x_5 = 0$
       $-2x_1 + 3x_2 - 5x_3 + 4x_5 = 0$
       $-x_1 + 2x_2 - x_3 + 2x_4 + 7x_5 = 0$
       $-3x_1 + 5x_2 - 6x_3 + 2x_4 + 11x_5 = 0$

7. For each of the following matrices,

    (a) show that it lacks one of the properties of a reduced matrix, and
    (b) specify an elementary row operation that changes it into a reduced matrix.

   (i) $\begin{bmatrix} 1 & 2 & 3 & 4 \\ 0 & 0 & 1 & 5 \end{bmatrix}$ (ii) $\begin{bmatrix} 1 & 0 & 3 & 4 & 5 \\ 0 & 2 & 6 & 8 & 4 \end{bmatrix}$

   (iii) $\begin{bmatrix} 0 & 0 & 0 & 1 \\ 0 & 0 & 1 & 0 \\ 0 & 1 & 0 & 0 \\ 0 & 0 & 0 & 0 \end{bmatrix}$ (iv) $\begin{bmatrix} 1 & 0 & 0 \\ 0 & 0 & 0 \\ 0 & 1 & 0 \end{bmatrix}$

8. What is the largest possible number of corners the reduced form R of a matrix A can have in the following cases?

   (a) A has three rows and five columns.
   (b) A has four rows and three columns.
   (c) A has four rows and four columns.

9. In the system

$$\begin{array}{rcrcrcl} x & + & 2y & + & z & = & 0 \\ 3x & + & 7y & + & 2z & = & 0 \\ 4x & + & 9y & + & kz & = & 0 \end{array} ,$$

   the coefficient of the variable z in the third equation is an unspecified constant k.

   (a) Determine a value of k for which the system has other solutions besides the zero solution.
   (b) With the value of k determined above, write down the general solution of the system in parametric form.

10. Find all choices of the variables x, y, z, w that are solutions of *both* of the following systems:

$$\begin{array}{ll} \text{(i)} \quad x + 2y + 3z + 4w = 0 & \text{(ii)} \quad x + 4y + 3z - 2w = 0 \\ \phantom{\text{(i)} \quad} x + 3y + 3z + w = 0 & \phantom{\text{(ii)} \quad} x + 3y + 2z + 2w = 0 \\ & \phantom{\text{(ii)} \quad} x + 2y + z + 6w = 0 . \end{array}$$

1.3 Consistency and Uniqueness

We have been using the method of matrix reduction to solve linear systems in which the right side of each equation is zero. The method, however, extends naturally to other linear systems as well. A new phenomenon is encountered; the case of linear systems having no solution. The possible forms of the set of solutions of a linear system and the conditions under which each form occurs are given in a summarizing theorem.

A linear system is said to be *homogeneous* if the right side of every equation is zero. Otherwise, it is said to be *nonhomogeneous*. In section 1.2, only the first example was nonhomogeneous:

Example 1

$$\begin{array}{rcrcl} x & + & 2y & = & 5 \\ 4x & + & y & = & 6 \end{array}$$

In the case of a nonhomogeneous system, the operations on equations have to be applied to the right sides of the equations as well. So we form the *augmented matrix* of the system, which has an extra column to contain the constants on the right sides of the equations of the system:

$$\left[\begin{array}{cc|c} 1 & 2 & 5 \\ 4 & 1 & 6 \end{array}\right] \quad .$$

The elementary row operations can then be applied to the augmented matrix:

$$R_2 - 4R_1 \left[\begin{array}{cc|c} 1 & 2 & 5 \\ 0 & -7 & -14 \end{array}\right] \quad R_1 + (\tfrac{2}{7})R_2 \left[\begin{array}{cc|c} 1 & 0 & 1 \\ 0 & -7 & -14 \end{array}\right] \quad \cdots$$

$$\cdots \quad (-\tfrac{1}{7})R_2 \left[\begin{array}{cc|c} 1 & 0 & 1 \\ 0 & 1 & 2 \end{array}\right] \quad .$$

Thus the unique solution, as obtained before, is:

$$\begin{aligned} x &= 1 \\ y &= 2 \quad , \end{aligned}$$

or $(x, y) = (1, 2)$.

Example 2

$$\begin{aligned} x_1 - x_2 - x_3 &= 2 \\ x_1 + x_2 + x_3 &= 1 \\ x_1 + 3x_2 + 3x_3 &= 2 \end{aligned}$$

The first few row operations on the augmented matrix are:

$$\left[\begin{array}{ccc|c} 1 & -1 & -1 & 2 \\ 1 & 1 & 1 & 1 \\ 1 & 3 & 3 & 2 \end{array}\right] \begin{array}{l} \\ R_2 - R_1 \\ R_3 - R_1 \end{array} \left[\begin{array}{ccc|c} 1 & -1 & -1 & 2 \\ 0 & 2 & 2 & -1 \\ 0 & 4 & 4 & 0 \end{array}\right] \quad \cdots$$

$$\cdots \quad \begin{array}{l} R_1 + \tfrac{1}{2}R_2 \\ \\ R_3 - 2R_2 \end{array} \left[\begin{array}{ccc|c} 1 & 0 & 0 & \tfrac{3}{2} \\ 0 & 2 & 2 & -1 \\ 0 & 0 & 0 & 2 \end{array}\right] \quad .$$

At this stage of the reduction process, the third row of the augmented matrix represents the equation $0 = 2$. Since this is impossible, it means that the system has no solution, or is *inconsistent*.

Inconsistency of a nonhomogeneous system is established as soon as a corner appears in the augmented column. Note that, by contrast, a homogeneous system is always consistent, since it always has at least one solution, namely the zero solution.

Example 3

$$\begin{aligned} x_1 + 2x_2 + 2x_3 + x_4 &= 14 \\ -x_1 - 2x_2 \quad\quad - 3x_4 &= -4 \\ 2x_1 + 4x_2 + 8x_3 - 2x_4 &= 48 \end{aligned}$$

The reduction of the augmented matrix proceeds as follows:

$$\begin{bmatrix} \underline{1} & 2 & 2 & 1 & \Big| & 14 \\ -1 & -2 & 0 & -3 & \Big| & -4 \\ 2 & 4 & 8 & -2 & \Big| & 48 \end{bmatrix} \begin{matrix} \\ R_2 + R_1 \\ R_3 - 2R_1 \end{matrix} \begin{bmatrix} \underline{1} & 2 & 2 & 1 & \Big| & 14 \\ 0 & 0 & \underline{2} & -2 & \Big| & 10 \\ 0 & 0 & 4 & -4 & \Big| & 20 \end{bmatrix} \cdots$$

$$\cdots \quad \begin{matrix} R_1 - R_2 \\ \\ R_3 - 2R_2 \end{matrix} \begin{bmatrix} \underline{1} & 2 & 0 & 3 & \Big| & 4 \\ 0 & 0 & \underline{2} & -2 & \Big| & 10 \\ 0 & 0 & 0 & 0 & \Big| & 0 \end{bmatrix} \tfrac{1}{2}R_2 \begin{bmatrix} \underline{1} & 2 & 0 & 3 & \Big| & 4 \\ 0 & 0 & \underline{1} & -1 & \Big| & 5 \\ 0 & 0 & 0 & 0 & \Big| & 0 \end{bmatrix} .$$

The reduced form of the augmented matrix represents the equations:

$$\begin{aligned} x_1 + 2x_2 \quad + 3x_4 &= 4 \\ x_3 - x_4 &= 5 \\ 0 &= 0 \quad . \end{aligned}$$

On moving the free variables x_2, x_4 to the right, we obtain the general solution in parametric form:

$$\begin{aligned} x_1 &= 4 - 2x_2 - 3x_4 \\ x_3 &= 5 \qquad + x_4 \quad . \end{aligned}$$

We can now summarize the conditions under which a system has a solution and under which the solution is unique. Clearly the placement of the corners in the coefficient matrix or augmented matrix of the system is decisive.

1. A system is *consistent* if there is no corner in the augmented column, and *inconsistent* if there is a corner there.

2. The solution of a consistent system is *unique* if all variables are corner variables, and *nonunique* if some variables are free variables (which serve as parameters).

In order to recast these observations into a formal theorem, we first introduce some terminology relating to corners.

The *rank* of a matrix A is defined to be the number of corners, or equivalently the number of nonzero rows, of its reduced form R. This number is denoted: rank (A).

Example 4 Compute the rank of $A = \begin{bmatrix} 1 & 4 & 7 \\ 2 & 5 & 8 \\ 3 & 6 & 9 \end{bmatrix}$.

Of course rank (A) is not usually obvious by inspection. At least part of the reduction process must be carried out before the number of corners can be counted. In this example, the necessary steps are:

$$\begin{bmatrix} \underline{1} & 4 & 7 \\ 2 & 5 & 8 \\ 3 & 6 & 9 \end{bmatrix} \begin{matrix} \\ R_2 - 2R_1 \\ R_3 - 3R_1 \end{matrix} \begin{bmatrix} \underline{1} & 4 & 7 \\ 0 & \underline{-3} & -6 \\ 0 & -6 & -12 \end{bmatrix} \begin{matrix} \\ \\ R_3 - 2R_2 \end{matrix} \begin{bmatrix} \underline{1} & 4 & 7 \\ 0 & \underline{-3} & -6 \\ 0 & 0 & 0 \end{bmatrix} .$$

At this point we can see, although the reduction process is incomplete, that there will be two corners, and so rank $(A) = 2$.

THEOREM 1.1

(Consistency and uniqueness)

(a) A homogeneous linear system is always *consistent*, since it always has the zero solution.

(b) A nonhomogeneous linear system is *consistent* when the ranks of its coefficient matrix and its augmented matrix are equal; otherwise it is *inconsistent*.

(c) Suppose a linear system in n variables, with coefficient matrix A, is consistent. Then the solution is *unique* when rank $(A) = n$ and otherwise is *nonunique*.

■**Proof** (a) Obvious.

(b) The coefficient matrix and augmented matrix, and therefore their reduced forms, are exactly the same except for the extra column carried by the augmented matrix. Thus the two matrices have the same number of corners, and therefore the same rank, if there is no corner in the extra column of the augmented matrix. But this is exactly the condition for consistency of the system.

(c) We have seen that a consistent linear system has a different solution for each choice of the parameters occurring in the parametric form of the general solution. So the solution is unique only when there are no parameters. Now, the parameters are the free variables, and therefore uniqueness occurs when there are no free variables. But then all n variables must be corner variables, and so there are n corners, and rank $(A) = n$.   Q.E.D.

COROLLARY 1.2

(Uniqueness and shape of the system)
If a consistent linear system has a unique solution, it must have at least as many equations as variables.

■**Proof** Uniqueness for a system with n variables implies the presence of n corners in the reduced form of the coefficient matrix. Since each corner occurs in a different row, the coefficient matrix has at least n rows; in other words, the system has at least n equations.

Q.E.D.

Note that of the consistent linear systems solved in this section, example 3, with three equations and four variables, did not have a unique solution, whereas example 1, with two equations and two variables, did.

Matrix Reduction and Geometry

Can we *see*, geometrically, how the matrix reduction method works? Take a simple case: that of a system in just two variables x, y. Then each equation in the system represents a line in the plane. For example, the first system we encountered in this chapter,

$$x + 2y = 5$$
$$4x + y = 6 \quad,$$

represents the intersection of the lines in Figure 1.2.

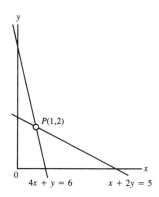

The given system

Figure 1.2

The first step in the reduction of the augmented matrix of the system is as follows:

$$\begin{bmatrix} 1 & 2 & | & 5 \\ 4 & 1 & | & 6 \end{bmatrix} \quad R_2 - 4R_1 \quad \begin{bmatrix} 1 & 2 & | & 5 \\ 0 & -7 & | & -14 \end{bmatrix} .$$

The new form of the augmented matrix represents the system

$$x + 2y = 5$$
$$-7y = -14 \quad.$$

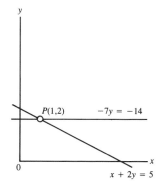

The system after one row operation

Figure 1.3

Thus the second line in the system has been changed from $4x + y = 6$ to $-7y = -14$. See Figure 1.3. Observe that the new line has a simpler form than the one it replaced; it is parallel to the x-axis. However, the new pair of lines intersect at the same point as the old pair did, and so the solution of the system is still the same.

Similarly, the next row operation leads to the augmented matrix representing the lines:

$$x \quad = \quad 1$$
$$-7y = -14 \quad.$$

Now the first line has been changed from $x + 2y = 1$ to $x = 1$ and is parallel to the y-axis. See Figure 1.4.

Of course the final row operation only results in the equation of the second line being simplified to $y = 2$; there is no change in the geometry of the system. See Figure 1.5. In the end, the system is still represented as a set of two lines, but they are now very simple lines: lines parallel to the coordinate axes.

For a system in three variables x, y, z, each equation can be shown to represent a plane in space (see section 2.4), and a solution of the system is therefore a point of intersection of these planes. By reasoning similar to the above, it can be shown that the matrix reduction process produces systems representing successively simpler sets of planes.

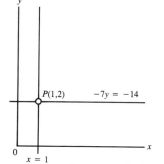

The system after two row operations

Figure 1.4

■ **Problem** The linear system

$$2x + y + z = 1$$
$$4x + 2y + 3z = q$$
$$-2x - y + z = 2$$

contains a constant q. For which choices of q is the system consistent? For which choices of q is there a unique solution?

■ **Solution** To analyze this problem, we develop the crossing line of the augmented matrix as far as possible.

$$\begin{bmatrix} \underline{2} & 1 & 1 & 1 \\ 4 & 2 & 3 & q \\ -2 & -1 & 1 & 2 \end{bmatrix} \begin{matrix} \\ R_2-2R_1 \\ R_3+R_1 \end{matrix} \begin{bmatrix} \underline{2} & 1 & 1 & 1 \\ 0 & 0 & \underline{1} & q-2 \\ 0 & 0 & 2 & 3 \end{bmatrix} \cdots$$

$$\begin{matrix} R_1-R_2 \\ \\ R_3-2R_2 \end{matrix} \begin{bmatrix} \underline{2} & 1 & 0 & 3-q \\ 0 & 0 & \underline{1} & q-2 \\ 0 & 0 & 0 & 7-2q \end{bmatrix}$$

At this point, we can see that there will be a further corner if the entry below the line in the augmented column is nonzero. In this case the system is inconsistent. In other words, the system is consistent only when the entry $7 - 2q$ is zero, or $q = \frac{7}{2}$.

To complete the study of the consistent case, note that the rank of the coefficient matrix is two, less than the number of variables, which is three. So the solution is not unique.

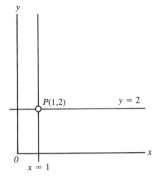

The reduced form of the system

Figure 1.5

■ **EXERCISES 1.3**

Using the augmented matrix method, find the general solution, in parametric form, of each of the following systems.

1. (a) $x + 4y = 5$
         $2x + 7y = 6$

 (b) $2x + 4y = -8$
     $6x + 12y = -24$

 (c) $x_1 + 3x_2 = 7$
       $2x_1 + 6x_2 = 12$

2. (a) $2x_1 + 3x_2 = 1$
       $4x_1 + 5x_2 = 6$

 (b) $x + 3y + 4z = 2$
      $3x + 8y + 12z = 5$

 (c) $3x + 2y - z = -2$
     $6x + 4y - 2z = 5$

 (d) $x_1 + 2x_2 + 3x_3 = 2$
      $4x_1 + 10x_2 + 6x_3 = 4$

 (e) $2x_1 + 3x_2 + x_3 = 0$
     $4x_1 + 6x_2 + 3x_3 = 8$

 (f) $x_1 + x_2 + x_3 = 3$
      $x_1 - x_2 + x_3 = 5$

3. (a)
$$x \qquad + 4z = 1$$
$$2x + 2y + 4z = 2$$
$$2x - 2y + 8z = 6$$

(b)
$$x + 2y + 4z = 1$$
$$3x + y + 7z = 5$$
$$5x + 5y + 15z = 9$$

(c)
$$3x + 4y + 2z = 1$$
$$6x + 12y + 5z = 4$$
$$3x + 8y + 3z = 3$$

(d)
$$x_1 + x_2 + x_3 = 2$$
$$x_1 + 2x_2 - x_3 = 4$$
$$x_1 + x_2 + 2x_3 = 4$$

(e)
$$x_1 + 3x_2 + 3x_3 = 5$$
$$2x_1 + 6x_2 + 3x_3 = 7$$
$$x_1 + 3x_2 - 3x_3 = 2$$

(f)
$$x_1 + 2x_2 - x_3 = 3$$
$$x_1 + 4x_2 + 2x_3 = 1$$
$$x_1 + 6x_2 + 4x_3 = 1$$

4. (a)
$$x + 3y + z + 2w = 1$$
$$-2x - 5y - z - 5w = 1$$

(b)
$$x_1 + 5x_2 + 3x_3 + 2x_4 + 3x_5 = 1$$
$$2x_1 + 10x_2 + 6x_3 + 5x_4 + 5x_5 = 4$$

5. (a)
$$x + 3y + 2z + w = 1$$
$$x + 3y + z + 2w = 1$$
$$x + 4y + z + 3w = 2$$

(b)
$$x + 3y + z + 7w = 5$$
$$3x + 9y + z + 11w = 17$$
$$2x + 6y + 2z + 14w = 10$$

(c)
$$x_1 + 2x_2 + 3x_3 + 4x_5 = 5$$
$$2x_1 + 4x_2 + 7x_3 + 5x_4 + x_5 = 8$$
$$3x_1 + 6x_2 + 10x_3 + 5x_4 + 7x_5 = 15$$

(d)
$$2x_1 + 4x_2 + 2x_3 + 2x_4 + x_5 = 18$$
$$4x_1 + 8x_2 + 6x_3 + 3x_5 = 44$$
$$-2x_1 - 4x_2 + 2x_3 - 10x_4 + 2x_5 = 2$$

6. (a)
$$x + y + z + w = 10$$
$$x + y - z - w = 4$$
$$x - y + z - w = 2$$
$$x - y - z + w = 0$$

(b)
$$x \qquad + z + 3w = 2$$
$$x + 2y + 3z + 2w = 4$$
$$3x + 2y + 7z \qquad = 1$$
$$x \qquad + 3z - 5w = 8$$

(c)
$$x + 2y + z + w = 1$$
$$x + 3y + z + w = 0$$
$$x + 2y + z + 2w = 1$$
$$x + y + z + 4w = 3$$

(d)
$$x_1 + 2x_2 + x_3 + 2x_4 + 3x_5 = 0$$
$$2x_1 + 5x_2 + 2x_3 + 5x_4 + 7x_5 = 1$$
$$x_1 + 3x_2 + x_3 + 3x_4 + 4x_5 = 1$$
$$x_1 + x_2 + x_3 + x_4 + x_5 = 0$$

7. Without doing any calculation, determine the correct placement of the crossing line for the matrix

$$A = \begin{bmatrix} 3 & 5 & 7 & 9 \\ 0 & 2 & 4 & 6 \end{bmatrix}$$

and deduce rank (A).

8. Each of the following linear systems contains an unspecified constant q. Determine for which values of q the system is consistent and for which of these values the solution of the system is unique.

(a) $\left[\begin{array}{ccc|c} 1 & 1 & 2 & 2 \\ 2 & 1 & 4 & 3 \\ 3 & 1 & q & 6 \end{array}\right]$ (b) $\left[\begin{array}{ccc|c} 1 & 0 & 1 & 2 \\ 2 & 1 & 1 & 3 \\ 1 & 4 & q & -2 \end{array}\right]$

(c) $\left[\begin{array}{ccc|c} 1 & 4 & 2 & 3 \\ 1 & 5 & q & 4 \\ 2 & 8 & 4 & 6 \end{array}\right]$ (d) $\left[\begin{array}{ccc|c} 1 & 2 & 3 & 2 \\ 2 & q & 5 & 4 \\ 2 & 4 & 4 & 2 \end{array}\right]$

9. (a) For which choices of the unspecified constants p and q will the linear system:

$$\begin{aligned} x \quad\quad + z \quad\quad &= q \\ y \quad\quad + 2w &= 0 \\ x \quad\quad + 2z + 3w &= 0 \\ 2y + 3z + pw &= 3 \end{aligned}$$

be (i) inconsistent
(ii) consistent, with a unique solution
(iii) consistent, with a nonunique solution?

(b) In case (iii), state the general solution of the system in parametric form.

10. Let $A = \begin{bmatrix} 1 & 1 & k \\ 1 & k & 1 \\ k & 1 & 1 \end{bmatrix}$.

Determine rank (A) in the cases:
(a) $k = 1$ (b) $k = -2$ (c) $k \neq 1$ or -2 .

1.4 Applications of Linear Systems

Linear systems occur naturally in practical and industrial problems.

Example 1 Suppose a mining company has a contract to deliver 8,000 oz gold, 78,000 oz silver, and 144,000 lb copper, which it plans to meet by milling ore from its North, Central, and South Mines. However, metal prices being low at the moment, the company does not want to have any metal left over after fulfilling the contract. Suppose that each ton of ore milled from the North Mine yields 1 oz gold, 10 oz silver, and 20 lb copper, and that the corresponding yields for the Central Mine are 2 oz, 18 oz, and 22 lb; and for the South Mine 1 oz, 12 oz, and 40 lb. Suppose that x, y, and z represent the numbers of tons of ore to be milled from the North, Central, and South Mines respectively. Then the quantities of the metals produced are:

gold $\qquad x + 2y + z = 8{,}000$ oz

silver $\qquad 10x + 18y + 12z = 78{,}000$ oz

copper $\qquad 20x + 22y + 40z = 144{,}000$ lb .

Now we solve the linear system for x, y, z.

$$\left[\begin{array}{ccc|c} 1 & 2 & 1 & 8{,}000 \\ 10 & 18 & 12 & 78{,}000 \\ 20 & 22 & 40 & 144{,}000 \end{array}\right] \begin{array}{c} \\ R_2 - 10R_1 \\ R_3 - 20R_1 \end{array} \left[\begin{array}{ccc|c} 1 & 2 & 1 & 8{,}000 \\ 0 & -2 & 2 & -2{,}000 \\ 0 & -18 & 20 & -16{,}000 \end{array}\right]$$

$$\begin{array}{c} R_1 + R_2 \\ \\ R_3 - 9R_2 \end{array} \left[\begin{array}{ccc|c} 1 & 0 & 3 & 6{,}000 \\ 0 & -2 & 2 & -2{,}000 \\ 0 & 0 & 2 & 2{,}000 \end{array}\right] \begin{array}{c} R_1 - (\frac{3}{2})R_3 \\ R_2 - R_1 \\ \\ \end{array} \left[\begin{array}{ccc|c} 1 & 0 & 0 & 3{,}000 \\ 0 & -2 & 0 & -4{,}000 \\ 0 & 0 & 2 & 2{,}000 \end{array}\right]$$

$$\begin{array}{c} \\ (-\frac{1}{2})R_2 \\ (\frac{1}{2})R_2 \end{array} \left[\begin{array}{ccc|c} 1 & 0 & 0 & 3{,}000 \\ 0 & 1 & 0 & 2{,}000 \\ 0 & 0 & 1 & 1{,}000 \end{array}\right] \quad , \quad \text{or} \quad \begin{array}{l} x = 3{,}000 \\ y = 2{,}000 \\ z = 1{,}000 \end{array} .$$

The company must therefore mill 3,000 tons of ore from the North Mine, 2,000 tons from the Central Mine, and 1,000 tons from the South Mine.

Example 2 Suppose that the ores described in the previous problem do not contain any copper and that the mining company has no contract to deliver copper. Then the quantities of ore to be milled from the three mines have to satisfy only the first two equations of the linear system constructed above. The augmented matrix and its reduced form are then easily seen to be:

$$\left[\begin{array}{ccc|c} 1 & 2 & 1 & 8{,}000 \\ 10 & 18 & 12 & 78{,}000 \end{array}\right] \quad \text{and} \quad R = \left[\begin{array}{ccc|c} 1 & 0 & 3 & 6{,}000 \\ 0 & 1 & -1 & 1{,}000 \end{array}\right] .$$

So the solution is nonunique:

$$x = 6{,}000 - 3z$$
$$y = 1{,}000 + z \quad .$$

The quantity, z, of ore milled from the South Mine can thus be chosen arbitrarily, except that it must not exceed 2,000 tons in order that the quantity, x, will be positive.

In order to select an optimal value for z, the company might decide to minimize its total production costs. If these costs, for the North, Central, and South Mines respectively, are \$200, \$300, and \$250 per ton of ore milled, then the total production cost is given by the *cost function*:

$$
\begin{aligned}
C(x, y, z) &= 200x + 300y + 250z \\
&= 200\,(6{,}000 - 3z) + 300\,(1{,}000 + z) + 250z \\
&= 1{,}500{,}000 - 50z \quad .
\end{aligned}
$$

So the cost is minimized when z is chosen as large as possible, that is, when $z = 2{,}000$.

The optimal solution is therefore that no ore is milled from the North Mine, 3,000 tons are milled from the Central Mine, and 2,000 tons from the South Mine.

Example 3   Suppose that in a fixed population of breakfast-food consumers, there are just three brands, A, B, and C, that share the market. Suppose that each year

10% of the brand A buyers shift to brand B and 20% to brand C,
20% of the brand B buyers shift to brand A and 20% to brand C, and
10% of the brand C buyers shift to brand A and none to brand B.

Suppose it is found that despite all this shifting, the market shares of the three brands are nevertheless unchanged from year to year. What are these market shares?

■ **Solution**   Let x, y, z represent the proportions of the market held by brands A, B, C in a given year.

The next year, A loses $10 + 20 = 30\%$ of its previous buyers, and so it keeps 70% of them. It also gains 20% of brand B's customers and 10% of brand C's. Thus its market becomes $\left(\frac{7}{10}\right)x + \left(\frac{2}{10}\right)y + \left(\frac{1}{10}\right)z$. Therefore if brand A's original market share is unchanged, we must have:

$$\left(\tfrac{7}{10}\right)x + \left(\tfrac{2}{10}\right)y + \left(\tfrac{1}{10}\right)z = x \quad .$$

Similar reasoning applied to brands B and C gives the further equations:

$$\left(\tfrac{1}{10}\right)x + \left(\tfrac{6}{10}\right)y \qquad\quad = y$$
$$\left(\tfrac{2}{10}\right)x + \left(\tfrac{2}{10}\right)y + \left(\tfrac{9}{10}\right)z = z \quad .$$

It is convenient to clear the fractions by multiplying each equation by 10. Then, on collecting terms, the equations assume the form:

$$-3x + 2y + z = 0$$
$$x - 4y \quad\;\; = 0$$
$$2x + 2y - z = 0 \;\;.$$

The coefficient matrix and its reduced form are then easily seen to be

$$\begin{bmatrix} -3 & 2 & 1 \\ 1 & -4 & 0 \\ 2 & 2 & -1 \end{bmatrix} \quad \text{and } R = \begin{bmatrix} 1 & 0 & -\frac{4}{10} \\ 0 & 1 & -\frac{1}{10} \\ 0 & 0 & 0 \end{bmatrix}.$$

The general solution is therefore: $x = \left(\frac{4}{10}\right)z,\; y = \left(\frac{1}{10}\right)z$.
Now brand A's proportion of the total market

$$= \left(\frac{\text{number of buyers of brand A}}{\text{total number of buyers}}\right) = \left(\frac{x}{x + y + z}\right)$$

$$= \frac{\left(\frac{4}{10}\right)z}{[\left(\frac{4}{10}\right)z + \left(\frac{1}{10}\right)z + z]} = \frac{4}{15}\;.$$

Similarly the market shares of brands B and C are $\frac{1}{15}$ and $\frac{10}{15}$ respectively.

Example 4 In the introductory section of this chapter, we derived a system of two equations satisfied by the quantities x, y, z of three grain types A, B, C needed to satisfy certain nutritional requirements.

The augmented matrix of this system and its reduced form are:

$$\begin{bmatrix} 1 & 4 & 2 & | & 270 \\ 1 & 3 & 1 & | & 190 \end{bmatrix} \quad \text{and} \quad R = \begin{bmatrix} 1 & 0 & -2 & | & -50 \\ 0 & 1 & 1 & | & 80 \end{bmatrix}.$$

The general solution of the system is therefore

$$x = -50 + 2z$$
$$y = \quad\; 80 - z \;\;.$$

From these formulas for x and y, we see that all components of the solution are positive as long as z is between 25 and 80.

An optimal solution can be isolated if we have the additional information that the grains of types A, B, C must be purchased at prices of, say, \$3, \$7, \$3 per bushel respectively. We then seek to minimize the total cost, given by the cost function:

$$C(x, y, z) = 3x + 7y + 3z$$
$$= 3(-50 + 2z) + 7(80 - z) + 3z$$
$$= 410 + 2z \;\;.$$

To minimize cost, we therefore choose z as small as possible, that is, $z = 25$. Then from the parametric formulas, $x = 0$, $y = 55$.

We conclude that the optimal solution is to purchase no grain of type A, 55 bushels of type B, and 25 bushels of type C.

■ **Problem**   Suppose that each day a tour group, composed of 40 people carrying one suitcase each, is driven across an island in taxis. The tour company can hire three kinds of taxis:

sedans,	which carry four passengers and eight suitcases,
limousines,	which carry eight passengers and four suitcases,
station wagons,	which carry six passengers and eighteen suitcases.

The tour organizers find that no matter what combination of taxis they rent, there is always some unused space in the taxis. They want to stop wasting space and are willing to change the size of the tour group somewhat if necessary. Is an adjustment essential, and if so by how much?

■ **Solution**   Suppose x sedans, y limousines, and z wagons are to be fully used by a tour group of n people. Then:

$$\text{passengers carried} = 4x + 8y + 6x = n$$
$$\text{suitcases carried} \;\;= 8x + 4y + 18z = n \;\;.$$

The augmented matrix of the system and its reduced form are:

$$\begin{bmatrix} 4 & 8 & 6 & \Big| & n \\ 8 & 4 & 18 & \Big| & n \end{bmatrix} \quad \text{and} \quad R = \begin{bmatrix} 1 & 0 & \tfrac{5}{2} & \Big| & \tfrac{n}{12} \\ 0 & 1 & -\tfrac{1}{2} & \Big| & \tfrac{n}{12} \end{bmatrix} \quad,$$

and so the general solution is

$$x = \tfrac{n}{12} - \tfrac{5z}{2} \;,$$
$$y = \tfrac{n}{12} + \tfrac{z}{2} \;;$$

or, for tour groups of 40,

$$x = \tfrac{10}{3} - \tfrac{5z}{2} \;,$$
$$y = \tfrac{10}{3} + \tfrac{z}{2} \;.$$

The number, z, of wagons rented can be chosen to be 0, 1, 2, etc. From the solution formulas, the corresponding number of sedans will be $x = \tfrac{10}{3}, \tfrac{5}{6}, -\tfrac{5}{3}$, etc. Thus the number of sedans is never a positive integer, as it must be for a valid solution. That is, there is no solution with "full use" when $n = 40$.

However, if the tour group size, n, is increased by 2 to 42, the general solution is

$$x = \tfrac{7}{2} - \tfrac{5z}{2} \;,$$
$$y = \tfrac{7}{2} + \tfrac{z}{2} \;.$$

Now if we choose $z = 1$, then $x = 1$ and $y = 4$; in other words a fleet of one sedan, four limousines, and one wagon will be fully used.

■ EXERCISES 1.4

1. Suppose types A, B, C of grain contain, respectively, one, two, and three units per bushel of nutrient I, and three, four, and four units per bushel of nutrient II, and cost, respectively, $4, $7, and $7 per bushel. If a total of 100 units of nutrient I and 150 units of nutrient II must be obtained, how many bushels of each type of grain must be ordered to minimize total cost?

2. What would be the solution to exercise 1 if the nutrient II content of grain C were only three units per bushel instead of four?

3. Suppose that 100 cartons, 60 drums, and 180 bales must be packed into steel containers for long-distance shipment. Suppose it is found that there are four efficient ways (modes) of packing a container:

    Mode A: pack 10 cartons, 10 drums, 10 bales
    Mode B: pack 8 cartons, 0 drums, 24 bales
    Mode C: pack 9 cartons, 6 drums, 18 bales
    Mode D: pack 9 cartons, 4 drums, 16 bales

   Is it possible for the shipment to consist entirely of fully packed containers? If so, in how many different ways can this be done?

4. Suppose that a second shipment like that of exercise 3 is to consist of 100 cartons, 24 drums, and 264 bales. Show that shipment in fully packed containers can be achieved in exactly one way.

5. Suppose a company has three mines, called the Pine, Cedar, and Rose Mines, which produce, respectively, 2, 1, and $\frac{3}{2}$ oz of gold and 12, 16, and 10 oz of silver per ton ore milled. Suppose the company is known to be using its full milling capacity of 1,000 tons of ore per day, and to be producing 1,700 oz of gold and 11,600 oz of silver per day. What percentage of the ore being milled is the company taking from each of the three mines?

6. Suppose the company of exercise 5 wants to reduce the amount of ore milled per day to the minimum possible while still maintaining the same output of gold and silver. What is the minimum possible milling level, and what percentage of the ore milled would then come from each mine?

7. Suppose the company of exercise 5 must, because of mechanical problems, reduce its milling rate to 500 tons of ore per day but must also, because of contract commitments, maintain gold production of at least 800 oz per day. What is the maximum rate of silver production that can be maintained under these conditions, and how can this rate be achieved?

8. Suppose that 400 trucks and 1,000 automobiles must be shipped by rail from a manufacturing plant to a regional storage lot. Three types of railroad flatcars, A, B, and C, can be used to carry the vehicles. Their rental costs and carrying capacity are given in Table 1.1.

Flatcar data		Flatcar Type		
		A	B	C
	Rental cost	$3,000	$4,000	$2,000
	Truck capacity	10	10	20
	Auto capacity	20	30	10

Table 1.1

To ship the vehicles at least cost, how many flatcars of each type should be rented?

9. Suppose that the owners of the type C flatcars of exercise 8 want to increase their rental price. How high can they raise it before it becomes less expensive for the manufacturer to use only type A and B flatcars?

10. As an alternative method of minimizing total cost in exercise 8, note that total cost, M, in thousands of dollars, satisfies $3x + 4y + 2z = M$, where x, y, z represent the numbers of flatcars of types A, B, C that are used.

    (a) Show that, if this equation is joined to the two equations for x, y, z used in the original method of solution, then a system with augmented matrix A and reduced form R is obtained as follows:

$$A = \begin{bmatrix} 1 & 1 & 2 & 40 \\ 2 & 3 & 1 & 100 \\ 3 & 4 & 2 & M \end{bmatrix}, \quad R = \begin{bmatrix} 1 & 0 & 0 & 5M - 680 \\ 0 & 1 & 0 & 440 - 3M \\ 0 & 0 & 1 & 140 - M \end{bmatrix}.$$

    (b) Find the smallest value of total cost M that results in x, y, z all being positive, and verify that this choice of M corresponds to the same values for x, y, z obtained in exercise 8.

11. Suppose that brands A, B, and C of a consumer staple are struggling for shares of a fixed market and that each year they gain consumers from each other as follows:

    Brand A gains 10% of B's, 30% of C's buyers.
    Brand B gains 20% of A's, 20% of C's buyers.
    Brand C gains none of A's, 40% of B's buyers.

    Find the proportion of the total market that each brand will have when the market reaches equilibrium, i.e., when market shares are unchanged from year to year.

*12. Suppose the people marketing brand B of the product in the previous exercise want to increase their market share to 40%, but they know that they cannot gain from brand A any faster than they are gaining now. How fast would they have to gain from brand C to achieve their goal? (Assume all other rates of shifting remain as given in the previous exercise.)

CHAPTER TWO

Vector Notation

Introduction

The equations in a linear system contain many coefficients, variables, and constants. In trying to keep track of all these quantities while solving a linear system, we have found it convenient to arrange the coefficients in a matrix and the variables in a sequential list, such as $x_1, \ldots, x_n$. Since matrices and sequential lists are encountered naturally at the beginning of the study of linear systems, it is reasonable to expect them to form basic building blocks in the further development of our subject. In order that they will be useful, however, we must first develop rules of calculation so that they can be combined and rearranged in mathematical equations. We must also introduce new and concise notations for these slightly bulky objects so that substantial amounts of calculation can be done with ease.

In the first section of this chapter, we interpret sequential lists as vectors of a certain kind, introduce vector notation, and discuss the algebraic properties of vectors. But in linear systems, matrices and vectors are mixed together. Therefore, in section 2.2 we present a way of combining matrices and vectors, called the matrix-vector product. The use of this product allows us to represent entire linear systems in just a few symbols. With the improved notation, it becomes feasible to interpret matrices in a new light: as objects that act on vectors and transform them into other vectors. The necessary technical framework for this viewpoint, the theory of transformations of vector spaces, is outlined in section 2.3.

The material in sections 2.1, 2.2, 2.3 suffices as preparation for the later chapters of this book. However, readers with time for a more leisurely treatment will profit by reading the starred section 2.4. This not only fills out the theory of vectors, but also provides an extensive geometric interpretation of linear systems that is of great intrinsic interest and importance.

2.1 Coordinate Vectors

In this section we present a brief introduction to the theory of vectors. The discussion is confined to those properties of vectors that are essential in the development of linear algebra. The streamlined treatment of vectors is intended to facilitate the most rapid possible approach to the methods of matrix algebra itself.

We regard a vector, in the first instance, as nothing more than a list of numbers. But methods of combining vectors are specified, and they are quickly organized into formal structures called vector spaces, which have many useful properties. Methods of constructing new vector spaces from the ones given initially are described, and a geometric interpretation of vectors is provided.

N-tuples

First, we define an *n-tuple* to be an ordered set of n numbers. For example, $(1,3,5,7)$ is a 4-tuple. Note that $(3,1,5,7)$ is a *different* 4-tuple, since the ordering is different. The numbers appearing in an *n*-tuple are called the *components* of the *n*-tuple. The symbol $\mathbf{x}$ represents the *n*-tuple with components $x_1, \ldots, x_n$ in order; or, $\mathbf{x} = (x_1, \ldots, x_n)$.

Given *n*-tuples $\mathbf{x} = (x_1, \ldots, x_n)$ and $\mathbf{y} = (y_1, \ldots, y_n)$, their *sum* $\mathbf{x} + \mathbf{y}$ is defined to be the new *n*-tuple

$$\mathbf{x} + \mathbf{y} = (x_1 + y_1, \ldots, x_n + y_n) \quad .$$

That is, the first component of $\mathbf{x} + \mathbf{y}$ is the sum of the first components of $\mathbf{x}$ and of $\mathbf{y}$, etc. So we can interpret the definition by saying that addition of *n*-tuples is done *componentwise*. For example,

$$(1,2) + (4,-8) = (1 + 4, 2 - 8) = (5,-6) \quad .$$

The *scalar product* $t\mathbf{x}$ of the *n*-tuple $\mathbf{x}$ and the number t is defined to be the new *n*-tuple

$$t\mathbf{x} = (tx_1, \ldots, tx_n) \quad .$$

That is, to multiply an *n*-tuple by t, multiply each component by t. For example,

$$3(4,5) = (3 \cdot 4, 3 \cdot 5) = (12,15) \quad .$$

Geometric Interpretations of N-tuples

Although it is not essential for the algebraic development that follows, geometric interpretation of *n*-tuples is helpful. A 2-tuple $\mathbf{x} = (x_1, x_2)$, for example, can of course be said to represent the point P in the plane whose Cartesian coordinates are x_1 and x_2. (See Figure 2.1.) It can also be said to represent the line segment from the origin O to the point P. This line segment is called the *position vector* of the point P, and denoted $\overrightarrow{OP}$.

Note that $t\mathbf{x} = (tx_1, tx_2)$ would then represent the position vector of the new point S with Cartesian coordinates tx_1 and tx_2. From a "similar triangles" argument of analytic geometry, it can easily be seen that the position vector $\overrightarrow{OS}$ is parallel to $\overrightarrow{OP}$ and t times as long (see Figure 2.2).

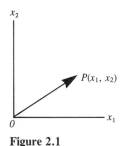

Figure 2.1

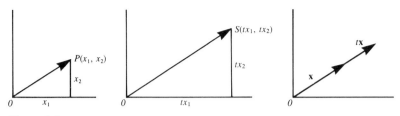

Figure 2.2

Similarly, if **x** and **y** represent the position vectors $\overrightarrow{OP}$ and $\overrightarrow{OQ}$, and if R is the fourth vertex of the parallelogram with sides OP and OQ (see Figure 2.3), then it can be shown (see exercise 1.8) that $\mathbf{x} + \mathbf{y}$ represents the position vector $\overrightarrow{OR}$ of vertex R.

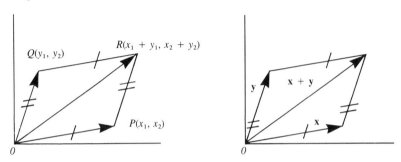

Figure 2.3

N-tuples as a Vector Space

Because operations on n-tuples are done componentwise, it is easy to see that they share many of the properties of ordinary numbers. For example, numbers have the *commutative* property that $x + y = y + x$, and so

$$\mathbf{x} + \mathbf{y} = (x_1 + y_1, \ldots, x_n + y_n)$$
$$= (y_1 + x_1, \ldots, y_n + x_n) = \mathbf{y} + \mathbf{x} \quad .$$

That is, addition of n-tuples is also commutative.

Furthermore, we can extend certain familiar notations of arithmetic to the case of n-tuples by writing **0** for the n-tuple $(0, \ldots, 0)$ and $-\mathbf{x}$ for the n-tuple $(-x_1, \ldots, -x_n)$.

It is then ridiculously easy to verify everything in the following list of properties of the set V of n-tuples.

I. Properties of addition

For any elements **x**, **y**, **z**, etc. in V:

1. $\mathbf{x} + \mathbf{y}$ is defined as an element of V (closure property).
2. $\mathbf{x} + \mathbf{y} = \mathbf{y} + \mathbf{x}$ (commutative law).

3. $(\mathbf{x} + \mathbf{y}) + \mathbf{z} = \mathbf{x} + (\mathbf{y} + \mathbf{z})$ (associative law).
4. $\mathbf{x} + \mathbf{0} = \mathbf{x}$ (existence of additive identity $\mathbf{0}$).
5. $\mathbf{x} + (-\mathbf{x}) = \mathbf{0}$ (existence of additive inverse).

II. Properties of scalar multiplication

For any elements $\mathbf{x}$, $\mathbf{y}$, etc. of V and any numbers s, t, etc.:

6. $t\mathbf{x}$ is defined as an element of V (closure property).
7. $(st)\mathbf{x} = s(t\mathbf{x})$ (associative law).
8. $(s + t)\mathbf{x} = s\mathbf{x} + t\mathbf{x}$ (distributive law one).
9. $t(\mathbf{x} + \mathbf{y}) = t\mathbf{x} + t\mathbf{y}$ (distributive law two).
10. $1\mathbf{x} = \mathbf{x}$ (existence of multiplicative identity 1).

The reason we set down this seemingly obvious list is that many other important mathematical systems, besides the n-tuples, happen to share the same set of properties. In fact, any *set* V that has all 10 properties is called a *vector space*, and the elements of the set V are then called *vectors*.

The set of n-tuples is often denoted R^n. Thus we can say that the set $V = R^n$ is a vector space, and that each n-tuple is a vector.

Subspaces

We will need to use only a few of the great variety of vector spaces that occur in mathematics. The easiest way to construct a new vector space is to choose the set V to be a suitable subset of R^n, instead of being all of R^n. Why should the subset V be a vector space in its own right? Because V "inherits" most of the properties of a vector space from the larger spaces R^n of which it is a part.

For example, since $\mathbf{x} + \mathbf{y} = \mathbf{y} + \mathbf{x}$ for *all* vectors $\mathbf{x}$, $\mathbf{y}$ in R^n, this property certainly holds for all $\mathbf{x}$, $\mathbf{y}$ in the smaller set V! By glancing over the list of 10 properties, we can see that all of them except #1 and #6 will remain valid by inheritance from R^n. Thus we only have to verify that V satisfies properties #1 and #6 to be sure that V is a vector space. This conclusion is simple but important; let us summarize it as a theorem:

THEOREM 2.1

A subset V of R^n is a vector space provided:

(a) whenever $\mathbf{x}$ and $\mathbf{y}$ are in V, $\mathbf{x} + \mathbf{y}$ is also in V, and

(b) whenever $\mathbf{x}$ is in V and t is a scalar, $t\mathbf{x}$ is also in V.

Example 1 Let V be the set of 2-tuples with two equal components; i.e., typical elements of V are $\mathbf{x} = (p,p)$, $\mathbf{y} = (q,q)$, etc. Let us check properties (a) and (b):

(a) $\mathbf{x} + \mathbf{y} = (p,p) + (q,q) = (p+q,p+q)$. So the two components of $\mathbf{x} + \mathbf{y}$ are both $p+q$, and $\mathbf{x} + \mathbf{y}$ is in V. Check.

(b) $t\mathbf{x} = t(p,p) = (tp,tp)$. So both components of $t\mathbf{x}$ are tp, and $t\mathbf{x}$ is in V. Check.

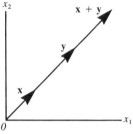

We conclude from Theorem 2.1 that V is a vector space.

The geometric interpretation of example 1 starts from the observation that the subset V of R^2 consists of position vectors of points in the plane with equal coordinates; that is, position vectors parallel to the line $x_1 = x_2$. When such vectors are added or multiplied by scalars, the results are still parallel to this line, and so are still in V. (See Figure 2.4).

Intuitively, one might say that, starting with vectors in V, there can be "no escape" from V by means of the vector space operations of addition and scalar multiplication. This is what makes V a vector space in its own right.

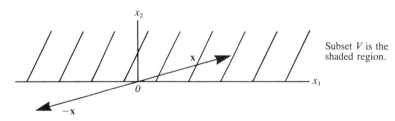

Figure 2.4

The above phenomenon of "nested" vector spaces, a smaller one inside a larger one, occurs frequently in practice. In fact, whenever we have two vector spaces V and W, and V is a subset of W, we say that V is a *subspace* of W. Thus the subset V of example 1 is a subspace of R^n.

Example 2 Let V be the set of 2-tuples $\mathbf{x} = (x_1,x_2)$ satisfying $x_2 > 0$. Geometrically, we can say that V consists of all position vectors pointing into the upper half plane. (See Figure 2.5.)

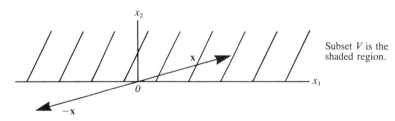

Subset V is the shaded region.

Figure 2.5

For example, $\mathbf{x} = (2,1)$ is evidently in V. However, if we choose $t = -1$, we get

$$t\mathbf{x} = (-1)(2,1) = (-2,-1) \quad ,$$

which has its second component negative and so points into the lower half plane. Thus we can escape from V, and therefore V is not a vector space. Technically, we say that condition (b) of Theorem 2.1 is violated.

Dot Product Notation

Given two n-tuples $\mathbf{x}$ and $\mathbf{y}$, the *dot product* $\mathbf{x} \cdot \mathbf{y}$ of $\mathbf{x}$ and $\mathbf{y}$ is defined to be the scalar quantity:

$$\mathbf{x} \cdot \mathbf{y} = x_1 y_1 + \ldots + x_n y_n \quad .$$

For example, if $\mathbf{x} = (1,2)$ and $\mathbf{y} = (3,2)$, then

$$\mathbf{x} \cdot \mathbf{y} = (1,2) \cdot (3,2) = 1 \cdot 3 + 2 \cdot 2 = 7 \quad .$$

This is a useful shorthand since the components of different vectors are often intermixed in formulas. A few simple properties of the dot product, easily checked by expanding all terms in components, are

(i) $\mathbf{x} \cdot \mathbf{y} = \mathbf{y} \cdot \mathbf{x}$ (commutative property).

(ii) $\mathbf{x} \cdot (\mathbf{y} + \mathbf{z}) = \mathbf{x} \cdot \mathbf{y} + \mathbf{x} \cdot \mathbf{z}$ (distributive property).

(iii) $t(\mathbf{x} \cdot \mathbf{y}) = (t\mathbf{x}) \cdot \mathbf{y} = \mathbf{x} \cdot (t\mathbf{y})$ associative property).

Dot Product and Geometry

The dot product, for 2-tuples and 3-tuples, is related to the concept of length from Euclidean geometry. In fact the Pythagorean formula gives the length, $|\mathbf{x}|$, of the plane position vector $\mathbf{x} = (x_1, x_2)$, as $|\mathbf{x}| = (x_1^2 + x_2^2)^{\frac{1}{2}}$, and that of the spatial position vector $\mathbf{x} = (x_1, x_2, x_3)$ as $|\mathbf{x}| = (x_1^2 + x_2^2 + x_3^2)^{\frac{1}{2}}$. But from the definition of the dot product for n-tuples, we have

$$\mathbf{x} \cdot \mathbf{x} = x_1^2 + \ldots + x_n^2 \quad .$$

So for 2- and 3-tuples, length can be expressed in terms of the dot product

$$|\mathbf{x}| = (\mathbf{x} \cdot \mathbf{x})^{\frac{1}{2}} \, (\textit{length formula}) \quad .$$

However, when n exceeds 3, our n-tuples no longer have an interpretation as position vectors in Euclidean geometry; so instead we use the length formula as the *definition of length of a vector* $\mathbf{x}$ *in* R^n.

It turns out that the angle θ between nonzero vectors $\mathbf{x}$ and $\mathbf{y}$ can also be expressed by the dot product (see Figure 2.6):

$$\cos \theta = \frac{\mathbf{x} \cdot \mathbf{y}}{|\mathbf{x}| \, |\mathbf{y}|} \quad (\textit{angle formula}) \quad ,$$

$$\mathbf{x} \cdot \mathbf{y} = |\mathbf{x}| \, |\mathbf{y}| \cos \theta \quad (\textit{dot product formula}) \quad .$$

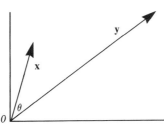
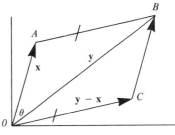

Figure 2.6

We can derive these formulas easily if we make use of the cosine law of trigonometry. Start by noting that $(\mathbf{y} - \mathbf{x}) + \mathbf{x} = \mathbf{y}$, so that in the parallelogram $OABC$ with sides $\overrightarrow{OA}$ and $\overrightarrow{OC}$ representing $\mathbf{x}$ and $\mathbf{y} - \mathbf{x}$, the diagonal $\overrightarrow{OB}$ represents the sum $\mathbf{y}$. For the sides of triangle OAB we have the lengths $|OA| = |\mathbf{x}|$, $|OB| = |\mathbf{y}|$, and also, since the opposite sides of a parallelogram have equal lengths, $|AB| = |OC| = |\mathbf{y} - \mathbf{x}|$.

Now the cosine law says that angle AOB, or θ, satisfies the equation:

$$|AB|^2 = |OA|^2 + |OB|^2 - 2|OA||OB|\cos\theta$$

or, $\quad |\mathbf{y} - \mathbf{x}|^2 = |\mathbf{x}|^2 + |\mathbf{y}|^2 - 2|\mathbf{x}||\mathbf{y}|\cos\theta$.

But we can derive another formula for $|\mathbf{y} - \mathbf{x}|^2$ as follows:

$$\begin{aligned}
|\mathbf{y} - \mathbf{x}|^2 &= (\mathbf{y} - \mathbf{x}) \cdot (\mathbf{y} - \mathbf{x}) \\
&= \mathbf{y} \cdot (\mathbf{y} - \mathbf{x}) - \mathbf{x} \cdot (\mathbf{y} - \mathbf{x}) \\
&= \mathbf{y} \cdot \mathbf{y} - \mathbf{y} \cdot \mathbf{x} - \mathbf{x} \cdot \mathbf{y} + \mathbf{x} \cdot \mathbf{x} \\
&= |\mathbf{x}|^2 - 2\mathbf{x} \cdot \mathbf{y} + |\mathbf{y}|^2 \quad .
\end{aligned}$$

Equating the two formulas for $|\mathbf{y} - \mathbf{x}|^2$ and cancelling, we get

$$-2|\mathbf{x}||\mathbf{y}|\cos\theta = -2\mathbf{x} \cdot \mathbf{y} \quad ,$$

which reduces to the stated formulas for $\cos\theta$ and $\mathbf{x} \cdot \mathbf{y}$.

Example 3 Let $\mathbf{x} = (3,4)$, $\mathbf{y} = (7,1)$. Then

$$\begin{aligned}
\cos\theta &= \frac{\mathbf{x} \cdot \mathbf{y}}{|\mathbf{x}||\mathbf{y}|} = \frac{(3,4) \cdot (7,1)}{(3^2 + 4^2)^{\frac{1}{2}}(7^2 + 1^2)^{\frac{1}{2}}} = \frac{21 + 4}{(25)^{\frac{1}{2}}(50)^{\frac{1}{2}}} \\
&= \left(\tfrac{25}{50}\right)^{\frac{1}{2}} = \frac{1}{\sqrt{2}} \quad .
\end{aligned}$$

Or, $\theta = \text{Arccos}\,\dfrac{1}{\sqrt{2}} = 45°$. Since both vectors point into the first quadrant, the angle between them should be less than $90°$, and it is.

Euclidean geometry no longer provides us with a concept of the angle between vectors once n is greater than 3. Instead, just as in the case of length, we use the formula for $\cos\theta$, or equivalently the formula

$$\theta = \cos^{-1}\left(\frac{\mathbf{x} \cdot \mathbf{y}}{|\mathbf{x}||\mathbf{y}|}\right) \quad ,$$

to *define* an angle between the n-tuples $\mathbf{x}$ and $\mathbf{y}$.

Sigma Notation

The usual symbolic method of writing the sum of the algebraic quantities x_2, x_3, x_4, x_5 is, of course, $x_2 + x_3 + x_4 + x_5$. This is inefficient since we keep writing the same symbol, x, with an index that increases by one unit at a time. For longer sums we have a shorthand form; for example, $x_2 + x_3 + \ldots + x_{75}$ denotes the sum of all the quantities x_i where the index i takes on all integer values between 2 and 75. This shorthand includes all the essential information, namely the symbol x and the lower and upper limits, 2 and 75, on the index. A still more compact way of specifying this information is the *sigma notation* for the sum:

$$\overset{i=75}{\underset{i=2}{\Sigma}} x_i \quad \text{or} \quad \overset{75}{\underset{i=2}{\Sigma}} x_i \quad .$$

This is read "sigma from i equals 2 to 75 of x sub i."

The dot product of the vectors $\mathbf{x}$ and $\mathbf{y}$ in R^3 would be written in sigma notation as:

$$\mathbf{x} \cdot \mathbf{y} = \overset{3}{\underset{i=1}{\Sigma}} x_i y_i \quad .$$

The discussion in this text minimizes the use of sigma notation, because it adds another layer of abstraction to the formulas. However, the compactness becomes essential in more advanced mathematics, where many indices may be summed over in the same formula, etc. We therefore provide practice in use of the sigma notation in many of the exercises, particularly the starred ones, throughout the book.

■ **Problem** Suppose $\mathbf{p}$ is a fixed, nonzero vector in R^3, and let V be the set of all vectors $\mathbf{x}$ in R^3 that are perpendicular to $\mathbf{p}$.

Show that V is a subspace of R^3, and give a description of the set V in terms of Euclidean geometry.

■ **Solution** The condition that $\mathbf{x}$ must be perpendicular to $\mathbf{p}$ can be expressed in terms of the dot product through the equation

$$\mathbf{p} \cdot \mathbf{x} = 0 \quad .$$

Suppose now that we start with two vectors, $\mathbf{x}$ and $\mathbf{y}$, which are in V; in other words,

$$\mathbf{p} \cdot \mathbf{x} = 0 \quad \text{and} \quad \mathbf{p} \cdot \mathbf{y} = 0 \quad .$$

Then
$$\mathbf{p} \cdot (\mathbf{x} + \mathbf{y}) = \mathbf{p} \cdot \mathbf{x} + \mathbf{p} \cdot \mathbf{y} = 0 + 0 = 0 \quad ,$$
and so $\mathbf{x} + \mathbf{y}$ is still in V .

Also,
$$\mathbf{p} \cdot (t\mathbf{x}) = t(\mathbf{p} \cdot \mathbf{x} = t0 = 0 \quad ,$$
and so $t\mathbf{x}$ is also in V.

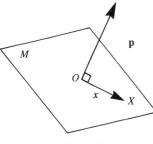

Figure 2.7

That is, there is "no escape" from V by means of the vector space operations; therefore V is indeed a subspace of R^3.

To interpret the subspace V geometrically, let M be the plane through the origin perpendicular to $\mathbf{p}$. (See Figure 2.7).

Then, by definition, V consists of those vectors $\mathbf{x}$ that lie in M. Or, on identifying a position vector $\mathbf{x} = OX$ with its endpoint X, we can say that V *is* the plane through the origin perpendicular to $\mathbf{p}$.

■ EXERCISES 2.1

1. Compute the following quantities:

(a) $(1,3) + (5,7)$

(b) $(7,-2) + (5,4)$

(c) $4(-2,3)$

(d) $6(1,2) + 8(-1,4)$

(e) $(1,-1,2) + (2,3,7)$

(f) $(8,0,4) + (0,1,0)$

(g) $3(1,2,3)$

(h) $2(4,1,2) - 3(1,0,2)$

(i) $7(1,3,2,-1)$

(j) $4(1,1,3,4) + 2(-2,1,-2,0)$

(k) $(1,3)\cdot(5,7)$

(l) $(1,5,5)\cdot(3,2,-1)$

(m) $(4,4,0,2)\cdot(0,2,4,-2)$

(n) $(3,1,2,-1)\cdot(1,2,4,2)$

2. Determine the vector **x** that satisfies each of the following equations:

(a) $\mathbf{x} + (5,2) = (4,3)$

(b) $3(\mathbf{x} + (1,2)) = (6,12)$

(c) $4\mathbf{x} + 5(1,2) = (-3,2)$

(d) $(2,3) - 2\mathbf{x} = (4,7)$

3. Let $\mathbf{x} = (1,3,2)$, $\mathbf{y} = (-1,2,5)$. Compute:

(a) $3\mathbf{x}$ (b) $\mathbf{x} + \mathbf{y}$ (c) $3\mathbf{x} - 5\mathbf{y}$.

(d) Find a vector **z** such that $\mathbf{x} - 4\mathbf{z} = \mathbf{y}$.

(e) Find a position vector **w** parallel to **x** but with second component equal to 9.

4. Determine whether or not the subset V is a subspace of R^2 in the following cases:

(a) V consists of those 2-tuples satisfying $x_1 = 3x_2$.

(b) V consists of those 2-tuples satisfying $x_1 = 0$.

(c) V consists of those 2-tuples satisfying $-1 \le x_1 \le 1$.

5. Show that n-tuple length has the properties:

(a) $|\mathbf{x}| > 0$ unless $\mathbf{x} = \mathbf{0}$.

(b) $|\mathbf{0}| = 0$.

(c) $|t\mathbf{x}| = |t|\,|\mathbf{x}|$.

6. Show that, for 2- and 3-tuples, the following inequalities hold:

(a) $|\mathbf{x} + \mathbf{y}| \le |\mathbf{x}| + |\mathbf{y}|$ (use triangle inequality from geometry).

(b) $|\mathbf{x}\cdot\mathbf{y}| \le |\mathbf{x}|\,|\mathbf{y}|$ (use cosine formula for $\mathbf{x}\cdot\mathbf{y}$).

7. Find the angles between the pairs of vectors:

(a) $\mathbf{x} = (5,12)$, $\mathbf{y} = (17,7)$.

(b) $\mathbf{x} = (11,2)$, $\mathbf{y} = (2,-1)$.

8. Show for 2- and 3-tuples, $\mathbf{x}\cdot\mathbf{y}$ is positive, zero, or negative according as the angle θ between **x** and **y** is an acute, right, or oblique angle.

9. Let V be the set of 3-tuples (x_1, x_2, x_3) satisfying the equation

$$2x_1 + 3x_2 + 6x_3 = 0 \quad .$$

Show, by reference to the problem preceding this set of exercises, or otherwise, that V is a subspace of R^3.

10. Suppose $\mathbf{u}$ is a fixed vector in a vector space W. Suppose the subset V of W consists of all scalar multiples of $\mathbf{u}$. That is, $\mathbf{x}$ is in V when for some scalar t, $\mathbf{x} = t\mathbf{u}$. Show that V is a subspace of W.

11. Let V be the set of all functions of one variable, so that typical elements of V are: $f(s) = s^2$, $g(s) = \sin s$, etc. Suppose sums and scalar multiples of functions are defined by the formulas:

$$(f+g)(s) = f(s) + g(s), \qquad (tf)(s) = t(f(s)) \quad .$$

For example, the sum of the functions f and g given above is the function $(f+g)(s) = s^2 + \sin s$, while the scalar 3 times the function f given above is the function $(3f)(s) = 3s^2$.
Show that V is a vector space.

12. With V as in the previous exercise, let $P(s)$ denote the subset of V consisting of all polynomials in s. A typical element of $P(s)$ is

$$p(s) = s^2 + 4s + 3 \quad .$$

Show that $P(s)$ is a subspace of V.

13. Let $P_N(s)$ denote the subset of $P(s)$ consisting of those polynomials of degree less than or equal to the positive integer N. That is, elements of $P_N(s)$ have the form:

$$p(s) = a_N s^N + a_{N-1} s^{N-1} + \cdots + a_1 s + a_0 \quad ,$$

where $a_0, a_1, \ldots, a_N$ are constants.
Show that $P_N(s)$ is a subspace of $P(s)$.

14. Prove that the parallelogram construction described in the text may be used to form the sum $\mathbf{x} + \mathbf{y}$ of the position vectors $\mathbf{x}$ and $\mathbf{y}$.
Hint: Use Figure 2.8 to show that the first coordinate of R is $x_1 + y_1$, as follows: first show that triangles OQA and PRB are congruent, then that $|CD| = y$, then that $|OD| = |OC| + |CD| = x_1 + y_1$.

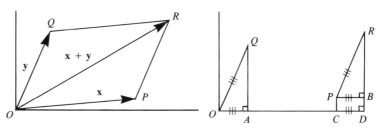

Figure 2.8

15. The inequality $|\mathbf{x} \cdot \mathbf{y}| \le |\mathbf{x}|\,|\mathbf{y}|$ holds in R^n and is called the Cauchy-Schwartz inequality.

(a) Prove this inequality by the following steps:

(i) Recall from the theory of quadratic equations that if the quadratic in q, $aq^2 + bq + c$, is never negative, then $b^2 \le 4ac$.

(ii) Reduce the nonnegative quadratic

$$\frac{1}{2} \sum_{i=1}^{n} (x_i q + y_i)^2 \text{ to the form:}$$

$$\left(\frac{1}{2} \sum_{i=1}^{n} x_i^2\right) q^2 + \left(\sum_{i=1}^{n} x_i y_i\right) q + \left(\frac{1}{2} \sum_{i=1}^{n} y_i^2\right) \ .$$

(iii) Deduce that $\left(\displaystyle\sum_{i=1}^{n} x_i y_i\right)^2 \le \left(\displaystyle\sum_{i=1}^{n} x_i^2\right)\left(\displaystyle\sum_{i=1}^{n} y_i^2\right) \ .$

(iv) Rewrite this as $(\mathbf{x} \cdot \mathbf{y})^2 \le |\mathbf{x}|^2 |\mathbf{y}|^2$ and deduce the required inequality.

(b) Prove that the triangle inequality $|\mathbf{x} + \mathbf{y}| \le |\mathbf{x}| + |\mathbf{y}|$ holds in R^n, by expanding $|\mathbf{x} + \mathbf{y}|^2 \ge 0$ as a dot product and using the Cauchy-Schwartz inequality.

(c) Check that the inverse cosine formula used in the text to define the angle θ between vectors $\mathbf{x}$, $\mathbf{y}$ in R^n makes sense for all nonzero choices of $\mathbf{x}$ and $\mathbf{y}$. (Use Cauchy-Schwartz inequality to show that the argument of the inverse cosine function will always be between -1 and $+1$.)

2.2 Linear Systems in Vector Notation

It is now clear that all the quantities that occur in a linear system can be organized conveniently into matrices and vectors. A truly efficient representation of a linear system should therefore amount to a concise, symbolic indication of how these matrices and vectors combine to satisfy mathematical equations.

But in a linear system, coefficients and variables, which are matrix entries and vector components respectively, are thoroughly mixed together. Accordingly, the essence of this section is a definition of a matrix-vector product that incorporates this complicated mixing of matrix and vector. By the use of this product, a few symbols suffice to represent a linear system.

We also streamline the presentation of the general solution of a linear system, recasting it in vector form. The improvements in notation developed here lead, in the next section, to a greater insight into the matrix concept; matrices then assume a new and more active role in relation to vectors.

Matrix-Vector Product

In general, a linear system of m equations in n variables takes the form

$$a_{11}x_1 + \cdots + a_{1n}x_n = c_1$$
$$a_{21}x_1 + \cdots + a_{2n}x_n = c_2$$
$$\ldots\ldots\ldots\ldots\ldots\ldots\ldots$$
$$a_{m1}x_1 + \cdots + a_{mn}x_n = c_m \quad.$$

The quantities appearing in these equations may be identified as follows:

On the right are the m constants $c_1, \ldots, c_m$ of the system.
On the left are the n variables $x_1, \ldots, x_n$ of the system.

The constants therefore form a vector $\mathbf{c}$ in R^m, and the variables a vector $\mathbf{x}$ in R^n. It is convenient to write those vectors in column form:

$$\mathbf{c} = \begin{bmatrix} c_1 \\ c_2 \\ \vdots \\ c_m \end{bmatrix} \quad, \quad \mathbf{x} = \begin{bmatrix} x_1 \\ x_2 \\ \vdots \\ x_n \end{bmatrix} \quad.$$

Also on the left are the *coefficients* $a_{11}, a_{12}, \ldots, a_{mn}$, which of course are the entries of the *coefficient matrix* A of the system, used in Chapter 1:

$$A = \begin{bmatrix} a_{11} & a_{12} & \cdots & a_{1n} \\ a_{21} & a_{22} & \cdots & a_{2n} \\ \ldots & \ldots & \ldots & \ldots \\ a_{m1} & a_{m2} & \cdots & a_{mn} \end{bmatrix} \quad.$$

This matrix has m rows and n columns and is therefore referred to as an $m \times n$ matrix. For example, a 2×3 coefficient matrix would correspond to a system with two equations and three unknowns.

Each coefficient a_{ij} carries a *row index i* and a *column index j*. For example, coefficient a_{21} appears in the second row and first column of the coefficient matrix A, or equivalently, in the second equation of the system as coefficient of the first variable x_1.

We have now designed shorthand notations, $\mathbf{c}$, $\mathbf{x}$, A, for all the quantities occurring in the linear system. The right side of the system itself can also be written in streamlined form: it is simply the column vector $\mathbf{c}$ in R^m. However, we do not yet have a streamlined notation for the mixture of a_{ij}'s and x_j's occurring on the left side of the system.

To obtain one, we simply define the *matrix-vector product*, $A\mathbf{x}$, by the formula

$$A\mathbf{x} = \begin{bmatrix} a_{11} & a_{12} & \cdots & a_{1n} \\ a_{21} & a_{22} & \cdots & a_{2n} \\ & \cdots \cdots \cdots \cdots \\ a_{m1} & a_{m2} & \cdots & a_{mn} \end{bmatrix} \begin{bmatrix} x_1 \\ x_2 \\ \cdot \\ \cdot \\ x_n \end{bmatrix} = \begin{bmatrix} a_{11}x_1 + \cdots + a_{1n}x_n \\ a_{21}x_1 + \cdots + a_{2n}x_n \\ \cdots \cdots \cdots \cdots \cdots \\ a_{m1}x_1 + \cdots + a_{mn}x_n \end{bmatrix}.$$

That is, the product of the $m \times n$ matrix A and the column vector $\mathbf{x}$ in R^n is the vector in R^m with ith component $a_{i1}x_1 + \cdots + a_{in}x_n$.

The general linear system then assumes the concise matrix-vector form:

$$A\mathbf{x} = \mathbf{c} \quad.$$

Calculation of Matrix-Vector Products

In order to use the improved notation effectively, we must first gain facility in numerical calculation.

The definition of the matrix product seems complicated, but actually it is easy to remember. For example, the first component of the product, $a_{11}x_1 + \ldots + a_{1n}x_n$, contains the n entries $a_{11}, a_{12}, \ldots, a_{1n}$ from the first row of the coefficient matrix. These are combined with the n variables $x_1, \ldots, x_n$ in the form of a dot product. Similar reasoning applies to the other components of the product $A\mathbf{x}$, and so we have the rule

$$(i\text{th component of } A\mathbf{x}) = (i\text{th row of } A) \cdot \mathbf{x}$$

Example 1 $A = \begin{bmatrix} 4 & 5 & 6 \\ 7 & 8 & 9 \end{bmatrix}$, $\mathbf{x} = \begin{bmatrix} 1 \\ 2 \\ 3 \end{bmatrix}$.

The components of $A\mathbf{x}$ are given by the dot products:

component 1: $(4,5,6) \cdot (1,2,3)$
component 2: $(7,8,9) \cdot (1,2,3)$.

Therefore,

$$A\mathbf{x} = \begin{bmatrix} 4 & 5 & 6 \\ 7 & 8 & 9 \end{bmatrix} \begin{bmatrix} 1 \\ 2 \\ 3 \end{bmatrix} = \begin{bmatrix} 4 \cdot 1 + 5 \cdot 2 + 6 \cdot 3 \\ 7 \cdot 1 + 8 \cdot 2 + 9 \cdot 3 \end{bmatrix} = \begin{bmatrix} 32 \\ 50 \end{bmatrix}.$$

It may help to visualize the computation by rotating the rows of the matrix into line with the column $\mathbf{x}$, and then multiplying corresponding components. The calculation then takes the form:

$$\begin{array}{ll} 4 \cdot 1 = 4 & 7 \cdot 1 = 7 \\ 5 \cdot 2 = 10 & 8 \cdot 2 = 16 \\ 6 \cdot 3 = \underline{18} & 9 \cdot 3 = \underline{27} \\ \text{total} = 32 & \text{total} = 50 \end{array}.$$

Example 2   Back in section 1.3, we studied a linear system $A\mathbf{x} = \mathbf{c}$ with
$$A = \begin{bmatrix} 1 & 2 \\ 4 & 1 \end{bmatrix} \quad \text{and} \quad \mathbf{c} = \begin{bmatrix} 5 \\ 6 \end{bmatrix} \quad \text{and found that } \mathbf{x} = (x,y) = (1,2)$$
was the unique solution. We can now check this result by computing $A\mathbf{x}$ (with $\mathbf{x}$ rewritten as a column vector, of course) and seeing whether the product equals $\mathbf{c}$ as it should:

$$A\mathbf{x} = \begin{bmatrix} 1 & 2 \\ 4 & 1 \end{bmatrix} \begin{bmatrix} 1 \\ 2 \end{bmatrix} = \begin{bmatrix} 1 \cdot 1 + 2 \cdot 2 \\ 4 \cdot 1 + 1 \cdot 2 \end{bmatrix} = \begin{bmatrix} 5 \\ 6 \end{bmatrix} = \mathbf{c}. \text{ Check.}$$

Example 3   The quantities in a matrix-vector product need not be given numerically. It is just as easy to compute a product such as

$$\begin{bmatrix} 1 & 2 \\ 3 & 4 \end{bmatrix} \begin{bmatrix} x \\ y \end{bmatrix} = \begin{bmatrix} 1 \cdot x + 2 \cdot y \\ 3 \cdot x + 4 \cdot y \end{bmatrix} = \begin{bmatrix} x + 2y \\ 3x + 4y \end{bmatrix} \quad .$$

There are dimensional restrictions on the matrix-vector products that are defined; the specified pattern is:

$$\begin{array}{ccc} A & \mathbf{x} & = \mathbf{c} \\ (m \times n) & (n) & (m) \end{array}$$
e.g. $(2 \times 3) \quad (3) \quad (2)$ in example 1.

For example, the product $\begin{bmatrix} 1 & 2 & 3 & 4 \\ 5 & 6 & 7 & 8 \end{bmatrix} \begin{bmatrix} 9 \\ 10 \\ 11 \end{bmatrix}$ is not defined because

in the pattern $(2 \times 4)(3)$ the values 4 and 3 for n imposed by the matrix and the vector do not match. We would discover the inconsistency during numerical computation because we would try to dot a row with a column of unequal length.

Vector Form of General Solution

In Chapter 1 we presented the general solutions to linear systems in parametric form; a typical result was example 3 in section 1.3, where the components of the solution vector $\mathbf{x}$ in R^4 were given by the parametric formulas

$$\begin{aligned} x_1 &= 4 - 2x_2 - 3x_4 \\ x_3 &= 5 \qquad\;\; + x_4 \quad . \end{aligned}$$

Let us develop a streamlined, vector form of the general solution, starting with the observation that on substituting for the corner variables x_1, x_3, in $\mathbf{x}$, we obtain:

$$\mathbf{x} = \begin{bmatrix} x_1 \\ x_2 \\ x_3 \\ x_4 \end{bmatrix} = \begin{bmatrix} 4 - 2x_2 - 3x_4 \\ x_2 \\ 5 \qquad + x_4 \\ x_4 \end{bmatrix} \quad .$$

Now, there is nothing to stop us from writing this formula for $\mathbf{x}$ as the sum of three vectors, one containing the numerical terms, the other two containing the terms with x_2 and x_4 respectively:

$$\mathbf{x} = \begin{bmatrix} 4 \\ 0 \\ 5 \\ 0 \end{bmatrix} + \begin{bmatrix} -2x_2 \\ x_2 \\ 0 \\ 0 \end{bmatrix} + \begin{bmatrix} -3x_4 \\ 0 \\ x_4 \\ x_4 \end{bmatrix} = \begin{bmatrix} 4 \\ 0 \\ 5 \\ 0 \end{bmatrix} + x_2 \begin{bmatrix} -2 \\ 1 \\ 0 \\ 0 \end{bmatrix} + x_4 \begin{bmatrix} -3 \\ 0 \\ 1 \\ 1 \end{bmatrix} .$$

Next we label the numerical vectors occurring in the last formula as follows:

$$\mathbf{v} = \begin{bmatrix} 4 \\ 0 \\ 5 \\ 0 \end{bmatrix} , \quad \mathbf{u}_1 = \begin{bmatrix} -2 \\ 1 \\ 0 \\ 0 \end{bmatrix} , \quad \mathbf{u}_2 = \begin{bmatrix} -3 \\ 0 \\ -1 \\ 1 \end{bmatrix} .$$

Then we can write the general solution more concisely as

$$\mathbf{x} = \mathbf{v} + x_2 \mathbf{u}_1 + x_4 \mathbf{u}_2 .$$

Furthermore, the free variables x_2, x_4 can be chosen arbitrarily: they are the parameters in the general solution. Therefore we can rename them b_1, b_2, so as to suggest their roles as first and second parameters in the general solution. We thereby obtain the *general solution in vector form*:

$$\mathbf{x} = \mathbf{v} + b_1 \mathbf{u}_1 + b_2 \mathbf{u}_2 .$$

Particular solutions are obtained from this general solution by assigning numerical values to the parameters. For example, if we select $b_1 = 0$, $b_2 = 0$, we obtain $\mathbf{x} = \mathbf{v}$. Thus $\mathbf{v}$ is a particular solution, and we call $\mathbf{v}$ the *anchor solution* of the system. The numerical vectors $\mathbf{u}_1$, $\mathbf{u}_2$ are called *generators*.

Let us summarize our new method of presenting the general solution:

THEOREM 2.2

The general solution of a consistent linear system $A\mathbf{x} = \mathbf{c}$ can always be represented in the vector form

$$\mathbf{x} = \mathbf{v} + b_1 \mathbf{u}_1 + \cdots + b_k \mathbf{u}_k ,$$

where $\mathbf{v}$ is a particular solution of the system, called the *anchor solution*; $\mathbf{u}_1, \ldots, \mathbf{u}_k$ are numerical vectors, called *generators*; and $b_1, \ldots, b_k$ are parameters, whose number k equals $n - \text{rank}\,(A)$.

■ Proof   In view of the above discussion, we need only check the formula for k. But the parameters $b_1, \ldots, b_k$ were just new names for the free variables, whose number is always $n - (\text{number of corner variables})$ or $n - \text{rank}\,(A)$.   Q.E.D.

Some special cases of this theorem should be noted.

1. *Case of uniqueness*. When rank $(A) = n$, the number of parameters $k = 0$. Then there are no generators, and the general solution is simply

   $$\mathbf{x} = \mathbf{v} \ .$$

   That is, the only solution is the anchor solution. This is in agreement with theorem 1.1, which specified rank $(A) = n$ as the condition for uniqueness.

2. *Case of homogeneity*. For the system $A\mathbf{x} = \mathbf{0}$, the constants form the vector $\mathbf{c} = \mathbf{0}$. Then the augmented matrix is not needed, and there are no constant terms in the parametric form of the general solution. Consequently the anchor solution $\mathbf{v}$ equals $\mathbf{0}$ as well, and the vector form of the general solution is:

   $$\mathbf{x} = b_1\mathbf{u}_1 + \cdots + b_k\mathbf{u}_k \ .$$

3. *Case of uniqueness and homogeneity*. When the system is $A\mathbf{x} = \mathbf{0}$, and rank $(A) = n$, the anchor solution is $\mathbf{x} = \mathbf{0}$ and there are no generators. So the general solution is

   $$\mathbf{x} = \mathbf{0} \ .$$

The following terminology is convenient: given any list of k vectors $\mathbf{z}_1, \ldots, \mathbf{z}_k$ in a vector space V, and any numbers $b_1, \ldots, b_k$, the vector $b_1\mathbf{z}_1 + \cdots + b_k\mathbf{z}_k$ in V is called a *linear combination* of $\mathbf{z}_1, \ldots, \mathbf{z}_k$.

We can say, for example, that each solution of the system $A\mathbf{x} = \mathbf{0}$ is a linear combination (l.c.) of the generators $\mathbf{u}_1, \ldots, \mathbf{u}_k$. Further, each solution of $A\mathbf{x} = \mathbf{c}$ is the sum of the anchor solution $\mathbf{v}$ plus an l.c. of the generators $\mathbf{u}_1, \ldots, \mathbf{u}_k$.

■ **Problem**   Suppose the augmented matrix of a linear system has the reduced form

$$\left[\begin{array}{cccc|c} 1 & 0 & 0 & 3 & 4 \\ 0 & 0 & 1 & 0 & 0 \end{array}\right] \ .$$

Find the general solution of the system in vector form.

■ **Solution**   Suppose the four variables of the system are $x_1, \ldots, x_4$. From the given reduced matrix, we can read off the general solution of the system in parametric form:

$$x_1 = 4 - 3x_4 \ .$$
$$x_3 = 0 \ .$$

The procedure of substitution for the corner variables in the general solution vector **x** must now be followed meticulously:

$$\mathbf{x} = \begin{bmatrix} x_1 \\ x_2 \\ x_3 \\ x_4 \end{bmatrix} = \begin{bmatrix} 4 - 3x_4 \\ x_2 \\ 0 \\ x_4 \end{bmatrix} = \begin{bmatrix} 4 \\ 0 \\ 0 \\ 0 \end{bmatrix} + x_2 \begin{bmatrix} 0 \\ 1 \\ 0 \\ 0 \end{bmatrix} + x_4 \begin{bmatrix} -3 \\ 0 \\ 0 \\ 1 \end{bmatrix}.$$

So the general solution is $\mathbf{x} = \mathbf{v} + b_1 \mathbf{u}_1 + b_2 \mathbf{u}_2$,

where the anchor solution is $\mathbf{v} = (4,0,0,0)$

and the generators are $\mathbf{u}_1 = (0,1,0,0)$ and $\mathbf{u}_2 = (-3,0,0,1)$.

Note that, although the column form of vectors is used in the above calculation, it is convenient and perfectly acceptable to list the results obtained in row form.

Note also that, in this example, one of the variables, x_2, does not appear explicitly in the parametric form of the general solution. It is nevertheless a free variable for which a value is chosen in each numerical solution, and it must be treated like any other free variable during the computation of the vector form of the general solution.

■ EXERCISES 2.2

1. Let $A = \begin{bmatrix} 2 & -1 & 3 \\ 0 & 4 & 1 \end{bmatrix}$, $B = \begin{bmatrix} 1 & 2 \\ 3 & 4 \end{bmatrix}$, $C = \begin{bmatrix} a & b \\ c & d \end{bmatrix}$,

   $\mathbf{x} = \begin{bmatrix} 1 \\ -1 \\ 2 \end{bmatrix}$, $\mathbf{y} = \begin{bmatrix} 3 \\ 1 \end{bmatrix}$, $\mathbf{z} = \begin{bmatrix} a \\ d \end{bmatrix}$.

   Compute those of the following matrix-vector products that are consistent:

   (a) $A\mathbf{x}$ (b) $A\mathbf{y}$ (c) $B\mathbf{x}$ (d) $B\mathbf{y}$ (e) $B(A\mathbf{x})$

   (f) $A(B\mathbf{y})$ (g) $B\mathbf{z}$ (h) $C\mathbf{z}$.

2. Assuming that the general solution of a linear system is as given below, rewrite the general solution in vector form.

   (a) ($\mathbf{x}$ in R^2) $x_1 = 4 + 3x_2$
   (b) ($\mathbf{x}$ in R^3) $x_1 = 5 + 3x_2$
                                $x_3 = 7$
   (c) ($\mathbf{x}$ in R^5) $x_1 = 1 + 4x_2 + 3x_4 + x_5$
                                $x_3 = 5 \qquad\quad - 2x_4 + 6x_5$
   (d) ($\mathbf{x}$ in R^3) $x_1 = -x_2 + 8x_3$

3. Rewrite in vector form the general solutions of the following systems from Chapter 1: section 1.2, #1(c), 2(b) through (f); section 1.3, #2(d), (e), (f), 4(a), (b).

4. Find the general solutions in vector form for the systems whose augmented matrices have the reduced forms:

(a) $\begin{bmatrix} 0 & 1 & 2 & 0 & | & 3 \\ 0 & 0 & 0 & 1 & | & 4 \end{bmatrix}$ (b) $\begin{bmatrix} 0 & 1 & 0 & | & 7 \\ 0 & 0 & 0 & | & 0 \end{bmatrix}$

(c) $\begin{bmatrix} 1 & 0 & | & 5 \\ 0 & 1 & | & 8 \\ 0 & 0 & | & 0 \\ 0 & 0 & | & 0 \end{bmatrix}$.

5. In exercise 2(a), suppose we introduce the new parameter $b_1' = b_1 - 2$. Show that the general solution changes

from $x = \begin{bmatrix} 4 \\ 0 \end{bmatrix} + b_1 \begin{bmatrix} 3 \\ 1 \end{bmatrix}$ to $x = \begin{bmatrix} 10 \\ 2 \end{bmatrix} + b_1' \begin{bmatrix} 3 \\ 1 \end{bmatrix}$.

Note: This example shows that the specification given in theorem 2.2 for the vector form of the general solution can be fulfilled in more than one way. In this case, only the anchor solution is altered, but in Chapter 6 we will see that different forms of the generators are also possible. It is important to realize that two forms of the general solution of a system may appear quite different and yet both be correct.

2.3 Transformations of Vector Spaces

Now that we have a matrix-vector product, we can read the equation $Ax = c$ as saying that the matrix A, applied to the vector x in R^n, produces the vector c in R^m. By this reading, a matrix assumes an active role: it transforms a vector x in one vector space to a new vector, c, in another vector space.

We recall that a similar role is played by ordinary functions in calculus. A function f maps each real number, x, to another real number, $y = f(x)$. But a real number may be regarded as a 1-tuple or element of R^1. So we can say that a function maps the set R^1 into R^1. Similarly we say that an $m \times n$ matrix transforms, or maps, R^n into R^m.

In one of the examples from Chapter 1, the components of the vector x in R^3 represented quantities of different grains purchased. The matrix A carried the information about the nutritional content of the various grains, and the components of the vector Ax represented the total amounts of each nutrient obtained. That is, the mapping of x to Ax transformed the given purchasing data into nutritional data. Similarly, in the mining problems we considered, a matrix served to transform data on ore milling into data on metal production.

The active interpretation of matrices, therefore, provides a mechanism for the transformation of one category of data into another. When fully developed in subsequent chapters, this mechanism will be seen to have great power in the analysis of many practical problems.

In this section, we set out the transformation, or mapping, interpretation of matrices. We first present the general concept of vector space transformations, then the specific case of matrix transformations; then we establish special properties of matrix transformations. Finally, the concepts of roots and range of a function are transferred to the vector context and renamed as the ''kernel'' and ''range'' of a transformation. Their immediate use is to improve our earlier descriptions of the set of solutions of a linear system.

Transformations

We are familiar with the concept of functions; that is, with mappings of real numbers into real numbers. For example, the mapping of each number x to its square x^2 is expressed in function notation as $f(x) = x^2$.

In the same way, mappings of one vector space, say R^n, into another vector space, say R^m, are called *transformations* and are usually denoted by capitals such as S, T, etc.

Example 1   A transformation T of R^2 into R^2 is given by the mapping of each vector (x_1, x_2) to the vector (x_1+3, x_2). Or, in function notation:

$$T((x_1,x_2)) = (x_1+3,x_2) \quad .$$

In particular, $T((0,0)) = (3,0), \qquad T((1,2)) = (4,2)$, etc.

Mappings of vector spaces cannot be conveniently graphed on paper as ordinary functions can. In fact, $\mathbf{x}$ and $T(\mathbf{x})$ together contain $n+m$ components, whereas there is room on paper for only two axes, or perhaps three if perspective drawing is used. So even a mapping of R^2 into R^2, involving $2 + 2 = 4$ components, cannot be sketched.

However, we can interpret the above example by noting that $\mathbf{x}$ and $T(\mathbf{x})$ are position vectors of points $P(x_1,x_2)$ and $Q(x_1+3,x_2)$ in the plane, and that Q is always three units to the right of P. We say that the transformation T *translates* each point three units to the right.

More generally, any transformation of R^n to R^n of the form

$$T(\mathbf{x}) = \mathbf{x} + \mathbf{a}$$

is called a *translation*. The above example is a translation with $\mathbf{a} = (3,0)$.

The collection of all transformations of R^n into R^m, like the collection of all real functions, is vast and unmanageable. We will therefore concentrate on a small but important subcategory of transformations with special properties.

Linear Transformations

Of primary interest to us are the *linear transformations* (or *linear mappings*), which are defined to be those satisfying the conditions

(i) $T(\mathbf{x} + \mathbf{y}) = T(\mathbf{x}) + T(\mathbf{y})$ for all vectors $\mathbf{x}$, $\mathbf{y}$

(ii) $T(t\mathbf{x}) = t(T(\mathbf{x}))$ for all vectors $\mathbf{x}$, numbers t.

In particular, condition (a) with $\mathbf{x} = \mathbf{y} = \mathbf{0}$ implies that for a linear transformation, $T(\mathbf{0}) = 2T(\mathbf{0})$, or $T(\mathbf{0}) = \mathbf{0}$. Thus the transformation in example 1 was *not* linear, since $T(\mathbf{0}) = (3,0)$.

Example 2   Let T be the transformation of R^2 into R^2 that reflects every vector across the x_1-axis (see Figure 2.9).

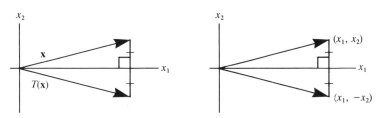

Figure 2.9

In terms of analytic geometry, this means that the x_1 coordinate is unchanged while the x_2 coordinate is reversed:

$$T((x_1, x_2)) = (x_1, -x_2) \quad .$$

To show that T is linear, we check properties (i) and (ii).

(i) $T(\mathbf{x} + \mathbf{y}) = T((x_1 + y_1, x_2 + y_2)) = (x_1 + y_1, -(x_2 + y_2))$
$$= (x_1, -x_2) + (y_1, -y_2)$$
$$= T(\mathbf{x}) + T(\mathbf{y}) \quad . \hspace{5cm} \text{Check.}$$

(ii) $T(t\mathbf{x}) = T((tx_1, tx_2)) = (tx_1, -tx_2) = t(x_1, -x_2)$
$$= tT(\mathbf{x}) \quad . \hspace{6cm} \text{Check.}$$

We could also have checked properties (i) and (ii) for this transformation by methods of Euclidean geometry but the above calculation is quite efficient.

Matrix Transformations

We now know that basic geometric operations such as translations and reflections can be placed within the framework of transformations. The next step is to show that the same framework can be used to describe the action of matrices on vectors.

Given an $m \times n$ matrix A, the mapping of each vector $\mathbf{x}$ in R^n to $A\mathbf{x}$ in R^m is a transformation of vector spaces and is called a *matrix transformation*.

Often the same symbol A is used to denote both the $m \times n$ matrix and the transformation of R^n to R^m that it induces. This overlapping of notation seems to be safe and convenient in practice.

Example 3   Let A be the 2×3 matrix $\begin{bmatrix} 1 & 2 & 3 \\ 4 & 9 & 10 \end{bmatrix}$.

$$\text{Then } A\mathbf{x} = \begin{bmatrix} 1 & 2 & 3 \\ 4 & 9 & 10 \end{bmatrix} \begin{bmatrix} x_1 \\ x_2 \\ x_3 \end{bmatrix} = \begin{bmatrix} x_1 + 2x_2 + 3x_3 \\ 4x_1 + 9x_2 + 10x_3 \end{bmatrix} .$$

That is, the matrix transformation A maps the vector $\mathbf{x} = (x_1, x_2, x_3)$ in R^3 to the vector

$$A\mathbf{x} = (x_1 + 2x_2 + 3x_3, 4x_1 + 9x_2 + 10x_3) \text{ in } R^2 .$$

In particular, A maps $(0,0,0)$ to $(0,0)$, $(1,-4,2)$ to $(-1,-12)$, etc.

The essential character of matrix transformations is given by the following theorem:

THEOREM 2.3

Every matrix transformation is linear. That is, if A is any $m\times n$ matrix, then,

(a) $A(\mathbf{x} + \mathbf{y}) = A\mathbf{x} + A\mathbf{y}$ for all $\mathbf{x}$, $\mathbf{y}$ in R^n .

(b) $A(t\mathbf{x}) = t(A\mathbf{x})$ for all $\mathbf{x}$ in R^n, all numbers t.

■Proof   Recall that from the definition of the matrix-vector product, the first component of $A\mathbf{x}$ is the dot product of the first row of A with the vector $\mathbf{x}$. Or, if we let $\mathbf{a}_1$ denote the first row of A, then $\mathbf{a}_1 \cdot \mathbf{x}$ denotes the first component of $A\mathbf{x}$.

Similarly, $\mathbf{a}_1 \cdot \mathbf{y}$ denotes the first component of $A\mathbf{y}$, and so the first component of $A\mathbf{x} + A\mathbf{y}$ is $\mathbf{a}_1 \cdot \mathbf{x} + \mathbf{a}_1 \cdot \mathbf{y}$, or simply $\mathbf{a}_1 \cdot (\mathbf{x} + \mathbf{y})$.

Now $\mathbf{a}_1 \cdot (\mathbf{x} + \mathbf{y})$ is also the first component of $A(\mathbf{x} + \mathbf{y})$, so we have shown that the vectors $A(\mathbf{x} + \mathbf{y})$ and $A\mathbf{x} + A\mathbf{y}$ have the same first component.

In exactly the same way we can show that they have the same second component, etc. Thus the entire vectors are equal, and identity (a) is proved.

As for identity (b), the first components on the left and right are $\mathbf{a}_1 \cdot (t\mathbf{x})$ and $t(\mathbf{a}_1 \cdot \mathbf{x})$. But since any number t can be factored from a dot product, these two quantities are the same. The first components are therefore equal, and so on.   Q.E.D.

■ Problem   Suppose A is a 2×3 matrix that maps $\mathbf{x} = (1,0,2)$ to $(5,3)$ and maps $\mathbf{y} = (0,1,3)$ to $(-1,-4)$. How does A map the vector $\mathbf{z} = (4,-2,2)$?

■ Solution If we can express **z** in terms of **x** and **y**, we can use the linearity properties of the matrix transformation A to solve the problem.

By inspection, $(4,-2,2) = 4(1,0,2) -2(0,1,3)$,

$$\text{or} \quad \mathbf{z} = 4\mathbf{x} - 2\mathbf{y} \ .$$

Therefore
$$\begin{aligned} A\mathbf{z} &= A(4\mathbf{x} - 2\mathbf{y}) = A(4\mathbf{x}) + A(-2\mathbf{y}) \\ &= 4(A\mathbf{x}) - 2(A\mathbf{y}) = 4(5,3) - 2(-1,-4) \\ &= (22,20) \quad . \end{aligned}$$

We have computed the vector $A\mathbf{z}$ as required, and without having to reconstruct the actual numerical matrix A. In fact, the information given does not suffice to determine A uniquely. Rather, it is an interesting exercise to construct two different matrices that map the given vectors **x** and **y** as specified for A.

The Kernel

To complete our development of vector notation in this chapter, we must transfer some terminology used for functions to vector spaces.

Recall that the *roots* of a function f are the numbers x, that f maps into 0, that is, numbers x such that $f(x) = 0$. For example, when $f(x) = x^2 - 4$, the roots of f are evidently $x = 2$ and $x = -2$.

Similarly, the *kernel* of a transformation T is defined to be the set of vectors **x** that T maps into **0**, that is, vectors **x** such that $T(\mathbf{x}) = \mathbf{0}$. We denote this set of vectors as $Ker(T)$.

Of course the actual computation of the roots of a function, or the kernel of a transformation, can be difficult. But in the case of most importance for us, that of a matrix transformation A, we already have a method of determining $Ker(A)$. From the definition of kernel, **x** is in $Ker(A)$ if $A\mathbf{x} = \mathbf{0}$. That is, **x** is in $Ker(A)$ if **x** is a solution of the linear system $A\mathbf{x} = \mathbf{0}$. But we know how to compute the general solution of a linear system! Therefore, we conclude:

Characterization of Ker(A) A listing of the vectors in the kernel of the matrix transformation A is given by the general solution, in vector form, of the linear system $A\mathbf{x} = \mathbf{0}$.

Example 4 To compute the kernel of the matrix $A = \begin{bmatrix} 1 & 3 \\ 2 & 6 \end{bmatrix}$, find reduced

form $R = \begin{bmatrix} 1 & 3 \\ 0 & 0 \end{bmatrix}$, and general solution $x_1 = -3x_2$, or

$$\mathbf{x} = \begin{bmatrix} -3x_2 \\ x_2 \end{bmatrix} = b_1 \begin{bmatrix} -3 \\ 1 \end{bmatrix} \ .$$

Therefore $Ker(A)$ consists of all scalar multiples of the vector $\mathbf{u} = (-3,1)$.

We can also say something substantial about the kernel for linear transformations in general.

THEOREM 2.4

Let T be a linear transformation of R^n into R^m.
Then $Ker(T)$ is a subspace of R^n.

■**Proof**   Since the set $V = Ker(T)$ is, by definition, a subset of R^n, we need only prove that V has properties #1 and #6 of vector spaces as given in section 2.1.

Suppose, then, that **x** and **y** are in V; in other words, $T(\mathbf{x}) = \mathbf{0}$ and $T(\mathbf{y}) = \mathbf{0}$. Then:

$$T(\mathbf{x} + \mathbf{y}) = T(\mathbf{x}) + T(\mathbf{y}) = \mathbf{0} + \mathbf{0} = \mathbf{0}$$

and $T(t\mathbf{x}) = t(T(\mathbf{x})) = t\mathbf{0} = \mathbf{0}$.

That is, $\mathbf{x} + \mathbf{y}$ and $t\mathbf{x}$ are still in V.

That is, properties #1 and #6 hold as required.   Q.E.D.

COROLLARY 2.5

Let A be an $m \times n$ matrix. Then the set of solutions of the linear system $A\mathbf{x} = \mathbf{0}$, is a subspace of R^n.

■**Proof**   The matrix transformation A is linear (theorem 2.3).
Therefore $Ker(A)$ is a subspace of R^n (theorem 2.4).
But $Ker(A)$ is the set of solutions of the linear system $A\mathbf{x} = \mathbf{0}$.   Q.E.D.

The Range

The range of a function is the set of all numbers that result from the mapping process. For example, consider again the function $y = f(x) = x^2 - 4$. The number that results from the mapping of x is $x^2 - 4$, which is always at least -4. So the range of f consists of the numbers y satisfying $y \geq -4$.

Similarly, the *range of a transformation* T is the set of all vectors that result from the mapping process. We denote this set of vectors as $Range(T)$.

Again, it is often difficult to compute the range of a function or a transformation. But in the case of a matrix transformation, the range concept can be interpreted in terms of the theory of linear systems. Suppose A is an $m \times n$ matrix. Then since A maps **x** in R^n to $A\mathbf{x}$ in R^m, the vectors resulting from the mapping process all have the form $A\mathbf{x}$. Therefore, a vector $\mathbf{c}$ in R^m is in the range of A if $A\mathbf{x} = \mathbf{c}$ for some choice of $\mathbf{x}$. That is, $\mathbf{c}$ is in $Range(A)$ if the linear system $A\mathbf{x} = \mathbf{c}$ is consistent. We can summarize our conclusions as follows.

Characterization of Range(A) The range of the matrix transformation A consists of those vectors $\mathbf{c}$ for which the linear system $A\mathbf{x} = \mathbf{c}$ is consistent.

We will delay the thorough study of $Range(A)$ until Chapter 6. However, for some matrices it is easy to see what the range must be.

Example 5 To find the range of the matrix $A = \begin{bmatrix} 1 & 3 \\ 2 & 6 \end{bmatrix}$, compute:

$$A\mathbf{x} = \begin{bmatrix} 1 & 3 \\ 2 & 6 \end{bmatrix} \begin{bmatrix} x_1 \\ x_2 \end{bmatrix} = \begin{bmatrix} x_1 + 3x_2 \\ 2x_1 + 6x_2 \end{bmatrix} = (x_1 + 3x_2) \begin{bmatrix} 1 \\ 2 \end{bmatrix} .$$

So $A\mathbf{x}$ is always a scalar multiple of the vector $(1,2)$. We conclude that $Range(A)$ consists of all scalar multiples of the vector $\mathbf{w} = (1,2)$.

The range of linear transformations in general also has some algebraic structure:

THEOREM 2.6

Let T be a linear transformation of R^n into R^m.
Then $Range(T)$ is a subspace of R^m.

■**Proof** Since the set $W = Range(T)$ is, by definition, a subset of R^m, we need only prove that W has properties #1 and #6 of vector spaces.

Suppose, then, that $\mathbf{z}$ and $\mathbf{w}$ are in $Range(T)$. Let $\mathbf{x}$ and $\mathbf{y}$ be vectors that T maps into $\mathbf{z}$ and $\mathbf{w}$ respectively: $\mathbf{z} = T(\mathbf{x})$ and $\mathbf{w} = T(\mathbf{y})$.

But then $\mathbf{z} + \mathbf{w} = T(\mathbf{x}) + T(\mathbf{y}) = T(\mathbf{x} + \mathbf{y})$
and $t\mathbf{z} = t(T(\mathbf{x})) = T(t\mathbf{x})$.

That is, $\mathbf{x} + \mathbf{y}$ maps into $\mathbf{z} + \mathbf{w}$ and $t\mathbf{x}$ maps into $t\mathbf{z}$. So $\mathbf{z} + \mathbf{w}$ and $t\mathbf{z}$ are still in the set W, and W has properties #1 and #6 as required.   Q.E.D.

COROLLARY 2.7

Let A be an $m \times n$ matrix. Then the set of those vectors $\mathbf{c}$ for which the system $A\mathbf{x} = \mathbf{c}$ is consistent, forms a subspace of R^m.

■ **Problem** Suppose the 2×2 matrix A has the following properties:

(a) $Ker(A)$ consists of all vectors parallel to $(1,8)$.

(b) $Range(A)$ consists of all vectors parallel to $(3,2)$.

(c) A maps $(4,5)$ to a vector whose first component is 162.

Does this information determine A uniquely?

■ **Solution** Let $A = \begin{bmatrix} a & b \\ c & d \end{bmatrix}$. From the information about $Ker(A)$,

$$A\begin{bmatrix} 1 \\ 8 \end{bmatrix} = \mathbf{0}, \text{ or } \begin{bmatrix} a & b \\ c & d \end{bmatrix}\begin{bmatrix} 1 \\ 8 \end{bmatrix} = \begin{bmatrix} 0 \\ 0 \end{bmatrix},$$

or $\begin{cases} a + 8b = 0 \\ c + 8d = 0 \end{cases}$, or $\begin{cases} a = -8b \\ c = -8d \end{cases}$.

Therefore A has the form $A = \begin{bmatrix} -8b & b \\ -8d & d \end{bmatrix}$.

Next, vectors in $Range(A)$ have the form $A\mathbf{x}$, or

$$\begin{bmatrix} -8b & b \\ -8d & d \end{bmatrix}\begin{bmatrix} x_1 \\ x_2 \end{bmatrix} = \begin{bmatrix} -8bx_1 + bx_2 \\ -8dx_1 + dx_2 \end{bmatrix} = (-8x_1 + x_2)\begin{bmatrix} b \\ d \end{bmatrix}.$$

But since these vectors must be parallel to $(3,2)$,

$(b,d) = m(3,2)$, or $b = 3m$, $d = 2m$.

Therefore A has the form $\begin{bmatrix} -24m & 3m \\ -16m & 2m \end{bmatrix}$.

Finally, from part (c) of the given information,

$$A\begin{bmatrix} 4 \\ 5 \end{bmatrix} = \begin{bmatrix} -24m & 3m \\ -16m & 2m \end{bmatrix}\begin{bmatrix} 4 \\ 5 \end{bmatrix} = \begin{bmatrix} -81m \\ -54m \end{bmatrix} = \begin{bmatrix} 162 \\ * \end{bmatrix}.$$

Therefore $-81m = 162$ or $m = -2$.

Thus A is uniquely determined as the matrix $\begin{bmatrix} 48 & -6 \\ 32 & -4 \end{bmatrix}$.

■ EXERCISES 2.3

1. Let T be the transformation of R^2 into R^2 given by

$T((x_1, x_2)) = ((x_1)^2, x_2)$.

(a) Compute $T((1,0))$ and $T((2,0))$.

(b) Deduce that $T(t\mathbf{x})$ does not always equal $t(T(\mathbf{x}))$ and that therefore T is not linear.

2. Show that the following transformations of R^2 into R^2 are linear:

(a) $P((x_1, x_2)) = (x_1, 0)$

(b) $T(\mathbf{x}) = 3\mathbf{x}$.

3. Let $A = \begin{bmatrix} 1 & 8 & 3 \\ 7 & 0 & 2 \end{bmatrix}$, $\mathbf{x} = \begin{bmatrix} 1 \\ 0 \\ 3 \end{bmatrix}$, $\mathbf{y} = \begin{bmatrix} 4 \\ 2 \\ 5 \end{bmatrix}$.

Compute $A(\mathbf{x} + \mathbf{y})$, $A\mathbf{x}$, $A\mathbf{y}$, $A\mathbf{x} + A\mathbf{y}$, and $A(t\mathbf{x})$, and thereby verify that $A(\mathbf{x} + \mathbf{y}) = A\mathbf{x} + A\mathbf{y}$ and that $A(t\mathbf{x}) = t(A\mathbf{x})$.

4. Suppose the 2×2 matrix A maps $(1,1)$ to $(4,7)$ and maps $(1,-1)$ to $(8,3)$. How does A map the following vectors?

(a) $(2,0)$ (b) $(1,0)$ (c) $(0,2)$

(d) $(0,1)$ (e) $(5,14)$ (f) $(9,3)$

5. Suppose the 3×2 matrix A maps $(1,7,0)$ to $(5,4)$ and maps $(0,3,1)$ to $(2,6)$. How does A map $(1,1,-2)$?

6. Let $A = \begin{bmatrix} 2 & 7 \\ -4 & -14 \end{bmatrix}$.

(a) Determine $Ker(A)$.

(b) Show that $Range(A)$ consists of those vectors parallel to $(1,-2)$.

7. Let $A = \begin{bmatrix} 1 & 8 \\ 2 & 5 \\ 3 & 7 \end{bmatrix}$, and let $\mathbf{z}_1 = \begin{bmatrix} 1 \\ 2 \\ 3 \end{bmatrix}$, $\mathbf{z}_2 = \begin{bmatrix} 8 \\ 5 \\ 7 \end{bmatrix}$.

(a) Show that $Ker(A)$ contains just the zero vector $\mathbf{0}$.

(b) Compute $A\mathbf{x}$. Deduce that $Range(A)$ consists of all linear combinations of $\mathbf{z}_1$ and $\mathbf{z}_2$; in other words all vectors of the form

$$\mathbf{w} = x_1 \mathbf{z}_1 + x_2 \mathbf{z}_2 .$$

8. Suppose that a 2×3 matrix and its three columns are denoted

$$A = \begin{bmatrix} a & b & c \\ d & e & f \end{bmatrix} , \quad \mathbf{z}_1 = \begin{bmatrix} a \\ d \end{bmatrix} , \quad \mathbf{z}_2 = \begin{bmatrix} b \\ e \end{bmatrix} , \quad \mathbf{z}_3 = \begin{bmatrix} c \\ f \end{bmatrix} .$$

(a) Show that $A\mathbf{x} = x_1 \mathbf{z}_1 + x_2 \mathbf{z}_2 + x_3 \mathbf{z}_3$.

(b) Deduce that $Range(A)$ consists of all linear combinations of the columns of A.

9. Suppose a 2×2 matrix A having the following properties is to be selected:

   (a) $Ker(A)$ consists of all vectors parallel to $(4,2)$.
   (b) $Range(A)$ consists of all vectors parallel to $(2,5)$.

   Determine all possible choices of A.

10. Is it possible for a 3×2 matrix to map $(1,2)$ to $(5,0,3)$; $(2,3)$ to $(2,1,4)$; and $(5,8)$ to $(9,2,12)$?

11. Let P be the transformation of R^2 into R^2 given by

    $$P((x_1, x_2)) = (0, x_2) \quad .$$

    (a) Show that $Ker(P)$ consists of all vectors parallel to $(1,0)$.
    (b) Show that $Range(P)$ consists of all vectors parallel to $(0,1)$.

12. (Alternative proof that $Ker(A)$ is a subspace.)

    (a) Show that if $\mathbf{q}_1, \ldots, \mathbf{q}_k$ are any k vectors in a vector space V, then the set of all linear combinations of $\mathbf{q}_1, \ldots, \mathbf{q}_k$ forms a subspace of V.
    (b) Use the fact that the general solution of $A\mathbf{x} = \mathbf{0}$ consists of all l.c.'s of the generators $\mathbf{u}_1, \ldots, \mathbf{u}_k$ to conclude that $Ker(A)$ is a subspace of R^n.

13. By using the ideas in exercises 8 and 10, construct an alternative proof of the theorem that $Range(A)$ is always a subspace of R^m.

*2.4 Lines and Planes

Introduction

So far we have emphasized the vector space R^n, the space of n-tuples. The construction of this space was independent of any geometric concepts, although the geometric *interpretation* of 2- and 3-tuples as position vectors was intuitively helpful. But now we are going to apply vector methods to a specifically geometric problem: the study of lines and planes in space. For this purpose, we first introduce a new kind of vector that is actually *defined* in terms of Euclidean geometry. These vectors, to be called "Euclidean vectors," are especially well suited to the problem at hand.

The study of lines and planes in space is, in fact, one of the most important and immediate applications of the theory of systems of linear equations.

Nevertheless, the development of matrix algebra presented in the chapters to follow does not depend on any knowledge of Euclidean vectors, lines, and planes; for that reason the rest of this chapter can be omitted on first reading.

Directed Line Segments

We begin with a concept from the Euclidean geometry of space (or the plane). Suppose A and B are two points (see Figure 2.10). Then we let AB denote the *line segment* joining points A and B.

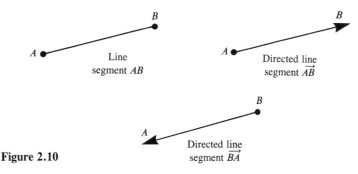

Figure 2.10

Of course, AB and BA denote the same line segment, i.e., the one joining A and B. However, we can distinguish between the two ends of a line segment by designating one, say A, as the *initial point* (or *point of application*) and the other, say B, as the *end point*. We then have a *directed line segment*, denoted $\overrightarrow{AB}$. In diagrams, an arrow is placed at the end point, but not at the initial point, of a directed line segment.

Observe that $\overrightarrow{BA}$ is a different directed line segment from $\overrightarrow{AB}$, because it has B as initial point and A as end point.

Consider the directed line segments $\overrightarrow{AB}$, $\overrightarrow{CD}$, and $\overrightarrow{EF}$, which lie on parallel lines. (See Figure 2.11.)

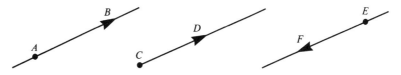

Figure 2.11

Since $\overrightarrow{AB}$ and $\overrightarrow{CD}$ point in the same direction, they are said to be *parallel*. But $\overrightarrow{AB}$ and $\overrightarrow{EF}$ point in opposite directions, and so are said to be *antiparallel*.

Directed line segments $\overrightarrow{AB}$ and $\overrightarrow{CD}$ are said to be *equivalent* when they are of equal length and parallel. (See Figure 2.12.) According to a theorem in Euclidean geometry (see exercises), this is so if and only if the quadrilateral $ABDC$ is a parallelogram.

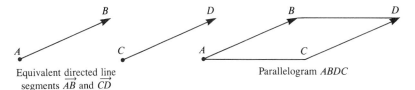

Equivalent directed line segments $\overrightarrow{AB}$ and $\overrightarrow{CD}$

Parallelogram *ABDC*

Figure 2.12

A *Euclidean vector* is a set of equivalent directed line segments. In other words, in order to obtain a Euclidean vector, choose one directed line segment $\overrightarrow{AB}$ and then take all directed line segments equivalent to $\overrightarrow{AB}$. (See Figure 2.13.) At first, a Euclidean vector seems to be a rather complicated object, consisting of so many distinct (though equivalent) directed line segments; but in practice we usually refer to only one of these segments at a time. The others can, when necessary, be deduced from the given one; and it turns out that the reserve supply of equivalent directed line segments greatly increases our power of calculation in geometric problems.

Figure 2.13

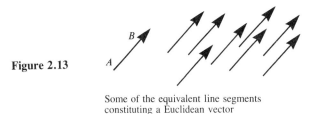

Some of the equivalent line segments constituting a Euclidean vector

The usual vector notation **u**, **v**, etc. is used for Euclidean vectors. Also, the formula $\mathbf{u} = \overrightarrow{AB}$ means that $\overrightarrow{AB}$ is *one* of the equivalent directed line segments constituting the Euclidean vector **u**.

Operations on Euclidean Vectors

Suppose $\mathbf{u} = \overrightarrow{AB}$ and $\mathbf{v} = \overrightarrow{BC}$ are Euclidean vectors. Then the *sum* of **u** and **v** is defined to be the vector $\mathbf{u} + \mathbf{v} = \overrightarrow{AC}$.

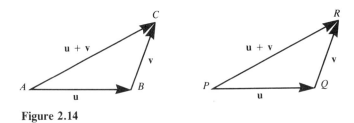

Figure 2.14

Note that if we represented the above vectors **u** and **v** by the directed line segments $\overrightarrow{PQ}$ and $\overrightarrow{QR}$ instead of $\overrightarrow{AB}$ and $\overrightarrow{BC}$, we would obtain for the sum the vector $\mathbf{u} + \mathbf{v} = \overrightarrow{PR}$, instead of $\overrightarrow{AC}$. However, it is easily seen that the triangles ABC and PQR would be congruent, with corresponding sides parallel. So $\overrightarrow{AC}$ and $\overrightarrow{PR}$ would be equivalent directed line segments and would represent the same Euclidean vector. That is, the same result $\mathbf{u} + \mathbf{v}$ is obtained as the sum of $\mathbf{u}$ and $\mathbf{v}$ no matter how the addition is performed.

Further, the scalar product of a vector $\mathbf{u} = \overrightarrow{AB}$ and a scalar $r \geq 0$ is defined to be the vector $r\mathbf{u}$ that is parallel to $\mathbf{u}$ and r times as long. (See Figure 2.15.) When $r < 0$, $r\mathbf{u}$ is defined to be antiparallel to $\mathbf{u}$ and $|r|$ times as long.

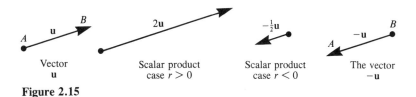

| Vector | Scalar product | Scalar product | The vector |
| $\mathbf{u}$ | case $r > 0$ | case $r < 0$ | $-\mathbf{u}$ |

Figure 2.15

Finally, the symbol $-\mathbf{u}$ denotes the vector $(-1)\mathbf{u}$. Evidently, if $\mathbf{u} = \overrightarrow{AB}$, then $-\mathbf{u} = \overrightarrow{BA}$.

Euclidean Vectors, Position Vectors, and N-tuples

Let us now relate Euclidean vectors to the kinds of vectors discussed earlier in this chapter. (See Figure 2.16.) Suppose that, as usual, O denotes the origin of a Cartesian coordinate system. Suppose $\mathbf{u} = \overrightarrow{AB}$ is a Euclidean vector, and let P be the fourth vertex of the parallelogram $OABP$.

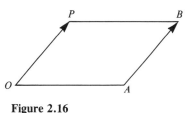

Figure 2.16

Then the directed line segment $\overrightarrow{OP}$ is equivalent to $\overrightarrow{AB}$, and so $\mathbf{u} = \overrightarrow{OP}$. That is, each Euclidean vector $\mathbf{u}$ can be represented by a position vector $\overrightarrow{OP}$.

But we can specify the Cartesian coordinates of the point P by writing $P(x_1, x_2)$ or $P(x_1, x_2, x_3)$ in the plane or space respectively. Thus the position vector $\overrightarrow{OP}$ corresponds to the n-tuple (x_1, x_2) or (x_1, x_2, x_3). Each Euclidean vector $\mathbf{u}$ therefore corresponds to a position vector $\overrightarrow{OP}$ *and* an n-tuple.

We can show right away that this correspondence holds up under the operations of addition and scalar multiplication.

Lemma 2.6 Suppose the Euclidean vectors $\mathbf{u},\mathbf{v}$ correspond to the n-tuples $(x_1, \ldots ,x_n)$ and $(y_1, \ldots ,y_n)$ respectively. Then

(a) the Euclidean vector $\mathbf{u} + \mathbf{v}$ corresponds to the n-tuple

$(x_1+y_1, \ldots ,x_n+y_n)$.

(b) the Euclidean vector $r\mathbf{u}$ corresponds to the n-tuple

$(rx_1, \ldots ,rx_n)$.

■**Proof**   We shall discuss the case $n = 2$, i.e., the Euclidean plane. The case $n = 3$, or Euclidean space, is exactly the same.

(a) Consider the points $P(x_1,x_2)$, $Q(y_1,y_2)$, and $S(x_1+y_1,x_2+y_2)$. (See Figure 2.17.) Then, as noted in section 2.1, OS is the diagonal of the parallelogram with adjacent sides OP and OQ.

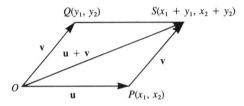

Figure 2.17

However, since $OPSQ$ is a parallelogram, $\mathbf{v} = \overrightarrow{OQ} = \overrightarrow{PS}$. Therefore,

$\mathbf{u} + \mathbf{v} = \overrightarrow{OP} + \overrightarrow{PS} = \overrightarrow{OS}$.

That is, $\mathbf{u} + \mathbf{v}$ corresponds to the position vector $\overrightarrow{OS}$ and therefore to the n-tuple (x_1+y_1,x_2+y_2), as required.

(b) We discuss the case $r > 0$. (The opposite case $r < 0$ is left as an exercise.) Let us plot the points O, $P(x_1,x_2)$, and $W(rx_1,rx_2)$. (See Figure 2.18.)

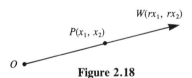

Figure 2.18

From analytic geometry, we know that $\overrightarrow{OP}$ and $\overrightarrow{OW}$ are parallel, and also that $\overrightarrow{OP}$ has length $(x_1^2 + x_2^2)^{\frac{1}{2}}$ whole $\overrightarrow{OW}$ has length $((rx_1)^2 + (rx_2)^2)^{\frac{1}{2}}$, or $r(x_1^2 + x_2^2)^{\frac{1}{2}}$. That is, $\overrightarrow{OW}$ is r times as long as $\overrightarrow{OP}$. But since $\mathbf{u}$ corresponds to $\overrightarrow{OP}$, $r\mathbf{u}$ corresponds to $\overrightarrow{OW}$, and therefore to the n-tuple (rx_1, rx_2).   Q.E.D.

Evidently there is a close structural relation between the Euclidean vectors and the n-tuples, despite the vastly different ways in which they are defined. Here is a concrete illustration of the linkage.

COROLLARY **2.7**

Let E^2 (or E^3) denote the set of all Euclidean vectors in the plane (or in space), with the addition and scalar multiplication operations defined above. Then E^2 and E^3 form vector spaces.

■**Proof**   To show that E^2, say, is a vector space, we need only verify that it has the 10 properties of a vector space as given in section 2.1. However, these properties can quickly be deduced from the corresponding properties of R^2, the space of 2-tuples. For example, to show that $\mathbf{u} + \mathbf{v} = \mathbf{v} + \mathbf{u}$, suppose that $\mathbf{u}$ and $\mathbf{v}$ correspond to the 2-tuples, (x_1, x_2) and (y_1, y_2) respectively. Then, by the previous lemma,

$$\mathbf{u} + \mathbf{v} \text{ corresponds to } (x_1+y_1, \ x_2+y_2)$$

while $\mathbf{v} + \mathbf{u}$ corresponds to $(y_1+x_1, y_2+x_2) = (x_1+y_1, \ x_2+y_2)$.

Thus $\mathbf{u} + \mathbf{v}$ and $\mathbf{v} + \mathbf{u}$ correspond to the same 2-tuple and so are the same Euclidean vector.

Verification of the other vector space properties is accomplished by similar methods, and can be left as an exercise. Q.E.D.

Example 1   Suppose we are given two points $A(2,1)$ and $B(3,4)$, and want to find the 2-tuple corresponding to the Euclidean vector $\mathbf{u} = \overrightarrow{AB}$. From Figure 2.19,

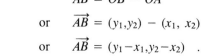

we have $\mathbf{w} = \mathbf{v} + \mathbf{u}$,   or $\mathbf{u} = \mathbf{w} - \mathbf{v}$.

Now $\mathbf{w}$ and $\mathbf{v}$ correspond to position vectors $\overrightarrow{OB}$ and $\overrightarrow{OA}$ respectively and, therefore, to n-tuples $(3,4)$ and $(2,1)$ respectively; and so $\mathbf{u}$ corresponds to the n-tuple $(3,4) - (2,1) = (1,3)$.

From now on we will streamline our terminology by writing

$$\mathbf{u} = \overrightarrow{OP} = (x_1, x_2)$$

to mean that the Euclidean vector $\mathbf{u}$ corresponds to the position vector $\overrightarrow{OP}$ and the n-tuple (x_1, x_2).

Figure 2.19

Example 2   Suppose we are given two arbitrary points $A(x_1, x_2)$ and $B(y_1, y_2)$ and want to find a formula for $\overrightarrow{AB}$ in terms of position vectors or n-tuples. From the method of the previous example, we have

$$\overrightarrow{AB} = \overrightarrow{OB} - \overrightarrow{OA}$$

or $\overrightarrow{AB} = (y_1, y_2) - (x_1, x_2)$

or $\overrightarrow{AB} = (y_1-x_1, y_2-x_2)$.

Example 3 Suppose we are given points $A(1,3,5)$ and $B(0,6,2)$ in space and want to find a vector $\mathbf{v}$ that is parallel to $\mathbf{u} = \overrightarrow{AB}$, but three times as long.

The calculation of the needed vector proceeds just as in the previous examples involving plane vectors, except that now there is a third component in all the n-tuples we use. First we compute $\mathbf{u}$.

$$\begin{aligned}\mathbf{u} = \overrightarrow{AB} &= \overrightarrow{OB} - \overrightarrow{OA} \\ &= (0,6,2) - (1,3,5) \\ &= (-1,3,-3) \quad .\end{aligned}$$

But we are given that $\mathbf{v} = 3\mathbf{u}$, that is:

$$\mathbf{v} = 3(-1,3,-3) = (-3,9,-9) \quad .$$

Example 4 Given two points $A(1,3)$ and $B(5,1)$, suppose we want to find the coordinates of the midpoint C of the line segment AB. From Figure 2.20, we have

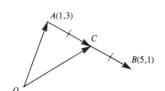

Figure 2.20

$$\begin{aligned}\overrightarrow{AC} &= (\tfrac{1}{2})\overrightarrow{AB} \\ \text{and} \quad \overrightarrow{OC} &= \overrightarrow{OA} + \overrightarrow{AC} \\ &= \overrightarrow{OA} + (\tfrac{1}{2})\overrightarrow{AB} \\ &= \overrightarrow{OA} + (\tfrac{1}{2})(\overrightarrow{OB} - \overrightarrow{OA}) \\ &= (\tfrac{1}{2})(\overrightarrow{OA} + \overrightarrow{OB}) \\ &= (\tfrac{1}{2})((1,3) + (5,1)) = (\tfrac{1}{2})(6,4) \\ \text{or} \quad \overrightarrow{OC} &= (3,2) \quad .\end{aligned}$$

That is, the required point is $C(3,2)$.

Example 5 Given two points $A(x_1, x_2, x_3)$ and $B(y_1,y_2,y_3)$, suppose we want to find the position vector of the point C that divides the line segment AB in the ratio $t : 1-t$, as in the diagram.

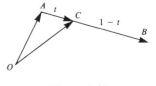

Figure 2.21

$$\begin{aligned}\text{Then} \quad \overrightarrow{OC} &= \overrightarrow{OA} + \overrightarrow{AC} = \overrightarrow{OA} + t(\overrightarrow{AB}) \\ &= \overrightarrow{OA} + t(\overrightarrow{OB} - \overrightarrow{OA}) \\ \text{or} \quad \overrightarrow{OC} &= (1-t)(\overrightarrow{OA}) + t(\overrightarrow{OB}) \\ \text{or} \quad \overrightarrow{OC} &= (1-t)(x_1,x_2,x_3) + t(y_1,y_2,y_3) \\ &= ((1-t)x_1+ty_1 \;,\; (1-t)x_2+ty_2 \;,\; (1-t)x_3+ty_3) \quad .\end{aligned}$$

The power of the method of Euclidean vectors in geometric problems is illustrated by the following proof of a classical theorem:

Example 6 A theorem from Euclidean geometry states that the medians of a triangle are concurrent and that their common point of intersection divides each median in a 2:1 ratio.

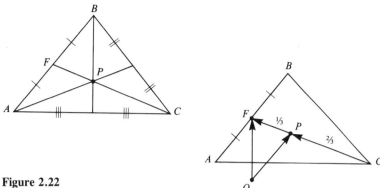

Figure 2.22

In triangle ABC in Figure 2.22, F is the midpoint of side AB, so that CF is one of the medians. In the right-hand diagram, P is the point that divides median CF in the 2:1 ratio. To match the formulation of example 5, we can express this ratio as $\frac{2}{3}:\frac{1}{3}$. From examples 4 and 5, we can see that

$$\overrightarrow{OF} = \left(\tfrac{1}{2}\right)(\overrightarrow{OA} + \overrightarrow{OB}),$$

while

$$\overrightarrow{OP} = \left(\tfrac{2}{3}\right)(\overrightarrow{OF}) + \left(\tfrac{1}{3}\right)(\overrightarrow{OC}) \quad.$$

Therefore,

$$\overrightarrow{OP} = \left(\tfrac{2}{3}\right)(\left(\tfrac{1}{2}\right)(\overrightarrow{OA} + \overrightarrow{OB})) + \left(\tfrac{1}{3}\right)(\overrightarrow{OC})$$

$$= \left(\tfrac{1}{3}\right)(\overrightarrow{OA} + \overrightarrow{OB}) + \left(\tfrac{1}{3}\right)(\overrightarrow{OC})$$

or

$$\overrightarrow{OP} = \left(\tfrac{1}{3}\right)(\overrightarrow{OA} + \overrightarrow{OB} + \overrightarrow{OC}) \quad.$$

Note that the formula is symmetric in A,B,C, and so we can see that if we compute the position vectors of the points dividing the other medians in 2:1 ratio, we must get the same result. Thus, the three medians are concurrent at P.

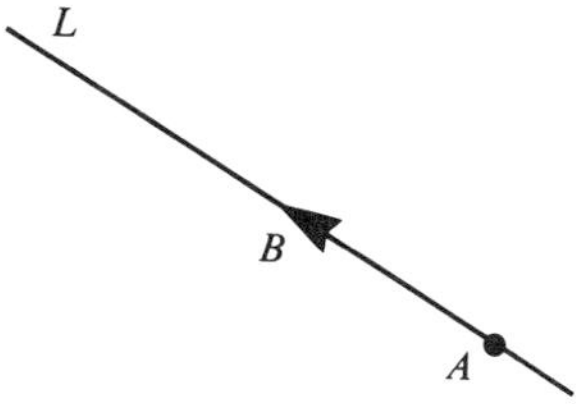

Equations of Lines

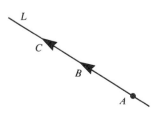

Figure 2.23

Figure 2.24

Let L be the line passing through two given points A and B. (See Figure 2.23). Then the Euclidian vector $\overrightarrow{AB}$ is parallel to L.

Similarly, an arbitrary point C is on L if and only if the Euclidean vector $\overrightarrow{AC}$ is parallel to L (see Figure 2.24), so that

$$\overrightarrow{AC} = t(\overrightarrow{AB}) \qquad \text{for some scalar } t.$$

But since $\overrightarrow{AC} = \overrightarrow{OC} - \overrightarrow{OA}$, we can restate the condition for C to be on L as

$$\overrightarrow{OC} = \overrightarrow{OA} + t(\overrightarrow{AB}) \qquad \text{for some scalar } t.$$

If we introduce the streamlined notations $\mathbf{a} = \overrightarrow{OA}$ and $\mathbf{u} = \overrightarrow{AB}$, our observations assume the following concise form.

Vector form of the equation of a line. Suppose the line L is parallel to the vector $\mathbf{u}$, and that $\mathbf{a}$ is the position vector of some point on L (see Figure 2.25). Then an arbitrary point C is on L if and only if its position vector has the form

$$\mathbf{x} = \mathbf{a} + t\mathbf{u} \qquad \text{for some scalar } t.$$

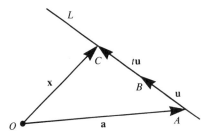

Example 1 To obtain the vector form of the equation of the line L through the points $A(6,1)$ and $B(9,0)$ in the plane, we compute

$$\mathbf{a} = \overrightarrow{OA} = (6,1)$$

and $\quad \mathbf{u} = \overrightarrow{AB} = \overrightarrow{OB} - \overrightarrow{OA} = (9,0) - (6,1) = (3,-1)$.

Then $\quad \mathbf{x} = (6,1) + t(3,-1)$.

Of course we can expand this formula as

$$(x_1, x_2) = (6+3t, 1-t)$$

or even rewrite it as a pair of equations:

$$\begin{cases} x_1 = 6 + 3t \\ x_2 = 1 - t \end{cases} .$$

For example, to obtain the intercept of L on the x_2-axis (see Figure 2.26), we write $x_1 = 6 + 3t = 0$, solve for $t = -2$, and on substitution obtain $x_2 = 1 - (-2) = 3$. Therefore the intercept is the point $K(0,3)$.

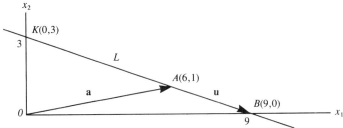

Figure 2.26

Note also that the parameter t can be eliminated between the equations:

since $\quad t = \dfrac{x_1 - 6}{3} \quad$ and $\quad t = \dfrac{x_2 - 1}{-1}$,

we have $\quad \dfrac{x_1 - 6}{3} = \dfrac{x_2 - 1}{-1}$,

or, on cross-multiplication, $x_1 + 3x_2 = 9$, which is the familiar *scalar form* of the equation of a line in the plane.

Example 2   To obtain the vector form of the equation of the line through the points $A(2,4,1)$ and $B(1,2,3)$ in space (see Figure 2.27), we compute

$$\mathbf{a} = \overrightarrow{OA} = (2,4,1) \quad ,$$
$$\mathbf{u} = \overrightarrow{OB} - \overrightarrow{OA} = (1,2,3) - (2,4,1) = (-1,-2,2) \quad ,$$

and obtain $\quad \mathbf{x} = (2,4,1) + t(-1,-2,2) \quad ,$

or $\quad \begin{cases} x_1 = 2 - t \\ x_2 = 4 - 2t \\ x_3 = 1 + 2t \quad . \end{cases}$

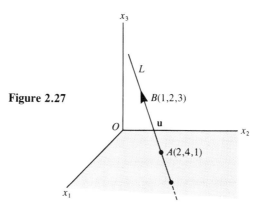

Figure 2.27

More generally, the line $\mathbf{x} = \mathbf{a} + t\mathbf{u}$ in space can be represented in the *parametric form*

$$\begin{cases} x_1 = a_1 + tu_1 \\ x_2 = a_2 + tu_2 \\ x_3 = a_3 + tu_3 \quad . \end{cases}$$

or, on solving each equation for t, in the *symmetric form*

$$\frac{x_1 - a_1}{u_1} = \frac{x_2 - a_2}{u_2} = \frac{x_3 - a_3}{u_3} \quad .$$

In particular, the line in example 2 has the symmetric form representation:

$$\frac{x_1 - 2}{-1} = \frac{x_2 - 4}{-2} = \frac{x_3 - 1}{2} \quad .$$

Example 3 Suppose we want to determine whether the lines

$$L_1: \quad \mathbf{x} = (1,2,7) + t(2,0,2)$$

and L_2: $\quad \mathbf{x} = (4,1,1) + t(1,2,3)$

intersect at some point P in space. If so, then P is the point on L_1 corresponding to a certain choice of the parameter t, say $t = r$; and P is also the point on L_2 corresponding to another choice of parameter, say $t = s$. Then we can equate the two representations of the point P:

$$(1,2,7) + r(2,0,2) = (4,1,1) + s(1,2,3) \quad .$$

The vector equation is equivalent to the scalar equations:

$$\begin{cases} 1 + 2r = 4 + s \\ 2 \quad\quad\;\; = 1 + 2s \\ 7 + 2r = 1 + 3s \end{cases} \quad \text{or} \quad \begin{cases} 2r - s = 3 \\ - 2s = -1 \\ 2r - 3s = -6 \end{cases} .$$

To solve for the unknowns r and s, we reduce the augmented matrix:

$$\begin{bmatrix} 2 & -1 & 3 \\ 0 & -2 & -1 \\ 2 & -3 & -6 \end{bmatrix} \longrightarrow \begin{bmatrix} 2 & -1 & 3 \\ 0 & -2 & -1 \\ 0 & -2 & -9 \end{bmatrix} \longrightarrow \begin{bmatrix} 2 & -1 & 3 \\ 0 & -2 & -1 \\ 0 & 0 & -8 \end{bmatrix} .$$

The system is inconsistent; thus there are no parameter values r and s that satisfy the original vector equation, and the given lines L_1 and L_2 do not intersect.

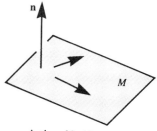

A plane M with normal vector $\mathbf{n}$

Figure 2.28

Equations of Planes

A vector $\mathbf{n}$ in space is said to be *normal* to a given plane M if $\mathbf{n}$ is orthogonal to every vector in M (see Figure 2.28).

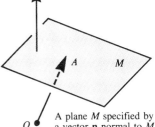

A plane M specified by a vector $\mathbf{n}$ normal to M and a point A in M

Figure 2.29

Conversely, the principles of three-dimensional Euclidean geometry imply that a plane M is uniquely specified by the choice of

(i) a point A in the plane M, with, say, position vector $\mathbf{a} = \overrightarrow{OA}$, and
(ii) a vector $\mathbf{n}$ normal to the plane M (see Figure 2.29).

Suppose, in fact, that a point C, with position vector $\mathbf{x}$, is in M (see Figure 2.30). Then $\overrightarrow{AC}$ is a vector in M, and so $\overrightarrow{AC}$ is orthogonal to $\mathbf{n}$.

That is, $\quad \mathbf{n} \cdot (\overrightarrow{AC}) = 0 \quad$,

and, since $\quad \overrightarrow{AC} = \overrightarrow{OC} - \overrightarrow{OA} = \mathbf{x} - \mathbf{a} \quad$,

we obtain the *normal form of the equation of a plane:*

$$\mathbf{n} \cdot (\mathbf{x} - \mathbf{a}) = 0 \quad .$$

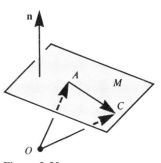

Figure 2.30

Example 4   The equation of the plane through the point $A(10,5,2)$ and with normal vector $\mathbf{n} = (3,4,5)$ is

$$(3,4,5)\cdot((x_1,\ x_2,\ x_3) - (10,5,2)) = 0$$

or $(3,4,5)\cdot(x_1-10,\ x_2-5,\ x_2-2) = 0$

or $3(x_1-10) + 4(x_2-5) + 5(x_3-2) = 0$

or $3x_1 + 4x_2 + 5x_3 - 60 = 0$.

More generally, the equation of a plane, $\mathbf{n}\cdot(\mathbf{x}-\mathbf{a}) = 0$, can be rewritten as $\mathbf{n}\cdot\mathbf{x} = \mathbf{n}\cdot\mathbf{a}$. But the dot product $\mathbf{n}\cdot\mathbf{a}$ is simply a scalar, say k. This gives us the *equation of a plane in scalar form*:

$$n_1x_1 + n_2x_2 + n_3x_3 = k .$$

Example 5   Suppose we have a plane M given by the scalar equation

$$12x_1 + 3x_2 + 4x_3 = 20 ,$$

and we want to represent M in normal form. First, we observe that the constant term, 20, in the scalar equation can be incorporated into one of the variable terms by writing

$$12x_1 + 3x_2 + 4(x_3-5) = 0 .$$

But the new equation is easily converted to normal form:

$$(12,3,4)\cdot(x_1-0,\ x_2-0,\ x_3-5) = 0$$

or $\mathbf{n}\cdot(\mathbf{x} - \mathbf{a}) = 0$,   where $\mathbf{n} = (12,3,4)$, $\mathbf{a} = (0,0,5)$.

Example 6   Suppose we must find the scalar form of the equation of the plane through the points $A(1,1,1)$, $B(2,3,1)$, and $C(3,3,4)$. We know that the required equation for M has the form

$$n_1x_1 + n_2x_2 + n_3x_3 = k ,$$

and on substituting the components of the three given points into this equation, we obtain a system of equations for the quantities n_1, n_2, n_3, k:

$$\begin{cases} n_1 + n_2 + n_3 = k \\ 2n_1 + 3n_2 + n_3 = k \\ 3n_1 + 3n_2 + 4n_3 = k . \end{cases}$$

Then we reduce the augmented matrix:

$$\begin{bmatrix} 1 & 1 & 1 & k \\ 2 & 3 & 1 & k \\ 3 & 3 & 4 & k \end{bmatrix} \longrightarrow \begin{bmatrix} 1 & 1 & 1 & k \\ 0 & 1 & -1 & -k \\ 0 & 0 & 1 & -2k \end{bmatrix} \longrightarrow \cdots$$

$$\cdots \begin{bmatrix} 1 & 0 & 2 & 2k \\ 0 & 1 & -1 & -k \\ 0 & 0 & 1 & -2k \end{bmatrix} \longrightarrow \begin{bmatrix} 1 & 0 & 0 & 6k \\ 0 & 1 & 0 & -3k \\ 0 & 0 & 1 & -2k \end{bmatrix}$$

and conclude that $\quad \mathbf{n} = (n_1, n_2, n_3) = (6k, -3k, -2k) = k(6, -3, -2) \quad$.

On choosing $k = 1$, we obtain $\mathbf{n} = (6, -3, -2)$ and a scalar equation for M:

$$6x_1 - 3x_2 - 2x_3 = 1 \quad .$$

Geometric Interpretation of Linear Systems

We know from analytic geometry (as well as from the earlier part of this section) that a linear equation in two variables, such as $x_1 + 2x_2 = 5$, represents a line in the plane. Thus a solution of the linear system:

$$\begin{cases} L_1 & x_1 + 2x_2 = 5 \\ L_2 & 4x_1 + x_2 = 6 \\ L_3 & 3x_1 + 7x_2 = 8 \end{cases}$$

is a point (x_1, x_2) at which the lines L_1, L_2, L_3 represented in the system intersect. Usually three (or more) lines in the plane have no common point (see Figure 2.31), and so we expect a system of three (or more) equations in two variables to be inconsistent. In particular, it is easily verified that the above system is inconsistent.

By contrast, two lines in the plane usually have a point of intersection, and so a system of two equations in two variables usually has a solution.

Clearly the vector methodology of this chapter is not necessary for geometric interpretation of systems in two variables. Now we increase the number of variables.

A linear equation in three variables, such as $x_1 + 2x_2 + x_3 = 4$, represents a plane in space. So a solution of the linear system

$$\begin{array}{ll} M_1 & x_1 + 2x_2 + x_3 = 4 \\ M_2 & 2x_1 + 5x_2 + x_3 = 9 \end{array}$$

is a point of intersection of the planes M_1, M_2 represented in the system. Geometric intuition suggests that two planes usually intersect along a line L. Let us see whether calculation confirms this expectation.

$$\begin{bmatrix} 1 & 2 & 1 & | & 4 \\ 2 & 5 & 1 & | & 9 \end{bmatrix} \quad \begin{bmatrix} 1 & 2 & 1 & | & 4 \\ 0 & 1 & -1 & | & 1 \end{bmatrix} \quad \begin{bmatrix} 1 & 0 & 3 & | & 2 \\ 0 & 1 & -1 & | & 1 \end{bmatrix}$$

or $\begin{cases} x_1 = 2 - 3x_3 \\ x_2 = 1 + x_3 \end{cases}$ or $\begin{cases} x = (2 - 3x_3, \ 1 + x_3, \ x_3) \\ \quad = (2, 1, 0) + x_3(-3, 1, 1) \end{cases}$.

On writing $\mathbf{a} = (2, 1, 0)$, $\mathbf{u} = (-3, 1, 1)$, and changing the parameter from x_3 to t, we obtain the general solution of the system in the form $\mathbf{x} = \mathbf{a} + t\mathbf{u}$. This is the equation, in vector form, of a line L through the point $A(2, 1, 0)$ and parallel to the vector $\mathbf{u}$. As expected.

Two planes are said to be parallel when they have no point of intersection. In this case the equations of the planes, in scalar form, must form an inconsistent linear system. For example, the system

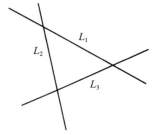

Typical placement of
three lines in the plane

Figure 2.31

$$x_1 + 2x_2 + x_3 = 4$$
$$3x_1 + 6x_2 + 3x_3 = 10$$

, with reduced matrix $\begin{bmatrix} 1 & 2 & 1 & 4 \\ 0 & 0 & 0 & -2 \end{bmatrix}$,

is inconsistent; and so the two planes represented in the system are parallel. Furthermore, the normal vectors to the two planes, $\mathbf{n} = (1,2,1)$ and $\mathbf{n}' = (3,6,3)$, are also parallel. In fact it is easily proved (see exercises) that in general, parallelism of planes is equivalent to that of their normal vectors (see Figure 2.32).

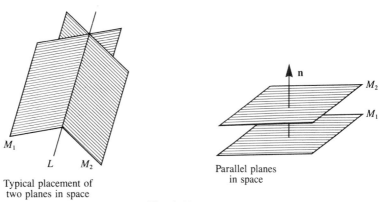

Typical placement of
two planes in space

Parallel planes
in space

Fig. 2.32

The general solution of a system of three equations in three variables gives us the points of intersection of three planes in space. Our experience in solving systems tells us that three equations in three variables usually have a unique solution, in harmony with our geometric intuition that three planes usually intersect in a single point. The most familiar example is that of the three coordinate planes, which of course intersect only at the origin (see Figure 2.33).

However, three planes do not always intersect in a unique point; for example the three panels of a revolving door intersect along the axis of the door (see Figure 2.34).

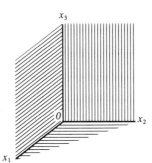

Fig. 2.33
Intersection of the three
coordinate planes at the
origin.

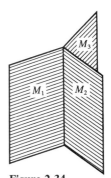

Figure 2.34
Revolving door: three
planes intersecting
along a line.

Example 7 For the system of three equations in three variables

$$\begin{cases} x_1 - 2x_2 - x_3 = 2 \\ x_1 - x_2 + 3x_3 = 4 \\ 2x_1 - 3x_2 + 2x_3 = 6 \end{cases},$$

the reduction sequence is

$$\begin{bmatrix} 1 & -2 & -1 & | & 2 \\ 1 & -1 & 3 & | & 4 \\ 2 & -3 & 2 & | & 6 \end{bmatrix} \rightarrow \begin{bmatrix} 1 & -2 & -1 & | & 2 \\ 0 & 1 & 4 & | & 2 \\ 0 & 1 & 4 & | & 2 \end{bmatrix} \rightarrow \begin{bmatrix} 1 & 0 & 7 & | & 6 \\ 0 & 1 & 4 & | & 2 \\ 0 & 0 & 0 & | & 0 \end{bmatrix},$$

and the general solution is $\begin{cases} x_1 = 6 - 7x_3 \\ x_2 = 2 - 4x_3 \end{cases}$

or $\quad \mathbf{x} = (6 - 7x_3, 2 - 4x_3, x_3) = (6,2,0) + x_3(-7, -4, 1)$,

which has the form, $\mathbf{x} = \mathbf{a} + t\mathbf{u}$, of a line. So the three planes represented in the system intersect along a line.

Consider now the general case of a linear system of m equations in n variables. We have seen that each equation in the system represents a line in the Euclidean plane when $n = 2$, and a plane in Euclidean space when $n = 3$. Classical geometry does not provide an interpretation for systems in more than three variables, but the imagery evoked by geometric language remains helpful to our understanding.

Thus the set of points $\mathbf{x}$ in R^n that satisfies a single linear equation in n variables is called a *hyperplane*. Also, the points given by a parametric formula of the form $\mathbf{x} = \mathbf{a} + t\mathbf{u}$ are said to form a *line* in R^n.

In particular, when the vector form of the general solution of a linear system contains just one generator, the general solution is a line in R^n. But a system of $n-1$ equations in n variables usually has $n-1$ corners and corner variables, and therefore one free variable and one generator.

We conclude that $n-1$ hyperplanes usually intersect in a line. Our earlier observation, that two planes in R^3 usually intersect in a line, is now seen to be a special case of this principle.

■ **Problem** Find all points of intersection of the line

$$\mathbf{x} = (5, -2, 7) + t(1, 2, -1)$$

and the plane

$$3x_1 + 4x_2 + 5x_3 = 60 \quad .$$

■ **Solution** The vector representation of the line can be converted to the set of parametric equations:

$$\begin{cases} x_1 = 5 + t \\ x_2 = -2 + 2t \\ x_3 = 7 - t \end{cases}.$$

On substitution of these formulas for the variables into the equation of the plane, we obtain

$$3(5+t) + 4(-2+2t) + 5(7-t) = 60$$

or $\qquad\qquad 42 + 6t \qquad\qquad = 60$

or $\qquad\qquad\qquad\qquad t = 3 \quad .$

But now substitution of the parameter value back into the equation of the line gives

$$\mathbf{x} = (5,-2,7) + 3(1,2,-1) = (8,4,4) \quad .$$

That is, the line and plane have a unique point of intersection, $C(8,4,4)$.

■ EXERCISES 2.4

1. Given points $A(1,4,2)$, $B(5,6,3)$, $C(9,7,4)$,
    (a) compute the vectors $\mathbf{u} = \vec{AB}$, $\mathbf{v} = \vec{BC}$, and $\mathbf{w} = \vec{AC}$;
    (b) verify that $\mathbf{u} + \mathbf{v}$ equals $\mathbf{w}$.

2. Given points $A(2,3,5)$ and $B(3,2,7)$,
    (a) compute $\mathbf{u} = \vec{AB}$;
    (b) find a vector $\mathbf{v}$ parallel to $\vec{AB}$ but 4 times as long;
    (c) find the point C such that $\mathbf{v} = \vec{AC}$.

3. Given the point $A(1,0,-2)$ and the Euclidean vectors $\vec{AC} = (1,3,2)$ and $\vec{BC} = (1,1,4)$, compute the points B and C.

4. Given the points $A(7,2,4)$ and $B(-13,12,-6)$,
    (a) find the midpoint P of the line segment AB;
    (b) find the point Q that divides the line segment AB in a 7:3 ratio;
    (c) find a point R colinear with A and B and such that B is between A and R and divides the line segment AR in a 2:1 ratio.

5. Find the point of intersection of the medians of the triangle whose vertices are $A(1,4,6)$, $B(2,7,-3)$, $C(3,-2,-9)$.

*6. Suppose $ABCD$ is a quadrilateral. Using vector methods, show that the midpoints of its sides are the vertices of a parallelogram.

7. Find the equation of the line through the points $A(2,3)$ and $B(5,7)$
    (a) in vector form   (b) in parametric form   (c) in scalar form.

8. Find the equation of the line through the points $A(7,1,4)$ and $B(3,2,5)$,

    (a) in vector form   (b) in parametric form   (c) in symmetric form.

9. Determine whether the lines $\mathbf{x} = (-1,6,-4) + t(1,-1,3)$ and $\mathbf{x} = (-3,6,7) + t(-4,2,5)$ intersect, and if so find all points of intersection.

10. Determine whether the lines $\mathbf{x} = (3,4,2) + t(1,1,8)$ and $\mathbf{x} = (-1,6,3) + t(2,-2,4)$ intersect, and if so find all points of intersection.

11. Suppose M is the plane through the point $A(1,7,8)$ with normal vector $\mathbf{n} = (2,1,3)$. Find the equation of M in scalar form.

12. Suppose M is the plane whose equation, in scalar form, is

$$7x_1 + 2x_2 - 4x_3 = 21 \quad .$$

    (a) Write down a normal vector to the plane M.
    (b) Find an equation for M in normal form.

13. Find an equation, in scalar form, for the plane M through the points $A(1,1,1)$, $B(2,3,9)$, $C(1,2,4)$.

14. Find an equation, in scalar form, for the plane M through the origin and the points $A(1,8,-3)$ and $B(3,4,-5)$.

15. Show that the lines $x_1 + 2x_2 = 3$, $x_1 - 4x_2 = -3$, and $x_1 + 6x_2 = 6$ have no common point.

16. Find all points of intersection of the line $\mathbf{x} = (1,4,2) + t(3,5,3)$ and the plane $2x_1 - 5x_2 + 3x_3 = 8$.

17. For which values of the parameter p will the line $\mathbf{x} = (4,0,1) + t(2,5,p)$ and the plane $x_1 + 2x_2 + 3x_3 = 8$ have no common point?

*18. Consider the system

$$b_1x_1 + b_2x_2 + b_3x_3 = h$$

$$c_1x_1 + c_2x_2 + c_3x_3 = k \quad .$$

    (a) Show that if the system is inconsistent, then $\mathbf{b}$ and $\mathbf{c}$ are parallel.
    (b) Show, conversely, that if $\mathbf{b}$ and $\mathbf{c}$ are parallel, then the planes represented by the two equations either coincide or have no common point.
    (c) Deduce that distinct planes are parallel if and only if their normal vectors are parallel.

CHAPTER THREE

Matrix Algebra

Introduction

We have now constructed two complementary interpretations of matrices. First they were introduced as independent mathematical objects, rectangular arrays of coefficients to which various techniques of calculation could be applied. Then they were "unleashed" to act on vectors, mapping one vector space into another. This latter, "active" interpretation of matrices as representing transformations is the one that we intend to emphasize both in the mathematical development of our subject and in the applications to real-life processes.

It is therefore natural to define sums and products of matrices by specifying the way in which they act on vectors. The familiar definitions of sums and composites of ordinary functions provide us with a guide, for they can be readily transferred to the context of transformations of vector spaces. In particular, the matrix product is defined in such a way as to represent the composite of two matrix transformations. We quickly find that some of the rules of elementary algebra no longer apply; for example, the order of factors in a matrix product is important.

Numerous applications of matrix algebra are presented. On the geometric side, we find that the net effect of a sequence of transformations can be expressed in a single transformation derived from a matrix product. For example, the difficulties of interpreting a geometric operation observed in a mirror are resolved by an easy matrix multiplication procedure.

In practical problems, repeated transformations of data from one category to another occur frequently. Matrix multiplication allows these multiple transformations to be collapsed into a single step. Section 3.2 is devoted to this method.

The important method of elementary matrices is presented in section 3.3. It allows us to do row operations on a matrix by introducing suitable factors into a matrix product. Matrix reduction can then be analyzed by the techniques of matrix algebra. This leads, in particular, to the rapid construction, in section 3.4, of the inverse matrix. Elementary matrices are also used systematically in the later development of theory in this text.

We show, in section 3.4, that by use of the inverse matrix the sequences of transformations of data in real-life processes can be reversed, and that observed outcomes can be used to infer earlier states of a process.

Finally, in section 3.5 we discuss matrix transposition and triangular matrices. These topics are presented simply as natural, intrinsic features of matrix algebra, but in later chapters they prove to constitute essential ideas.

3.1 Matrix Addition and Multiplication

Our intention is to define sums and products of matrices by specifying how they act on vectors. This method works well, for it turns out to be easy to construct a matrix that acts in the required way. In fact, the only problem is to make sure that the construction can be done in exactly one way so that sums and products are uniquely defined objects. We resolve this point at the beginning, with the simple but effective lemma 3.1. The section then proceeds through the definitions, examples, geometric illustrations, and some preliminary study of our fledgling matrix algebra.

Matrix Action on Vectors

First we introduce notation for a few important vectors. From analytic geometry we recognize the vectors $\mathbf{e}_1 = (1,0)$, $\mathbf{e}_2 = (0,1)$ as the unit vectors along the coordinate axes (see Figure 3.1).

Similarly, in R^n, the *standard vector* $\mathbf{e}_j$ is defined as having the jth coordinate equal to 1 and all other coordinates equal to 0. In R^3, for example,

$$\mathbf{e}_1 = (1,0,0), \quad \mathbf{e}_2 = (0,1,0), \quad \mathbf{e}_3 = (0,0,1).$$

Consider the action of a 2×3 matrix A on $\mathbf{e}_2$, say:

$$A\mathbf{e}_2 = \begin{bmatrix} a_{11} & a_{12} & a_{13} \\ a_{21} & a_{22} & a_{23} \end{bmatrix} \begin{bmatrix} 0 \\ 1 \\ 0 \end{bmatrix} = \begin{bmatrix} a_{12} \\ a_{22} \end{bmatrix} = \text{column 2 of } A.$$

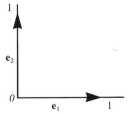

Figure 3.1

Similarly, $A\mathbf{e}_j = \text{column } j$ of A, and it is easy to see that this result holds for any $m \times n$ matrix A acting on the standard vectors in R^n. To summarize:

Lemma 3.1   An $m \times n$ matrix can be reconstructed from its action on the standard vectors $\mathbf{e}_1, \ldots, \mathbf{e}_n$. In fact, if $A\mathbf{e}_j = \mathbf{q}_j$, then the columns of A are $\mathbf{q}_1, \ldots, \mathbf{q}_n$.

An implication of the above lemma is that no two matrices can act on every vector in exactly the same way, and so it is feasible to define matrices by specifying how they act on vectors.

Matrix Operations

Recall that the sum $f+g$ of functions f and g is defined by specifying how it maps each number x:

$$(f+g)(x) = f(x) + g(x).$$

For example, if $f(x) = x^2$ and $g(x) = 2x$, then $(f+g)(x) = x^2 + 2x$. For matrices, we adapt this definition as follows:

Let A and B be $m \times n$ matrices. Then their *sum*, $A+B$, is defined to be the $m \times n$ matrix such that for each vector $\mathbf{x}$ in R^n

$$(A+B)\mathbf{x} = A\mathbf{x} + B\mathbf{x}.$$

To reconstruct the entries of $A+B$ from those of A and B, let us examine the defining formula $A\mathbf{x} + B\mathbf{x}$:

$$
A\mathbf{x} + B\mathbf{x} = \begin{bmatrix} a_{11}x_1 + \cdots + a_{1n}x_n \\ \cdots\cdots\cdots\cdots\cdots \\ a_{m1}x_1 + \cdots + a_{mn}x_n \end{bmatrix} + \begin{bmatrix} b_{11}x_1 + \cdots + b_{1n}x_n \\ \cdots\cdots\cdots\cdots\cdots \\ b_{m1}x_1 + \cdots + b_{mn}x_n \end{bmatrix}
$$

$$
= \begin{bmatrix} (a_{11}+b_{11})x_1 + \cdots + (a_{1n}+b_{1n})x_n \\ \cdots\cdots\cdots\cdots\cdots\cdots\cdots \\ (a_{m1}+b_{m1})x_1 + \cdots + (a_{mn}+b_{mn})x_n \end{bmatrix}
$$

$$
= \begin{bmatrix} a_{11}+b_{11} \cdots a_{1n}+b_{1n} \\ \cdots\cdots\cdots\cdots\cdots \\ a_{m1}+b_{m1} \cdots a_{mn}+b_{mn} \end{bmatrix} \mathbf{x} \quad .
$$

That is, the required action of $A+B$ on $\mathbf{x}$ is provided by the matrix with entries $a_{11} + b_{11}$, etc. So we conclude that matrix addition is done *entrywise*:

Matrix sum $\quad (A + B)_{ij} = a_{ij} + b_{ij} \quad .$

For example, $\begin{bmatrix} 1 & 2 \\ 3 & 4 \end{bmatrix} + \begin{bmatrix} 5 & 6 \\ 7 & 8 \end{bmatrix} = \begin{bmatrix} 1+5 & 2+6 \\ 3+7 & 4+8 \end{bmatrix} = \begin{bmatrix} 6 & 8 \\ 10 & 12 \end{bmatrix} \quad .$

Recall also that if t is a scalar and $f(x)$ a function, then the product function tf acts on numbers according to the formula

$$(tf)(x) = t[f(x)] \quad .$$

For example, with $f(x) = x^2$, $(3f)(x) = 3x^2$. The matrix adaptation of this definition is as follows:

The product of an $m \times n$ matrix A and a number t is defined to be the $m \times n$ matrix tA such that, for each vector $\mathbf{x}$ in R^n,

$$(tA)\mathbf{x} = t(A\mathbf{x}) \quad .$$

But $t(A\mathbf{x}) = t \begin{bmatrix} a_{11}x_1 + \cdots + a_{1n}x_n \\ \cdots \cdots \cdots \cdots \cdots \\ a_{m1}x_1 + \cdots + a_{mn}x_n \end{bmatrix} = \begin{bmatrix} ta_{11}x_1 + \cdots + ta_{1n}x_n \\ \cdots \cdots \cdots \cdots \cdots \cdots \\ ta_{mn}x_1 + \cdots + ta_{mn}x_n \end{bmatrix}$

$= \begin{bmatrix} ta_{11} \cdots ta_{1n} \\ ta_{m1} \cdots ta_{mn} \end{bmatrix} \mathbf{x}$

From this we read off the *entrywise* formula for tA:

Scalar product $(tA)_{ij} = ta_{ij}$.

For example, $3 \begin{bmatrix} 1 & 2 \\ 3 & 4 \end{bmatrix} = \begin{bmatrix} 3 \cdot 1 & 3 \cdot 2 \\ 3 \cdot 3 & 3 \cdot 4 \end{bmatrix} = \begin{bmatrix} 3 & 6 \\ 9 & 12 \end{bmatrix}$.

$M \times N$ Matrices as a Vector Space

We have shown that addition and scalar multiplication are done entrywise for matrices just as they are done componentwise for n-tuples. In fact, since $m \times n$ matrices have mn entries, they may be thought of as mn-tuples arranged in a slightly different way from the usual row or column form.

Therefore, the set of $m \times n$ matrices forms a vector space, which is essentially R^{mn}. Special matrices include

the *zero matrix* $O = \begin{bmatrix} 0 & \cdots & 0 \\ \cdots \cdots \cdots \\ 0 & \cdots & 0 \end{bmatrix}$

and the *additive inverse* $-A = \begin{bmatrix} -a_{11} & \cdots & -a_{1n} \\ \cdots \cdots \cdots \cdots \cdots \\ -a_{m1} & \cdots & -a_{mn} \end{bmatrix}$.

All 10 vector space properties automatically hold. Besides the closure properties, those are

2. $A + B = B + A$ 7. $(st)A = s(tA)$

3. $(A + B) + C = A + (B + C)$ 8. $(s+t)A = sA + tA$

4. $A + O = A$ 9. $t(A + B) = tA + tB$

5. $A + (-A) = O$ 10. $1A = A$.

However, the vector space interpretation of matrices does not provide any guide for forming products of matrices, because the vector space structure in itself provides no rule for multiplying vectors. It is here that the parallel between matrices and functions proves most fruitful.

Matrix Multiplication

Recall that the composite of the functions f and g is the function $f \circ g$ satisfying the condition

$(f \circ g)(x) = f(g(x))$.

That is, starting with the number x, the mapping g is applied first, then f. For example, if $f(x) = x^2$, $g(x) = 2x$, then $(f \circ g)(x) = (2x)^2 = 4x^2$. Note, however, that $(g \circ f)(x) = 2x^2$, and so $f \circ g \neq g \circ f$ in general. So we are forewarned that the matrix version of the composition operation may have novel algebraic properties.

Let A and B be $m \times k$ and $k \times n$ matrices respectively. Then the *product* AB of A and B is defined to be the $m \times n$ matrix such that, for all $\mathbf{x}$ in R^n:

$$(AB)\mathbf{x} = A(B\mathbf{x}) \quad .$$

Note that the defining formula $A(B\mathbf{x})$ represents a composite transformation applied to $\mathbf{x}$:

First, B, as a transformation of R^n into R^k, maps $\mathbf{x}$ from R^n to $B\mathbf{x}$ in R^k;

then, A, as a transformation of R^k into R^m, maps $B\mathbf{x}$ from R^k to $A(B\mathbf{x})$ in R^m.

So the composite is indeed a map of R^n into R^m. Let us calculate this map, selecting the case when A and B are both 2×2 matrices to save writing:

$$A(B\mathbf{x}) = \begin{bmatrix} a_{11} & a_{12} \\ a_{21} & a_{22} \end{bmatrix} \begin{bmatrix} b_{11}x_1 + b_{12}x_2 \\ b_{21}x_1 + b_{22}x_2 \end{bmatrix}$$

$$= \begin{bmatrix} a_{11}(b_{11}x_1 + b_{12}x_2) + a_{12}(b_{21}x_1 + b_{22}x_2) \\ a_{21}(b_{11}x_1 + b_{12}x_2) + a_{22}(b_{21}x_1 + b_{22}x_2) \end{bmatrix}$$

$$= \begin{bmatrix} (a_{11}b_{11} + a_{12}b_{21})x_1 + (a_{11}b_{12} + a_{12}b_{22})x_2 \\ (a_{21}b_{11} + a_{21}b_{12})x_1 + (a_{21}b_{12} + a_{22}b_{22})x_2 \end{bmatrix}$$

or, $\quad (AB)\mathbf{x} = \begin{bmatrix} a_{11}b_{11} + a_{12}b_{21} & a_{11}b_{12} + a_{12}b_{22} \\ a_{21}b_{11} + a_{21}b_{12} & a_{21}b_{12} + a_{22}b_{22} \end{bmatrix} \mathbf{x} \quad .$

The last line spells out the matrix AB, but is hardly the sort of formula one wants to memorize! Order begins to emerge out of chaos when we study, say, the upper right entry of AB:

$$(AB)_{12} = a_{11}b_{12} + a_{12}b_{22}$$
$$= (a_{11}, a_{12}) \cdot (b_{12}, b_{22})$$
$$= (\text{row 1 of } A) \cdot (\text{column 2 of } B) \quad .$$

Observe that the indices 1, 2 on AB match the specification of row 1 of A, column 2 of B in the dot product on the right side. It is easy to check that the same pattern applies to every entry of AB, and so we obtain the "row-column rule":

Row-column rule: $\quad (AB)_{ij} = (\text{row } i \text{ of } A) \cdot (\text{column } j \text{ of } B) \quad .$

As an exercise, the above calculations could be repeated to confirm that the row-column rule continues to hold in the general case where A, B are $m \times k$, $k \times n$ matrices respectively.

Example 1 $A = \begin{bmatrix} 1 & 2 \\ 3 & 4 \end{bmatrix}$, $B = \begin{bmatrix} 5 & 6 \\ 7 & 8 \end{bmatrix}$.

Then $AB = \begin{bmatrix} 1 & 2 \\ 3 & 4 \end{bmatrix} \begin{bmatrix} 5 & 6 \\ 7 & 8 \end{bmatrix} = \ldots$

$$\ldots \begin{bmatrix} (1,2)\cdot(5,7) & (1,2)\cdot(6,8) \\ (3,4)\cdot(5,7) & (3,4)\cdot(6,8) \end{bmatrix} = \begin{bmatrix} 19 & 22 \\ 43 & 50 \end{bmatrix} .$$

Example 2 $A = [3 \quad 4 \quad 1]$, $B = \begin{bmatrix} 1 & 2 \\ 0 & 3 \\ 2 & 1 \end{bmatrix}$.

Then $AB = [3 \quad 4 \quad 1] \begin{bmatrix} 1 & 2 \\ 0 & 3 \\ 2 & 1 \end{bmatrix} = [(3,4,1)\cdot(1,0,2) \quad (3,4,1)\cdot(2,3,1)]$

$= [5 \quad 19]$, which is a 1×2 matrix.

The dimensions of a matrix product can be predicted from the scheme

$$\begin{array}{ccc} (A) & (B) & (AB) \\ (m \times k) & (k \times n) & = & (m \times n) \end{array} .$$

In example 1, $(2 \times 2) (2 \times 2) = (2 \times 2)$.

In example 2, $(1 \times 3) (3 \times 2) = (1 \times 2)$.

In example 2, the product in reverse order, BA, would not make sense, since the pattern $(3 \times 2)(1 \times 3)$ imposes the inconsistent demands $k = 2$ and $k = 1$.

Suppose now that we select an integer n and restrict our attention to $n \times n$ matrices (also called *square* matrices). Then the products always make sense and are always $n \times n$ matrices, in view of the scheme $(n \times n)(n \times n) = (n \times n)$.

In example 1 above, A and B were both 2×2 matrices, and so was the product AB. We can also compute BA:

$$BA = \begin{bmatrix} 5 & 6 \\ 7 & 8 \end{bmatrix} \begin{bmatrix} 1 & 2 \\ 3 & 4 \end{bmatrix} = \ldots$$

$$\ldots \begin{bmatrix} (5,6)\cdot(1,3) & (5,6)\cdot(2,4) \\ (7,8)\cdot(1,3) & (7,8)\cdot(2,4) \end{bmatrix} = \begin{bmatrix} 23 & 34 \\ 31 & 46 \end{bmatrix} .$$

Even for square matrices, then, a product depends on the order of the factors:

$AB \neq BA$.

We say that matrix multiplication is *noncommutative*.

Example 3 $A = \begin{bmatrix} 1 & 2 & 3 \\ 4 & 5 & 6 \end{bmatrix}$, $B = \begin{bmatrix} 1 \\ 0 \\ 7 \end{bmatrix}$.

Then $AB = \begin{bmatrix} 1 & 2 & 3 \\ 4 & 5 & 6 \end{bmatrix} \begin{bmatrix} 1 \\ 0 \\ 7 \end{bmatrix} = \begin{bmatrix} (1,2,3)\cdot(1,0,7) \\ (4,5,6)\cdot(1,0,7) \end{bmatrix} = \begin{bmatrix} 22 \\ 46 \end{bmatrix}$.

Here the 3×1 matrix B is similar in form to a column vector $\mathbf{x}$, and the row-column rule for computing the matrix product coincides with the formula for the matrix-vector product $A\mathbf{x}$. Conversely, it is sometimes convenient to reinterpret a matrix-vector product as a matrix product in which the second factor is a matrix with only one column.

First Applications of the Matrix Product

In order to obtain some hint of the mathematical power of matrix multiplication, let us give some further thought to geometrically defined transformations of vector spaces. When two transformations S and T are applied in succession, the result can always be regarded as a single transformation, which is denoted $S \circ T$ and called the composite of S and T. The formula defining the composite is then

$(S \circ T)(\mathbf{x}) = S(T(\mathbf{x}))$.

However, it is often difficult to obtain a simple formula for, or an intuitively satisfying description of, the composite $S \circ T$, even when the individual transformations S and T are well understood.

The difficulty disappears when S and T have matrix representations A and B, say; for then $S \circ T$ is represented by the product matrix AB. Let us consider a few examples.

Example 1   Suppose the vectors in R^2 are reflected across the x_1-axis twice in succession. To obtain a description of the composite transformation, note that the reflection F is represented by the matrix

$$A = \begin{bmatrix} 1 & 0 \\ 0 & -1 \end{bmatrix} ,$$

since $A\mathbf{x} = \begin{bmatrix} 1 & 0 \\ 0 & -1 \end{bmatrix} \begin{bmatrix} x_1 \\ x_2 \end{bmatrix} = \begin{bmatrix} x_1 \\ -x_2 \end{bmatrix}$,

in agreement with the formula $F((x_1, x_2)) = (x_1, -x_2)$.

So the two consecutive reflections are represented by the composite $F \circ F$, or by the product matrix

$$I = AA = \begin{bmatrix} 1 & 0 \\ 0 & -1 \end{bmatrix} \begin{bmatrix} 1 & 0 \\ 0 & -1 \end{bmatrix} = \begin{bmatrix} 1 & 0 \\ 0 & 1 \end{bmatrix} .$$

To interpret I, observe that:

$$I\mathbf{x} = \begin{bmatrix} 1 & 0 \\ 0 & 1 \end{bmatrix} \begin{bmatrix} x_1 \\ x_2 \end{bmatrix} = \begin{bmatrix} x_1 \\ x_2 \end{bmatrix} = \mathbf{x} ;$$

and so the matrix I maps every vector to itself. Thus the result of reflecting across the x_1-axis twice in succession is to map every vector to itself. Of course, this conclusion agrees with our geometric intuition that two successive reflections cancel each other.

The matrix encountered, and denoted I, in the example above is very useful in matrix algebra. Let us make further note of it.

The *identity matrix I*, or I_m, is the square $m \times m$ matrix:

$$I = \begin{bmatrix} 1 & 0 & \dots & 0 \\ 0 & 1 & \dots & 0 \\ \dots & \dots & \dots & \dots \\ 0 & 0 & \dots & 1 \end{bmatrix} .$$

In general, the *diagonal* of a matrix consists of entries a_{11}, a_{22}, and so on. Thus the identity matrix has ones on the diagonal and zeros elsewhere.

It is easily verified by direct multiplication that the identity matrix has the properties:

$$I_n \mathbf{x} = \mathbf{x} \qquad \text{for all } x \text{ in } R^n$$

and $\quad I_m B = B$ and $B I_n = B \quad$ for all $m \times n$ matrices B.

Example 2 Suppose each point $P(x_1, x_2)$ is shifted to the right, or left, by ax_2 units. This gives us a transformation of position vectors in R^2:

$$S_a((x_1, x_2)) = (x_1 + ax_2, x_2) .$$

The mapping S_a can be implemented by a matrix; in fact

$$\begin{bmatrix} 1 & a \\ 0 & 1 \end{bmatrix} \begin{bmatrix} x_1 \\ x_2 \end{bmatrix} = \begin{bmatrix} x_1 + ax_2 \\ x_2 \end{bmatrix} , \quad \text{as required.}$$

We can therefore determine the composite $S_b \circ S_a$ of two such shifts: it is represented by the matrix product:

$$\begin{bmatrix} 1 & b \\ 0 & 1 \end{bmatrix} \begin{bmatrix} 1 & a \\ 0 & 1 \end{bmatrix} = \begin{bmatrix} 1 & a+b \\ 0 & 1 \end{bmatrix} .$$

That is, $S_b \circ S_a = S_{a+b}$. We conclude that the net effect of the two consecutive shifts is additive.

Example 3 (A mirrored transformation) Suppose we want to implement the shift S_a of example 2 "in a mirror." In precise geometric terms, we do so by first applying the reflection transformation F (which takes us "through the looking glass"), then applying the shift S_a, and finally applying the reflection F again in order to return from the looking glass world.

The composite of the first two steps is represented by the matrix product

$$\begin{bmatrix} 1 & a \\ 0 & 1 \end{bmatrix} \begin{bmatrix} 1 & 0 \\ 0 & -1 \end{bmatrix} = \begin{bmatrix} 1 & -a \\ 0 & -1 \end{bmatrix} .$$

The composite of the third step with the earlier ones is then given by the further product:

$$\begin{bmatrix} 1 & 0 \\ 0 & -1 \end{bmatrix} \begin{bmatrix} 1 & -a \\ 0 & -1 \end{bmatrix} = \begin{bmatrix} 1 & -a \\ 0 & 1 \end{bmatrix} \ .$$

That is, the full three-step transformation is simply S_{-a}. We conclude that doing the shift in a mirror reverses its direction.

The application of matrix algebra to geometry is developed extensively in Chapter 4.

Matrix Algebra

Matrix algebra is not like ordinary algebra: we have already seen, for example, that the order of factors in a matrix product matters. Consider now the cancellation law of ordinary algebra: if $b \neq 0$ but $ab = 0$, we can cancel b and obtain $a = 0$. However, if we let A and B be the nonzero matrices

$$A = \begin{bmatrix} 1 & 2 \\ 4 & 8 \end{bmatrix} \ , \quad B = \begin{bmatrix} -6 & 4 \\ 3 & -2 \end{bmatrix} \ ,$$

then $\quad AB = \begin{bmatrix} 1 & 2 \\ 4 & 8 \end{bmatrix} \begin{bmatrix} -6 & 4 \\ 3 & -2 \end{bmatrix} = \begin{bmatrix} 0 & 0 \\ 0 & 0 \end{bmatrix} = O \ .$

But we cannot cancel the nonzero matrix B from the matrix equation $AB = O$, because this would give us the false result $A = O$. So matrix algebra has *no cancellation law.*

At this stage it becomes essential to establish what we *can* do in matrix algebra.

THEOREM **3.2**

The following laws of matrix multiplication are valid whenever the stated matrix products are defined.

1. $(AB)C = A(BC)$ (associative law),
2. $A(B+C) = AB + AC$ (right distributive law),
3. $(A+B)C = AC + BC$ (left distributive law),
4. $t(AB) = (tA)B = A(tB)$ (scalar factorization law).

Dimensional note. In equation (1), for example, the dimensional patterns of A, B, C must be $m \times k$, $m \times 1$, $1 \times n$. Then AB is $m \times 1$, and $(AB)C$ is $m \times n$; while BC is $k \times n$, and $A(BC)$ is $m \times n$. All products in the equation are defined.

■Proof   Equation (1).

Both sides of the first equation act on vectors **x** in R^n.

(i) *Left side:* $((AB)C)\mathbf{x}$ means that first C acts on **x**, and then AB acts on it. But the second step, AB acting on x, can be subdivided into first B and then A acting on x. So there are three steps: C, then B, then A acting in turn.

(ii) *Right side:* $(A(BC))\mathbf{x}$ means that first BC acts on **x** and then A acts. But the first step, BC acting, can be subdivided as C acting and then B acting. So there are three steps: C, then B, then A acting in turn.

Therefore, $((AB)C)\mathbf{x} = (A(BC))\mathbf{x}$ for all **x** in R^n. It follows, from lemma 3.1, that $(AB)C$ and $A(BC)$ are the same matrix.

Equation (2). This time the pattern is that A is $m \times k$, while B and C are $k \times n$. Then for all **x** in R^n,

$$(B+C)\mathbf{x} \qquad = B\mathbf{x} + C\mathbf{x} \text{ (definition of sum of matrices),}$$
$$A((B+C)\mathbf{x}) \qquad = A(B\mathbf{x}) + A(C\mathbf{x}) \text{ (linearity of transformation } A),$$
$$\text{or} \quad (A(B+C))\mathbf{x} = (AB)\mathbf{x} + (AC)\mathbf{x} \text{ (definition of matrix products),}$$
$$\text{or} \quad (A(B+C))\mathbf{x} = (AB+AC)\mathbf{x} \text{ (definition of sum of matrices).}$$

It follows, from lemma 3.1 again, that $A(B+C)$ and $AB + AC$ are the same matrix.

Equation (3). Proof similar to equation (2).

Equation (4). Proof based on definition of scalar product tA; do this as an easy exercise.   Q.E.D.

From the associative law in particular, we see that the product ABC of three matrices can be written *without bracketing*, since it doesn't matter whether we multiply AB first or BC.

Another shorthand applies to square, $n \times n$ matrices: a product of p factors of A, or $AA \cdots A$, can be written as A^p.

Example 1 Let us expand the product $(A+B)(A-B)$ and compare our result with what would be obtained in ordinary algebra:

$$(A+B)(A-B) = A(A-B) + B(A-B) \text{ (left distributive law)}$$
$$= AA - AB + BA - BB \text{ (right distributive law, applied twice)}$$
$$= A^2 - AB + BA - B^2 \text{ (shorthand).}$$

But here the calculation stops, since for matrices, AB and BA are different. The product does not reduce to the familiar $A^2 - B^2$.

Example 2 Simplify the product $ABABAAB$. Although the bracketing doesn't matter, the order of factors does, and we cannot collect the four factors of A. Only the two adjacent factors of A can be combined, and so the simplest form of the product is $ABABA^2B$. Note that all the B factors are isolated from each other.

■ **Problem** Let $B = \begin{bmatrix} 0 & 2 \\ 3 & 0 \end{bmatrix}$. Show that if the 2×2 matrix A satisfies the matrix equation $AB = BA$, then there are constants p and q such that $A = pB + qI$.

■ **Solution** Let $A = \begin{bmatrix} a & b \\ c & d \end{bmatrix}$.

Then $AB = \begin{bmatrix} a & b \\ c & d \end{bmatrix} \begin{bmatrix} 0 & 2 \\ 3 & 0 \end{bmatrix} = \begin{bmatrix} 3b & 2a \\ 3d & 2c \end{bmatrix}$,

and $BA = \begin{bmatrix} 0 & 2 \\ 3 & 0 \end{bmatrix} \begin{bmatrix} a & b \\ c & d \end{bmatrix} = \begin{bmatrix} 2c & 2d \\ 3a & 3b \end{bmatrix}$.

Now if the two matrix products are equal, they must agree entry-by-entry. This gives us the four scalar equations:

$3b = 2c \qquad 2a = 2d$
$3d = 3a \qquad 2c = 3b$.

Two of these equations simplify to the same conclusion, $a=d$, and we can let the constant q denote the common value of a and d: $q=a=d$. Similarly, the other two equations both read $3b=2c$, and we can conveniently introduce a second constant p through the formula $6p=3b=2c$. This is equivalent to the formulas $b=2p$ and $c=3p$.

Therefore $A = \begin{bmatrix} q & 2p \\ 3p & q \end{bmatrix} = \begin{bmatrix} 0 & 2p \\ 3p & 0 \end{bmatrix} + \begin{bmatrix} q & 0 \\ 0 & q \end{bmatrix}$

$= p \begin{bmatrix} 0 & 2 \\ 3 & 0 \end{bmatrix} + q \begin{bmatrix} 1 & 0 \\ 0 & 1 \end{bmatrix}$,

or $\qquad A = pB + qI$.

■ EXERCISES 3.1

1. Compute the following matrix products:

(a) $\begin{bmatrix} 1 & 0 \\ 3 & 5 \end{bmatrix} \begin{bmatrix} 0 & 3 \\ 1 & 4 \end{bmatrix}$

(b) $\begin{bmatrix} 1 & 2 & 0 \\ 0 & -1 & 1 \end{bmatrix} \begin{bmatrix} 2 & 1 \\ 0 & 1 \\ 3 & 0 \end{bmatrix}$

(c) $\begin{bmatrix} 1 & 2 & 3 \\ 0 & 0 & 1 \\ 1 & 0 & 0 \end{bmatrix} \begin{bmatrix} 0 & 4 & 3 \\ 2 & 1 & 1 \\ 8 & 0 & 4 \end{bmatrix}$

(d) $\begin{bmatrix} 1 & 0 \\ 2 & 3 \\ 5 & 0 \end{bmatrix} \begin{bmatrix} 1 & 0 & 3 & 5 \\ 0 & 1 & 1 & 2 \end{bmatrix}$

(e) $[3 \quad 1 \quad 0] \begin{bmatrix} 2 & 3 \\ 3 & 0 \\ 0 & 5 \end{bmatrix}$ (f) $[2 \quad 3] \begin{bmatrix} 4 \\ 5 \end{bmatrix}$

(g) $\begin{bmatrix} 2 & 1 \\ 0 & 4 \end{bmatrix} \begin{bmatrix} 0 & 1 \\ 1 & 2 \end{bmatrix} \begin{bmatrix} -1 & 0 & 1 \\ 2 & 1 & 0 \end{bmatrix}$

(h) $\begin{bmatrix} 2 & -1 & 0 \\ 0 & 0 & 3 \end{bmatrix} \begin{bmatrix} 3 & 3 & 5 \\ 1 & 0 & 0 \\ 0 & 2 & 0 \end{bmatrix} \begin{bmatrix} 3 & 2 \\ 1 & 0 \\ 2 & 0 \end{bmatrix}$

(i) $\begin{bmatrix} 1 & 3 \\ 5 & 4 \end{bmatrix} \begin{bmatrix} a & b \\ c & d \end{bmatrix}$ (j) $\begin{bmatrix} x & y \\ 2 & 3 \end{bmatrix} \begin{bmatrix} a & b \\ -1 & 4 \end{bmatrix}$

2. Let $A = \begin{bmatrix} 1 & 2 & 3 \\ 4 & 5 & 6 \end{bmatrix}$, $B = \begin{bmatrix} 1 & 0 & 2 \\ -3 & 1 & 4 \end{bmatrix}$.

   (a) Compute $2B$, $A + 2B$, and $3(A + 2B)$.
   (b) Compute $3A$, $6B$, and $3A + 6B$.
   (c) Why are the last results in parts (a) and (b) the same?

3. Let $A = \begin{bmatrix} 1 & 2 \\ 3 & -1 \end{bmatrix}$, $B = \begin{bmatrix} 2 & 6 & 1 \\ 1 & -1 & 0 \end{bmatrix}$, $C = \begin{bmatrix} 3 & 2 \\ 1 & 2 \\ 4 & 1 \end{bmatrix}$

   (a) Compute AB, $(AB)C$, BC, $A(BC)$ and verify that the bracketing of the product ABC doesn't matter.
   (b) Can the order of the factors in the product ABC be changed?

4. Let $A = \begin{bmatrix} -1 & 3 \\ 2 & 1 \end{bmatrix}$, $B = \begin{bmatrix} 0 & 4 \\ -3 & 2 \end{bmatrix}$. Compute:

   (a) $A^2 + 2AB + B^2$ (b) $(A+B)^2$ (c) $AB - BA$, and (d) verify that result (c) equals result (a) minus result (b).

5. Scalars satisfy the identity: $a^3 - b^3 = (a-b)(a^2+ab+b^2)$. For $n \times n$ matrices A, B, find a formula for the difference between $A^3 - B^3$ and $(A-B)(A^2+AB+B^2)$.

6. Let $P = \begin{bmatrix} g & h \\ h & -g \end{bmatrix}$.

   (a) Show that $P^2 = (g^2+h^2)I$.
   (b) Deduce that $P^4 = (g^2+h^2)^2 I$.
   (c) Deduce that $P^{2q} = (g^2+h^2)^q I$.
   (d) Compute $\begin{bmatrix} 1 & 2 \\ 2 & -1 \end{bmatrix}^{12}$.

7. Let $B = \begin{bmatrix} 1 & 0 \\ 0 & -1 \end{bmatrix}$, $C = \begin{bmatrix} 0 & 1 \\ 1 & 0 \end{bmatrix}$.

    (a) Suppose M commutes with B. That is, $MB = BM$. Show that M is a diagonal matrix; that is, M has the form $M = \begin{bmatrix} a & 0 \\ 0 & d \end{bmatrix}$.

    (b) Suppose that M commutes with C as well as with B. Show that M has the form $M = aI$.

8. (a) Find a 2×2 matrix $A \neq O$, satisfying $A^2 = O$.

    (b) Find a 2×2 matrix A, other than O or I, satisfying $A^2 = A$.

       Note: parts (a) and (b) can both be solved by using matrices with only one nonzero entry.

    (c) Find a 2×2 matrix A, with real numbers as entries, satisfying $A^2 = -I$. (There are solutions with only two nonzero entries.)

9. (a) Suppose the 2×2 matrix A satisfies $A\mathbf{e}_1 = 4\mathbf{e}_2$.

       Show that A has the form $A = \begin{bmatrix} 0 & b \\ 4 & d \end{bmatrix}$.

    (b) Suppose that, in addition, A satisfies the matrix equation $A^2 = I$. Compute A.

*10. (The matrix product as a recombination of rows or columns.) Suppose that A and B are $m \times k$ and $k \times n$ matrices respectively.

    (a) Show that if the row vector $\mathbf{b}_j$ represents row j of B, then row i of the product AB can be expressed as the following linear combination of the rows of B:

$$\sum_{j=1}^{k} a_{ij}\mathbf{b}_j \quad .$$

    (b) Show that if the column vector $\mathbf{q}_i$ represents column i of A, then column j of AB can be expressed as the following linear combination of the columns of A:

$$\sum_{i=1}^{k} b_{ij}\mathbf{q}_i \quad .$$

*11. (a) Prove that the following rules apply to the sigma notation:

       (i) $\displaystyle\sum_{p=1}^{k} (cf_p + dg_p) = c\left(\sum_{p=1}^{k} f_p \right) + d\left(\sum_{p=1}^{k} g_p \right)$.

       (ii) $\displaystyle\sum_{p=1}^{k} \left(\sum_{q=1}^{l} h_{pq} \right) = \sum_{q=1}^{l} \left(\sum_{p=1}^{k} h_{pq} \right)$.

(b) Show that the row-column rule for the product AB of the $m \times k$ matrix A and $k \times n$ matrix B can be expressed in terms of the sigma notation as:

$$(AB)_{ij} = \sum_{p=1}^{k} a_{ip}b_{pj} \quad .$$

(c) Suppose that the row-column rule, as given in part (b), had been used to *define* the matrix product. Using the rules in part (a), show that the four laws of matrix multiplication given in theorem 3.2 would still hold.

(d) Deduce, as a special case of the associative law of matrix multiplication, that for all vectors $\mathbf{x}$,

$$A(B\mathbf{x}) = (AB)\mathbf{x} \quad ,$$

so that the "composition" definition of the matrix product can be recovered from the row-column rule.

*12. Let $B = \begin{bmatrix} 1 & 0 \\ 0 & -1 \end{bmatrix}$, $C = \begin{bmatrix} 0 & 1 \\ 1 & 0 \end{bmatrix}$.

(a) Show that the pair of matrices B, C satisfy the "anticommutation relation" $BC + CB = O$, and that $B^2 = C^2 = I$.
(b) Find a third matrix D such that the pairs B, D and C, D both satisfy the anticommutation relation, and $D^2 = I$. (Complex entries will be needed.)
Note: The matrices B, C, D are called the *Pauli spin matrices* in physics and are used in the study of electron spin.

*13. Find *all* solutions of each of the 2×2 matrix equations given in problem 8.

3.2 Applications of the Matrix Product

Example 1 Let us reformulate the model, introduced in example 1 of section 1.4, of the production of three metals, #1, gold; #2, silver; #3, copper; from three mines, #1, North Mine; #2, Central Mine; #3, South Mine. Suppose we let x_1, x_2, x_3 represent tons of ore taken from the three mines respectively; let:

b_{jk} = quantity of metal #j produced per ton of ore from mine #k ;

and let y_1, y_2, y_3 represent the total quantities of the three metals produced. Then the equations derived earlier linking the x's and y's take the form

$$y_1 = b_{11}x_1 + b_{12}x_2 + b_{13}x_3$$
$$y_2 = b_{21}x_1 + b_{22}x_2 + b_{23}x_3$$
$$y_3 = b_{31}x_1 + b_{32}x_2 + b_{33}x_3 \quad,$$

or, in matrix-vector notation, $\mathbf{y} = B\mathbf{x}$. Furthermore, the data on production rates given in section 1.4 specify the numerical form of B as:

$$B = \begin{bmatrix} 1 & 2 & 1 \\ 10 & 18 & 12 \\ 20 & 22 & 40 \end{bmatrix} \cdot$$

Now, suppose the unit selling prices of the three metals are \$400, \$10, and \$1 respectively, while the royalties that must be paid to the government per unit of production of the metals are \$50, \$1, and \$0.10 respectively. Then, if z_1, z_2 represent total sales revenue and total royalty costs respectively, we have

$$z_1 = 400y_1 + 10y_2 + \quad y_3$$
$$z_2 = \ 50y_1 + \quad y_2 + (\cdot 1)y_3 \quad,$$

or, in matrix-vector notation, $\mathbf{z} = A\mathbf{y}$, where

$$A = \begin{bmatrix} 400 & 10 & 1 \\ 50 & 1 & 0.1 \end{bmatrix} \cdot$$

The dollar-denominated vector $\mathbf{z}$ carrying the revenue and royalty data for the total operation can be linked directly to the ore tonnage vector $\mathbf{x}$. In fact,

$$\mathbf{z} = A\mathbf{y} = A(B\mathbf{x}) = (AB)\mathbf{x} \quad,$$

the last step being justified as the definition of the product matrix AB. Since we have the numerical entries of A and B, we can compute AB:

$$AB = \begin{bmatrix} 400 & 10 & 1 \\ 50 & 1 & 0.1 \end{bmatrix} \begin{bmatrix} 1 & 2 & 1 \\ 10 & 18 & 12 \\ 20 & 22 & 40 \end{bmatrix} = \begin{bmatrix} 520 & 1002 & 560 \\ 62 & 120.2 & 66 \end{bmatrix} \cdot$$

For example, with ore tonnages 3,000, 2,000, and 1,000, as found in section 1.4, the dollar vector is

$$\mathbf{z} = \begin{bmatrix} 520 & 1002 & 560 \\ 62 & 120.2 & 66 \end{bmatrix} \begin{bmatrix} 3000 \\ 2000 \\ 1000 \end{bmatrix} = \begin{bmatrix} 4,124,000 \\ 492,400 \end{bmatrix} \cdot$$

That is, sales are \$4,124,000 and royalties \$492,400.

As this first application suggests, the matrix product, as we have defined it, is perfectly adapted to the study of the multistep processes that occur everywhere in science, commerce, and so on. Each step of the process is represented as a matrix transformation, and the whole process is the composite of all these transformations; in other words, it is the product of all the representing matrices.

Example 2   Suppose that in a fixed population of breakfast cereal consumers, there are just two brands, Crunch and Munch, that share the market. Suppose that during year n, Crunch and Munch have proportions x_n and y_n of the market, so that their market shares can be represented as a column vector, $\mathbf{x}_n = \begin{bmatrix} x_n \\ y_n \end{bmatrix}$. Suppose that each year 20% of the Crunch buyers shift to Munch, while 30% of the Munch buyers shift to Crunch.

We can then predict the market shares of the brands in all future years through a concise matrix formula as follows. First, in year $n+1$, Crunch retains 80% of last year's buyers and acquires 30% of Munch's, and so Crunch's new market share is (see also example 3 of section 1.4 for a previous example of this type of reasoning):

$$x_{n+1} = \left(\tfrac{80}{100}\right)x_n + \left(\tfrac{30}{100}\right)y_n \quad .$$

By similar arguments Munch's new market share is:

$$y_{n+1} = \left(\tfrac{20}{100}\right)x_n + \left(\tfrac{70}{100}\right)y_n \quad ;$$

or, in matrix-vector terms,

$$\begin{bmatrix} x_{n+1} \\ y_{n+1} \end{bmatrix} = \begin{bmatrix} \tfrac{80}{100} & \tfrac{30}{100} \\ \tfrac{20}{100} & \tfrac{70}{100} \end{bmatrix} \begin{bmatrix} x_n \\ y_n \end{bmatrix} \quad ,$$

or $\mathbf{x}_{n+1} = B\mathbf{x}_n$, where $B = \begin{bmatrix} \tfrac{8}{10} & \tfrac{3}{10} \\ \tfrac{2}{10} & \tfrac{7}{10} \end{bmatrix}$.

For example, if in year 0, Munch has 60% of the market, so that $\mathbf{x}_0 = (40/100, 60/100)$, then in year 1,

$$\mathbf{x}_1 = \begin{bmatrix} \tfrac{8}{10} & \tfrac{3}{10} \\ \tfrac{2}{10} & \tfrac{7}{10} \end{bmatrix} \begin{bmatrix} \tfrac{40}{100} \\ \tfrac{60}{100} \end{bmatrix} = \begin{bmatrix} \tfrac{1}{2} \\ \tfrac{1}{2} \end{bmatrix} \quad ;$$

that is, the market shares are now 50% each.

Now, we can also determine the market shares in years 2, 3, and so on, by computing $\mathbf{x}_2 = B\mathbf{x}_1$, $\mathbf{x}_3 = B\mathbf{x}_2$, etc.; but those steps can be combined, since $\mathbf{x}_2 = B\mathbf{x}_1 = B(B\mathbf{x}_0) = B^2\mathbf{x}_0$, and similarly $\mathbf{x}_k = B^k\mathbf{x}_0$. Similarly, if we start from year n with market share vector $\mathbf{x}_n$, the market shares k *years later* are given by the *market share predictor* formula

$$\mathbf{x}_{n+k} = B^k\mathbf{x}_n \quad .$$

To predict k years ahead, simply apply the power B^k to the present market share vector. In our example,

$$B^2 = \left(\tfrac{1}{10}\right)^2 \begin{bmatrix} 8 & 3 \\ 2 & 7 \end{bmatrix}^2 = \left(\tfrac{1}{100}\right) \begin{bmatrix} 70 & 45 \\ 30 & 55 \end{bmatrix} = \left(\tfrac{1}{20}\right) \begin{bmatrix} 14 & 9 \\ 6 & 11 \end{bmatrix} \quad ,$$

and

$$B^4 = (B^2)^2 = \left(\tfrac{1}{400}\right) \begin{bmatrix} 250 & 255 \\ 150 & 175 \end{bmatrix} = \left(\tfrac{1}{16}\right) \begin{bmatrix} 10 & 9 \\ 6 & 7 \end{bmatrix} \quad .$$

So after four years, the market shares are

$$\mathbf{x}_4 = \left(\tfrac{1}{16}\right) \begin{bmatrix} 10 & 9 \\ 6 & 7 \end{bmatrix} \begin{bmatrix} \frac{40}{100} \\ \frac{60}{100} \end{bmatrix} = \begin{bmatrix} \frac{47}{80} \\ \frac{33}{80} \end{bmatrix} .$$

Crunch's market share is now almost 59%.

Example 3 Suppose a taxicab company uses the following strategy to maintain a fleet of fixed size: on December 31 it sells one-fifth of the cars that are one year old (and have proved to be lemons) and all the cars that are three years old. The next day, January 1, it buys enough new cars to replace those sold the day before.

To describe the process of fleet renewal, we let $x_1(n)$, $x_2(n)$, $x_3(n)$ denote the number of cars in first, second, and third year of service, respectively, during calendar year n. This gives us a three-component fleet vector $\mathbf{x}(n)$ describing the age distribution of the fleet during year n. Next year's fleet vector, $\mathbf{x}(n+1)$, is deduced from the chart of company strategy (see Figure 3.2).

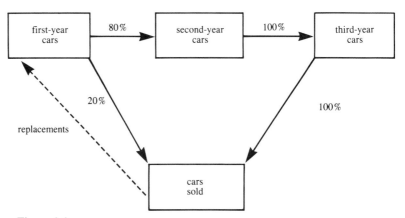

Figure 3.2

At the end of year n, cars sold include one-fifth of the first-year cars and all of the third-year cars, or $\tfrac{1}{5}x_1(n) + x_3(n)$. But cars sold at the end of this year equal cars bought at the beginning of next year, and so the number of first-year cars in year $n + 1$, or $x_1(n+1)$, is

$$x_1(n+1) = \tfrac{1}{5}x_1(n) + x_3(n).$$

Next year's second-year cars are just 80% of this year's first-year cars, or

$$x_2(n+1) = \tfrac{4}{5}x_1(n),$$

and next year's third-year cars are just this year's second-year cars

$$x_3(n+1) = x_2(n).$$

The three equations combine into the matrix-vector equation

$$\mathbf{x}(n+1) = B\mathbf{x}(n), \text{ where } B = \begin{bmatrix} \frac{1}{5} & 0 & 1 \\ \frac{4}{5} & 0 & 0 \\ 0 & 1 & 0 \end{bmatrix}.$$

Just as in example 2, we can predict the fleet k years ahead with the *fleet predictor* formula:

$$\mathbf{x}(n+k) = B^k\mathbf{x}(n).$$

For example, if the company starts in year n with a fleet consisting entirely of q new cars, then $\mathbf{x}(n) = (q,0,0) = q\mathbf{e}_1$, and we can easily compute the fleet vectors for the next few years from the formula $\mathbf{x}(n+k) = B^k\mathbf{x}(n) = qB^k\mathbf{e}_1$:

$$\mathbf{x}(n+1) = q(.2000, \ .8000, \ 0)$$
$$\mathbf{x}(n+2) = q(.0400, \ .1600, \ .8000)$$
$$\mathbf{x}(n+3) = q(.8080, \ .0320, \ .1600)$$
$$\mathbf{x}(n+4) = q(.3216, \ .6464, \ .0320).$$

The extreme fluctuations in the fleet vector from year to year make us wonder whether the age distribution of the cars ever stabilizes. In Chapter 8, the method of matrix diagonalization will be applied to this type of problem.

■ EXERCISES 3.2

1. In the mining example in this section, suppose we wish to incorporate the calculation of "net revenue" from the metal production into the matrix formulation. That is, the formula: net revenue $r = $ sales $-$ royalties $= z_1 - z_2$, is to be interpreted as a third step in a composite process.

   (a) Express r in the form $r = N\mathbf{z}$, where N is a 1×2 matrix.
   (b) Show that $r = NAB\mathbf{x}$.
   (c) Compute the numerical matrix NAB.
   (d) Deduce the numerical value of r when the numerical vector $\mathbf{x}$ is as given in the text discussions.

2. Suppose that the taxicab company described in example 3 of this section alters its fleet renewal procedures somewhat, so that the strategy chart changes (see Figure 3.3).

   (a) Find the 3×3 matrix B that connects the fleet vector in successive years through the formula $\mathbf{x}(n+1) = B\mathbf{x}(n)$.
   (b) Suppose the company starts on a shoestring, with a fleet consisting entirely of third-year cars. What proportion of the fleet will consist of third-year cars in each of the following four years?

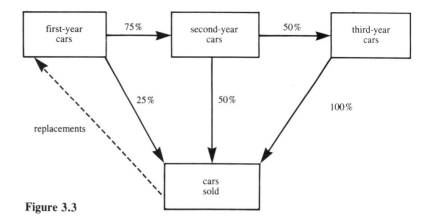

Figure 3.3

(c) Show that, if the fleet vector is the same in two successive years, in other words, $\mathbf{x}(n+1) = \mathbf{x}(n)$, then $\mathbf{x}(n)$ is a solution of the equation $B\mathbf{x} = \mathbf{x}$, or, $(B-I)\mathbf{x} = \mathbf{0}$. Find such a fleet vector.

(d) Write down a 3×3 matrix C that moves the fleet vector two years ahead, that is, $\mathbf{x}(n+2) = C\mathbf{x}(n)$.

(e) Are there any fleet vectors that are not the same every year but do repeat themselves after a period of two years?

3. Suppose a competing taxi company also has an annual fleet renewal scheme, which is based on a simpler classification of its fleet into *new* cars and *old* cars. The company strategy is that, at the turn of the year, all the new cars become old cars, while half the old cars are sold and half remain as old cars for another year. Then a new car is bought to replace each car sold.

(a) Write down a strategy chart for the company.

(b) Find a 2×2 matrix B that gives the transformation of the two-component fleet vector from one year to the next.

(c) Show that $B^{10} = \left(\frac{1}{1,024}\right) \begin{bmatrix} 342 & 341 \\ 682 & 683 \end{bmatrix}$.

(d) Deduce that, no matter what the initial proportions of new and old cars in the fleet, after 10 years the proportion of new cars will fall between 33% and 34%.

4. In example 3 of section 1.4, a competitive situation among three breakfast cereal brands was described.

(a) Find the 3×3 matrix transforming the three-component market share vector from one year to the next.

(b) Assume that brand A starts with 100% of the market and, by computing the first few powers of the transforming matrix found in (a), find out how many years elapse before brand A's market share falls below 50%.

5. Suppose just two brands, Crunch and Munch, have the market, as in example 2 of this section, but that each year the proportion of its own buyers that Crunch keeps, and the proportion of Munch's buyers that it wins over, are exactly the same, say p. Show that no matter what the initial market shares of Crunch and Munch are, their market shares a year later and in all subsequent years will be p and $1 - p$ respectively.

6. Suppose that a leasing firm rents out both large and small cars on one-year contracts. Suppose it finds that for each of three consecutive years, 20% of the large-car renters shift to small cars while 30% of the small-car renters shift to large cars.

   However, in the final year of a four-year cycle, during which fuel prices rise sharply, the above percentages shift abruptly to 50% and 10% respectively.

   (a) Construct a predictor formula of the form $x(4(n+1)) = Gx(4n)$ that indicates the leasing pattern a full cycle ahead.

   (b) Suppose that fuel price escalation and its consequences occur in both the second and fourth years of the four-year cycle. What would the predictor formula then be?

7. Suppose that one year the leasing firm of the previous exercise rents out equal numbers of large and small cars. The proportion of large cars rented out eight years later will depend on the cyclical pattern in force during the two intervening four-year cycles.

   What is the difference in the ultimate proportion of large cars resulting from the two cyclical patterns just studied?

■3.3 The Elementary Matrices

One of the most important questions about a transformation or mapping is, can it be reversed? For the mappings represented by ordinary functions, such as $y = f(x) = 2x + 5$, the problem is familiar. We solve for x as a function of y, obtaining $x = g(y) = \frac{1}{2}(y-5)$ in this instance. Then the new function g maps y back to the starting point x. We say that the mapping g is the inverse of f. In terms of composite notation, we can write:

$$(g \circ f)(x) = g(f(x)) = g(y) = x \quad .$$

Consider now the case of a matrix transformation B that maps x to y, that is, $Bx = y$. To reverse this mapping we must, given y, recover x. Our procedure for doing so, as developed in section 1.3, is to set up the augmented matrix $[B|\mathbf{y}]$ and reduce it by row operations. This method, however, reverses the mapping for only one vector at a time; given a numerical vector $\mathbf{y}$, we are able to calculate its antecedent numerical vector $\mathbf{x}$. We do not obtain a new transformation, say A, that reverses the effect of the given transformation B.

It stands to reason that, in order to solve the equation $B\mathbf{x} = \mathbf{y}$ for $\mathbf{x}$ and obtain a result of the form $\mathbf{x} = A\mathbf{y}$, we must be able to apply the methods of matrix algebra to the given equation $B\mathbf{x} = \mathbf{y}$. The augmented matrix method does gradually modify B, but only through a sequence of symbolically specified row operations on B. These coded instructions take us outside the structure of matrix algebra itself in which the operations are matrix addition and multiplication.

The solution to the problem is this: given a row operation, such as $R_2 - 3R_1$, we are going to find a matrix Q, called an elementary matrix, such that applying the given row operation to B is the same as changing B to QB. Each step of the matrix reduction procedure can then be expressed within the framework of matrix algebra as an act of matrix multiplication.

In this section, we introduce the elementary matrices and establish their basic algebraic properties. This paves the way for the construction, in the following section, of the inverse matrix transformation.

The Elementary Matrices

The row-column rule seems to mix the factors of a matrix product rather thoroughly. Yet the following example shows that some traces of the original factors endure in the product.

Example 1

$$\begin{bmatrix} 0 & k & 0 \\ 0 & 0 & 0 \\ 0 & 0 & 0 \end{bmatrix} \begin{bmatrix} b_{11} & b_{12} \\ b_{21} & b_{22} \\ b_{31} & b_{32} \end{bmatrix} = \begin{bmatrix} kb_{21} & kb_{22} \\ 0 & 0 \\ 0 & 0 \end{bmatrix} \ .$$

This product has the form, $S \quad B = B'$, with the dimensional pattern $(m \times m)(m \times n) = (m \times n)$.

The left factor, S, is a square matrix, while the right factor B and the product B' have the same dimensions. Furthermore,

(row 1 of B') = k times (row 2 of B) .

The row indices, $1, 2$, appearing in this equation are the same as those of the sole nonzero entry of S: $s_{12} = k$.

By checking another example or two, we can convince ourselves that if the only nonzero entry of S is $s_{ij} = k$, then

(row i of B') = k times (row j of B) .

This suggests that row operations on B can be done by multiplying B on the left by suitable square matrices. We develop this idea in the following examples.

Example 2 (*Type* E *row operations.*) To exchange rows 1 and 2 of B, let us select nonzero entries for S from Table 3.1:

Row movement Row $j \rightarrow$ row i	Nonzero entry of S: s_{ij}
row 1 $\rightarrow$ row 2	$s_{21} = 1$
row 2 $\rightarrow$ row 1	$s_{12} = 1$
row 3 $\rightarrow$ row 3	$s_{33} = 1$

Table 3.1

Let's see if this works by multiplying out SB:

$$SB = \begin{bmatrix} 0 & 1 & 0 \\ 1 & 0 & 0 \\ 0 & 0 & 1 \end{bmatrix} \begin{bmatrix} b_{11} & b_{12} \\ b_{21} & b_{22} \\ b_{31} & b_{32} \end{bmatrix} = \begin{bmatrix} b_{21} & b_{22} \\ b_{11} & b_{12} \\ b_{31} & b_{32} \end{bmatrix} \cdots$$

$$= \begin{bmatrix} \text{row 2 of } B \\ \text{row 1 of } B \\ \text{row 3 of } B \end{bmatrix} . \quad \text{Check.}$$

More generally, suppose the movement from B to SB exchanges rows i and j of B. Then the square matrix S is called a *type* E *elementary matrix*, and denoted $E(i,j)$.

The entries of $E(i,j)$ can be constructed easily from the row movement table, but they can also be spelled out algebraically as follows:

$s_{ij} = s_{ji} = 1$; $s_{pp} = 1$, when p differs from i and j; all other entries are 0.

Example 3 (*Type* A *row operations.*) Suppose we want to add four times row 2 to row 3. In this situation, we are just adding on to the rows already present, so the first three statements in the row movement table are there to keep the existing rows in place. The last line of Table 3.2 adds on the specified multiple $k = 4$ of row 2 to row 3:

Row $j \rightarrow$ Row i	s_{ij}
row 1 $\rightarrow$ row 1	$s_{11} = 1$
row 2 $\rightarrow$ row 2	$s_{22} = 1$
row 3 $\rightarrow$ row 3	$s_{33} = 1$
row 2 $\rightarrow$ row 3	$s_{32} = 4$

Table 3.2

To check our result, we compute:

$$SB = \begin{bmatrix} 1 & 0 & 0 \\ 0 & 1 & 0 \\ 0 & 4 & 1 \end{bmatrix} \begin{bmatrix} b_{11} & b_{12} \\ b_{21} & b_{22} \\ b_{31} & b_{32} \end{bmatrix} = \begin{bmatrix} b_{11} & b_{12} \\ b_{21} & b_{22} \\ 4b_{21} + b_{31} & 4b_{22} + b_{32} \end{bmatrix} \quad \text{. Check.}$$

More generally, suppose the movement from B to SB adds k times row j of B to row i of B. Then the square matrix S is called a *type* A *elementary matrix*, and denoted $A(k;j,i)$.

The algebraic specification of $A(k;j,i)$ is: $s_{pp} = 1$ for all p; $s_{ij} = k$, all other entries 0.

Example 4 (*Type* M *row operations.*) Suppose row 1 of B is to be multiplied by 5. Then the row movements are:

Row j → Row i	s_{ij}
row 1 → row 1	$s_{11} = 5$
row 2 → row 2	$s_{22} = 1$
row 3 → row 3	$s_{33} = 1$

Table 3.3

$$SB = \begin{bmatrix} 5 & 0 & 0 \\ 0 & 1 & 0 \\ 0 & 0 & 1 \end{bmatrix} \begin{bmatrix} b_{11} & b_{12} \\ b_{21} & b_{22} \\ b_{31} & b_{32} \end{bmatrix} = \begin{bmatrix} 5b_{11} & 5b_{12} \\ b_{21} & b_{22} \\ b_{31} & b_{32} \end{bmatrix} \quad \text{. Check.}$$

More generally, suppose the movement from B to SB multiplies row i of B by k. Then the square matrix S is called a *type* M *elementary matrix* and denoted $M(k;i)$.

The algebraic specification of $M(k;i)$ is: $s_{ii} = k$; $s_{pp} = 1$ for $p \neq i$; all other entries are 0.

Properties of Elementary Matrices

The following small corollary of lemma 3.1 facilitates the study of matrix products.

COROLLARY **3.3**

Suppose two $m \times k$ matrices, C and D, multiply every matrix in exactly the same way; that is, if

$CB = DB$ whenever the products are defined,

then $C = D$.

■Proof   Our assumption implies, in particular, that C and D multiply $k \times 1$ matrices, or column vectors, in the same way. But then lemma 3.1 implies $C = D$.   Q.E.D.

Let us apply this observation to elementary matrices.

If rows i and j of a matrix B are exchanged twice in succession, the outcome is simply the original matrix B. In symbols, we have

$$E(i,j)E(i,j)B = B, \text{ or } E(i,j)^2 B = IB \quad.$$

It follows from the corollary that

$$E(i,j)^2 = I \quad. \quad (\textit{elementary matrix identity}),$$

and we can check this result by direct calculation. For example,

$$E(1,2)^2 = E(1,2)E(1,2) = \begin{bmatrix} 0 & 1 & 0 \\ 1 & 0 & 0 \\ 0 & 0 & 1 \end{bmatrix} \begin{bmatrix} 0 & 1 & 0 \\ 1 & 0 & 0 \\ 0 & 0 & 1 \end{bmatrix} = \ldots$$

$$\ldots \begin{bmatrix} 1 & 0 & 0 \\ 0 & 1 & 0 \\ 0 & 0 & 1 \end{bmatrix} = I \quad.$$

To find a similar identity for type A elementary matrices, note that if we operate twice on a matrix B by first adding k times row j to row i, and then subtracting k times row j from row i, we are back to the initial matrix B. So for all B, $A(-k;j,i)A(k;j,i)B = B$, and we deduce that

$$A(-k;j,i)A(k;j,i) = I = (\textit{elementary matrix identity}).$$

A specific example of this identity is:

$$A(-k;2,1)A(k;2,1) = \begin{bmatrix} 0 & -k & 1 \\ 0 & 1 & 0 \\ 0 & 0 & 1 \end{bmatrix} \begin{bmatrix} 1 & k & 0 \\ 0 & 1 & 0 \\ 0 & 0 & 1 \end{bmatrix} = \begin{bmatrix} 1 & 0 & 0 \\ 0 & 1 & 0 \\ 0 & 0 & 1 \end{bmatrix} = I \quad.$$

As for type M's, we cancel the effect of multiplying a row by k by multiplying the same row by $\frac{1}{k}$. This leads to the identity:

$$M(\tfrac{1}{k};i)M(k;i) = I \quad (\textit{elementary matrix identity}).$$

Let us summarize our new approach to row operations in a theorem.

THEOREM 3.4

(a) All elementary row operations on a matrix B may be done by left multiplication of B by the square elementary matrices of types E, A, and M constructed above.

(b) To every elementary matrix Q, there is an elementary matrix Q' of the same type as Q, such that $Q'Q = I$.

Since the process of reducing a matrix is done by a sequence of, say, h consecutive row operations, the process may be summarized as follows:

COROLLARY 3.5

Suppose the $m \times n$ matrix B has reduced form R. Then there are elementary matrices $Q_1, \ldots, Q_h$ such that:

$$Q_h Q_{h-1} \ldots Q_2 Q_1 B = R .$$

Also, if $Q'_1, \ldots, Q'_h$ are the elementary matrices such that $Q'_i Q_i = I$, then

$$B = Q'_1 Q'_2 \ldots Q'_{h-1} Q'_h R .$$

■Proof   The first equation simply affirms what we already know: the elementary row operations that reduce B to R can be done through left multiplication of B by elementary matrices $Q_1, \ldots, Q_h$ in turn.

Suppose now we multiply this equation by Q'_h. Then we have

$$Q'_h Q_h Q_{h-1} \ldots Q_2 Q_1 B = Q'_h R ,$$

or, since $Q'_h Q_h = I$,

$$Q_{h-1} \ldots Q_2 Q_1 B = Q'_h R .$$

Similarly, multiplying in turn by Q'_{h-1} etc., we finally obtain

$$B = Q'_1 Q'_2 \ldots Q'_{h-1} Q'_h R \quad \text{as required.}$$

Q.E.D.

■ Problem   Let $B = \begin{bmatrix} 0 & 2 \\ 1 & -6 \end{bmatrix}$.

(a) Find elementary matrices $Q_1, \ldots, Q_h$ such that $Q_h \ldots Q_1 B = I$.

(b) Find elementary matrices $Q'_1, \ldots, Q'_h$ such that $B = Q'_1 \ldots Q'_h$.

■ Solution (a) The reduction of B by row operations is a familiar process:

$$B = \begin{bmatrix} 0 & 2 \\ 1 & -6 \end{bmatrix} \overset{R_2}{\underset{R_1}{\longrightarrow}} \begin{bmatrix} 1 & -6 \\ 0 & 2 \end{bmatrix} \overset{R_1 + 3R_2}{\longrightarrow} \begin{bmatrix} 1 & 0 \\ 0 & 2 \end{bmatrix} \overset{\frac{1}{2}R_2}{\longrightarrow} \begin{bmatrix} 1 & 0 \\ 0 & 1 \end{bmatrix} = I .$$

These row operations are implemented by elementary matrices as follows:

Step 1. Row exchange, implemented by $Q_1 = E(1,2) = \begin{bmatrix} 0 & 1 \\ 1 & 0 \end{bmatrix}$.

      Check: $Q_1 B = \begin{bmatrix} 0 & 1 \\ 1 & 0 \end{bmatrix} \begin{bmatrix} 0 & 2 \\ 1 & -6 \end{bmatrix} = \begin{bmatrix} 1 & -6 \\ 0 & 2 \end{bmatrix}$.

Step 2. Type A operation, implemented by $Q_2 = A(3;2,1) = \begin{bmatrix} 1 & 3 \\ 0 & 1 \end{bmatrix}$.

      Check: $Q_2(Q_1 B) = \begin{bmatrix} 1 & 3 \\ 0 & 1 \end{bmatrix} \begin{bmatrix} 1 & -6 \\ 0 & 2 \end{bmatrix} = \begin{bmatrix} 1 & 0 \\ 0 & 2 \end{bmatrix}$.

Step 3. Type M operation, implemented by $Q_3 = M(2;\frac{1}{2}) = \begin{bmatrix} 1 & 0 \\ 0 & \frac{1}{2} \end{bmatrix}$.

      Check: $Q_3(Q_2 Q_1 B) = \begin{bmatrix} 1 & 0 \\ 0 & \frac{1}{2} \end{bmatrix} \begin{bmatrix} 1 & 0 \\ 0 & 2 \end{bmatrix} = \begin{bmatrix} 1 & 0 \\ 0 & 1 \end{bmatrix}$.

(b) In view of the preceding corollary 3.5, it suffices to find matrices Q_i such that $Q_i' Q_i = I$. But the elementary matrix identities imply that:

    for $Q_1 = E(1,2)$, $Q_1' = E(1,2) = \begin{bmatrix} 0 & 1 \\ 1 & 0 \end{bmatrix}$;

    for $Q_2 = A(3;2,1)$, $Q_2' = A(-3;2,1) = \begin{bmatrix} 1 & -3 \\ 0 & 1 \end{bmatrix}$;

    for $Q_3 = M(2;\frac{1}{2})$, $Q_3' = M(2;2) = \begin{bmatrix} 1 & 0 \\ 0 & 2 \end{bmatrix}$.

   Check: $Q_1' Q_2' Q_3' = \begin{bmatrix} 0 & 1 \\ 1 & 0 \end{bmatrix} \begin{bmatrix} 1 & -3 \\ 0 & 1 \end{bmatrix} \begin{bmatrix} 1 & 0 \\ 0 & 2 \end{bmatrix} = \ldots$

          $\ldots \begin{bmatrix} 0 & 1 \\ 1 & 0 \end{bmatrix} \begin{bmatrix} 1 & -6 \\ 0 & 2 \end{bmatrix} = \begin{bmatrix} 0 & 2 \\ 1 & -6 \end{bmatrix} = B$.

■ EXERCISES 3.3

1. Decide whether each of the following matrices is elementary; and if so, identify it as, for example, $M(8;2)$.

(a) $\begin{bmatrix} 1 & 0 \\ 0 & 8 \end{bmatrix}$ (b) $\begin{bmatrix} 1 & 0 \\ 4 & 1 \end{bmatrix}$ (c) $\begin{bmatrix} 2 & 0 \\ 0 & 3 \end{bmatrix}$

(d) $\begin{bmatrix} 1 & 1 \\ 0 & 2 \end{bmatrix}$ (e) $\begin{bmatrix} 0 & 0 & 1 \\ 0 & 1 & 0 \\ 1 & 0 & 0 \end{bmatrix}$ (f) $\begin{bmatrix} 1 & 0 & 0 \\ 0 & 1 & -5 \\ 0 & 0 & 1 \end{bmatrix}$

(g) $\begin{bmatrix} 1 & 0 & 0 \\ 0 & 0 & 3 \\ 0 & 0 & 1 \end{bmatrix}$ (h) $\begin{bmatrix} 1 & 0 & 0 \\ 0 & -1 & 0 \\ 0 & 0 & 1 \end{bmatrix}$

2. Let $B = \begin{bmatrix} 1 & 3 \\ 2 & 3 \end{bmatrix}$. Find elementary matrices Q_1, Q_2, Q_3 such that $Q_3 Q_2 Q_1 B = I$. Also find elementary matrices Q'_1, Q'_2, Q'_3 such that $B = Q'_1 Q'_2 Q'_3$.

3. Let $C = \begin{bmatrix} 0 & 6 & 13 \\ 0 & 2 & 4 \\ 1 & 0 & 0 \end{bmatrix}$. Find elementary matrices Q_1 through Q_4 such that $Q_4 \ldots Q_1 C = I$. Also find elementary matrices Q'_1 through Q'_4 such that $C = Q'_1 \ldots Q'_4$.

4. (a) Show that an elementary matrix differs from the identity matrix I in at most two rows.

    (b) Deduce that $P = \begin{bmatrix} 0 & 1 & 0 \\ 0 & 0 & 1 \\ 1 & 0 & 0 \end{bmatrix}$ is not an elementary matrix.

    (c) Show that $P^2 \neq I$, but $P^3 = I$.

    (d) Show that P is the product of two type E elementary matrices.

5. (a) Show that $E(1,2)$ and $E(3,4)$ commute.

    (b) Show that $E(1,2)$ and $E(2,3)$ do not commute.

    (c) Show that any two $n \times n$ diagonal matrices (matrices in which all nondiagonal entries are zero) commute.

    (d) Deduce that any two $n \times n$ type M elementary matrices commute.

6. Let $A = \begin{bmatrix} 2 & 4 & 5 \\ 0 & 1 & 3 \\ 6 & 10 & 9 \end{bmatrix}$, $B = \begin{bmatrix} 2 & 0 & -7 \\ 0 & 1 & 3 \\ 2 & 2 & -1 \end{bmatrix}$.

    Find elementary matrices Q_1 and Q_2 such that $B = Q_2 Q_1 A$.

7. Suppose A and B are any two $m \times n$ matrices with the same reduced form R. Show that there are $m \times m$ elementary matrices $S_1, \ldots, S_r$ such that $B = S_r \ldots S_1 A$.

8. Let $A = \begin{bmatrix} 3 & 5 \\ 7 & 5 \end{bmatrix}$, $B = \begin{bmatrix} -2 & 0 \\ 10 & 10 \end{bmatrix}$.

    Find elementary matrices Q_1, Q_2 such that $Q_2 Q_1 A = B$.

9. Let $A = \begin{bmatrix} 1 & 2 \\ 3 & 4 \end{bmatrix}$, $B = \begin{bmatrix} 3 & 4 \\ 2 & 4 \end{bmatrix}$.

    Find elementary matrices Q_1, Q_2 such that $Q_1 A = Q_2 B$.

10. (a) Suppose the $m \times n$ matrix B is multiplied on the *right* by the $n \times n$ elementary matrix $M(k;i)$. Show that in the move from B to $BM(k;i)$, column i of B is multiplied by k.

(b) Interpret the multiplication of B on the right by types E and A elementary matrices as "column operations" on B as well.

*11. A *permutation matrix* is a square matrix P with a single 1 in each row and column, and all other entries zero.

    (a) Show that an $m \times m$ matrix P is a permutation matrix if and only if, for any $m \times n$ matrix B, the move from B to PB simply rearranges (or "permutes") the rows of B.

    (b) Show that a square matrix P is a permutation matrix if and only if it is a product of type E elementary matrices.

    (c) Show that the product of $m \times m$ permutation matrices is another $m \times m$ permutation matrix.

    (d) Show that there are exactly $m!$ (recall that $m!$, or "m factorial" is the product of all integers from 1 to m) different $m \times m$ permutation matrices.

∎3.4 The Inverse Matrix

Introduction

Suppose B is a square, $n \times n$ matrix, mapping R^n into R^n. Then an $n \times n$ matrix A is called an *inverse of B* if $AB = I$.

For example, suppose $B = \begin{bmatrix} 3 & 7 \\ 2 & 5 \end{bmatrix}$. Then $A = \begin{bmatrix} 5 & -7 \\ -2 & 3 \end{bmatrix}$ is an inverse of B, simply because

$$AB = \begin{bmatrix} 5 & -7 \\ -2 & 3 \end{bmatrix} \begin{bmatrix} 3 & 7 \\ 2 & 5 \end{bmatrix} = \begin{bmatrix} 1 & 0 \\ 0 & 1 \end{bmatrix} = I \ .$$

The inverse matrix, when it can be found, greatly facilitates computation. Suppose, for example, that we want to solve $Bx = c$ for various choices of c without having to reduce the augmented matrix all over again for every different c. Instead, if we have an inverse A, just multiply the equation by A:

$$ABx = Ac \ ,$$
or $\quad Ix = Ac \ ,$
or $\quad x = Ac \ .$

That is, we compute the solution to the system by applying the inverse matrix A to the vector c.

For example, to solve the equation $Bx = c$ for x, with $B = \begin{bmatrix} 3 & 7 \\ 2 & 5 \end{bmatrix}$ as above, and for several choices of c, say $c = (4,1)$, $c = (2,6)$, $c = (4,8)$, etc., we just compute the matrix-vector product Ax for the given choices of c:

$$\mathbf{x} = A\mathbf{c} = \begin{bmatrix} 5 & -7 \\ -2 & 3 \end{bmatrix} \begin{bmatrix} 4 \\ 1 \end{bmatrix} = \begin{bmatrix} 13 \\ -5 \end{bmatrix} \quad,$$

$$\mathbf{x} = A\mathbf{c} = \begin{bmatrix} 5 & -7 \\ -2 & 3 \end{bmatrix} \begin{bmatrix} 2 \\ 6 \end{bmatrix} = \begin{bmatrix} -32 \\ 14 \end{bmatrix} \quad, \quad \text{etc.}$$

Alternatively, the solution to $B\mathbf{x} = \mathbf{c}$ can be expressed directly in terms of the components of $\mathbf{c}$ through the formula

$$\mathbf{x} = A\mathbf{c} = \begin{bmatrix} 5 & -7 \\ -2 & 3 \end{bmatrix} \begin{bmatrix} c_1 \\ c_2 \end{bmatrix} = \begin{bmatrix} 5c_1 - 7c_2 \\ -2c_1 + 3c_2 \end{bmatrix} \quad.$$

Thus for $\mathbf{c} = (4,8)$, $\mathbf{x} = (5(4)-7(8), -2(4)+3(8)) = (-36, 16)$, etc.

Existence of the Inverse

However, not every $n \times n$ matrix can have an inverse. In fact the above calculation shows that when there is an inverse, the system $B\mathbf{x} = \mathbf{c}$ has the unique solution $\mathbf{x} = A\mathbf{c}$. And uniqueness is only possible when rank $(B) = n$, according to theorem 1.1 in Chapter 1.

The converse holds as well. For if the $n \times n$ matrix B has rank n, its reduced form R has a corner in each of the n columns, and so $R = I$. Therefore, the reduction of B to R through the elementary matrices Q_1, Q_2, etc., given in corollary 3.5, takes the form

$$Q_h \ldots Q_1 B = I \quad.$$

That is, $AB = I$, where $A = Q_h \ldots Q_1$, and so A is an inverse of B. Thus we have proved the following theorem.

THEOREM 3.6

The $n \times n$ matrix B has an inverse if and only if rank $(B) = n$.

Computation of the Inverse

The method of elementary matrices also leads to an efficient method of computing the inverse of a given numerical matrix B. We will illustrate the method with a 2×2 matrix.

Example 1 $B = \begin{bmatrix} 1 & 3 \\ 2 & 5 \end{bmatrix}$

First we write the matrices B and I side by side, thereby forming an $n \times 2n$ matrix that will be denoted:

$$[B|I] \quad \text{or} \quad \left[\begin{array}{cc|cc} 1 & 3 & 1 & 0 \\ 2 & 5 & 0 & 1 \end{array} \right] \quad.$$

This matrix can be thought of as being *partitioned* into two $n \times n$ blocks, B and I.

The first row operation of the reduction sequence for B is, of course, row 2 minus twice row 1. We apply this row operation not only to B but also to the right block, I, obtaining:

Step 1. $[Q_1B|Q_1I]$ or

$$R_2 - 2R_1 \quad \begin{bmatrix} 1 & 3 & | & 1 & 0 \\ 0 & \boxed{-1} & | & -2 & 1 \end{bmatrix} \quad .$$

Similarly, we apply the remaining row operations of the reduction sequence for B to both the left and the right blocks together:

Step 2. $[Q_2Q_1B|Q_2Q_1I]$ or $R_1 + 3R_2$ $\begin{bmatrix} 1 & 0 & | & -5 & 3 \\ 0 & \boxed{-1} & | & -2 & 1 \end{bmatrix}$.

Step 3. $[Q_3Q_2Q_1B|Q_3Q_2Q_1I]$ or $-R_2$ $\begin{bmatrix} 1 & 0 & | & -5 & 3 \\ 0 & 1 & | & 2 & -1 \end{bmatrix}$.

On observing that the left block of the numerical matrix is now I, we conclude that $Q_3Q_2Q_1B = I$. Therefore, according to the preceding discussion of the inverse, the right block $A = Q_3Q_2Q_1I = Q_3Q_2Q_1$ is the inverse of B. That is, we now have

$$[I|A] \quad \text{or} \quad A = \begin{bmatrix} -5 & 3 \\ 2 & -1 \end{bmatrix} \quad .$$

It is a good strategy to detect possible numerical errors by checking that $AB = I$:

$$\begin{bmatrix} -5 & 3 \\ 2 & -1 \end{bmatrix} \begin{bmatrix} 1 & 3 \\ 2 & 5 \end{bmatrix} = \begin{bmatrix} 1 & 0 \\ 0 & 1 \end{bmatrix} = I \quad . \quad \text{Check.}$$

Let us summarize this important method in a theorem.

THEOREM 3.7

Suppose that B and I are $n \times n$ matrices and that by a sequence of elementary row operations the $n \times 2n$ matrix $[B|I]$ can be changed to $[I|A]$.
   Then A is an inverse of B.

Example 2 Suppose we want to determine, first, whether $B = \begin{bmatrix} 2 & 4 \\ 4 & 8 \end{bmatrix}$ has an

inverse; and second, what the inverse is if it exists. We could answer the first question by checking directly whether rank $(B) = n$; but this is not necessary. Instead, we can start the reduction of $[B|I]$ right away:

$$[B|I] = \begin{bmatrix} 2 & 4 & | & 1 & 0 \\ 4 & 8 & | & 0 & 1 \end{bmatrix} \quad R_2 - 2R_1 \quad \begin{bmatrix} 2 & 4 & | & 1 & 0 \\ 0 & 0 & | & -2 & 1 \end{bmatrix} \quad .$$

At this stage, we can see that the left block, B, is reducing, not to I, but to another reduced matrix, R, with fewer than $n = 2$ corners. So we learn during the reduction process that B is not invertible, and discontinue the computation at this point.

Example 3 Suppose we want to invert the 3×3 matrix

$$B = \begin{bmatrix} 3 & 1 & 4 \\ 4 & 1 & 6 \\ 1 & 0 & 1 \end{bmatrix} .$$

We can see that reduction of $[B|I]$ by the standard sequence of row operations set out in Chapter 1 is going to involve fractions from the beginning, because of the entry $b_{11} = 3$. Fortunately, our construction of the inverse, as summarized in theorem 3.7, does not at any stage call for the use of this standard sequence. All that is required is that the sequence of row operations chosen results in the identity matrix I ultimately appearing in the left block.

In this example, it is best to start the operations on $[B|I]$ with the interchange of rows 1 and 3. This will result in the first corner being the number 1 and will at least postpone the appearance of fractions:

$$[B|I] = \left[\begin{array}{ccc|ccc} 3 & 1 & 4 & 1 & 0 & 0 \\ 4 & 1 & 6 & 0 & 1 & 0 \\ 1 & 0 & 1 & 0 & 0 & 1 \end{array} \right] \quad \begin{array}{c} R_3 \\ \\ R_1 \end{array} \left[\begin{array}{ccc|ccc} 1 & 0 & 1 & 0 & 0 & 1 \\ 4 & 1 & 6 & 0 & 1 & 0 \\ 3 & 1 & 4 & 1 & 0 & 0 \end{array} \right]$$

$$\begin{array}{c} \\ R_2 - 4R_1 \\ \\ \end{array} \left[\begin{array}{ccc|ccc} 1 & 0 & 1 & 0 & 0 & 1 \\ 0 & 1 & 2 & 0 & 1 & -4 \\ 0 & 1 & 1 & 1 & 0 & -3 \end{array} \right] \quad \begin{array}{c} \\ \\ R_3 - R_2 \end{array} \left[\begin{array}{ccc|ccc} 1 & 0 & 1 & 0 & 0 & 1 \\ 0 & 1 & 2 & 0 & 1 & -4 \\ 0 & 0 & -1 & 1 & -1 & 1 \end{array} \right]$$

$$\begin{array}{c} R_1 + R_3 \\ R_2 + 2R_3 \\ \\ \end{array} \left[\begin{array}{ccc|ccc} 1 & 0 & 0 & 1 & -1 & 2 \\ 0 & 1 & 0 & 2 & -1 & -2 \\ 0 & 0 & -1 & 1 & -1 & 1 \end{array} \right]$$

$$\begin{array}{c} \\ \\ -R_3 \end{array} \left[\begin{array}{ccc|ccc} 1 & 0 & 0 & 1 & -1 & 2 \\ 0 & 1 & 0 & 2 & -1 & -2 \\ 0 & 0 & 1 & -1 & 1 & -1 \end{array} \right] = [I|A] .$$

That is, the inverse is $A = \begin{bmatrix} 1 & -1 & 2 \\ 2 & -1 & -2 \\ -1 & 1 & -1 \end{bmatrix}$. (Check that $AB = I$.)

Of course, in many problems the matrix entries are much more difficult numbers than integers to begin with, and the ideal computational strategies are based on quite different principles from those above. See Chapter 10 on numerical methods. Nevertheless, the simpler matrices with integer entries are also important in both theory and practice, and the computational short-cuts discussed here are significant.

Algebraic Properties of the Inverse

It turns out that the matrix inverse can be used with facility in algebraic computation, provided that due attention is given to the ordering of factors. It is even possible to recover a form of the cancellation law. We start with a "uniqueness" property that simplifies things by allowing us to talk about *the* inverse.

Lemma 3.8 Suppose an $n \times n$ matrix B has an inverse. Then,
   (a) the inverse, denoted B^{-1}, is unique; and
   (b) the matrix and its inverse commute, that is,

$$B^{-1}B = BB^{-1} = I \quad .$$

■ Proof We know from earlier discussion in this section that if rank $(B) = n$, then B has at least one inverse A, where A and B have the elementary matrix representations:

$$A = Q_h Q_{h-1} \cdots Q_2 Q_1, \quad \text{and} \quad B = Q'_1 Q'_2 \cdots Q'_{h-1} Q'_h \quad .$$

In the product

$$BA = Q'_1 Q'_2 \cdots Q'_{h-1} Q'_h Q_h Q_{h-1} \cdots Q_2 Q_1 \quad ,$$

the adjacent factors Q'_h, Q_h satisfy $Q'_h Q_h = I$. These factors can therefore be deleted, and the product simplifies to

$$BA = Q'_1 Q'_2 \cdots Q'_{h-1} \ Q_{h-1} \cdots Q_2 Q_1 \quad .$$

Evidently the product $Q'_{h-1} Q_{h-1}$ can now be deleted and so on. Eventually the simplifications reach the conclusion

$$BA = Q'_1 Q_1 = I \quad .$$

Suppose now that B has another inverse, C, so that $CB = I$ and $AB = I$ both hold. Subtracting equations, we get $(C-A)B = O$; or, on multiplying this equation in turn by A, $(C-A)BA = O$; or, since $BA = I, C - A = O$; or, $C = A$. The new inverse, C, is therefore the same as the old one, A. That is, there is just one inverse, $A = B^{-1}$, and since we know that $AB = BA = I$ we have $B^{-1}B = BB^{-1} = I$.   Q.E.D.

Example We saw that $B = \begin{bmatrix} 1 & 3 \\ 2 & 5 \end{bmatrix}$ has the inverse $B^{-1} = \begin{bmatrix} -5 & 3 \\ 2 & -1 \end{bmatrix} \quad .$

We can verify the commutative rule for the inverse by direct computation:

$$BB^{-1} = \begin{bmatrix} 1 & 3 \\ 2 & 5 \end{bmatrix} \begin{bmatrix} -5 & 3 \\ 2 & -1 \end{bmatrix} = \begin{bmatrix} 1 & 0 \\ 0 & 1 \end{bmatrix} = I \quad . \quad \text{Check.}$$

> **THEOREM 3.9**
>
> Matrix products and inverses satisfy the following rules:
>
> First, modified cancellation laws hold. If B is an invertible square matrix, then
>
> (a) $CB = DB$ implies $C = D$
>
> (b) $BC = BD$ implies $C = D$.
>
> Also, when B, B_1, B_2, etc. are invertible $n \times n$ matrices, then
>
> (c) $B^{-1}B = BB^{-1}$
>
> (d) $(B_1 B_2)^{-1} = B_2^{-1} B_1^{-1}$
>
> (e) $(B_1 \cdots B_k)^{-1} = B_k^{-1} \cdots B_1^{-1}$
>
> (f) $(B^p)^{-1} = (B^{-1})^p$.

■**Proof**   All the statements are easy to prove.

(a) $CB = DB$ implies $CBB^{-1} = DBB^{-1}$ or $CI = DI$ or $C = D$.

(b) Similar to (a).

(c) This is a restatement of the previous lemma.

(d) $(B_2^{-1} B_1^{-1})(B_1 B_2) = B_2^{-1}(B_1^{-1}B_1)B_2 = B_2^{-1}IB_2 = B_2^{-1}B_2 = I$. This equation says precisely that $(B_2^{-1}B_1^{-1})$ is the inverse of $(B_1 B_2)$.

(e) Same method as (d).

(f) This is just (e), with p factors, all equal to B.   Q.E.D.

Special Formulas for the Inverse

So far we have been able to compute inverses of specific numerical matrices by row operations. However, there are special cases in which an algebraic formula for B^{-1} can be obtained directly in terms of B.

Example 1   Suppose the $n \times n$ matrix B is known to satisfy the polynomial matrix equation $6B^3 + 4B + 2I = O$. Then we can shift the I term to the right side and factor B from the rest of the terms:

$$-3B^2 - 2B = I \quad \text{or} \quad (-3B - 2I)B = I \quad .$$

But the last equation says simply that the inverse of B is

$$B^{-1} = -3B - 2I \quad .$$

Example 2 Consider the general 2×2 matrix $B = \begin{bmatrix} a & b \\ c & d \end{bmatrix}$. The following matrix identity happens to result from reshuffling the entries of B:

$$\begin{bmatrix} d & -b \\ -c & a \end{bmatrix} \begin{bmatrix} a & b \\ c & d \end{bmatrix} = \begin{bmatrix} ad-bc & 0 \\ 0 & ad-bc \end{bmatrix} = (ad - bc)I \ .$$

This implies that, when $ad - bc \neq 0$, the inverse is

$$B^{-1} = \frac{1}{ad - bc} \begin{bmatrix} d & -b \\ -c & a \end{bmatrix} \ .$$

That is, to invert the 2×2 matrix B, the steps are the following:

(a) exchange the diagonal entries of B,

(b) reverse the signs of the off-diagonal entries,

(c) divide the matrix by $ad-bc$.

An example of the steps is

$$B = \begin{bmatrix} 2 & 3 \\ 4 & 5 \end{bmatrix} \ , \quad \begin{bmatrix} 5 & 3 \\ 4 & 2 \end{bmatrix} \ , \quad \begin{bmatrix} 5 & -3 \\ -4 & 2 \end{bmatrix} \ , \ldots$$

$$\ldots (-\tfrac{1}{2}) \begin{bmatrix} 5 & -3 \\ -4 & 2 \end{bmatrix} = \begin{bmatrix} -\tfrac{5}{2} & \tfrac{3}{2} \\ 2 & -1 \end{bmatrix} = B^{-1} \ .$$

Since for 2×2 matrices we can express the entries of B^{-1} algebraically in terms of those of B, we might naturally demand a similar formula in the $n\times n$ case. In fact, such a formula, known as "Cramer's Rule", is derived in Appendix B. It is important in theoretical and algebraic contexts but is inefficient for numerical computation when n is large.

■ **Problem** Suppose that brand I of megavitamins used to monopolize a certain regional market until brand II was gradually placed in stores in various towns throughout the region. Suppose that a market researcher is trying to find out when brand II was put on the market in one particular town. He knows that the present market shares there are 72% for brand I and 28% for brand II. He also knows that, whenever the two brands are in competition, brand II takes 14% of brand I's customers each year while brand I takes 6% of brand II's. Can he reconstruct the sales histories of the two brands in the town?

■ **Solution** Let the proportions of the market held by brands I and II during year n be represented by the components $x_1(n), x_2(n)$ of the "market share vector" $\mathbf{x}(n)$. The given information about the dynamics of competition implies the market share predictor formula

$$\mathbf{x}(n+1) = B\mathbf{x}(n) \ , \quad \text{where } B = (\tfrac{1}{100}) \begin{bmatrix} 86 & 6 \\ 14 & 94 \end{bmatrix} \ .$$

But from the formula for the 2×2 matrix inverse,

$$B^{-1} = \left(\frac{(100)^2}{(86)(94) - (06)(14)} \right) \begin{bmatrix} 94 & -6 \\ -14 & 86 \end{bmatrix} = \ldots$$

$$\ldots = \left(\tfrac{1}{80} \right) \begin{bmatrix} 94 & -6 \\ -14 & 86 \end{bmatrix} = \left(\tfrac{1}{40} \right) \begin{bmatrix} 47 & -3 \\ -7 & 43 \end{bmatrix} .$$

Then multiplication of the predictor formula by B^{-1} gives $B^{-1}\mathbf{x}(n+1) = \mathbf{x}(n)$, or, on introducing a new time parameter $p = n+1$,

$$\mathbf{x}(p-1) = B^{-1}\mathbf{x}(p) \quad , \quad \text{which is the } history \; formula.$$

For the present year, year p, the market share vector $\mathbf{x}(p)$ is known to be $\mathbf{x}(p) = \left(\tfrac{72}{100}, \tfrac{28}{100} \right)$. Application of the history formula then gives last year's shares as

$$\mathbf{x}(p-1) = B^{-1}\mathbf{x}(p) =$$

$$\left(\tfrac{1}{40} \right) \begin{bmatrix} 47 & -3 \\ -7 & 43 \end{bmatrix} \left(\tfrac{1}{100} \right) \begin{bmatrix} 72 \\ 28 \end{bmatrix} = \left(\tfrac{1}{40} \right) \begin{bmatrix} 33 \\ 7 \end{bmatrix}$$

Similarly, $\mathbf{x}(p-2) = B^{-1}\mathbf{x}(p-1) =$

$$\left(\tfrac{1}{40} \right) \begin{bmatrix} 47 & -3 \\ -7 & 43 \end{bmatrix} \left(\tfrac{1}{40} \right) \begin{bmatrix} 33 \\ 7 \end{bmatrix} = \left(\tfrac{1}{100} \right) \begin{bmatrix} 153 \\ 7 \end{bmatrix} .$$

and $\mathbf{x}(p-3) = B^{-2}\mathbf{x}(p-2) =$

$$\left(\tfrac{1}{40} \right) \begin{bmatrix} 47 & -3 \\ -7 & 43 \end{bmatrix} \left(\tfrac{1}{160} \right) \begin{bmatrix} 153 \\ 7 \end{bmatrix} = \left(\tfrac{1}{640} \right) \begin{bmatrix} 717 \\ -77 \end{bmatrix}$$

Thus brand II's market shares in years p, $p-1$ and $p-2$ respectively are $x_2(p) = \tfrac{28}{100}$ or 28%, $x_2(p-1) = \tfrac{7}{40}$ or 17.5%, $x_2(p-2) = \tfrac{7}{160}$, or 4.4%.

However, our calculations give $x_2(p-3) = -\tfrac{77}{640}$, which is negative.

Thus $x_2(p-3)$ does not represent a market share! Now, if the competition between brands I and II had really started three or more years ago, the history formula would have retrieved the (positive) market shares of three years ago. The market researcher can therefore conclude that competition between brands had not yet started three years ago. Brand II must have gone on the market between two and three years ago.

■ EXERCISES 3.4

1. For each of the following matrices, compute the inverse or determine that none exists by reducing the block form $[B|I]$. Check your answers by multiplying out BB^{-1}.

(a) $\begin{bmatrix} 1 & 4 \\ 2 & 7 \end{bmatrix}$ (b) $\begin{bmatrix} 2 & 5 \\ 4 & 8 \end{bmatrix}$ (c) $\begin{bmatrix} 3 & 4 \\ 4 & -3 \end{bmatrix}$

(d) $\begin{bmatrix} 5 & -12 \\ 12 & 5 \end{bmatrix}$ (e) $\begin{bmatrix} 1 & 2 & -1 \\ 2 & 5 & 3 \\ 1 & 3 & 9 \end{bmatrix}$ (f) $\begin{bmatrix} 3 & 7 & 0 \\ 2 & 3 & 4 \\ 4 & 8 & 2 \end{bmatrix}$

(g) $\begin{bmatrix} 1 & 2 & 3 \\ 2 & 3 & 4 \\ 3 & 4 & 6 \end{bmatrix}$ (h) $\begin{bmatrix} 3 & 3 & 8 \\ 5 & 6 & 8 \\ 1 & 1 & 2 \end{bmatrix}$ (i) $\begin{bmatrix} 5 & 0 & 4 \\ 6 & 1 & 0 \\ 1 & 0 & 1 \end{bmatrix}$

(j) $\begin{bmatrix} 1 & 2 & 3 \\ 0 & 1 & 2 \\ 0 & 0 & 1 \end{bmatrix}$ (k) $\begin{bmatrix} 1 & 1 & 1 \\ 1 & 2 & 3 \\ 1 & 4 & 9 \end{bmatrix}$ (l) $\begin{bmatrix} 1 & 2 & 3 \\ 4 & 5 & 6 \\ 7 & 8 & 9 \end{bmatrix}$

(m) $\begin{bmatrix} 0 & 0 & 1 & 0 \\ 0 & 0 & 0 & 1 \\ 0 & 2 & 0 & 0 \\ 3 & 0 & 0 & 0 \end{bmatrix}$

2. Compute the inverses of matrices (a) through (d) in exercise 1 by using the special formula for the 2×2 inverse given in this section.

3. The matrix $B = \begin{bmatrix} -2 & 1 & p \\ 0 & -1 & 1 \\ 1 & 2 & 0 \end{bmatrix}$ contains a parameter p.

   (a) For which values of the parameter p is B invertible?
   (b) Show that, for those values of p for which the inverse exists, there is a formula for B^{-1} in terms of p.

4. Show that, for all values of the parameters p, q, r, the matrix

   $$B = \begin{bmatrix} 1 & p & q \\ 0 & 1 & r \\ 0 & 0 & 1 \end{bmatrix}$$

   has an inverse. Compute this inverse.

5. Suppose the $n \times n$ matrix B satisfies the equation $B^7 - 3B + I = 0$. Show that B is invertible and find a formula for B^{-1} in terms of B.

6. Suppose that the $n \times n$ matrix A satisfies $A^g = 0$ for some positive integer g.

   (a) Show that A cannot have an inverse.
   (b) Show that for any constant a, $I - aA$ has the inverse

   $$(I - aA)^{-1} = I + aA + a^2 A^2 + \cdots + a^{g-1} A^{g-1} \quad .$$

7. For the taxi company described in example 3 of section 3.2, the matrix

$B = \begin{bmatrix} \frac{1}{5} & 0 & 1 \\ \frac{4}{5} & 0 & 0 \\ 0 & 1 & 0 \end{bmatrix}$ was found to predict next year's taxi fleet vector,

x_{n+1}, from this year's, x_n, through the formula $x_{n+1} = Bx_n$.

(a) Compute B^{-1}.

(b) Given that this year's fleet vector is $x_n = (609, 36, 80)$, use the formula $x_{n-1} = B^{-1}x_n$ to calculate the fleet vectors of last year and the year before that.

(c) By computing and inspecting the vector $x_{n-3} = B^{-1}x_{n-2}$, show that the given fleet renewal strategy could not have been in effect as long as three years ago.

8. For the breakfast cereals Crunch and Munch described in example 2 of section 2.2, the market share vectors in years n and $n+1$ were found to be related by the formulas

$$x_{n+1} = Bx_n, \text{ or } x_n = B^{-1}x_{n+1}, \text{ where } B = \left(\frac{1}{10}\right) \begin{bmatrix} 8 & 3 \\ 2 & 7 \end{bmatrix} .$$

(a) Compute B^{-1}.

(b) Assuming that this year Crunch has 75% of the market, what was its market share last year?

9. (a) To generalize from the previous exercise, suppose Crunch and Munch lose proportions p and q, respectively, of their buyers to the opposite brand each year. Show that the formula for B becomes

$$B = \begin{bmatrix} 1-p & q \\ p & 1-q \end{bmatrix} .$$

(b) Show that B is invertible unless $p + q = 1$.

(c) In the special case $p + q = 1$, can last year's market shares be deduced in *any* way from this year's?

10. Develop the 2×2 inverse formula directly by reducing

$$[A|I] = \begin{bmatrix} a & b & | & 1 & 0 \\ c & d & | & 0 & 1 \end{bmatrix} ,$$

starting with the row operation $R_2 - \left(\frac{c}{a}\right)R_1$.

*11. Suppose $m > n$, and let b be an $m \times n$ matrix with rank $(B) = n$. By following the steps outlined below, show that B has a *left inverse*, that is, an $n \times m$ matrix A such that

$$AB = I_n = n \times n \text{ identity matrix.}$$

(a) Let $O_{m,n}$ represent the $m \times n$ zero matrix, and let F and G be, respectively, the $n \times m$ and $m \times n$ matrices with block representations

$$F = [I_n | O_{n, m-n}] \quad , \quad G = \left[\begin{array}{c} I_n \\ \hline O_{m-n, n} \end{array} \right] .$$

Show that F is a left inverse of G.

(b) Show that $[B | I_m]$ reduces to $[G | S]$ so that the $m \times m$ matrix S satisfies $SB = G$.

(c) Show that $A = FS$ is a left inverse for B.

(d) Find a left inverse for $B = \begin{bmatrix} 1 & 3 \\ 2 & 6 \\ 3 & 12 \end{bmatrix}$.

(e) *Nonuniqueness.* Show that if C is any $n \times (m-n)$ matrix, then $F' = [I_n | C]$ is also a left inverse of G, and so $A' = F'S$ is also a left inverse of B.

*12. (a) Show that when the $m \times n$ matrix B has rank $(B) < n$, then B cannot have a left inverse.

(b) Suppose now that $m < n$ (the opposite case from the previous exercise). It is then possible for the $m \times n$ matrix B to satisfy rank $(B) = m$. Develop a theory of the *right inverse* in this case.

*13. (a) There is only a finite number of $n \times n$ permutation matrices (see section 3.3, starred exercise). Deduce that, given a permutation matrix P, there must be two different integers q, r such that $P^q = P^r$.

(b) Show that for every permutation matrix P, there is a positive integer s such that $P^s = I$.

(c) Find a 5×5 permutation matrix P such that $s = 6$ is the least positive integer for which $P^s = I$.

*14. Suppose B is an invertible $n \times n$ matrix, and suppose the column vectors $\mathbf{d}_1, \ldots, \mathbf{d}_k$ are the solutions of the k linear systems $B\mathbf{x} = \mathbf{c}_1, \ldots, B\mathbf{x} = \mathbf{c}_k$. That is, the augmented matrices $[B|\mathbf{c}_1], \ldots, [B|\mathbf{c}_k]$ reduce to $[I|\mathbf{d}_1], \ldots, [I|\mathbf{d}_k]$ respectively.

(a) Suppose the k columns $\mathbf{c}_1, \ldots, \mathbf{c}_k$ are adjoined to B, forming the $n \times (n+k)$ matrix $J = [B|\mathbf{c}_1| \ldots |\mathbf{c}_k]$. Show that J reduces to $R = [I|\mathbf{d}_1| \ldots |\mathbf{d}_k]$, and deduce that the k linear systems given above can all be solved with a single reduction process.

(b) Show that in the special case when the $\mathbf{c}_i$'s are the standard vectors $\mathbf{e}_1, \ldots, \mathbf{e}_n$, then $J = [B|I]$ and the reduced form $R = [I|B^{-1}]$. That is, show that the matrix reduction method is a special case of this procedure.

3.5 Triangular Matrices

Introduction

The matrices used in many of the illustrative problems in earlier sections represented tables of data. For example, the entry a_{ij} of a matrix might represent the concentration of the ith mineral in ore from the jth mine. In this case, all the information concerning the ith mineral is set out along the ith row of the table. But we can easily imagine problems in which it is necessary to list the information about each mineral down a column. Then we have to transpose the table of data, or matrix, by converting its rows into columns.

Furthermore, transpositions must sometimes be done in the middle of a calculation when matrices are entangled in complicated formulas. Since we cannot foresee the precise contexts in which transpositions will be needed, it seems wise to work out the general algebraic properties of the operation for use as needed. It turns out later (see Chapters 7 and 8) that transposition has an essential place in the analysis of geometric transformations, and even in the further development of matrix algebra itself.

This section also includes a discussion of triangular matrices, which are packed solidly with zeros below (or above) the diagonal. These matrices are often easier to work with than matrices in general, and many matrix calculations are designed to produce triangular matrices. For example, reduced matrices are always triangular.

Triangular matrices are of importance in the theory of numerical computation and are applied to this purpose in Chapter 10.

The Transpose Matrix

Suppose the entries of the $m \times n$ matrix A and the $n \times m$ matrix G are related by the formula $g_{ij} = a_{ji}$. Then G is called the *transpose* of A and is denoted A^t.

In particular, column j of G consists of the entries $g_{1j}, g_{2j}, \ldots, g_{nj}$, which are the same as $a_{j1}, \ldots, a_{jn}$, which in turn constitute row j of A. That is,

column j of A^t = row j of A.

Example 1   The rows of the 2×3 matrix $A = \begin{bmatrix} 1 & 2 & 3 \\ 4 & 5 & 6 \end{bmatrix}$ are $(1,2,3)$ and $(4,5,6)$; therefore the columns of the 3×2 matrix A^t are $\begin{bmatrix} 1 \\ 2 \\ 3 \end{bmatrix}$ and $\begin{bmatrix} 4 \\ 5 \\ 6 \end{bmatrix}$.

That is, $A^t = \begin{bmatrix} 1 & 4 \\ 2 & 5 \\ 3 & 6 \end{bmatrix}$.

Similar arguments show that

row i of A^t = column i of A,

and so we can say that the transposition operation on A *interchanges* the rows and columns of A.

The *diagonal* of a matrix A consists of the entries a_{11}, a_{22}, ... Now, from the definition of $G = A^t$, it follows that $g_{ii} = a_{ii}$, and so transposition leaves the diagonal of a matrix unchanged. In particular, A and A^t both have diagonal entries 1,5 in example 1 above.

In practice, we often transpose a table of data to get it the "right way around," and we will see that transposition plays an important role in the discussion of metric concepts in Chapter 7. For now, we simply note some algebraic properties of the transpose. There is a surprising similarity to the properties of the inverse.

THEOREM 3.10

(a) $(A^t)^t = A$.   (b) $(AB)^t = B^t A^t$.

(c) For an invertible $n \times n$ matrix A, $(A^t)^{-1} = (A^{-1})^t$.

■**Proof**   (a) ijth entry of $(A^t)^t$
$= ji$th entry of A^t (transpose law)
$= ij$th entry of A (transpose law).

(b) ijth entry of $(AB)^t$
$= ji$th entry of AB (transpose law)
$= (\text{row } j \text{ of } A) \cdot (\text{column } i \text{ of } B)$ (row-column rule)
$= (\text{column } i \text{ of } B) \cdot (\text{row } j \text{ of } A)$ (dot product commutes)
$= (\text{row } i \text{ of } B^t) \cdot (\text{column } j \text{ of } A^t)$ (transpose law)
$= ij$th entry of $(B^t A^t)$ (row-column rule).

(c) $(A^{-1})^t A^t = (AA^{-1})^t$ (from part (b))
$= I^t$
$= I$.

But the equation $(A^{-1})^t A^t = I$ says precisely that $(A^{-1})^t$ is the inverse of A^t. That is, $(A^{-1})^t = (A^t)^{-1}$.   Q.E.D

Example 2 Suppose the unknown matrix C satisfies the matrix equation
$C^t A = B$, where $A = \begin{bmatrix} 1 & 2 \\ 0 & 1 \end{bmatrix}$, $B = \begin{bmatrix} 1 & 3 \\ 2 & 2 \\ 1 & 5 \end{bmatrix}$. Since A is inver-

tible, and in fact $A^{-1} = \begin{bmatrix} 1 & -2 \\ 0 & 1 \end{bmatrix}$, we can recover C as follows:

$C^t = BA^{-1}$, or $C = (BA^{-1})^t$.

Now $BA^{-1} = \begin{bmatrix} 1 & 3 \\ 2 & 2 \\ 1 & 5 \end{bmatrix} \begin{bmatrix} 1 & -2 \\ 0 & 1 \end{bmatrix} = \begin{bmatrix} 1 & 1 \\ 2 & -2 \\ 1 & 3 \end{bmatrix}$,

and so $C = \begin{bmatrix} 1 & 2 & 1 \\ 1 & -2 & 3 \end{bmatrix}$.

Triangular Matrices

The *diagonal* of a matrix A consists of the entries $a_{11}, a_{22}, \ldots$. We now define three kinds of *triangular matrices* by reference to the diagonal.

  (a) A matrix U is *upper triangular* if all entries below the diagonal are zero.

  (b) A matrix L is *lower triangular* if all entries above the diagonal are zero.

  (c) A matrix D is *diagonal* if all entries off the diagonal are zero.

Examples are: $D = \begin{bmatrix} 2 & 0 \\ 0 & 3 \end{bmatrix}$, $U = \begin{bmatrix} 1 & 3 \\ 0 & 1 \end{bmatrix}$, $L = \begin{bmatrix} 2 & 0 \\ 4 & 6 \end{bmatrix}$.

The categories of triangular matrices overlap; in fact a diagonal matrix is simply one that is both upper and lower triangular.

Among elementary matrices, type M's are of course diagonal, whereas type E's are not triangular at all. However, type A's are upper triangular when they add a multiple of a row to an earlier row, and lower triangular when they add it to a later row. For example, the matrix U above adds three times row 2 to row 1, and is upper triangular.

As for matrix operations on triangular matrices, consider the product of upper triangular matrices:

$$UU' = \begin{bmatrix} a & b \\ 0 & d \end{bmatrix} \begin{bmatrix} a' & b' \\ 0 & d' \end{bmatrix} = \begin{bmatrix} aa' & ab' + bd' \\ 0 & dd' \end{bmatrix} ,$$

which is again upper triangular. More generally, it is easy to check that the product of two $n \times n$ triangular matrices of the same kind is again a triangular matrix of the same kind.

The next examples show that triangular matrices are not always invertible:

$$U = \begin{bmatrix} 2 & 4 & 5 \\ 0 & 4 & 2 \\ 0 & 0 & 1 \end{bmatrix} \qquad U' = \begin{bmatrix} 2 & 3 & 7 \\ 0 & 0 & 2 \\ 0 & 0 & 1 \end{bmatrix} \quad .$$

The diagonal entries of U are nonzero, and so each forms a corner and U is invertible. But the second diagonal entry of U' is a zero, and since there are only zero entries below it, there can be no corner in column 2. So U' is not invertible. In general, it is easily checked that an $n \times n$ triangular matrix is invertible if and only if all its diagonal entries are nonzero.

As the following examples show, the inverse of a triangular elementary matrix is a triangular elementary matrix of the same kind:

$$D = \begin{bmatrix} 2 & 0 \\ 0 & 1 \end{bmatrix} \quad , \quad D^{-1} = \begin{bmatrix} \frac{1}{2} & 0 \\ 0 & 1 \end{bmatrix} \quad ,$$

$$L = \begin{bmatrix} 1 & 0 \\ 4 & 1 \end{bmatrix} \quad , \quad L^{-1} = \begin{bmatrix} 1 & 0 \\ -4 & 1 \end{bmatrix} \quad .$$

To extend this result to invertible triangular matrices in general, consider the 3×3 invertible upper triangular matrix U given above. In reducing U to the identity, we first eliminate the entries above the corners by means of upper triangular type A elementary matrices U_1, U_2, U_3, and then change the corner entries to ones by means of diagonal type M elementary matrices D_1, D_2. Therefore,

$$D_2 D_1 U_3 U_2 U_1 U = I, \quad \text{or} \quad U^{-1} = D_2 D_1 U_3 U_2 U_1 \quad .$$

That is, U^{-1} is the product of upper triangular matrices, and therefore it is upper triangular. Similar arguments can be used to show that the inverse of a triangular matrix can only be a triangular matrix of the same kind. Let us summarize our conclusions in a lemma.

Lemma 3.11 Suppose U and V are $n \times n$ triangular matrices of the same kind (diagonal, upper triangular, or lower triangular). Then

(a) UV is also a triangular matrix of the same kind as U and V.

(b) U^{-1} exists if and only if all diagonal entries of U are nonzero.

(c) U^{-1}, if it exists, is a triangular matrix of the same kind as U.

■ **Problem** Suppose A is an upper triangular 3×3 matrix satisfying the conditions $a_{12} = 2$, $a_{23} = 4$. Suppose it is also known that

$$A^t A = \begin{bmatrix} * & 2 & 3 \\ * & * & 14 \\ * & * & 25 \end{bmatrix} \quad , \quad \text{where each star denotes an unknown entry.}$$

Can we recover the matrix A from this fragmentary information?

■ **Solution** From the first part of the given information, we have

$$A = \begin{bmatrix} a & 2 & d \\ 0 & b & 4 \\ 0 & 0 & c \end{bmatrix}.$$

Therefore $A\,^tA =$

$$\begin{bmatrix} a & 0 & 0 \\ 2 & b & 0 \\ d & 4 & c \end{bmatrix} \begin{bmatrix} a & 2 & d \\ 0 & b & 4 \\ 0 & 0 & c \end{bmatrix} = \begin{bmatrix} a^2 & 2a & ad \\ 2a & 4+b^2 & 2d+4b \\ ad & 2d+4b & d^2+16+c^2 \end{bmatrix} = \begin{bmatrix} * & 2 & 3 \\ * & * & 14 \\ * & * & 25 \end{bmatrix}.$$

From row 1 of this matrix equation, we have $2a = 2$ and $ad = 3$, which implies $a = 1$ and $d = 3$. From row 2, we obtain the further equation $2d+4b = 14$, and so $b = 2$. From row 3 finally, $d^2+16+c^2 = 25$, or $c^2 = 0$, or $c = 0$.

We conclude that $A = \begin{bmatrix} 1 & 2 & 3 \\ 0 & 2 & 4 \\ 0 & 0 & 0 \end{bmatrix}.$

■ EXERCISES 3.5

1. Let $A = \begin{bmatrix} 3 & 2 \\ 5 & 4 \end{bmatrix}$, $B = \begin{bmatrix} 2 & 3 & -1 \\ 4 & 0 & 7 \end{bmatrix}.$

   Verify that $(AB)^t = B\,^tA\,^t$ by computing both the left and the right sides of the equation.

2. With A as above, verify that $(A^t)^{-1} = (A^{-1})^t$ by computing both sides of the equation.

3. With A and B as in exercise 1, solve the following matrix equations for the unknown matrix C:

   (a) $AC^t = B$ (b) $CA = B^t$.

4. A square matrix A is said to be *symmetric* if $A^t = A$, and *antisymmetric* if $A^t = -A$.

   (a) Determine whether A is necessarily symmetric, necessarily antisymmetric, or neither, when A has each of the following forms:
       (i) $A = B + B^t$ (ii) $A = BB^t$
       (iii) $A = B - B^t$ (iv) $A = B\,^tB$.

   (b) Show that every square matrix is the sum of a symmetric matrix and an antisymmetric matrix.

5. (a) Find the inverse of the triangular matrix $U = \begin{bmatrix} 2 & 9 \\ 0 & 3 \end{bmatrix}.$

(b) Show that if a square invertible triangular matrix A has diagonal entries $a_{11}, \ldots, a_{nn}$, then its inverse has diagonal entries $a_{11}^{-1}, \ldots, a_{nn}^{-1}$. (That is, $(A^{-1})_{ii} = a_{ii}^{-1}$.)

6. Suppose a matrix has the form $B = \begin{bmatrix} 1 & 8 & * \\ 0 & 2 & * \\ 0 & 0 & 3 \end{bmatrix}$, and let $A = B^{-1}$.

    Determine the entries $a_{11}, a_{12}, a_{32}, a_{33}$ of A.

7. Suppose $AB = D$, where A, B, and D are respectively lower triangular, upper triangular, and diagonal $n \times n$ matrices. Assume D is invertible. Show that A and B must in fact both be diagonal matrices.

8. Show that the matrix $E = \begin{bmatrix} 0 & 1 \\ 1 & 0 \end{bmatrix}$ cannot be expressed as the product of an upper triangular and a lower triangular matrix.

*9. Find a formula for A^n , where $A = \begin{bmatrix} 1 & a & b \\ 0 & 1 & a \\ 0 & 0 & 1 \end{bmatrix}$.

CHAPTER FOUR

Matrices and Geometry

Introduction

The perspective on $m \times n$ matrices that we have emphasized is that each one represents a mapping of the vector space R^n into R^m. Furthermore, we have defined the matrix product AB in such a way that it represents the composite of the mappings given by A and B individually.

Suppose now that we start with geometrically defined mappings rather than matrices. For example, suppose we transform each vector in R^2 by first rotating it counterclockwise through $90°$ and then reflecting the resulting vector across the x_1-axis. The geometric interpretation of this composite mapping is not immediately obvious. However, if we can find matrices A and B that represent the reflection and the rotation respectively, then we know that the composite is represented by AB. If, further, we can extract from the matrix transformation AB a geometric interpretation, then this is the interpretation of the composite transformation with which we started.

Is this way of studying geometric operations by the methods of matrix algebra feasible? Obviously, it will work only if we can find a substantial class of geometric transformations that have matrix representations and we are able to move back and forth easily between transformations and their representing matrices. The geometric transformations to be studied will have to be linear, since according to theorem 2.3 all matrix transformations are linear.

We have, in fact, already used our proposed new method — immediately after introducing the matrix product in section 3.1 — but only to interpret a few specific composite mappings. It turns out that we can do much more. The surprisingly strong theoretical result in section 4.1 tells us that the linearity condition on a mapping is also sufficient to ensure the existence of a matrix representation for it. Equally important, it also provides us with a very efficient and convenient method for actually computing the representing matrix for a given linear transformation.

A considerable collection of geometric transformations and their representing matrices is developed in section 4.2, along with illustrative geometric problems solved by methods of matrix algebra.

Apart from their intrinsic interest, the ideas presented in this chapter recur later on. Scaling, shearing, and reflection transformations are used to construct the geometric interpretation of the determinant in section 5.3, while rotations and reflections figure prominently in the study of orthogonal transformations in section 7.2.

4.1 Matrix Representation of Linear Transformations

There are a few geometric transformations whose matrix representations can be constructed almost by inspection. For example, in section 2.3 we introduced reflection across an axis,

$$F((x_1,x_2)) = (x_1,-x_2) \quad ,$$

and showed the mapping to be linear. In section 3.1, we quoted its matrix representation,

$$F = \begin{bmatrix} 1 & 0 \\ 0 & -1 \end{bmatrix} \quad .$$

This matrix could easily have been constructed by the following argument. If we write vectors in column form, we can say that

$$F \text{ maps } \begin{bmatrix} x_1 \\ x_2 \end{bmatrix} \text{ to } \begin{bmatrix} x_1 \\ -x_2 \end{bmatrix} \quad .$$

But this transformation is recognizable as a row operation on 2×1 matrices: row 2 is multiplied by -1, and so the transformation is represented by the type M elementary matrix $M(-1;2) = \begin{bmatrix} 1 & 0 \\ 0 & -1 \end{bmatrix}$.

Improvised reasoning of this kind is effective only in a few simple cases, of course. Thus it is not at all easy to see what matrix represents reflection across a line not parallel to an axis! We therefore proceed to develop systematic methods.

Linear Transformations and Matrices

A feature of all linear transformations from R^n to R^m is that once we know how they map the standard vectors $e_1, \ldots, e_n$, we can deduce how they map every vector $\mathbf{x}$ in R^n.

First we express the vector $\mathbf{x}$ itself in terms of the standard vectors:

$$\begin{bmatrix} x_1 \\ \cdot \\ \cdot \\ \cdot \\ x_n \end{bmatrix} = x_1 \begin{bmatrix} 1 \\ 0 \\ \cdot \\ \cdot \\ 0 \end{bmatrix} + \cdots + x_n \begin{bmatrix} 0 \\ \cdot \\ \cdot \\ 0 \\ 1 \end{bmatrix} \quad ,$$

or $\quad \mathbf{x} = x_1 \mathbf{e}_1 + \cdots + x_n \mathbf{e}_n$.

Now, if we apply a linear transformation T to this equation, we obtain

$$T(\mathbf{x}) = x_1 T(\mathbf{e}_1) + \cdots + x_n T(\mathbf{e}_n) \quad .$$

Moreover, since any matrix transformation, say A, is linear, we can also write

$$A\mathbf{x} = x_1 A\mathbf{e}_1 + \cdots + x_n A\mathbf{e}_n \quad .$$

From these formulas for $A\mathbf{x}$ and $T(\mathbf{x})$, we can see that A and T will act on every vector $\mathbf{x}$ in the same way, provided that they act on each of the standard vectors $\mathbf{e}_1, \ldots, \mathbf{e}_n$ in the same way, that is, provided

$$A\mathbf{e}_j = T(\mathbf{e}_j) \quad , \quad \text{for } j = 1, \ldots, n \quad .$$

Recall, however, from section 3.1, that $A\mathbf{e}_j$ always equals column j of the matrix A. So if we define A to be the $m \times n$ matrix whose jth column is $T(\mathbf{e}_j)$, then automatically $A\mathbf{e}_j = T(\mathbf{e}_j)$, and so $A\mathbf{x} = T(\mathbf{x})$ for every vector $\mathbf{x}$ in R^n. Let us summarize this important observation in a theorem.

THEOREM 4.1

Suppose T is a linear transformation of R^n to R^m, and let A be the $m \times n$ matrix whose jth column is $T(\mathbf{e}_j)$. Then

$$T(\mathbf{x}) = A\mathbf{x} \text{ for all } \mathbf{x} \text{ in } R^n \quad ,$$

and the matrix transformation A is the same as T.

Example 1 Let P be the transformation of R^2 into R^2 that projects every vector along the x_1-axis.

Geometrically (see Figure 4.1), this means that we drop a perpendicular from $\mathbf{x}$ to the x_1-axis, and the foot of this perpendicular marks $P(\mathbf{x})$:

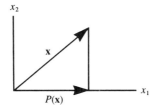 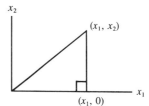

Figure 4.1

In terms of analytic geometry, the mapping is easily seen to be

$$P((x_1, x_2)) = (x_1, 0) \quad .$$

Now it is easily proved, just as in the case of reflection, that the projection transformation is linear, and so we can apply theorem 4.1 to obtain a matrix representation of P. The columns of the representing matrix are

$$P\mathbf{e}_1 = P((1,0)) = (1,0) = \mathbf{e}_1$$

and $\quad P\mathbf{e}_2 = P((0,1)) = (0,0) = \mathbf{0} \quad .$

On filling in the columns e_1, 0 of the 2×2 representing matrix, we obtain

$$P = \begin{bmatrix} 1 & 0 \\ 0 & 0 \end{bmatrix} \quad .$$

Note that $P^2 = \begin{bmatrix} 1 & 0 \\ 0 & 0 \end{bmatrix}^2 = \begin{bmatrix} 1 & 0 \\ 0 & 0 \end{bmatrix} = P$, so that projecting a vector twice is no different from projecting it just once. Geometrically, this is reasonable since the first projection map gets the vector to the x_1-axis, leaving nothing more for the second projection to do.

Example 2 Suppose the transformation T of R^3 into R^3 maps each vector $x = (x_1, x_2, x_3)$ as follows:

$$T((x_1, x_2, x_3)) = (x_1 + 2x_2, \; x_2 - 2x_1 + 3x_3, \; x_3 - x_2) \quad .$$

To obtain a 3×3 matrix A representing T, we compute

$$\begin{aligned}
T(e_1) &= T((1,0,0)) = (1, -2, 0) \\
T(e_2) &= T((0,1,0)) = (2, 1, -1) \\
T(e_3) &= T((0,0,1)) = (0, 3, 1) \quad .
\end{aligned}$$

Then $A = \begin{bmatrix} 1 & 2 & 0 \\ -2 & 1 & 3 \\ 0 & -1 & 1 \end{bmatrix} \quad .$

■ **Problem** Suppose S and T are transformations of R^2 into R^2, and we know that

$$S((x_1, x_2)) = (5x_1 + 4x_2, \; 6x_1 + 5x_2)$$

and $\qquad S \circ T = I \quad .$

Can we reconstruct the transformation T?

■ **Solution** Suppose S and T have matrix representations A and B respectively. Then $S \circ T$ is represented by AB, and so $AB = I$ or $B = A^{-1}$.

Also, $\quad S(e_1) = (5, 6)$

and $\quad S(e_2) = (4, 5) \quad .$

Therefore $A = \begin{bmatrix} 5 & 4 \\ 6 & 5 \end{bmatrix} \quad ,$

and, from the formula for the 2×2 matrix inverse,

$$A^{-1} = \begin{bmatrix} 5 & -4 \\ -6 & 5 \end{bmatrix} \quad .$$

Then $T(\mathbf{x}) = B\mathbf{x} = A^{-1}\mathbf{x} = \begin{bmatrix} 5 & -4 \\ -6 & 5 \end{bmatrix} \begin{bmatrix} x_1 \\ x_2 \end{bmatrix} = \ldots$

$$\ldots \begin{bmatrix} 5x_1 - 4x_2 \\ -6x_1 + 5x_2 \end{bmatrix} ,$$

or, $T((x_1,x_2)) = (5x_1 - 4x_2, \ -6x_1 + 5x_2)$.

■ EXERCISES 4.1

1. Find matrices representing the following transformations:

   (a) $T((x_1,x_2)) = (x_1 + 2x_2, x_2 - x_1, x_1 + x_2)$
   (b) $S((x_1,x_2,x_3)) = (x_1 - x_2 - x_3, 2x_2 + 5x_3 - 3x_1)$.

   By using matrix products, deduce that

   (c) $S \cdot T$ is the identity transformation of R^2 into R^2 .
   (d) $(T \cdot S)((x_1,x_2,x_3)) = (-5x_1 + 3x_2 + 9x_3 , \ -4x_1$
   $+ \ 3x_2 + 6x_3 \ , -2x_1 + x_2 + 4x_3)$.

2. Suppose $S((x_1,x_2)) = (3x_1 + 10x_2, 2x_1 + 7x_2)$
   and $U((x_1,x_2)) = (8x_1 + 9x_2, 2x_1 + 2x_2)$.

   Find matrix transformations T of R^2 into R^2 such that

   (a) $T = S \circ U$ (b) $S \circ T = I$ (c) $S \circ T = U$.

3. (a) Using theorem 4.1, construct the matrices representing the trans-
       formations F', P' of reflection across the x_2-axis and projection
       along the x_2-axis, respectively.

   (b) Show that $F'F = -I$, while $P'P = O$.

   (c) Give geometric interpretations of the formulas in (b).

4. Suppose the linear transformation T of R^2 into R^2 satisfies

   $T((1,1)) = (2,3)$ and $T((-1,1)) = (4,5)$.

   (a) Using the linearity of T, compute $T((1,0))$ and $T((0,1))$.

   (b) Deduce the matrix representing T.

5. Reflection and projection are linked by the matrix identity $I + F = 2P$.
   Give a geometric interpretation of this identity by showing that $\mathbf{x}$, $F\mathbf{x}$,
   and $2P\mathbf{x}$ are two adjacent sides and a diagonal of a parallelogram.

6. (Projection and reflection in R^3)

   (a) For each point H in R^3, let H' be the foot of the perpendicular from
       H to the x_1x_2 plane. Suppose we define projection P onto the x_1x_2

plane to be the transformation of R^3 into R^3, which sends H to
H'. Find the 3×3 matrix representing P.

(b) Also, let H'' be the point such that H' is the midpoint of the line
segment HH'', and define reflection F across the x_1x_2 plane to be
the map that sends H to H''. Find the 3×3 matrix representing F.

(c) Show that the formulas $F^2 = I$, $P^2 = P$, $I + F = 2P$ still hold.

*7. Show that the two-dimensional reflection matrix F has no real square
root, that is, show that there is no matrix A with real entries satisfying
$A^2 = F$.

*8. Show that all 2×2 solutions of $A^2 = I$, except $A = \pm I$, can be writ-
ten in the form $A = JFJ^{-1}$, where J is an invertible matrix.

4.2 Geometric Transformations and Their Matrix Representations

We can now consider the most frequently used geometric transformations
of R^2 and, by use of theorem 4.1, construct matrix representations for them. In
doing so, we begin to implement the method that was proposed in the introduction to this chapter.

Rotations form the first category of transformations to be considered, and
the brevity of the construction of their representing matrices gives an immediate indication of the efficiency of matrix methods in geometry. Comparable theories of rotation of axes developed in trigonometry texts are much
more cumbersome, as some readers may recall.

An interesting feature of our method is that rotation matrices are also
used in constructing the matrix representations of other, quite different,
transformations. For example, any reflection involves an axis of reflection;
and the reflection is easily described when that axis happens to be the x_1-axis.
But by use of rotations, any axis of reflection can be brought into alignment
with the x_1-axis and the required matrix representation deduced from the
easy case.

Of course, matrix algebra and geometry inevitably illuminate each other
once the linkage between them is established. For instance, we all have an
intuitive understanding of how the order in which geometric operations are
performed affects the result. And so we can see that the matrix product,
which represents the composite of two operations, must depend on the order of the factors. In other words, the great descriptive power of matrix
algebra automatically entails the loss of commutativity.

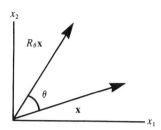

Figure 4.2

Rotations in the Plane

Let R, or R_θ, denote the geometric transformation that rotates every position vector in R^2 counterclockwise through angle θ (see Figure 4.2).

It is easy to prove by geometry that the transformation R_θ is linear, and we leave this to the exercises.

In order to obtain a matrix representation for R_θ, we need only compute $R_\theta e_1$ and $R_\theta e_2$. Observe that e_1 and e_2 are unit vectors (see Figure 4.3) and, since rotation does not change the length of a vector, therefore Re_1, Re_2 are also unit vectors. Now a point on the unit circle, at angle ω counterclockwise from the positive x_1-axis, has coordinates $(\cos \omega, \sin \omega)$. So when the vector e_1, which is along the positive x_1-axis, is rotated counterclockwise through θ degrees, the result is $R_\theta e_1 = (\cos \theta, \sin \theta)$.

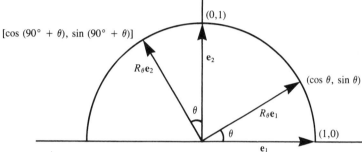

Figure 4.3

The vector e_2, however, is along the positive x_2-axis, and so is already $90°$ counterclockwise from the positive x_1-axis. Rotation of e_2 through the further angle θ counterclockwise results in the vector $R_\theta e_2$ at the total angle $\omega = 90° + \theta$ counterclockwise from the positive x_1-axis. That is, $R_\theta e_2 = (\cos(90° + \theta), \sin(90° + \theta)) = (-\sin \theta, \cos \theta)$.

Having computed the columns $R_\theta e_1$, $R_\theta e_2$ of the matrix representing the transformation R_θ, we can write

$$R_\theta = \begin{bmatrix} \cos \theta & -\sin \theta \\ \sin \theta & \cos \theta \end{bmatrix} .$$

Rotation Counter clockwise

Example 1   In order to rotate the vector $x = (2,1)$ counterclockwise through $90°$, we first compute the rotation matrix $R_{90°}$:

$$R_{90°} = \begin{bmatrix} \cos 90° & -\sin 90° \\ \sin 90° & \cos 90° \end{bmatrix} = \begin{bmatrix} 0 & -1 \\ 1 & 0 \end{bmatrix} ;$$

we then apply it to the given vector:

$$R_{90°}x = \begin{bmatrix} 0 & -1 \\ 1 & 0 \end{bmatrix} \begin{bmatrix} 2 \\ 1 \end{bmatrix} = \begin{bmatrix} -1 \\ 2 \end{bmatrix} .$$

We can check this result by noting that $R_{90°}\mathbf{x} = (-1,2)$ and $\mathbf{x} = (2,1)$ have dot product 0, as vectors separated by 90° must.

Note also that $R_{90°}$ satisfies the matrix equation $A^2 = -I$. To see why this should be so, observe that $(R_{90°})^2$ amounts to rotation by 90° twice, or rotation by 180°, which of course reverses every vector just as the transformation $-I$ does.

Matrix Algebra and Geometry

It is easily seen through examples that the composite of two geometric transformations depends on the order in which they are done. For example, let us track the progress of the vector $\mathbf{x} = (1,1)$ when it is first rotated counterclockwise 90°, and then reflected across the x_1-axis.

But when the order of implementation of the operations is reversed,

The final result is quite different in the two cases. The matrices representing the rotation and reflection respectively are

$$R_{90°} = \begin{bmatrix} 0 & -1 \\ 1 & 0 \end{bmatrix} \quad , \quad F = \begin{bmatrix} 1 & 0 \\ 0 & -1 \end{bmatrix} .$$

So the two composite transformations are, respectively,

$$FR_{90°} = \begin{bmatrix} 1 & 0 \\ 0 & -1 \end{bmatrix} \begin{bmatrix} 0 & -1 \\ 1 & 0 \end{bmatrix} = \begin{bmatrix} 0 & -1 \\ -1 & 0 \end{bmatrix} ,$$

$$R_{90°}F = \begin{bmatrix} 0 & -1 \\ 1 & 0 \end{bmatrix} \begin{bmatrix} 1 & 0 \\ 0 & -1 \end{bmatrix} = \begin{bmatrix} 0 & 1 \\ 1 & 0 \end{bmatrix} .$$

Therefore $FR_{90°} \neq R_{90°}F$, confirming the geometric observations.

At this stage we can see why matrix algebra *cannot* be a commutative algebra. After all, the matrix product was defined as a composite transformation; and since the composite of, say, geometric transformations clearly depends on the order of implementation, so the matrix product must also depend on the order of the factors.

Conversely, a commutative algebra could not have the descriptive powers of matrix algebra.

Example 2 For which choices of angle of rotation θ does rotation commute with reflection across the x_1-axis? We can solve this problem algebraically by computing the matrix products:

$$FR_\theta = \begin{bmatrix} 1 & 0 \\ 0 & -1 \end{bmatrix} \begin{bmatrix} \cos\theta & -\sin\theta \\ \sin\theta & \cos\theta \end{bmatrix} = \begin{bmatrix} \cos\theta & -\sin\theta \\ -\sin\theta & -\cos\theta \end{bmatrix} .$$

$$R_\theta F = \begin{bmatrix} \cos\theta & -\sin\theta \\ \sin\theta & \cos\theta \end{bmatrix} \begin{bmatrix} 1 & 0 \\ 0 & -1 \end{bmatrix} = \begin{bmatrix} \cos\theta & \sin\theta \\ \sin\theta & -\cos\theta \end{bmatrix} .$$

The products are the same if $\sin\theta = 0$, that is, precisely when the angle θ of rotation is 0 or 180 degrees.

General Reflections and Projections

From a geometric point of view, we can just as easily reflect vectors across an arbitrary line ℓ through the origin as across an axis.

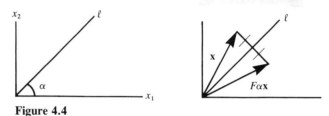

Figure 4.4

However, much more effort would be required to compute this transformation in terms of analytic geometry than before. Fortunately we can deduce the representing matrix from transformations we already know. The steps are as follows (see Figure 4.5).

Step 1. Suppose the line ℓ is positioned counterclockwise from the positive x_1-axis by the angle α. Then, if we rotate the line ℓ and the vector **x** clockwise through angle α, line ℓ then runs along the x_1-axis. This first step is done by the transformation $R_{-\alpha}$.

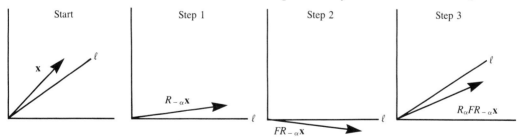

Figure 4.5

Step 2. Reflection across line ℓ is then the same as reflection across the x_1-axis, and is implemented by F.

Step 3. To return line ℓ to its original position, we rotate back through angle α counterclockwise, using R_α.

The reflection F_α across the line ℓ at angle α to the positive x_1-axis is therefore a composite of these mappings:

$$
\begin{aligned}
F_\alpha &= R_\alpha F R_{-\alpha} \\
&= \begin{bmatrix} \cos\alpha & -\sin\alpha \\ \sin\alpha & \cos\alpha \end{bmatrix} \begin{bmatrix} 1 & 0 \\ 0 & -1 \end{bmatrix} \begin{bmatrix} \cos(-\alpha) & -\sin(-\alpha) \\ \sin(-\alpha) & \cos(-\alpha) \end{bmatrix} \\
&= \begin{bmatrix} \cos\alpha & \sin\alpha \\ \sin\alpha & -\cos\alpha \end{bmatrix} \begin{bmatrix} \cos\alpha & \sin\alpha \\ -\sin\alpha & \cos\alpha \end{bmatrix} \\
&= \begin{bmatrix} \cos^2\alpha - \sin^2\alpha & 2\sin\alpha\cos\alpha \\ 2\sin\alpha\cos\alpha & \sin^2\alpha - \cos^2\alpha \end{bmatrix}
\end{aligned}
$$

or, on looking up or remembering the trigonometric "double angle" formulas (see also the exercises of this section),

$$
F_\alpha = \begin{bmatrix} \cos 2\alpha & \sin 2\alpha \\ \sin 2\alpha & -\cos 2\alpha \end{bmatrix} . \qquad \text{Reflection}
$$
$$
F_\alpha^2 = I
$$

Example 3 In order to reflect the vector $(2,3)$ across the line ℓ at angle $\alpha = 45°$ to the x_1-axis, we compute:

$$
F_{45°} = \begin{bmatrix} \cos 90° & \sin 90° \\ \sin 90° & -\cos 90° \end{bmatrix} = \begin{bmatrix} 0 & 1 \\ 1 & 0 \end{bmatrix} ,
$$

and $\quad F_{45°} \begin{bmatrix} 2 \\ 3 \end{bmatrix} = \begin{bmatrix} 0 & 1 \\ 1 & 0 \end{bmatrix} \begin{bmatrix} 2 \\ 3 \end{bmatrix} = \begin{bmatrix} 3 \\ 2 \end{bmatrix} .$

More generally, the reflection $F_{45°}$ will map the vector (x_1, x_2) to (x_2, x_1), exchanging $\mathbf{e}_1$ and $\mathbf{e}_2$ and so on.

Note that $F_{45°}$ happens to be an elementary matrix:

$$
F_{45°} = E(1,2) ,
$$

and this gives us a geometric interpretation of type E elementary matrices as reflections.

Note also that the case studied previously of the reflection F across the x_1-axis is simply the case $\alpha = 0$; so we can write F_0 instead of F.

Projection P_α along an arbitrary line ℓ through the origin is also geometrically equivalent to a special case discussed previously, projection P along the x_1-axis (see Figure 4.6).

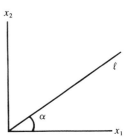

The matrix representing P_α is derived from that representing P exactly as in the case of reflection:

$$
\begin{aligned}
P_\alpha &= R_\alpha P R_{-\alpha} \\
&= \begin{bmatrix} \cos\alpha & -\sin\alpha \\ \sin\alpha & \cos\alpha \end{bmatrix} \begin{bmatrix} 1 & 0 \\ 0 & 0 \end{bmatrix} \begin{bmatrix} \cos(-\alpha) & -\sin(-\alpha) \\ \sin(-\alpha) & \cos(-\alpha) \end{bmatrix} \\
&= \begin{bmatrix} \cos\alpha & 0 \\ \sin\alpha & 0 \end{bmatrix} \begin{bmatrix} \cos\alpha & \sin\alpha \\ -\sin\alpha & \cos\alpha \end{bmatrix}
\end{aligned}
$$

or $\quad P_\alpha = \begin{bmatrix} \cos^2\alpha & \cos\alpha\sin\alpha \\ \cos\alpha\sin\alpha & \sin^2\alpha \end{bmatrix} . \qquad \text{Projection}$

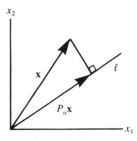

Figure 4.6

Example 4   From the formula, $P_{45°} = \begin{bmatrix} (\frac{1}{\sqrt{2}})^2 & (\frac{1}{\sqrt{2}})(\frac{1}{\sqrt{2}}) \\ (\frac{1}{\sqrt{2}})(\frac{1}{\sqrt{2}}) & (\frac{1}{\sqrt{2}})^2 \end{bmatrix} = \begin{bmatrix} \frac{1}{2} & \frac{1}{2} \\ \frac{1}{2} & \frac{1}{2} \end{bmatrix}$,

and so $P_{45°} \begin{bmatrix} x_1 \\ x_2 \end{bmatrix} = \begin{bmatrix} \frac{1}{2} & \frac{1}{2} \\ \frac{1}{2} & \frac{1}{2} \end{bmatrix} \begin{bmatrix} x_1 \\ x_2 \end{bmatrix} = \begin{bmatrix} \left(\dfrac{x_1 + x_2}{2}\right) \\ \left(\dfrac{x_1 + x_2}{2}\right) \end{bmatrix} = \dots$

$$\left(\dfrac{(x_1 + x_2)}{2}\right) \begin{bmatrix} 1 \\ 1 \end{bmatrix} .$$

That is, projection along the line inclined at 45° always results in a vector with equal components. This agrees with the fact that the line ℓ in this case has the equation $x_2 = x_1$.

Example 5   Suppose the line ℓ is given in the form $ax_1 + bx_2 = 0$. To construct P_α, note that $x_1 = -b, x_2 = a$ satisfies the equation, so that $(-b, a)$ is a point on ℓ. A unit position vector along ℓ is therefore

$$\mathbf{u} = \frac{-b}{(a^2 + b^2)^{\frac{1}{2}}} , \frac{a}{(a^2 + b^2)^{\frac{1}{2}}}$$

and, since this position vector is on the unit circle, it also has the form $\mathbf{u} = (\cos \alpha, \sin \alpha)$. That is,

$$\cos \alpha = \frac{-b}{(a^2 + b^2)^{\frac{1}{2}}}$$

$$\sin \alpha = \frac{a}{(a^2 + b^2)^{\frac{1}{2}}} .$$

For instance, if ℓ is the line $4x_1 - 3x_2 = 0$, then $(a^2 + b^2)^{1/2} = (16 + 9)^{1/2} = 5$, $\cos \alpha = \frac{3}{5}$, $\sin \alpha = \frac{4}{5}$, and

$$P_\alpha = \begin{bmatrix} \frac{9}{25} & \frac{12}{25} \\ \frac{12}{25} & \frac{16}{25} \end{bmatrix} = \left(\frac{1}{25}\right) \begin{bmatrix} 9 & 12 \\ 12 & 16 \end{bmatrix} .$$

Note that it is not necessary to compute α in this procedure. All we needed was a unit vector $\mathbf{u} = (\frac{3}{5}, \frac{4}{5})$ along ℓ. In fact, an alternative notation for the projection based directly on the unit vector $\mathbf{u} = (u_1, u_2)$ along ℓ is sometimes used; we write

$$P_\mathbf{u} = \begin{bmatrix} u_1^2 & u_1 u_2 \\ u_1 u_2 & u_2^2 \end{bmatrix} .$$

Projection on unit vector
eg $4u_2 - 3u_1$

Example 6   To construct the matrix F_α for reflection across the above line, $4x_1 - 3x_2 = 0$, we can use the same method as for P_α. But this is not necessary, because F_α and P_α are algebraically related. In fact, since

$$(x_1, x_2) + (x_1, -x_2) = (2x_1, 0) ,$$

therefore $\qquad$ $\mathbf{x} + F_0\mathbf{x} = 2P_0\mathbf{x}$ for all $\mathbf{x}$,

or $\qquad$ $(I + F_0)\mathbf{x} = 2P_0\mathbf{x}$ for all $\mathbf{x}$,

or $\qquad$ $I + F_0 = 2P_0$.

Then $\qquad$ $R_\alpha(I + F_0)R_{-\alpha} = R_\alpha(2P_0)R_{-\alpha}$,

or $\qquad$ $R_\alpha R_{-\alpha} + R_\alpha F_0 R_{-\alpha} = 2R_\alpha P_0 R_{-\alpha}$,

or $\qquad$ $I + F_\alpha = 2P_\alpha$,

or, $\qquad$ $F_\alpha = 2P_\alpha - I$ *(link between reflection and projection)*.

In our example, $\qquad$ $F_\alpha = 2 \begin{bmatrix} \frac{9}{25} & \frac{12}{25} \\ \frac{12}{25} & \frac{16}{25} \end{bmatrix} - \begin{bmatrix} 1 & 0 \\ 0 & 1 \end{bmatrix}$,

or $\qquad$ $F_\alpha = F_\mathbf{u} = F_{(\frac{3}{5},\frac{4}{5})} = \begin{bmatrix} \frac{-7}{25} & \frac{24}{25} \\ \frac{24}{25} & \frac{7}{25} \end{bmatrix}$.

We may want to do some checking of our calculation at this point, making sure, for example, that $F_\alpha^2 = I$, as expected for a reflection; and that $F_\alpha\mathbf{u} = \mathbf{u}$, since a vector along ℓ shouldn't be moved by the reflection.

Scaling and Shearing

Think of stretching the plane like a sheet of rubber along the x_1-axis, so that the distance of each point from the x_2-axis is multiplied by the same factor k. The unit square, for example, stretches as shown in Figure 4.7.

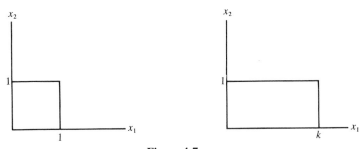

Figure 4.7

Since the x_1 coordinate of each point is multiplied by k, we can interpret the transformation as a rescaling of the coordinate system, and write down the *scaling transformation*

$T_0((x_1, x_2)) = (kx_1, x_2)$.

Evidently this has the matrix representation

$T_0 = \begin{bmatrix} k & 0 \\ 0 & 1 \end{bmatrix}$, Scaling in 2 dim

which we recognize as the type M elementary matrix $M(k;1)$.

Stretching in an arbitrary direction (parallel to the line ℓ inclined at angle α to the positive x_1-axis, as usual) can of course be derived from $M(k;1)$ through the formula $T_\alpha = R_\alpha M(k;1)R_{-\alpha}$, or, by easy calculation (see exercises),

$$T_\alpha = (k-1)P_\alpha + I \ . \qquad \textit{Scaling arbitral direction}$$

Think, now, of a force pushing the top of the unit square to the right while the base stays fixed, distorting the square and causing the x_1-coordinate of each of its points to increase by an amount proportional to the height of the point above the base (see Figure 4.8).

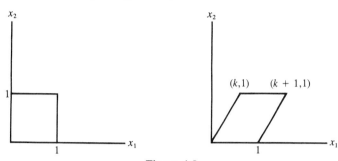

Figure 4.8

This is called a *shearing* transformation and can be written

$$S_0((x_1,x_2)) = (x_1 + kx_2, x_2) \ .$$

The matrix representing S_0 is clearly

$$S_0 = \begin{bmatrix} 1 & k \\ 0 & 1 \end{bmatrix}, \qquad \textit{Shearing in 2 dim.}$$

which we remember as the type A elementary matrix $A(k;2,1)$.

Shearing along an arbitrary direction is, of course, expressed by the formula $S_\alpha = R_\alpha S_0 R_{-\alpha}$. $\qquad \textit{Shearing arbitrary direction}$

Example 7 Suppose shearing from upper left to lower right along the line $5x + 12y = 0$ occurs with "shearing factor" $k = 2$, and we want to know whether the point $(2,1)$ is pushed into the lower half plane. Then we compute

$$(a^2 + b^2)^{\frac{1}{2}} = (5^2 + 12^2)^{\frac{1}{2}} = 13,\ \cos\alpha = -\tfrac{12}{13},\ \sin\alpha = \tfrac{5}{13}\ ;$$

$$S_\alpha = R_\alpha S_0 R_{-\alpha} = \begin{bmatrix} -\tfrac{12}{13} & -\tfrac{5}{13} \\ \tfrac{5}{13} & -\tfrac{12}{13} \end{bmatrix} \begin{bmatrix} 1 & 2 \\ 0 & 1 \end{bmatrix} \begin{bmatrix} -\tfrac{12}{13} & \tfrac{5}{13} \\ -\tfrac{5}{13} & -\tfrac{12}{13} \end{bmatrix}$$

$$= \left(\tfrac{1}{169}\right) \begin{bmatrix} 289 & 288 \\ -50 & 49 \end{bmatrix} \ ,$$

and so $S_\alpha \begin{bmatrix} 2 \\ 1 \end{bmatrix} = \left(\tfrac{1}{169}\right) \begin{bmatrix} 289 & 288 \\ -50 & 49 \end{bmatrix} \begin{bmatrix} 2 \\ 1 \end{bmatrix} = \left(\tfrac{1}{169}\right) \begin{bmatrix} 866 \\ -51 \end{bmatrix} \ .$

That is, point $(2,1)$ is moved to a point with negative x_2 coordinate $\tfrac{-51}{169}$, in the lower half plane.

In the course of these discussions, we have obtained the geometric interpretations for all three types of elementary matrices:

Type A: shearing transformation
Type M: scaling transformation
Type E: reflection.

We have concentrated on 2×2 case, but since elementary matrices never change more than two components of a vector, the geometric content of elementary matrix transformations is fully covered.

■ **Problem** The composite transformation $T = F_0 R_{90°}$ was discussed in the introduction to this chapter, and its matrix representation,

$$T = \begin{bmatrix} 0 & -1 \\ -1 & 0 \end{bmatrix} ,$$

was computed in this section.

In accordance with the method proposed in the introduction to this chapter, reconstruct the geometric interpretation of the transformation T from its matrix representation.

■ **Solution** We try to identify the above matrix as an instance of one of the categories of matrix representation that we have studied in this chapter. For example, if we guess that T may be a reflection, we write the equation

$$T = F_\alpha \quad \text{or} \quad \begin{bmatrix} 0 & -1 \\ -1 & 0 \end{bmatrix} = \begin{bmatrix} \cos 2\alpha & \sin 2\alpha \\ \sin 2\alpha & -\cos 2\alpha \end{bmatrix} .$$

The matrix equation amounts to the two independent scalar equations:

$$\cos 2\alpha = 0 \quad , \quad \sin 2\alpha = -1 \quad .$$

That is, $2\alpha = -90°$, or $\alpha = -45°$. Thus $T = F_{-45°}$.

Therefore, T is indeed a reflection transformation. See section 7.2 for further discussion of products of reflections and rotations.

In exercises 2 and 3 below, a somewhat more systematic method for going from matrices back to geometric transformations is suggested.

■ EXERCISES 4.2

1. (a) Compute the matrices $R_{45°}$, $F_{60°}$, $P_{30°}$.
   (b) Compute the matrices $R_{-30°}$, $F_{120°}$, $P_{-90°}$.
   (c) By multiplying out the product $R_{45°} F_{60°} R_{45°}$, show that it equals $F_{60°}$.

(d) Find the matrices representing projection along, and reflection across, the line $2x + y = 0$.

2. Let $A = \begin{bmatrix} a & b \\ c & d \end{bmatrix}$. Show that

   (a) if A is a projection, P_α, then $a + d = 1$ and $b = c$.
   (b) if A is a reflection, F_α, then $a = -d$ and $b = c$.
   (c) if A is a rotation, R_θ, then $a = d$ and $b = -c$.

3. Identify the geometric transformation of R^2 represented by each of the following matrices:

   (a) $\begin{bmatrix} 0 & 1 \\ -1 & 0 \end{bmatrix}$ (b) $\begin{bmatrix} \frac{1}{2} & -\frac{1}{2} \\ -\frac{1}{2} & \frac{1}{2} \end{bmatrix}$ (c) $\begin{bmatrix} 0 & -1 \\ -1 & 0 \end{bmatrix}$.

4. For which choices of angle α does the equation $F_\alpha P_0 = P_0 F_\alpha$ hold?

5. (a) Find all projection matrices that map $(1,1)$ to $(0,1)$.
   (b) Find all rotation and reflection matrices that map $(1,1)$ to $(-1,1)$.

6. Suppose the transformation T in R^2 acts on each vector $\mathbf{x}$ as follows: first $\mathbf{x}$ is rotated counterclockwise by $90°$, and then the resulting vector is projected along the line $y = 3x$.

   (a) Find the matrix representing T.
   (b) Determine whether this matrix is a projection matrix, P_α.
   (c) Compute $T \circ T$.

7. (a) Show that a type E elementary row operation is equivalent to a sequence of row operations of types A and M by verifying the formula:
   $$E(1,2) = M(-1;2)A(1;2,1)A(-1; 1,2)A(1; 2,1) \quad .$$

   (b) Track the movement of the unit square as the four type A and M operations are successively applied to it, and verify that its ultimate displacement is the same as if just $E(1,2)$ had been applied.

8. (a) Show that the scaling transformation T_0 is linked to projection by the formula $T_0 = (k-1)P_0 + I$.
   (b) Deduce the formula $T_\alpha = (k-1)P_\alpha + I$.

9. Suppose we write $S_\alpha(k)$ to specify the ''shearing factor'' k in the shearing transformation S_α.

   (a) Show that $S_0(k)\, S_0(k') = S_0(k + k')$.
   (b) Deduce that $S_\alpha(k)\, S_\alpha(k') = S_\alpha(k + k')$.

*10. (a) Show, by multiplying out the product, that $R_\theta F_0 R_\theta = F_0$.

(b) Deduce that $R_\theta F_\alpha R_\theta = F_\alpha$. (Exercise 1(c) was a special case of this.)

(c) Give a geometric interpretation of result (b).

*11. Show that an alternative formula for the projection map $P_\mathbf{u}$ is given in terms of the dot product as follows:

$$P_\mathbf{u} \mathbf{x} = (\mathbf{x} \cdot \mathbf{u}) \mathbf{u} \quad .$$

*12. It is geometrically clear that rotation through angle θ, followed by another rotation through angle ω, is the same as a single rotation through angle $\theta + \omega$. That is,

$$R_\omega R_\theta = R_{\theta + \omega} \quad .$$

By multiplying out the left side of this identity, prove the trigonometric formulas:

(a) $\cos \omega \cos \theta - \sin \omega \sin \theta = \cos (\omega + \theta)$

(b) $\sin \omega \cos \theta + \cos \omega \sin \theta = \sin (\omega + \theta)$.

*13. Find all rotation and reflection matrices that map $(8,1)$ to $(7,4)$.

CHAPTER FIVE

Determinants

Introduction

Two of the most important questions concerning a square $n \times n$ matrix A are:

1. Does A have an inverse?
2. Does the linear system $A\mathbf{x} = \mathbf{0}$ have a nonzero solution?

The two questions always have opposite answers. According to theorem 1.1, the nonzero solution exists when rank$(A) < n$; while from theorem 3.6, the inverse exists when rank$(A) = n$.

In principle, it is easy to determine which of the two properties a given square matrix has. We have only to reduce it, count the number of corners in its reduced form, and thereby determine its rank. This procedure works well for a specific, numerical matrix.

In later chapters, however, we will often encounter matrices whose entries depend on parameters. As early as sections 1.2 and 1.3, we saw that reduction of such matrices is difficult and often branches into many distinct cases. An alternative approach to the general problem is therefore needed, and one has already been introduced. Recall that in section 3.4, we observed that a 2×2 matrix

$$A = \begin{bmatrix} a & b \\ c & d \end{bmatrix}$$

has an inverse only when $ad - bc \neq 0$. That is, we found an algebraic condition on the entries of A that tells us whether A^{-1} exists.

For example, the parameter λ occurs in two of the entries of the matrix

$$A = \begin{bmatrix} \lambda+3 & 4 \\ 3 & \lambda+2 \end{bmatrix} .$$

The algebraic condition for invertibility is then

$$(\lambda+3)(\lambda+2) - (4)(3) \neq 0 \quad \text{or} \quad \lambda^2 + 5\lambda - 6 \neq 0$$

or $\quad (\lambda - 1)(\lambda + 6) \neq 0 \quad .$

So A is invertible unless the parameter λ is 1 or -6.

Our main objective in this chapter is to obtain a comparable algebraic condition for invertibility of a square matrix of any size. Starting from the entries of a square matrix A, we will construct an algebraic formula called the "determinant of A," or det A, such that A is invertible unless det $A = 0$.

Sections 5.1 and 5.2 are devoted to the construction of the determinant and demonstration of its essential properties. The crucial step is at the beginning, in lemma 5.1, after which everything unfolds naturally. The use of the method of elementary matrices facilitates the rapid development of a complete theory.

It turns out that the determinant also has an important geometric interpretation; it describes the way in which matrix transformations modify areas and volumes. This interpretation is presented in section 5.3.

5.1 Determinants and Row Operations

Our main objective in this chapter has been outlined in the introduction. Given a square matrix A, we want to construct a scalar quantity, to be called the determinant of A, or det A, in such a way that A^{-1} exists unless det $A = 0$.

Let us review what we know about inverses up to now. Of course a scalar quantity, a, has an inverse, a^{-1}, unless $a = 0$. But since the algebra of 1×1 matrices is the same as that of scalars, we can say that

the 1×1 matrix, $A = [a_{11}]$, has an inverse unless $a_{11} = 0$.

In the next dimension up, the corresponding result, as noted above, is as follows:

the 2×2 matrix, $A = \begin{bmatrix} a_{11} & a_{12} \\ a_{21} & a_{22} \end{bmatrix}$, has an inverse unless

$a_{11}a_{22} - a_{12}a_{21} = 0$.

Accordingly, we can define the determinant in dimensions one and two as follows:

$$\det \begin{bmatrix} a_{11} \end{bmatrix} = a_{11} \quad , \quad \det \begin{bmatrix} a_{11} & a_{12} \\ a_{21} & a_{22} \end{bmatrix} = a_{11}a_{22} - a_{12}a_{21} \quad .$$

Our strategy will be to construct a determinant in successively higher dimensions in turn. That is, the 3×3 determinant will be defined in terms of 2×2 determinants; then the 4×4 determinant will be defined in terms of 3×3 determinants, and so on.

The concept that allows us to move from one dimension to the next is simple: when row i and column j of an $n \times n$ matrix A are deleted, the result is an $(n-1) \times (n-1)$ matrix, which is called the *ijth minor of A* and is denoted M_{ij}.

Example 1 Let $A = \begin{bmatrix} 3 & 4 \\ 7 & 8 \end{bmatrix}$.

To compute M_{11}, delete row 1 and column 1 and obtain

$\begin{bmatrix} 3 & 4 \\ 7 & 8 \end{bmatrix}$ or $M_{11} = [8]$

To compute M_{12}, delete row 1 and column 2 and obtain

$\begin{bmatrix} 3 & 4 \\ 7 & 8 \end{bmatrix}$ or $M_{12} = [7]$

Example 2 Let $A = \begin{bmatrix} 1 & 2 & 3 \\ 4 & 5 & 6 \\ 7 & 8 & 9 \end{bmatrix}$.

To compute M_{11}, delete row 1 and column 1 and obtain

$\begin{bmatrix} 1 & 2 & 3 \\ 4 & 5 & 6 \\ 7 & 8 & 9 \end{bmatrix}$ or $M_{11} = \begin{bmatrix} 5 & 6 \\ 8 & 9 \end{bmatrix}$.

To compute M_{23}, delete row 2 and column 3 and obtain

$\begin{bmatrix} 1 & 2 & 3 \\ 4 & 5 & 6 \\ 7 & 8 & 9 \end{bmatrix}$ or $M_{23} = \begin{bmatrix} 1 & 2 \\ 7 & 8 \end{bmatrix}$.

When row 1 and column 1 of a 2×2 matrix A are deleted, we obtain the 1×1 minor $M_{11} = [a_{22}]$. Similarly, if row 1 and column 2 are deleted, the result is the minor $M_{12} = [a_{21}]$.

The formula for the 2×2 determinant can therefore be rewritten as

$\det A = a_{11} \det M_{11} - a_{12} \det M_{12}$.

This form of det A is said to be the *Laplace expansion of det A across row one*, since the entries of the first row of A appear as factors in each term.

Using this expansion as a model, we define the *determinant of a 3×3 matrix A* by the formula

$\det A = a_{11} \det M_{11} - a_{12} \det M_{12} + a_{13} \det M_{13}$.

Example 3 $\det \begin{bmatrix} 1 & 2 & 3 \\ 4 & 5 & 6 \\ 7 & 8 & 9 \end{bmatrix} = 1 \det \begin{bmatrix} 5 & 6 \\ 8 & 9 \end{bmatrix} - 2 \det \begin{bmatrix} 4 & 6 \\ 7 & 9 \end{bmatrix} + 3 \det \begin{bmatrix} 4 & 5 \\ 7 & 8 \end{bmatrix}$

$= 1\,(45 - 48) - 2\,(36 - 42) + 3\,(32 - 35)$
$= -3 + 12 - 9 = 0$.

Note that the jth term in the expansion of det A can also be written $(-1)^{j+1}a_{1j}$ det M_{1j}, since the factor $(-1)^{j+1}$ makes the signs preceding successive terms of the expansion alternate between plus and minus.

With these notations, the *determinant of the $n \times n$ matrix A* is defined by the expansion

$$\det A = a_{11} \det M_{11} - a_{12} \det M_{12} + \ldots + (-1)^{n+1}a_{1n} \det M_{1n} \quad .$$

Example 4   The Laplace expansion of a typical 4×4 determinant is

$$\det \begin{bmatrix} 1 & 3 & 5 & 7 \\ 2 & 4 & 6 & 8 \\ 4 & 3 & 2 & 1 \\ 2 & 3 & 4 & 5 \end{bmatrix} =$$

$$1 \det \begin{bmatrix} 4 & 6 & 8 \\ 3 & 2 & 1 \\ 3 & 4 & 5 \end{bmatrix} - 3 \det \begin{bmatrix} 2 & 6 & 8 \\ 4 & 2 & 1 \\ 2 & 4 & 5 \end{bmatrix} + 5 \det \begin{bmatrix} 2 & 4 & 8 \\ 4 & 3 & 1 \\ 2 & 3 & 5 \end{bmatrix} \cdots$$

$$\ldots - 7 \det \begin{bmatrix} 2 & 4 & 6 \\ 4 & 3 & 2 \\ 2 & 3 & 4 \end{bmatrix} \quad .$$

But the calculation of even a 4×4 determinant may be lengthy: it expands into four 3×3 determinants, each of which in turn expands into three 2×2 determinants, for a total of twelve 2×2 determinants. We must work not only to show that the $n \times n$ determinant has the same theoretical significance as the 2×2 determinant, but also to find more efficient methods of calculation. The solution to these problems flows from the following lemma:

Lemma 5.1   If rows 1 and 2 of the $n \times n$ matrix A are exchanged, then det A is multiplied by -1.

■**Proof**   The argument is conveniently illustrated by the 4×4 matrix with rows given as row vectors **p**, **q**, **r**, **s**; that is, the matrix

$$A = \begin{bmatrix} p_1 & p_2 & p_3 & p_4 \\ q_1 & q_2 & q_3 & q_4 \\ r_1 & r_2 & r_3 & r_4 \\ s_1 & s_2 & s_3 & s_4 \end{bmatrix} \quad .$$

Now, det A will be expressed as a sum of twelve 2×2 determinants formed from the **r** and **s** rows once we have expanded across the **p** and **q** rows in turn. For example, the terms containing p_3 and p_4 respectively at the first expansion are

$$(-1)^{3+1}p_3 \det \begin{bmatrix} q_1 & q_2 & q_4 \\ r_1 & r_2 & r_4 \\ s_1 & s_2 & s_4 \end{bmatrix} \quad , \quad (-1)^{4+1}p_4 \det \begin{bmatrix} q_1 & q_2 & q_3 \\ r_1 & r_2 & r_3 \\ s_1 & s_2 & s_3 \end{bmatrix} \quad .$$

When the 3×3 determinants are expanded in turn across the **q** row, the terms containing $p_3 q_4$ and $p_4 q_3$ respectively are

$$(-1)^{3+1} p_3 (-1)^{3+1} q_4 \det \begin{bmatrix} r_1 & r_2 \\ s_1 & s_2 \end{bmatrix} \quad , \quad \ldots$$

$$\ldots (-1)^{4+1} p_4 (-1)^{3+1} q_3 \det \begin{bmatrix} r_1 & r_2 \\ s_1 & s_2 \end{bmatrix} \quad .$$

The arrangement of signs in these Laplace expansions is mostly predictable. Since p_3 and q_3 start off as column 3 entries, we expect them to carry the sign $(-1)^{3+1}$, and similarly p_4 and q_4 factors should carry the sign $(-1)^{4+1}$. But there is a break in this pattern in the $p_3 q_4$ term. Because q_4 was to the *right* of p_3 in A, the deletion of the column containing p_3 shifted q_4 one column to the left. The q_4 factor in the $p_3 q_4$ term therefore carried the sign $(-1)^{3+1}$ instead of the expected $(-1)^{4+1}$.

In consequence, the $p_3 q_4$ and $p_4 q_3$ terms necessarily carry opposite signs. On combining these two terms, we see that the coefficient of det $\begin{bmatrix} r_1 & r_2 \\ s_1 & s_2 \end{bmatrix}$ is $(-1)^{4+3}(p_4 q_3 - p_3 q_4)$.

But if rows **p** and **q** were exchanged, the coefficient would instead be $(-1)^{4+3}(q_4 p_3 - q_3 p_4)$, exactly -1 times its original value. Since the same applies to every 2×2 determinant in the sum representing det A, the determinant would be multiplied by -1. Q.E.D.

Using this result, we can compute the following 4×4 determinant by exploiting the strings of zeros in row 2.

Example 5 $\quad \det \begin{bmatrix} 1 & 1 & 0 & 3 \\ 0 & 1 & 0 & 0 \\ 1 & 2 & 4 & 0 \\ 1 & 1 & 0 & 1 \end{bmatrix} = -\det \begin{bmatrix} 0 & 1 & 0 & 0 \\ 1 & 1 & 0 & 3 \\ 1 & 2 & 4 & 0 \\ 1 & 1 & 0 & 1 \end{bmatrix} \quad \ldots$

$$\ldots \quad = -\left(0 - 1 \det \begin{bmatrix} 1 & 0 & 3 \\ 1 & 4 & 0 \\ 1 & 0 & 1 \end{bmatrix} + 0 - 0 \right)$$

$$= 1 \det \begin{bmatrix} 4 & 0 \\ 0 & 1 \end{bmatrix} - 0 + 3 \det \begin{bmatrix} 1 & 4 \\ 1 & 0 \end{bmatrix} = (1)(4) + (3)(-4) = -8 \quad .$$

THEOREM 5.2

(a) If any two rows of the $n \times n$ matrix A are exchanged, then det A is multiplied by -1.

(b) If any two rows of the $n \times n$ matrix A are the same, then det $A = 0$.

■ Proof   (a) For 2×2 matrices, rows 1 and 2 are the only rows, and so the preceding lemma applies directly. The proof can then be extended to successively higher dimensions. Once we have proved it in dimension $n-1$, say, we reason in dimension n as follows.

When rows $i+1$ and $j+1$ of the $n \times n$ matrix A are exchanged, rows i and j of the minors used in expanding det A are also exchanged. Since the minors are $(n-1) \times (n-1)$ matrices, we know that the determinant of each minor is multiplied by -1. It follows from the expansion

$$\det A = a_{11} \det M_{11} - a_{12} \det M_{12} + \ldots$$

that det A is also multiplied by -1. This disposes of all cases where row 1 of A is not moved.

As for the exchange of row 1 and row j of A, let the rows $1, 2$, and j of A be the vectors $\mathbf{p}, \mathbf{q}, \mathbf{w}$ respectively. Suppose we reshuffle rows $1, 2$, and j as follows:

$\mathbf{p}, \mathbf{q}, \mathbf{w}$ change to $\mathbf{q}, \mathbf{p}, \mathbf{w}$, which change to $\mathbf{q}, \mathbf{w}, \mathbf{p}$, which change to $\mathbf{w}, \mathbf{q}, \mathbf{p}$.

That is, we exchange first rows 1 and 2, then rows 2 and j, then rows 1 and 2 again. At the first and third steps, det A is multiplied by -1, in accordance with lemma 5.1. At the second step, det A is also multiplied by -1, in accordance with the earlier part of the present lemma. Altogether, det A is multiplied by $(-1)^3$ or -1, as required.

(b) If rows i and j of A are the same, then A is unchanged when these rows are exchanged, and so det A is also unchanged. But from part (a) we know that det A is multiplied by -1. Thus det $A = -\det A$ or det $A = 0$.   Q.E.D.

Example 6   The above theorem helps us to shorten the calculation of the determinant by exploiting the zeros in the matrix wherever they occur. In the following determinant, for example, a zero occurs in the row $(0, 8, 9)$; so we move this row to the top by exchanging it with the rows above it in turn, and then we expand:

$$\det \begin{bmatrix} 1 & 2 & 3 \\ 4 & 5 & 6 \\ 0 & 8 & 9 \end{bmatrix} = (-1)\det \begin{bmatrix} 1 & 2 & 3 \\ 0 & 8 & 9 \\ 4 & 5 & 6 \end{bmatrix} = (-1)^2\det \begin{bmatrix} 0 & 8 & 9 \\ 1 & 2 & 3 \\ 4 & 5 & 6 \end{bmatrix}$$

$$= (-1)^2 \left\{ \left(0 \det \begin{bmatrix} 2 & 3 \\ 5 & 6 \end{bmatrix} - 8 \det \begin{bmatrix} 1 & 3 \\ 4 & 6 \end{bmatrix} \ldots \right. \right.$$

$$\left. \left. \ldots + 9 \det \begin{bmatrix} 1 & 2 \\ 4 & 5 \end{bmatrix} \right) \right\}$$

$$= (-8)(-6) + (9)(-3) = 21 \ .$$

Note that the expansion in 2×2 matrices can be interpreted as an expansion of the original matrix across row 3. In fact, it is exactly

$$(-1)^2 (a_{31} \det M_{31} - a_{32} \det M_{32} + a_{33} \det M_{33}) \ ,$$

since M_{31} is the minor $\begin{bmatrix} 2 & 3 \\ 5 & 6 \end{bmatrix}$ obtained by deleting row 3 and column 1 of the original matrix, etc.

We can see that the factor $(-1)^2$ occurs in front of the expansion because the expansion row, row 3, is exactly two rows from the top. Similarly, the extra factor $(-1)^{i-1}$ must occur when we expand across row i. The validity of the following theorem is therefore clear.

THEOREM 5.3

For any $n \times n$ matrix A, det A may be expanded across row i as follows:

$$\det A = (-1)^{i-1}(a_{i1} \det M_{i1} - a_{i2} \det M_{i2} + \ldots) .$$

Example 7   In the next example, we first expand across the row containing the maximum number of zeros, row 2:

$$\det \begin{bmatrix} 1 & 2 & 3 & 4 \\ 0 & 0 & 0 & 3 \\ 0 & 0 & 7 & 5 \\ 4 & 3 & 2 & 6 \end{bmatrix} = (-1)^{2-1} \left\{ 0 - 0 + 0 - 3 \det \begin{bmatrix} 1 & 2 & 3 \\ 0 & 0 & 7 \\ 4 & 3 & 2 \end{bmatrix} \right\}$$

and now, expanding the 3×3 determinant across its second row, we get

$$3 \, (-1)^{2-1} \left\{ 0 - 0 + 7 \det \begin{bmatrix} 1 & 2 \\ 4 & 3 \end{bmatrix} \right\} = (3)(-1)(7)(-5) = 105 .$$

Example 8   It is convenient to expand the determinant of an upper triangular matrix across the last row; for instance;

$$\det \begin{bmatrix} 2 & 3 & 4 \\ 0 & 5 & 6 \\ 0 & 0 & 7 \end{bmatrix} = 0 - 0 + 7 \det \begin{bmatrix} 2 & 3 \\ 0 & 5 \end{bmatrix} \cdots$$

$$\cdots = 7 \, ((2)(5)) = (2)(5)(7) .$$

That is, the determinant is the product of the diagonal entries. Clearly this result would be obtained for any triangular matrix. This gives us a useful corollary.

COROLLARY 5.4

(a) The determinant of a triangular $n \times n$ matrix A is the product of its diagonal entries:

$$\det A = a_{11} a_{22} \ldots a_{nn} .$$

(b) For the $n \times n$ identity matrix I, $\det I = 1$.

The next theorem prepares the ground for the systematic use of row operations in the rest of the chapter.

THEOREM 5.5

When an elementary row operation is applied to an $n \times n$ matrix A, det A changes as follows.

Type A: a multiple of one row is added to another; det A is unchanged.
Type M: a row is multiplied by a constant k; det A is multiplied by k.
Type E: two rows are exchanged; det A is multiplied by -1.

■**Proof**   For type E, this is just theorem 5.2. For type M, suppose that row i is multiplied by k; then the expansion of det A across row i is altered to

$$(-1)^{i-1}(ka_{i1} \det M_{i1} - ka_{i2} \det M_{i2} + \ldots) \quad ,$$

which is clearly k times det A, as required.

For type A, suppose that the rows of the matrix A are **p**, **q**, **r**, etc., and suppose, for example, that k times row 2 is added to row 1. Then row 1 is changed to **p** + **kq**, and the expansion of det A across row 1 is changed to

$$(p_1+kq_1) \det M_{11} - (p_2+kq_2) \det M_{12} + \ldots$$
$$= (p_1 \det M_{11} - p_2 \det M_{12} + \ldots) + k(q_1 \det M_{11} - q_2 \det M_{12} + \ldots).$$

Now of the two bracketed expressions, the first is just the expansion of the original matrix A across row 1, while the second is the expansion across row 1 of the matrix with rows **q**, **q**, **r**, etc. This latter matrix has two rows the same, and so its determinant is 0. So in fact the quantity added to the original det A is 0, as required.

Similarly, if row i, say, is acted on by the type A row operation, we expand across row i and obtain the same result.   Q.E.D.

Now we can easily show that the $n \times n$ determinant has the same theoretical significance noted for the 2×2 determinant at the beginning of the chapter.

THEOREM 5.6

The $n \times n$ matrix A has an inverse, A^{-1}, unless det $A = 0$.

■**Proof**   The matrix A is changed to reduced form R by a sequence of elementary row operations. Each of these row operations multiplies det A by a nonzero factor, according to the previous theorem, and so det A and det R are both zero or both nonzero.

Now if A is invertible, its reduced form is $R = I$, and det R = det I = 1, and det A is nonzero also. But if A is not invertible, the last row of its reduced form R is all zeros. So by expanding R along its last row, we get det $R = 0$, and therefore det A is also zero.   Q.E.D.

For example, the determinant of the matrix A at the right was found by computation to be 0 at the beginning of the chapter, and so by theorem 5.6, A has no inverse. As an exercise, the matrix inversion algorithm of Chapter 3 can be used to confirm that A^{-1} does not exist.

$$A = \begin{bmatrix} 1 & 2 & 3 \\ 4 & 5 & 6 \\ 7 & 8 & 9 \end{bmatrix}.$$

■ **Problem**   Given the matrix $A = \begin{bmatrix} 2 & 2 & 3 & 4 \\ 7 & 4 & 6 & 2 \\ 8 & 8 & 8 & 7 \\ 1 & 1 & 1 & 1 \end{bmatrix}$, compute det A.

■ **Solution**   Since there are no zero entries in the matrix, repeated Laplace expansion will eventually result in 12 2×2 determinants. To reduce the amount of calculation and the chance of numerical error it is important to try to create some zeros by means of row operations.

In this example, it happens that row 3 is "almost" eight times row 4. So if we subtract eight times row 4 from row 3, we will get several zeros in row 3. Note that the row operation is type A, and doesn't change the determinant at all. Therefore,

$$\det A = \det \begin{bmatrix} 2 & 2 & 3 & 4 \\ 7 & 4 & 6 & 2 \\ 0 & 0 & 0 & -1 \\ 1 & 1 & 1 & 1 \end{bmatrix}.$$

Now, of course, we expand across row 3 and obtain

$$\det A = (-1)^{3-1} \left(0 - 0 + 0 - (-1) \det \begin{bmatrix} 2 & 2 & 3 \\ 7 & 4 & 6 \\ 1 & 1 & 1 \end{bmatrix} \right)$$

$$= \det \begin{bmatrix} 2 & 2 & 3 \\ 7 & 4 & 6 \\ 1 & 1 & 1 \end{bmatrix}.$$

At this point, a further shortcut is provided by the type A row operation $R_1 - 2R_3$, and we obtain

$$\det A = \det \begin{bmatrix} 0 & 0 & 1 \\ 7 & 4 & 6 \\ 1 & 1 & 1 \end{bmatrix} = (-1)^{1-1} \left(0 - 0 + 1 \det \begin{bmatrix} 7 & 4 \\ 1 & 1 \end{bmatrix} \right)$$

$$= 7 - 4 = 3 \quad.$$

■ EXERCISES 5.1

1. Compute the determinants of the following matrices.

(a) $\begin{bmatrix} 1 & 2 \\ 3 & 4 \end{bmatrix}$ (b) $\begin{bmatrix} 1 & -1 \\ -3 & 2 \end{bmatrix}$ (c) $\begin{bmatrix} 2 & 1 & 3 \\ 1 & 2 & 5 \\ 5 & 0 & 4 \end{bmatrix}$

(d) $\begin{bmatrix} 1 & 1 & 5 \\ 0 & 1 & 3 \\ 1 & 0 & 2 \end{bmatrix}$ (e) $\begin{bmatrix} -4 & -6 & 2 \\ 2 & -3 & 1 \\ 1 & 1 & 1 \end{bmatrix}$ (f) $\begin{bmatrix} 5 & 0 & 3 \\ 2 & 0 & 4 \\ 3 & 0 & 7 \end{bmatrix}$

(g) $\begin{bmatrix} 1 & 1 & 1 \\ 4 & 3 & 1 \\ 16 & 9 & 1 \end{bmatrix}$ (h) $\begin{bmatrix} 2 & 0 & 0 & 3 \\ 4 & 1 & 1 & 5 \\ 1 & 0 & 1 & 3 \\ 2 & 1 & 0 & 2 \end{bmatrix}$ (i) $\begin{bmatrix} 0 & 0 & 1 & 2 \\ 0 & 0 & 3 & 4 \\ 5 & 6 & 0 & 0 \\ 7 & 8 & 0 & 0 \end{bmatrix}$

(j) $\begin{bmatrix} 1 & 1 & 1 & 1 \\ 2 & 3 & 3 & 3 \\ 8 & 4 & 5 & 5 \\ 9 & 8 & 6 & 7 \end{bmatrix}$ (k) $\begin{bmatrix} 1 & 0 & 1 & 1 \\ 0 & 1 & 1 & 0 \\ 1 & 0 & 3 & 1 \\ 0 & 1 & 0 & 3 \end{bmatrix}$

2. Let $M = \begin{bmatrix} 1 & 1 & 2 \\ 1 & 0 & 5 \\ p & 2 & 4 \end{bmatrix}$.

   Compute det M, and determine for which values of the parameter p the matrix M has an inverse.

3. Show that the matrix $\begin{bmatrix} a & 1 & 1 \\ 1 & a & 1 \\ 1 & 1 & a \end{bmatrix}$ is invertible unless $a = 1$ or -2.

4. Let $A = \begin{bmatrix} 1 & 2 \\ 0 & 1 \end{bmatrix}$, $B = \begin{bmatrix} 2 & 0 \\ 1 & 3 \end{bmatrix}$.

   (a) Compute the matrices $C = AB$ and $D = A + B$.
   (b) Compute the determinants of $A, B, C,$ and D .
   (c) Verify that $\det(AB) = (\det A)(\det B)$,
       but that $\det(A + B) \neq (\det A) + (\det B)$.

5. Suppose that $\det \begin{bmatrix} a & b & c \\ d & e & f \\ g & h & i \end{bmatrix} = k$. Using row operations, show that

   (a) $\det \begin{bmatrix} d & e & f \\ 4a & 4b & 4c \\ 3g & 3h & 3i \end{bmatrix} = -12k$.

(b) $\det \begin{bmatrix} a+5g & b+5h & c+5i \\ g & h & i \\ -2d & -2e & -2f \end{bmatrix} = 2k$.

6. Suppose $\det \begin{bmatrix} a & b \\ c & d \end{bmatrix} = 3$.

Show that $\det \begin{bmatrix} 2a & 2b \\ 2c & 2d \end{bmatrix} = 12$.

7. Let $A = \begin{bmatrix} 1 & 1 & 1 \\ a & b & c \\ a^2 & b^2 & c^2 \end{bmatrix}$. Compute $\det A$ as follows:

(a) Find a type A row operation that changes A to

$$B = \begin{bmatrix} 1 & 1 & 1 \\ a-b & 0 & c-b \\ a^2 & b^2 & c^2 \end{bmatrix} .$$

(b) By Laplace expansion of $\det B$ across row 2, deduce that

$$\det A = (b-a)(c^2-b^2) - (c-b)(b^2-a^2) .$$

(c) Deduce that $\det A = (b-a)(c-b)(c-a)$.

8. Suppose $\det \begin{bmatrix} 3 & 2 & p \\ 0 & p & 1 \\ 1 & 0 & 2 \end{bmatrix} = 10$.

What are the possible values of the parameter p?

9. Suppose $A = \begin{bmatrix} 1 & 3 & 5 & * \\ 0 & 4 & 0 & 6 \\ 0 & 1 & 0 & 2 \\ 3 & * & 7 & 8 \end{bmatrix}$, where the stars denote

unknown entries. Find all possible values of $\det A$.

10. (a) Show that, if A is the 3×3 matrix whose rows are the vectors **a**, **b**, **c**, then
$$\det A = a_1(b_2c_3-b_3c_2) + a_2(b_3c_1-b_1c_3) + a_3(b_1c_2-b_2c_1) .$$

*(b) Suppose that A is a 3×3 matrix, and that rank $(A) < 3$.
Show that the vector **x** whose components are

$$x_1 = \det M_{31} , \quad x_2 = -\det M_{32} , \quad x_3 = \det M_{33} ,$$

is in the kernel of A.

(c) Use part (b) to find a vector **x** in the kernel of $A = \begin{bmatrix} 1 & 2 & 3 \\ 4 & 5 & 6 \\ 7 & 8 & 9 \end{bmatrix}$.

*11. (a) Suppose that the $n \times n$ matrix A has entries a_{ij}, and let $N(i,j)$ denote the $(n-2) \times (n-2)$ matrix obtained by deleting rows 1 and 2 and columns i and j of A.

Show that the two-row expansion formula used in the proof of lemma 5.1 can be expressed in the form

$$\det A = \Sigma \quad (-1)^{1+i+j} \det \begin{bmatrix} a_{1i} & a_{1j} \\ a_{2i} & a_{2j} \end{bmatrix} \det N(i,j) \quad .$$

(b) Find a similar formula for $\det A$ based on expansion across rows h and k instead of rows 1 and 2.

(c) Use the formula in part (a) to evaluate $\det \begin{bmatrix} 0 & 0 & 1 & 2 \\ 0 & 0 & 3 & 4 \\ 5 & 6 & 0 & 0 \\ 7 & 8 & 0 & 0 \end{bmatrix}$.

(Compare problem 1(i) above.)

*12. A scalar valued function $f(\mathbf{a}_1, \mathbf{a}_2, \ldots, \mathbf{a}_n)$, whose n variables $\mathbf{a}_1$ through $\mathbf{a}_n$ are all vectors in R^n, is said to be an *antisymmetric multilinear form* if it satisfies the following two properties:

(i) $f(\mathbf{a}_1, \ldots, \mathbf{a}_i, \ldots, \mathbf{a}_j, \ldots, \mathbf{a}_n)$
$\quad = -f(\mathbf{a}_1, \ldots, \mathbf{a}_j, \ldots, \mathbf{a}_i, \ldots, \mathbf{a}_n)$
(ii) $f(\mathbf{a}_1, \ldots, h\mathbf{a}_i + k\mathbf{b}_i, \ldots, \mathbf{a}_n)$
$\quad = hf(\mathbf{a}_1, \ldots, \mathbf{a}_i, \ldots, \mathbf{a}_n) + kf(\mathbf{a}_1, \ldots, \mathbf{b}_i, \ldots, \mathbf{a}_n) \quad .$

That is, f must change sign when two variables are interchanged and must be linear in each variable separately.

(a) Let $\mathbf{a}_1, \ldots, \mathbf{a}_n$ denote the n rows of the $n \times n$ matrix A, and rewrite the determinant function $\det A$ as $\det (\mathbf{a}_1, \ldots, \mathbf{a}_n)$. Show that det is an antisymmetric multilinear form in the n rows $\mathbf{a}_1$ through $\mathbf{a}_n$ of A.

(b) Show that det satisfies the additional condition $\det (\mathbf{e}_1, \ldots, \mathbf{e}_n)$
$= 1$.

(c) Show that any antisymmetric multilinear form f is a scalar multiple of det, and that

$$f(\mathbf{a}_1, \ldots, \mathbf{a}_n) = f(\mathbf{e}_1, \ldots, \mathbf{e}_n) \det (\mathbf{a}_1, \ldots, \mathbf{a}_n) \quad .$$

5.2 Algebra of Determinants

We have seen that large determinants can be evaluated quickly when enough zero entries are present, or can be created, along a row. But no reliable means of creating these zeros has yet appeared. Columns of zeros, however, can readily be created by row operations, as in the matrix reduction process. Our computational powers will therefore be greatly enhanced if we can show that Laplace expansion of determinants along columns as well as rows is valid. Now the transpose operation converts rows to columns, so we need only show that transposition doesn't change determinants. Results of this sort form the content of this section.

We used elementary matrices as building blocks in developing matrix algebra and will do so again in presenting the algebra of determinants. We start by discussing the determinants of the elementary matrices themselves.

Lemma 5.7 Suppose Q is an elementary $n \times n$ matrix. Then

(a) det $(QA) = ($det $Q)($det $A)$, for any $n \times n$ matrix A.
(b) det $(Q^t) = $ det Q.

◼**Proof** (a) Theorem 5.5 says that applying an elementary row operation to A changes det A by a factor, c, which depends only on the row operation, not on A. That is, if Q is an elementary matrix, det $(QA) = c($det $A)$. To determine the scalar c, substitute $A = I$. Then we get det $Q = c($det $I) = c$. Substituting det Q for c in the previous formula, we get

det $(QA) = ($det $Q)($det $A)$ as required.

(b) For types M and E elementary matrices, $Q^t = Q$, and so the determinants of Q^t and Q are necessarily equal. Suppose now that Q is a type A elementary matrix. Typically,

$$Q = \begin{bmatrix} 1 & 0 & 3 \\ 0 & 1 & 0 \\ 0 & 0 & 1 \end{bmatrix} \quad , \qquad Q^t = \begin{bmatrix} 1 & 0 & 0 \\ 0 & 1 & 0 \\ 3 & 0 & 1 \end{bmatrix} \quad .$$

Clearly Q^t is also type A, and by repeated expansion we easily compute det $Q = $ det $Q^t = 1$.
\hfill Q.E.D.

THEOREM 5.8

For any $n \times n$ matrix A,
(a) det $A^t = $ det A;
(b) det A can be expanded along column j according to the formula

$$\det A = (-1)^{j-1}(a_{1j} \det M_{1j} - a_{2j} \det M_{2j} + \ldots) \quad .$$

Proof (a) From matrix algebra, we know that if one of A and A^t lacks an inverse, so does the other, and thus $\det A = \det A^t = 0$.

Consider now the other case, when A and A^t both have inverses. Then from matrix algebra again, A and A^t are both products of elementary matrices, say

$$A = Q_1 Q_2 \ldots Q_h \; , \qquad A^t = Q_h^t \ldots Q_2^t Q_1^t \; .$$

Using the previous lemma, we compute

$$\det A = \det((Q_1)(Q_2 \ldots Q_h)) = (\det Q_1)(\det(Q_2 \ldots Q_h)) \; ,$$

and by splitting off successive factors Q_2 etc., we get

$$\det A = (\det Q_1)(\det Q_2) \ldots (\det Q_h) \; .$$

Similarly we compute

$$\det A_t = (\det Q_h^t) \ldots (\det Q_2^t)(\det Q_1^t)$$
$$= (\det Q_h) \ldots (\det Q_2)(\det Q_1) \text{ by the previous lemma.}$$

The formulas we have obtained for $\det A$ and $\det A^t$ are both products of scalars and they are obviously equal as required.

(b) Let B denote A^t, so that $b_{ij} = a_{ji}$ and $\det A = \det B$. Also let N_{ij} denote the ijth minor of B.

For example, if $A = \begin{bmatrix} 1 & 2 & 3 \\ 4 & 5 & 6 \\ 7 & 8 & 9 \end{bmatrix}$, then $B = \begin{bmatrix} 1 & 4 & 7 \\ 2 & 5 & 8 \\ 3 & 6 & 9 \end{bmatrix}$,

and $M_{13} = \begin{bmatrix} 4 & 5 \\ 7 & 8 \end{bmatrix}$, $N_{31} = \begin{bmatrix} 4 & 7 \\ 5 & 8 \end{bmatrix}$, so $N_{31} = M_{13}^t$,

and similarly $N_{ji} = M_{ij}^t$ in general.

But on expanding $\det B$ along row j, we get

$$\det B = (-1)^{j-1}(b_{j1} \det N_{j1} - b_{j2} \det N_{j2} + \ldots)$$
$$= (-1)^{j-1}(a_{1j} \det M_{1j}^t - a_{2j} \det M_{2j}^t + \ldots)$$
$$= (-1)^{j-1}(a_{1j} \det M_{1j} - a_{2j} \det M_{2j} + \ldots) \; ,$$

which is the stated expansion along column j of A.   Q.E.D.

Example 1 In the following determinant, it is convenient to expand along column 2:

$$\det \begin{bmatrix} 1 & 3 & 1 & 1 \\ 2 & 0 & 1 & 5 \\ 2 & 0 & 2 & 4 \\ 5 & 0 & 2 & 0 \end{bmatrix} = (-1)^{2-1} \left\{ 3 \det \begin{bmatrix} 2 & 1 & 5 \\ 2 & 2 & 4 \\ 5 & 2 & 0 \end{bmatrix} - 0 + 0 - 0 \right\}$$

$$= -3 \left\{ (-1)^{3-1}(5 \det \begin{bmatrix} 2 & 2 \\ 5 & 2 \end{bmatrix} - 4 \det \begin{bmatrix} 2 & 1 \\ 5 & 2 \end{bmatrix} + 0) \right\}$$

$$= -3 \{(5)(-6) - (4)(-1)\} = 78 \; .$$

Even more important is the possibility of creating a column of zeros by applying type A row operations, which do not change the determinant, and then expanding down that column. This is done in the next example.

Example 2 $\det \begin{bmatrix} 1 & 1 & 1 & 1 \\ 1 & 1 & 2 & 2 \\ 1 & 1 & 3 & 4 \\ 1 & 2 & 3 & 9 \end{bmatrix} = \det \begin{bmatrix} 1 & 1 & 1 & 1 \\ 0 & 0 & 1 & 1 \\ 0 & 0 & 2 & 3 \\ 0 & 1 & 2 & 8 \end{bmatrix}$

$$= (-1)^{1-1} \left\{ 1 \det \begin{bmatrix} 0 & 1 & 1 \\ 0 & 2 & 3 \\ 1 & 2 & 8 \end{bmatrix} - 0 + 0 - 0 \right\}$$

$$= (-1)^{1-1} \left\{ 0 - 0 + 1 \det \begin{bmatrix} 1 & 1 \\ 2 & 3 \end{bmatrix} \right\}$$

$$= 3 - 2 = 1 \quad .$$

Clearly the combination of type A row operations with row or column expansions will be the most efficient means of tackling large determinants without many zeros.

Our final algebraic result is of great theoretical importance.

THEOREM 5.9

(Product rule) For any $n \times n$ matrices A and B,
 $\det (AB) = (\det A)(\det B)$.

■Proof Suppose first that A is invertible, so that A is a product of elementary matrices:

$A = Q_1 Q_2 \ldots Q_h$ and $\det A = (\det Q_1)(\det Q_2) \ldots (\det Q_h)$.

Then

$\det (AB) = \det (Q_1 Q_2 \ldots Q_h B) = (\det Q_1)(\det Q_2) \ldots (\det Q_h)(\det B)$
$\qquad\qquad = (\det A)(\det B),$ as required.

If A is not invertible, then $\det A = 0$, and A reduces to a matrix R with a zero row at the bottom. That is,

$A = Q_1 Q_2 \ldots Q_h R$ and $AB = Q_1 Q_2 \ldots Q_h (RB)$.

But by the row-column rule of matrix multiplication, RB also has a zero row at the bottom, and so $\det (RB) = 0$ also. Therefore,

$\det (AB) = (\det Q_1)(\det Q_2) \ldots (\det Q_h)(\det (RB)) = 0$.   Q.E.D.

> **COROLLARY 5.10**
>
> Let A be an $n \times n$ matrix. Then
>
> (a) $\det (A^{-1}) = (\det A)^{-1}$, whenever A^{-1} exists.
> (b) $\det (A^m) = (\det A)^m$, for any positive integer m; and if A^{-1} exists, the formula holds for negative integers also.
> (c) $\det (cA) = c^n \det A$, for any scalar c.

■**Proof** (a) Since $A^{-1}A = I$, therefore $\det (A^{-1}A) = \det I$, or $(\det (A^{-1}))(\det A) = 1$ as required.

(b) $\det (A^2) = \det (AA) = (\det A)(\det A) = (\det A)^2$. Similarly, $\det (A^3) = \det (AA^2) = (\det A)(\det (A^2)) = (\det A)^3$, etc. Also, $\det (A^{-m}) = \det ((A^{-1})^m) = (\det (A^{-1}))^m = ((\det A)^{-1})^m = (\det A)^{-m}$.

(c) Note that $cA = (cI)(A)$, and of course the diagonal matrix

$$cI = \begin{bmatrix} c & & \\ & \ddots & \\ & & c \end{bmatrix}$$

has determinant c^n. Therefore

$$\det (cA) = (\det (cI))(\det A) = c^n \det A \quad . \qquad \text{Q.E.D.}$$

Example 3 We can see at a glance that

$$\det \left(\begin{bmatrix} 6 & 3 \\ 15 & 9 \end{bmatrix}^{-7} \right) = \left(\det \begin{bmatrix} 6 & 3 \\ 15 & 9 \end{bmatrix} \right)^{-7} = \dots$$

$$\dots \det \left(3 \begin{bmatrix} 2 & 1 \\ 5 & 3 \end{bmatrix}^{-7} \right)$$

$$= \left(3^2 \det \begin{bmatrix} 2 & 1 \\ 5 & 3 \end{bmatrix} \right)^{-7} = ((9)(1))^{-7} = 3^{-14} \quad .$$

■ **Problem** Evaluate $\det A$, where $A = \begin{bmatrix} a & b & c \\ b & c & a \\ c & a & b \end{bmatrix}$.

■ **Solution** Laplace expansion would give an algebraic formula for $\det A$ in terms of a, b, c; but this might be bulky and unusable. Instead, recombination of the rows of A may provide a common factor in one or more rows and thereby simplify the calculation of $\det A$. The method uses a matrix equation of the form $JA = K$, which in this case is

$$\begin{bmatrix} 1 & 1 & 1 \\ 0 & 1 & 0 \\ 0 & 0 & 1 \end{bmatrix} \begin{bmatrix} a & b & c \\ b & c & a \\ c & a & b \end{bmatrix} = \begin{bmatrix} a+b+c & a+b+c & a+b+c \\ b & c & a \\ c & a & b \end{bmatrix} \quad .$$

Observe that the numerical matrix J, applied to A, replaces row 1 of A by the sum of all three rows. In other words, it recombines the rows of A. Now det K is relatively easy to compute:

$$\det K = (a+b+c) \quad \det \begin{bmatrix} 1 & 1 & 1 \\ b & c & a \\ c & a & b \end{bmatrix}$$

$$= (a+b+c) \, ((bc-a^2) - (b^2-ac) + (ab-c^2))$$
$$= (a+b+c) \, ((ab-bc+ca) - (a^2+b^2+c^2))$$

This result has the interesting alternative form

$$\det K = -\tfrac{1}{2}(a+b+c) \, ((a-b)^2 + (b-c)^2 + (c-a)^2).$$

Then, by the product rule for the determinant,

$$\det A = \left(\frac{\det K}{\det J}\right) = \left(\frac{\det K}{1}\right) = \det K \quad .$$

In particular, the formula we have obtained for det A allows us to see that det $A = 0$ only when a, b, c are all the same, or $a+b+c = 0$.

■ EXERCISES 5.2

1. Compute the determinants of the following matrices.

   (a) $\begin{bmatrix} 1 & 0 & 2 & 1 \\ 1 & 3 & 7 & 4 \\ 1 & 4 & 2 & 6 \\ 1 & 5 & 0 & 6 \end{bmatrix}$ (b) $\begin{bmatrix} 1 & 1 & 2 & 2 \\ 2 & 2 & 1 & 1 \\ 1 & 2 & 1 & 2 \\ 1 & 2 & 2 & 2 \end{bmatrix}$ (c) $\begin{bmatrix} 3 & 5 & 7 & 1 \\ 4 & 1 & 2 & 2 \\ 1 & 0 & 3 & 1 \\ 2 & 1 & 1 & 1 \end{bmatrix}$

   (d) $\begin{bmatrix} 3 & 1 & -2 & 4 \\ 2 & 0 & -5 & 1 \\ 1 & -1 & 2 & 6 \\ -2 & 3 & -2 & 3 \end{bmatrix}$ (e) $\begin{bmatrix} 0 & 1 & 2 & 3 \\ 2 & 0 & 3 & 5 \\ 4 & 5 & 0 & 6 \\ 6 & 3 & 9 & 0 \end{bmatrix}$ (f) $\begin{bmatrix} 1 & 1 & 1 & 1 \\ 1 & -1 & 2 & -2 \\ 1 & 1 & 4 & 4 \\ 1 & -1 & 8 & -8 \end{bmatrix}$

2. Show that det $\begin{bmatrix} a & b & 0 & a \\ 0 & a & a & b \\ b & a & a & 0 \\ a & 0 & b & a \end{bmatrix} = b^2(4a^2 - b^2)$.

3. Suppose that the 3×3 matrix A satisfies det $A = 5$.

   (a) Show that det $(4A) = 4^3 \cdot 5 = 320$.
   (b) Compute det (A^2).
   (c) Compute det $(4A^2)$.

4. Suppose A is a 6×6 matrix and $A^4 = 2A$.

    (a) Show that $(\det A)^4 = 64 \det A$.

    (b) Deduce that $\det A = 0$ or 4.

5. Show that if the square matrix A satisfies $A^t A = I$, then $\det A = \pm 1$.

6. For an $n \times n$ matrix A, show that $\det(-A) = (-1)^n \det A$.

7. An $n \times n$ matrix A is said to be skew-symmetric if $A^t = -A$. Show that if A is skew-symmetric and $\det A$ is not zero, then the dimension n must be even.

8. Suppose A and B are $n \times n$ matrices and that $\det A = 2$ and $\det B = -1$. Compute $\det(ABA^t B^{-3} A^2 B^t)$.

9. Suppose $\det A = 0$. Show that $\det(A^3 + 5A^2 + 6A) = 0$.

10. Using the row recombination method described in the problem at the end of this section, show that

$$\det \begin{bmatrix} a & a & c \\ b & c & a \\ c & b & b \end{bmatrix} = (a+b+c)(a-c)(c+b) \quad .$$

11. Show that $\det \begin{bmatrix} 1 & a & a^2 & a^3 \\ a & a^2 & a^3 & 1 \\ a^2 & a^3 & 1 & a \\ a^3 & 1 & a & a^2 \end{bmatrix} = (a^4 - 1)^3 \quad .$

12. Let $J = \begin{bmatrix} 1 & 1 & 1 & 1 \\ 1 & -1 & 1 & -1 \\ 1 & 0 & -1 & 0 \\ 0 & 1 & 0 & -1 \end{bmatrix}$, $A = \begin{bmatrix} a & b & c & d \\ b & c & d & a \\ c & d & a & b \\ d & a & b & c \end{bmatrix}$,

$K = \begin{bmatrix} p & p & p & p \\ q & -q & q & -q \\ r & s & -r & -s \\ s & -r & -s & r \end{bmatrix}$

    (a) Show that $\det J = -8$.

    (b) Show that $\det K = 8pq(r^2 + s^2)$.

    (c) Show that $JA = K$, for suitable choice of p, q, r, s.

    (d) Deduce that
$$\det A = -(a+b+c+d)(a-b+c-d)((a-c)^2 + (b-d)^2) \quad .$$

13. Show that $\det \begin{bmatrix} 1 & 1 & 1 & 1 \\ a & b & b & b \\ c & d & e & e \\ f & g & h & k \end{bmatrix} = (b-a)(e-d)(k-h)$.

*14. (a) Suppose D is a 2×3 matrix.
Show that $D^t D$ is a 3×3 matrix and that $\det (D^t D) = 0$.
(Note that since D and D^t are not square matrices, the product rule for determinants cannot be used.)

(b) Is it necessarily true that $\det (DD^t) = 0$?

*15. The *cross product* $\mathbf{b} \times \mathbf{c}$ of vectors $\mathbf{b}$ and $\mathbf{c}$ in R^3 is defined to be the vector

$$\mathbf{b} \times \mathbf{c} = (_2c_3 - b_3c_2, \; b_3c_1 - b_1c_3, \; b_1c_2 - b_2c_1) \quad .$$

Suppose now that the rows of the 3×3 matrix A are the vectors $\mathbf{a}, \mathbf{b}, \mathbf{c}$ in R^3.

(a) Show that the formula for $\det A$ in exercise 10(a), section 5.1, can be rewritten as $\det A = \mathbf{a} \cdot (\mathbf{b} \times \mathbf{c})$.

(b) Show that, if A is invertible, then

$$A^{-1} = \frac{[|\mathbf{b} \times \mathbf{c}|\mathbf{c} \times \mathbf{a}|\mathbf{a} \times \mathbf{b}]}{(\mathbf{a} \cdot (\mathbf{b} \times \mathbf{c}))}$$

16. The n dimensional *Van der Monde determinant* is defined to be

$$V_n(p_1, \ldots, p_n) = \det \begin{bmatrix} 1 & 1 & \ldots & 1 \\ p_1 & p_2 & \ldots & p_n \\ p_1^2 & p_2^2 & \ldots & p_n^2 \\ \cdot & \cdot & & \cdot \\ \cdot & \cdot & & \cdot \\ \cdot & \cdot & & \cdot \\ p_1^{n-1} & p_2^{n-1} & \ldots & p_n^{n-1} \end{bmatrix} \quad .$$

(a) Show that $P(x) = V_n(p_1, \ldots, p_{n-1}, x)$ is a polynomial in x of degree $n-1$ with coefficient of x^{n-1} equal to $V_{n-1}(p_1, \ldots, p_{n-1})$.

(b) Show that p_1 through p_{n-1} are all roots of $P(x)$ and that therefore
$$V_n(p_1, \ldots, p_{n-1}, x) = V_{n-1}(p_1, \ldots, p_{n-1}) (x-p_1)(x-p_2)$$
$$\ldots (x - p_{n-1}).$$

*(c) Show that $V_n(p_1, \ldots, p_n) = \prod_{i<j} (p_j - p_i)$.

(Compare problem 7 in section 5.1.)

5.3 Geometric Interpretation of the Determinant

Matrices are linked to geometry because they are used to represent transformations of vector spaces, and a geometric interpretation of the determinant develops naturally from this point of view.

A 2×2 matrix, say, specifies a transformation of R^2 into R^2. That is, it maps each point in the plane to another point in the plane. Accordingly, it maps each set S of points in the plane to another set S' of points in the plane. For some matrix transformations, the sets S and S' will be closely related. If the transformation is a rotation, for example, S and S' will always be congruent. Of course we cannot expect anything like this to hold for a general matrix transformation A, but it turns out that there is a very strong result. In fact, the areas of the sets S and S' are proportional; and the constant of proportionality is exactly $|\det A|$.

Our intention is to sketch a proof of this excellent theorem, although omitting many details. We must remember that even the precise definition of area for a general plane region is a rather difficult problem in calculus, requiring approximation of an area by the sum of many small squares, etc. Essentially, we content ourselves with tracing the progress of a square region under a matrix transformation. This is the part of the theory that is based on linear algebra.

We begin by studying the action of elementary matrices on areas.

The 2×2 type M elementary matrix $Q_1 = \begin{bmatrix} 3 & 0 \\ 0 & 1 \end{bmatrix}$ represents a mapping of the xy plane, R^2, into itself. In fact it maps the point (x,y) to the point $(3x,y)$, and so it is easy to see that it maps the unit square, $0 < x < 1$, $0 < y < 1$, into the rectangle $0 < x < 3, 0 < y < 1$, which is a rectangle of base 3 and height 1. That is, a square of area 1 is mapped into a rectangle of area 3. Note that, since $\det Q_1 = 3$, we can say that the original square is mapped into a region of area $|\det Q_1|$ times as great. It is easily seen that that area would be multiplied by the same factor no matter what the size of the original square, and no matter whether one of its corners happened to be at the origin or not.

Now, in calculus it is shown that areas in the plane are approximated by fitting as many squares as possible inside and then taking a limit as smaller and smaller squares are used. So it can be demonstrated that the mapping Q_1 multiplies the area of every plane region by the same factor $|\det Q_1|$. Clearly this conclusion applies to every type M elementary matrix.

In fact the same conclusion applies to every elementary matrix. For example, the type E matrix $Q_2 = \begin{bmatrix} 0 & 1 \\ 1 & 0 \end{bmatrix}$ maps the point (x,y) to (y,x), and so it just reflects regions across the line $y = x$, and does not change areas at all. That is, it multiplies areas by 1. This agrees with the fact that $|\det Q_2| = |-1| = 1$.

The type A matrix $Q_3 = \begin{bmatrix} 1 & k \\ 0 & 1 \end{bmatrix}$ has a different geometric action. It maps the vertices of the unit square to the points $(0,0)$, $(1,0)$, $(k,1)$, $(k+1,1)$, which are the vertices of a parallelogram of base 1 and height 1, that is, a parallelogram of unit area. So area is again multiplied by 1, in agreement with the fact that det $Q_3 = 1$ (see Figure 5.1).

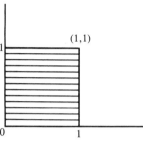

Unit square: base 1,
height 1, area 1.

Figure 5.1

Unit square transformed
to parallelogram: same
base, height, area.

In order to compute the factors by which 3×3 elementary matrices multiply volumes in R^3, note that 3×3 elementary matrices act on only two dimensions at a time. For example, $Q = \begin{bmatrix} 1 & k & 0 \\ 0 & 1 & 0 \\ 0 & 0 & 1 \end{bmatrix}$ maps the base of the unit cube into a parallelogram in the xy plane of area 1 but leaves the z coordinate unchanged. So the cube is replaced by a volume with vertical walls and uniform height 1 on a new base with the same area 1. So the new volume is again 1, just as for a 2×2 type A elementary matrix. The rest of the arguments are just as in dimension two.

Let us summarize our conclusions so far in a lemma.

Lemma 5.11 The linear transformation of R^2 (or R^3) represented by a 2×2 (or 3×3) elementary matrix Q multiplies areas (or volumes) by the factor $|\det Q|$.

We now enjoy a smooth passage from elementary matrices to the general case.

THEOREM 5.12

The linear transformation of R^2 (or R^3) represented by a 2×2 (or 3×3) matrix A multiplies areas (or volumes) by the factor $|\det A|$.

■ Proof   Suppose first that A happens to be a reduced 2×2 matrix R. Either

$$R = \begin{bmatrix} 1 & 0 \\ 0 & 1 \end{bmatrix} = I, \quad \det R = 1, \quad \text{and} \quad R\mathbf{x} = \mathbf{x} \ ,$$

or

$$R = \begin{bmatrix} a & b \\ 0 & 0 \end{bmatrix} \ , \quad \det R = 0, \quad \text{and} \quad R\mathbf{x} = \begin{bmatrix} ax_1 + bx_2 \\ 0 \end{bmatrix} \ .$$

In the first case, R multiplies areas by 1. In the second case, R maps the entire plane onto the x_1 axis, that is, onto a region of area 0, and so it multiplies areas by 0.

In both cases, the transformation R multiplies areas by $|\det R|$.

More generally, any 2×2 matrix A can be represented as a product

$$A = Q_k \ldots Q_2 Q_1 R \ ,$$

where R is a reduced matrix and Q_1 through Q_k are elementary matrices. That is, the transformation represented by A is the composite of the transformations represented by $R, Q_1, \ldots, Q_k$ acting in turn. But these transformations multiply areas by the factors $|\det R|, |\det Q_1|, \ldots, |\det Q_k|$ in turn. So the composite transformation A multiplies areas by the product of these factors, which is

$$\begin{aligned} |\det Q_k| \ldots |\det Q_1||\det R| &= |(\det Q_k) \ldots (\det Q_1)(\det R)| \\ &= |\det (Q_k \ldots Q_1 R)| \\ &= |\det A| \ . \end{aligned}$$

Exactly similar arguments apply in the 3×3 case.   Q.E.D.

■ EXERCISES 5.3

1. Suppose the invertible 3×3 matrix A satisfies $\det A = 3$. By what factors do the following matrices multiply volumes in R^3?

   (a) A (b) A^2 (c) A^{-1} (d) $2A$

2. By what factor does the 2×2 matrix A multiply areas when A is

   (a) a rotation   (b) a reflection   (c) a projection?

3. Give an example of a matrix transformation of R^2 into R^2 that does not preserve lengths of vectors but does preserve areas.

4. Let A be the 3×3 matrix whose columns are the vectors $\mathbf{a}, \mathbf{b}, \mathbf{c}$ in R^3. That is, $A = [\mathbf{a}|\mathbf{b}|\mathbf{c}]$.

   (a) By expanding $\det A$ along column 1, show that $\det A = \mathbf{a} \cdot (\mathbf{b} \times \mathbf{c})$. (Compare exercise 15(a), section 5.2.)

(b) Show that A maps the unit cube into the parallelepiped with adjacent edges **a**, **b**, **c**.

(c) Deduce that the volume of a parallelepiped with adjacent edges **a**, **b**, **c** must be $|\mathbf{a} \cdot (\mathbf{b} \times \mathbf{c})|$. (Compare exercise 15(a), section 5.2.)

*5. Prove the following theorem:

THEOREM (Geometric properties of the cross product)

Let **b**, **c** be vectors in R^3, and let θ be the angle between them.
Then   (i) $\mathbf{b} \times \mathbf{c}$ is perpendicular to both **b** and **c**.
    (ii) $|\mathbf{b} \times \mathbf{c}| = |\mathbf{b}||\mathbf{c}| \sin \theta$.

Method of proof: Let $\mathbf{a} = \dfrac{|\mathbf{b}| \times |\mathbf{c}|}{|\mathbf{b} \times \mathbf{c}|}$,   and let A be the 3×3 matrix with the columns **a**, **b**, **c**.

(a) Show that $\mathbf{b} \cdot (\mathbf{b} \times \mathbf{c}) = 0$.

(b) Deduce that $\mathbf{b} \times \mathbf{c}$ is perpendicular to **b** and to **c**.

(c) Using exercise 4(a), show that $\det A = |\mathbf{b} \times \mathbf{c}|$.

(d) Using exercises 4(a) and 4(c) and the fact that **a** is a unit vector perpendicular to **b** and **c**, show that

   $|\det A|$ = area of parallelogram with adjacent edges **b** and **c**.

(e) Deduce from parts (c) and (d) that if θ represents the angle between **b** and **c**, then

   $|\mathbf{b} \times \mathbf{c}| = |\mathbf{b}||\mathbf{c}| \sin \theta$.

CHAPTER SIX

Basis and Dimension

Introduction

At this stage we have a fairly good grasp of such vector spaces as R^2 and R^3, which we can visualize as the plane and space respectively. More elusive are the various subspaces of R^n, which seem to float somewhere within R^n, anchored only by the requirement that subspaces always contain the zero vector. Let us try to get a better grip on vector spaces in general. Our method will be to represent the contents of vector spaces in terms of much smaller, more manageable sets of vectors.

For example, suppose we can express every vector $\mathbf{x}$ in a subspace V of R^3 (see Figure 6.1) as a linear combination of just two nonparallel vectors $\mathbf{u}_1$ and $\mathbf{u}_2$:

$$\mathbf{x} = a_1\mathbf{u}_1 + a_2\mathbf{u}_2 \quad .$$

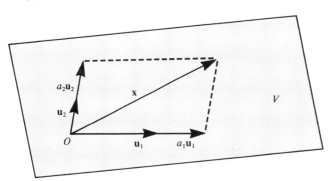

A vector space V interpreted as a plane in space.

Figure 6.1

Then $\mathbf{x}$ is the diagonal of the parallelogram with adjacent sides $a_1\mathbf{u}_1$ and $a_2\mathbf{u}_2$, and so, in terms of Euclidean geometry, $\mathbf{x}$ is always in the same plane as the vectors $\mathbf{u}_1$ and $\mathbf{u}_2$. We conclude that V is a plane within the Euclidean space R^3.

Usually it is easy to find a finite set of vectors in terms of which all the vectors in a given space V can be expressed. Such a set is called a "spanning set." In the case of a vector space of the type $V = \text{Ker}(A)$, in particular, the set of generators for the system $A\mathbf{x} = \mathbf{0}$ serves the purpose.

However, a given spanning set may be uneconomical; it may contain ten elements when two would suffice and thereby convey an inaccurate impression of the size of the space V. An essential objective is therefore to obtain a "minimal" set of vectors that serve as a spanning set for the vector space. Such a set is called a "basis" for V, and the number of elements in a basis can, as it turns out, justifiably be regarded as the "dimension" of V. Furthermore, the constants $a_1, \ldots, a_k$, used to express a vector $\mathbf{x}$ in terms of a basis $\mathbf{u}_1, \ldots, \mathbf{u}_k$, prove to be uniquely determined and can therefore serve as a label for $\mathbf{x}$. In effect, the vectors $\mathbf{x}$ in the difficult space V are indexed by the vectors $\mathbf{a} = (a_1, \ldots, a_k)$ in the simple space R^k.

We begin by studying the intrinsic properties of small sets of vectors in section 6.1. Here the algebraic concept of "linear dependence" sharpens the intuitive notion of a "minimal set" and is supported by a geometric interpretation. An efficient method of determining the dependence or independence of a given set of vectors is presented.

In sections 6.2 and 6.3 the concepts of basis and dimension are developed and applied to the vector spaces of most importance to us. The "Main Dimension Theorem," linking the sizes of the spaces $\text{Ker}(A)$ and Range (A), caps the development of the theory.

The description of vector space transformations in relation to different bases and the use of matrices in such descriptions are the subject of section 6.4. Although of considerable theoretical importance, this last section is not essential for understanding the developments in later chapters, and may be skipped on first reading.

6.1 Linear Independence

As mentioned in the introduction, our objective is to describe many vectors, such as the contents of a vector space V, in terms of a few vectors, say $\mathbf{u}_1, \ldots, \mathbf{u}_k$. And once we have such a set, we want to make sure it is as small as possible while still serving its purpose. Interestingly enough, one can determine whether the set is of minimal size by studying the intrinsic algebraic properties of the set itself, without further reference to the full vector space V.

We therefore begin by discussing the algebraic properties of small (finite) sets of vectors.

Linear Independence

Which linear combinations of the vectors $\mathbf{u}_1 = (2,0)$ and $\mathbf{u}_2 = (0,3)$ add up to $\mathbf{0}$? In other words, for which choices of the constants c_1, c_2 is it true that $c_1\mathbf{u}_1 + c_2\mathbf{u}_2 = \mathbf{0}$? In this case, the equation can be written as

$$c_1 \begin{bmatrix} 2 \\ 0 \end{bmatrix} + c_2 \begin{bmatrix} 0 \\ 3 \end{bmatrix} = \begin{bmatrix} 0 \\ 0 \end{bmatrix} \quad \text{or} \quad \begin{bmatrix} 2c_1 \\ 3c_2 \end{bmatrix} = \begin{bmatrix} 0 \\ 0 \end{bmatrix} \quad,$$

and so we see that $c_1 = 0$, $c_2 = 0$ is the only choice.

But if we take another pair of vectors, $\mathbf{v}_1 = (1,2)$ and $\mathbf{v}_2 = (2,4)$, we can see at a glance that

$$2 \begin{bmatrix} 1 \\ 2 \end{bmatrix} - \begin{bmatrix} 2 \\ 4 \end{bmatrix} = \begin{bmatrix} 0 \\ 0 \end{bmatrix} \quad;$$

therefore the non-zero choice of constants, $c_1 = 2$, $c_2 = -1$, is also valid. This algebraic distinction between different sets of vectors turns out to be of great importance, and is pinpointed in the following definitions.

The vectors $v_1, \ldots, \mathbf{v}_k$ in a vector space V are said to be *linearly independent* (l.i.) if the linear combination $c_1\mathbf{v}_1 + \cdots + c_k\mathbf{v}_k$ equals $\mathbf{0}$ *only* when *all* of the constants $c_1, \ldots, c_k$ equal zero.

Thus, the vectors (2,0) and (0,3) form a linearly independent set.

Also, vectors $\mathbf{v}_1, \ldots, \mathbf{v}_k$ in a vector space V are said to be *linearly dependent* (l.d.) when they are not linearly independent.

The vectors (1,2) and (2,4) furnish an example of an l.d. set, since we found that they do not satisfy the condition for being l.i.

More generally, when vectors $\mathbf{v}_1, \ldots, \mathbf{v}_k$ are l.d., it means that the equation $c_1\mathbf{v}_1 + \cdots + c_k\mathbf{v}_k = \mathbf{0}$ holds for a set of constants $c_1, \ldots, c_k$, at least one of which is not zero. Suppose, for example, that $c_1 \neq 0$. Then we can solve the equation for $\mathbf{v}_1$:

$$\mathbf{v}_1 = \left(\frac{-c_2}{c_1}\right) \mathbf{v}_2 + \cdots + \left(\frac{-c_k}{c_1}\right) \mathbf{v}_k \quad.$$

That is, one of the vectors in an l.d. set can be expressed as a linear combination of the others. Conversely, whenever one vector in a set $\mathbf{v}_1, \ldots, \mathbf{v}_k$ can be expressed as a linear combination of the others, say,

$$\mathbf{v}_1 = d_2\mathbf{v}_2 + \cdots d_k\mathbf{v}_k \quad,$$

then $\quad 1\,\mathbf{v}_1 + (-d_2)\mathbf{v}_2 + \cdots + (-d_k)\mathbf{v}_k = \mathbf{0} \quad,$

and therefore $\mathbf{v}_1, \ldots, \mathbf{v}_k$ form an l.d. set. Let us summarize these observations in a lemma.

Lemma 6.1 (Algebraic interpretation of linear dependence)
Vectors $\mathbf{v}_1, \ldots, \mathbf{v}_k$ in a vector space V are linearly dependent if and only if one of them is a linear combination of the others.

Geometric Interpretation of Dependence

For sets of vectors in R^2 and R^3, geometric interpretations of linear dependence are readily obtained. By lemma 6.1, a set of two vectors, $\mathbf{v}_1$ and $\mathbf{v}_2$, is dependent if $\mathbf{v}_1 = d_2\mathbf{v}_2$ or $\mathbf{v}_2 = d_1\mathbf{v}_1$. This means that $\mathbf{v}_1$ and $\mathbf{v}_2$ are parallel.

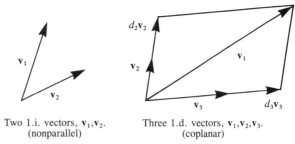

Two 1.i. vectors, $\mathbf{v}_1,\mathbf{v}_2$.
(nonparallel)

Three l.d. vectors, $\mathbf{v}_1,\mathbf{v}_2,\mathbf{v}_3$.
(coplanar)

Figure 6.2

Dependence of three vectors is equivalent to an equation of the type $\mathbf{v}_1 = d_2\mathbf{v}_2 + d_3\mathbf{v}_3$, which means that $\mathbf{v}_1$ is a diagonal of a parallelogram with sides $d_2\mathbf{v}_2$ and $d_3\mathbf{v}_3$ (see Figure 6.2). This implies that $\mathbf{v}_1,\mathbf{v}_2,\mathbf{v}_3$ are all in the plane of this parallelogram and so are certainly in the same plane, or are *coplanar*. The converse, that any three coplanar vectors are dependent, is given as a problem in Euclidean geometry in exercises 6.1. To summarize,

Lemma 6.2 (Geometric interpretation of linear dependence)
Let $V = R^2$ or R^3. Then

(a) two vectors in V form a linearly dependent set if and only if they are parallel;

(b) three vectors in V form a linearly dependent set if and only if they are coplanar.

Independence Test

The dependence or independence of a given set of numerical vectors can be determined quite rapidly by the method provided by the following theorem:

THEOREM 6.3

Suppose the vectors $\mathbf{v}_1, \ldots, \mathbf{v}_n$ in R^m form the columns of the $m \times n$ matrix $A = [\mathbf{v}_1|\ldots|\mathbf{v}_n]$ and that the constants $c_1, \ldots, c_n$ form the vector $\mathbf{c}$ in R^n. Then

(a) linear combinations of the vectors $\mathbf{v}_1, \ldots, \mathbf{v}_n$ can be rewritten in matrix-vector form with the formula

$$c_1\mathbf{v}_1 + \cdots + c_n\mathbf{v}_n = A\mathbf{c} \quad ;$$

(b) the vectors $\mathbf{v}_1, \ldots, \mathbf{v}_n$ are l.i. when rank $(A) = n$
and l.d. when rank $(A) < n$.

Proof (a) Recall, from the beginning of Chapter 4, that each vector can be expressed in terms of the standard vectors by the formula

$$\mathbf{c} = c_1\mathbf{e}_1 + \cdots + c_n\mathbf{e}_n$$

and that, by linearity of matrix transformations,

$$A\mathbf{c} = c_1A\mathbf{e}_1 + \cdots + c_nA\mathbf{e}_n \quad .$$

Recall also, from the beginning of Chapter 3, that

$$A\mathbf{e}_j = (\text{column } j \text{ of } A = \mathbf{v}_j) \quad .$$

Therefore, $A\mathbf{c} = c_1\mathbf{v}_1 + \cdots + c_n\mathbf{v}_n \quad .$

(b) Linear independence means that

$$c_1\mathbf{v}_1 + \cdots + c_n\mathbf{v}_n = \mathbf{0}$$

only when all of the constants $c_1, \ldots, c_n$ are zero.

This is equivalent, by part (a), to the condition

$$A\mathbf{c} = \mathbf{0} \text{ only when } \mathbf{c} = \mathbf{0} \quad .$$

But this in turn is equivalent, in view of the theory of linear equations of Chapter 1, to the condition rank $(A) = n$. Q.E.D.

Example 1 To determine whether the vectors $\mathbf{v}_1 = (1,2,3)$ and $\mathbf{v}_2 = (4,5,6)$ are l.i. or l.d., we set them up as the columns of a matrix A and reduce A far enough to determine its rank:

$$A = [\mathbf{v}_1|\mathbf{v}_2] = \begin{bmatrix} 1 & 4 \\ 2 & 5 \\ 3 & 6 \end{bmatrix} \begin{matrix} \\ R_3 - 2R_1 \\ R_3 - 3R_1 \end{matrix} \begin{bmatrix} 1 & 4 \\ 0 & -3 \\ 0 & -6 \end{bmatrix} \quad .$$

At this stage we can see that rank $(A) = 2 = n$. So $\mathbf{v}_1$, $\mathbf{v}_2$ are an independent set.

Example 2 The vectors $\mathbf{v}_1 = (1,2)$, $\mathbf{v}_2 = (3,7)$, $\mathbf{v}_3 = (5,8)$ are the columns of the matrix

$$A = [\mathbf{v}_1|\mathbf{v}_2|\mathbf{v}_3] = \begin{bmatrix} 1 & 3 & 5 \\ 2 & 7 & 8 \end{bmatrix} \quad ,$$

which we can see at a glance satisfies the condition that rank $(A) = 2 < n = 3$.

Therefore $\mathbf{v}_1$, $\mathbf{v}_2$, $\mathbf{v}_3$ are dependent.

Note that, when the number of vectors given, n, exceeds the number of components they have, m, then necessarily rank $(A) \leq m < n$, and so the vectors are automatically dependent. This reasoning could have been applied to example 2, where we had three vectors with only two components.

Example 3   Suppose a set of vectors in R^m is dependent. To find all sets of constants $c_1, \ldots, c_n$ such that $c_1\mathbf{v}_1 + \cdots + c_n\mathbf{v}_n = \mathbf{0}$, we have only to find all solutions $\mathbf{c}$ of the equivalent equation $A\mathbf{c} = \mathbf{0}$. That is, we simply compute $\text{Ker}(A)$. For the dependent vectors of example 2, the steps are as follows.

$$A = \begin{bmatrix} 1 & 3 & 5 \\ 2 & 7 & 8 \end{bmatrix} \xrightarrow{R_2 - 2R_1} \begin{bmatrix} 1 & 3 & 5 \\ 0 & 1 & -2 \end{bmatrix} \xrightarrow{R_1 - 3R_2} \begin{bmatrix} 1 & 0 & 11 \\ 0 & 1 & -2 \end{bmatrix}.$$

In parametric form, the solution is $\begin{aligned} c_1 &= -11c_3, \\ c_2 &= 2c_3, \end{aligned}$

$$\text{or} \quad \mathbf{c} = \begin{bmatrix} c_1 \\ c_2 \\ c_3 \end{bmatrix} = \begin{bmatrix} -11c_3 \\ 2c_3 \\ c_3 \end{bmatrix} = c_3 \begin{bmatrix} -11 \\ 2 \\ 1 \end{bmatrix}, \quad \text{or } \mathbf{c} = a(-11,2,1).$$

So we must choose $c_1 = -11a$, $c_2 = 2a$, $c_3 = a$, and can then conclude that

$$(-11a)(1,2) + (2a)(3,7) + a(5,8) = \mathbf{0} \quad.$$

The calculation should be checked, of course, by verifying that the left side does add up to $\mathbf{0}$.

Example 4   Let $\mathbf{v}_1 = (1,2,3)$, $\mathbf{v}_2 = (1,3,5)$, $\mathbf{v}_3 = (1,3,6)$, $\mathbf{v}_4 = (2,3,4)$, and suppose we want to determine which, if any, of these vectors can be expressed as a linear combination of the others.

In fact, since we have four vectors in R^3, we know from the start that they must be dependent, and therefore that at least one of them is an l.c. of the others. For more precise information, the essential step is to find all l.c.'s of the four vectors that add to $\mathbf{0}$. We follow the method of the previous example, abbreviating the specification of the row operations to arrows between successive matrices.

$$A = [\mathbf{v}_1|\mathbf{v}_2|\mathbf{v}_3|\mathbf{v}_4] = \begin{bmatrix} 1 & 1 & 1 & 2 \\ 2 & 3 & 3 & 3 \\ 3 & 5 & 6 & 4 \end{bmatrix} \rightarrow \begin{bmatrix} 1 & 1 & 1 & 2 \\ 0 & 1 & 1 & -1 \\ 0 & 2 & 3 & -2 \end{bmatrix}$$

$$\rightarrow \begin{bmatrix} 1 & 0 & 0 & 3 \\ 0 & 1 & 1 & -1 \\ 0 & 0 & 1 & 0 \end{bmatrix} \rightarrow \begin{bmatrix} 1 & 0 & 0 & 3 \\ 0 & 1 & 0 & -1 \\ 0 & 0 & 1 & 0 \end{bmatrix}, \quad \text{or} \quad \begin{cases} c_1 = -3c_4 \\ c_2 = c_4 \\ c_3 = 0 \end{cases}.$$

Therefore $\mathbf{c} = a(-3,1,0,1)$, or $c_1 = -3a$, $c_2 = a$, $c_3 = 0$, $c_4 = a$, and the most general l.c. of the given vectors adding to $\mathbf{0}$ is

$$(-3a)\mathbf{v}_1 + a\mathbf{v}_2 + a\mathbf{v}_4 = \mathbf{0} \quad.$$

We conclude that any of $\mathbf{v}_1$, $\mathbf{v}_2$, or $\mathbf{v}_4$ can be expressed in terms of the other $\mathbf{v}_i$'s, for example, $\mathbf{v}_2 = 3\mathbf{v}_1 - \mathbf{v}_4$. But $\mathbf{v}_3$ is *not* an l.c. of $\mathbf{v}_1$, $\mathbf{v}_2$, $\mathbf{v}_4$.

■ **Problem**   For which choices of the parameter p do the vectors

$$\mathbf{v}_1 = (1,2,3), \quad \mathbf{v}_2 = (2,6,9,), \quad \mathbf{v}_3 = (p,2,4)$$

form a linearly independent set?

■ **Solution** The matrix whose columns are the given vectors will necessarily contain the parameter p. However, we are accustomed,
from Chapter 1, to dealing with the reduction of such matrices.

$$A = [\mathbf{v}_1|\mathbf{v}_2|\mathbf{v}_3] = \begin{bmatrix} 1 & 2 & p \\ 2 & 6 & 2 \\ 3 & 9 & 4 \end{bmatrix} \rightarrow \begin{bmatrix} 1 & 2 & p \\ 0 & 2 & 2-2p \\ 0 & 3 & 4-3p \end{bmatrix} \rightarrow \begin{bmatrix} 1 & 2 & p \\ 0 & 2 & 2-2p \\ 0 & 0 & 1 \end{bmatrix}.$$

Thus rank $(A) = 3 = n$ regardless of the value of p. So the vectors $\mathbf{v}_1$, $\mathbf{v}_2$, $\mathbf{v}_3$
are always l.i.

■ EXERCISES 6.1

1. Determine whether each of the following sets of vectors is linearly independent or linearly dependent. In the case of a dependent set, find all
   linear combinations of the vectors that add to **0**.

   (a) $\mathbf{v}_1 = (1,2,1,4)$, $\mathbf{v}_2 = (2,1,8,0)$, $\mathbf{v}_3 = (3,5,1,1)$
   (b) $\mathbf{v}_1 = (-4,4,1)$, $\mathbf{v}_2 = (1,-1,2)$, $\mathbf{v}_3 = (2,-2,1)$
   (c) $\mathbf{v}_1 = (1,2)$, $\mathbf{v}_2 = (3,5)$
   (d) $\mathbf{v}_1 = (1,1)$, $\mathbf{v}_2 = (2,3)$, $\mathbf{v}_3 = (3,4)$
   (e) $\mathbf{v}_1 = (1,0,1)$, $\mathbf{v}_2 = (0,1,2)$, $\mathbf{v}_3 = (2,0,3)$
   (f) $\mathbf{v}_1 = (2,0,4,3)$, $\mathbf{v}_2 = (4,0,8,6)$

2. Show that, for any choice of the parameter k, the vectors

   $$\mathbf{v}_1 = (1,-2,1), \mathbf{v}_2 = (-2,4,3), \mathbf{v}_3 = (1,1,k)$$

   form a linearly independent set.

3. (a) Show that there is just one choice of the parameter p for which the
       vectors

   $$\mathbf{v}_1 = (2,3,5,-1), \mathbf{v}_2 = (1,1,4,0), \mathbf{v}_3 = (2,p,-1,-3)$$

   are linearly independent.
   (b) For the above choice of p, express $\mathbf{v}_3$ as a linear combination of $\mathbf{v}_1$
       and $\mathbf{v}_2$.

4. Show that if the vectors $\mathbf{v}_1$, $\mathbf{v}_2$, $\mathbf{v}_3$ in a vector space V form an l.i. set,
   then the vectors $\mathbf{v}_1$, $\mathbf{v}_2$ also form an l.i. set.

5. (a) Show that if the vectors $\mathbf{v}_1, \ldots, \mathbf{v}_k$ in R^n form an l.d. set and A is
       an $m \times n$ matrix, then the vectors $A\mathbf{v}_1, \ldots, A\mathbf{v}_n$ also form an l.d.
       set.
   (b) Show that if the vectors $\mathbf{v}_1, \ldots, \mathbf{v}_k$ in R^n form an l.i. set and A is
       an invertible $n \times n$ matrix, then the vectors $A\mathbf{v}_1, \ldots, A\mathbf{v}_n$ also form
       an l.i. set.

6. Show that if the vectors **u**,**v** in a vector space V form an l.i. set, then so do **u**, **u** + **v**.

7. Show that if the vector $\mathbf{u}_0 \neq \mathbf{0}$ in a vector space V can be expressed as a linear combination of the vectors $\mathbf{v}_1$, $\mathbf{v}_2$ in V, and also as a linear combination of the vectors $\mathbf{v}_3$, $\mathbf{v}_4$ in V, then $\mathbf{v}_1$, $\mathbf{v}_2$, $\mathbf{v}_3$, $\mathbf{v}_4$ form an l.d. set.

6.2 Spanning Sets and Bases

The vector spaces R^n are relatively easy to understand, and one way of explaining this is to say that their vectors are linear combinations (l.c.'s) of the rather simple vectors $\mathbf{e}_1, \ldots, \mathbf{e}_n$. Our next objective is to obtain comparable representations for other vector spaces, and to make these representations just as convenient and useful as in the exemplary case of R^n.

In describing the contents of an arbitrary subspace V of R^n, we cannot expect to be able to use the vectors $\mathbf{e}_1, \ldots, \mathbf{e}_n$. In most cases, these vectors will not even be in the space V. Thus the plane V within R^3 sketched in the chapter introduction tilts obliquely in space and may very well not contain any of the coordinate axes along which the vectors $\mathbf{e}_1$, $\mathbf{e}_2$, $\mathbf{e}_3$ point.

Instead, we have to find vectors within V that take over the role that $\mathbf{e}_1, \ldots, \mathbf{e}_n$ play in the case of the space R^n. The role to be filled is spelled out in the definition of ''spanning set.'' In various examples, we find that these sets are easy to obtain. We then proceed to identify a distinguished category of spanning sets, called bases, which are at once economical and induce a systematic and convenient labeling of the contents of a space.

Spanning Set

A set of vectors $\mathbf{v}_1, \ldots, \mathbf{v}_k$ in a vector space V is said to *span* V if every vector **x** in V is a linear combination of them, that is,

$$\mathbf{x} = c_1\mathbf{v}_1 + \cdots + c_k\mathbf{v}_k \quad,$$

for some set of constants $c_1, \ldots, c_k$.

Example 1 The vector space R^n is spanned by the standard vectors $\mathbf{e}_1, \ldots, \mathbf{e}_n$. In fact, given **x** in R^3, say, we have

$$\mathbf{x} = (x_1,x_2,x_3) = x_1 (1,0,0) + x_2 (0,1,0) + x_3 (0,0,1)$$

$$= x_1\mathbf{e}_1 + x_2\mathbf{e}_2 + x_3\mathbf{e}_3 \quad.$$

Therefore **x** is an l.c. of $\mathbf{e}_1$, $\mathbf{e}_2$, $\mathbf{e}_3$, and the constants in the l.c. are x_1, x_2, x_3.

Example 2   Let V be the kernel of an $m \times n$ matrix transformation A, or $V = \text{Ker}(A)$. In Chapter 2, V was shown to be a vector space and the elements of V were shown to have the form

$$\mathbf{x} = b_1 \mathbf{u}_1 + \cdots + b_k \mathbf{u}_k \quad ,$$

where $\mathbf{u}_1, \ldots, \mathbf{u}_k$ are the generating solutions of the system $A\mathbf{x} = \mathbf{0}$. These generating solutions therefore span $\text{Ker}(A)$.

Example 3   As a specific instance of the previous example, consider the following matrix and its reduced form R:

$$A = \begin{bmatrix} 1 & 3 & 0 & 5 \\ 0 & 0 & 1 & 2 \\ 0 & 0 & 2 & 4 \end{bmatrix} \quad, \quad R = \begin{bmatrix} 1 & 3 & 0 & 5 \\ 0 & 0 & 1 & 2 \\ 0 & 0 & 0 & 0 \end{bmatrix} \quad.$$

The general solution is, in parametric form,

$$x_1 = -3x_2 - 5x_4$$

$$x_3 = \qquad -2x_4 \quad,$$

or, in vector form,

$$\mathbf{x} = \begin{bmatrix} x_1 \\ x_2 \\ x_3 \\ x_4 \end{bmatrix} = \begin{bmatrix} -3x_2 - 5x_4 \\ x_2 \\ -2x_4 \\ x_4 \end{bmatrix} = x_2 \begin{bmatrix} -3 \\ 1 \\ 0 \\ 0 \end{bmatrix} + x_4 \begin{bmatrix} -5 \\ 0 \\ -2 \\ 1 \end{bmatrix} \quad.$$

That is, the generating solutions are $\mathbf{u}_1 = (-3,1,0,0)$ and $\mathbf{u}_2 = (-5,0,-2,1)$, the general solution is $\mathbf{x} = b_1 \mathbf{u}_1 + b_2 \mathbf{u}_2$, and the generating solutions span $\text{Ker}(A)$.

Example 4   Let V be the range of an $m \times n$ matrix transformation A, or $V = \text{Range}(A)$. This was shown to be a vector space in Chapter 2. Now the range of A consists of all vectors of the form $A\mathbf{x}$, and by linearity of A,

$$A\mathbf{x} = x_1 A\mathbf{e}_1 + \cdots + x_n A\mathbf{e}_n \quad.$$

In other words, every vector of the form $A\mathbf{x}$ is an l.c. of the vectors $A\mathbf{e}_1, \ldots, A\mathbf{e}_n$, which therefore span $\text{Range}(A)$. But since $A\mathbf{e}_j$ is simply column j of the matrix A, as shown in lemma 3.1, we conclude that $\text{Range}(A)$ is spanned by the columns of A.

Example 5   As a concrete case of example 4, consider again the 3×4 matrix

$$A = \begin{bmatrix} 1 & 3 & 0 & 5 \\ 0 & 0 & 1 & 2 \\ 0 & 0 & 2 & 4 \end{bmatrix} \quad.$$

Here we can say that the vector space $V = \text{Range}(A)$ is spanned by the four columns of A:

$$\mathbf{q}_1 = (1,0,0), \ \mathbf{q}_2 = (3,0,0), \ \mathbf{q}_3 = (0,1,2), \ \mathbf{q}_4 = (4,2,4) \quad.$$

Basis

We have seen that the contents of a vector space are conveniently described within the framework of a spanning set. The next step is to find a systematic way of labeling individual vectors in the space.

To illustrate the objective, let $V = $ Range (A), where A is the 3×4 numerical matrix just studied in example 5. The vector $\mathbf{x} = (3,4,8)$ is in V; in fact:

$$(3,4,8) = 3(1,0,0) + 4(0,1,2) = 3\mathbf{q}_1 + 4\mathbf{q}_3 \quad,$$

or $\quad \mathbf{x} = 3\mathbf{q}_1 + 0\mathbf{q}_2 + 4\mathbf{q}_3 + 0\mathbf{q}_4 \quad.$

So we would like to use the set of constants $3,0,4,0$ in the linear combination as a label for the vector $\mathbf{x}$ in V. We encounter a problem, though, in that $\mathbf{x}$ can also be written as

$$(3,4,8) = -7(1,0,0) + 2(5,2,4) = -7\mathbf{q}_1 + 2\mathbf{q}_4 \quad,$$

or $\quad \mathbf{x} = -7\mathbf{q}_1 + 0\mathbf{q}_2 + 0\mathbf{q}_3 + 2\mathbf{q}_4 \quad.$

That is, the alternative set of constants, $-7,0,0,2$, would have just as good a claim to be the label for $\mathbf{x}$.

It is better to arrange things so that multiple labels do not occur. The next definition accomplishes this.

A set of vectors $\mathbf{v}_1, \ldots, \mathbf{v}_k$ in a vector space V is said to form a *basis* for V if every vector $\mathbf{x}$ in V can be expressed as a linear combination of them in exactly one way. That is,

$$\mathbf{x} = c_1\mathbf{v}_1 + \cdots + c_k\mathbf{v}_k$$

for a unique choice of the constants $c_1, \ldots, c_k$.

In this case, the constants $c_1, \ldots, c_k$ are said to be the *coordinates of* $\mathbf{x}$ *relative to the basis* $\mathbf{v}_1, \ldots, \mathbf{v}_k$.

Example 6   To see that the spanning set $\mathbf{e}_1, \ldots, \mathbf{e}_n$ forms a basis for R^n, suppose that

$$\mathbf{x} = g_1\mathbf{e}_1 + \cdots + g_n\mathbf{e}_n = h_1\mathbf{e}_1 + \cdots h_n\mathbf{e}_n \quad.$$

Then $\mathbf{x} = (g_1, \ldots, g_n) = (h_1, \ldots, h_n)$, and so $g_1 = h_1, \ldots, g_n = h_n$. The choice of constants in the representation of $\mathbf{x}$ is therefore unique, as was required.

We will refer to the basis $\mathbf{e}_1, \ldots, \mathbf{e}_n$ as the *standard basis* for R^n, and each element $\mathbf{e}_i$ will now be called a *standard basis vector*.

The most efficient method for determining whether a spanning set forms a basis is derived from the following theorem.

THEOREM 6.4

Suppose the vectors $\mathbf{v}_1, \ldots, \mathbf{v}_k$ span the vector space V. Then these vectors form a basis for V if and only if they are linearly independent.

■Proof First, suppose $\mathbf{v}_1, \ldots, \mathbf{v}_k$ form a basis for V. Since the *unique* representation of the $\mathbf{0}$ vector must be

$$\mathbf{0} = 0\mathbf{v}_1 + \cdots + 0\mathbf{v}_k \quad,$$

therefore, $\mathbf{0} \neq c_1\mathbf{v}_1 + \cdots + c_k\mathbf{v}_k$

when any of the constants c_i is nonzero. That is, the vectors $\mathbf{v}_1, \ldots, \mathbf{v}_k$ are independent.

Conversely, suppose $\mathbf{v}_1, \ldots, \mathbf{v}_k$ are independent. Then if we have two representations of $\mathbf{x}$, such as

$$\mathbf{x} = g_1\mathbf{v}_1 + \cdots + g_k\mathbf{v}_k$$

and $\mathbf{x} = h_1\mathbf{v}_1 + \cdots + h_k\mathbf{v}_k \quad,$

subtraction of equations gives us

$$\mathbf{0} = (g_1 - h_1)\mathbf{v}_1 + \cdots + (g_k - h_k)\mathbf{v}_k \quad.$$

The independence of the vectors $\mathbf{v}_1, \ldots, \mathbf{v}_k$ implies that $g_i - h_i = 0$ for each index i, or $g_i = h_i$, and so the two representations of $\mathbf{x}$ are in fact the same. That is, the representation of each vector $\mathbf{x}$ is unique, and $\mathbf{v}_1, \ldots, \mathbf{v}_k$ form a basis for V. Q.E.D.

A shorthand summary of this theorem would be

span + l.i. = basis.

Example 7 We found, in example 3, that the kernel of a specific 3×4 matrix A was spanned by the generating solutions $\mathbf{u}_1 = (-3,1,0,0)$, $\mathbf{u}_2 = (-5,0,-2,1)$. To check on their linear independence, we set them up as the columns of a matrix:

$$B = [\mathbf{u}_1|\mathbf{u}_2] = \begin{bmatrix} -3 & -5 \\ 1 & 0 \\ 0 & -2 \\ 0 & 1 \end{bmatrix} \rightarrow \begin{bmatrix} -3 & -5 \\ 0 & \frac{5}{3} \\ 0 & -2 \\ 0 & 1 \end{bmatrix} \quad.$$

Since rank $(B) = 2 = n$, therefore $\mathbf{u}_1$, $\mathbf{u}_2$ are l.i., and, by theorem 6.4, form a basis for Ker(A).

Example 8 Let A be the matrix of example 3 once again. Suppose we want to find the coordinates of the vector $\mathbf{u} = (3,4,6,-3)$ in Ker(A) relative to the basis $\mathbf{u}_1$, $\mathbf{u}_2$ just found for Ker(A). This amounts to finding constants c_1, c_2 such that

$$c_1\mathbf{u}_1 + c_2\mathbf{u}_2 = \mathbf{u} \quad,$$

or, with $B = [\mathbf{u}_1|\mathbf{u}_2]$, and $\mathbf{c} = (c_1,c_2)$,

$$B\mathbf{c} = \mathbf{u} \quad.$$

We solve for **c** by reducing the augmented matrix:

$$[B|\mathbf{u}] = [\mathbf{u}_1|\mathbf{u}_2|\mathbf{u}] = \begin{bmatrix} -3 & -5 & 3 \\ 1 & 0 & 4 \\ 0 & -2 & 6 \\ 0 & 1 & -3 \end{bmatrix} \rightarrow \begin{bmatrix} 1 & 0 & 4 \\ -3 & -5 & 3 \\ 0 & -2 & 6 \\ 0 & 1 & -3 \end{bmatrix}$$

$$\rightarrow \begin{bmatrix} 1 & 0 & 4 \\ 0 & -5 & 15 \\ 0 & -2 & 6 \\ 0 & 1 & -3 \end{bmatrix} \rightarrow \begin{bmatrix} 1 & 0 & 4 \\ 0 & 1 & -3 \\ 0 & -2 & 6 \\ 0 & 1 & -3 \end{bmatrix} \rightarrow \begin{bmatrix} 1 & 0 & 4 \\ 0 & 1 & -3 \\ 0 & 0 & 0 \\ 0 & 0 & 0 \end{bmatrix}, \text{ or } \begin{matrix} c_1 = 4 \\ c_2 = -3 \end{matrix}.$$

To check, we compute $4\mathbf{u}_1 - 3\mathbf{u}_2$:

$$4(-3,1,0,0) - 3(-5,0,-2,1) = (3,4,6,-3) = \mathbf{u} \quad.$$

Bases for R^n

The standard basis for R^2 consists of the two unit vectors $\mathbf{e}_1$, $\mathbf{e}_2$ along the Cartesian coordinate axes used in analytic geometry. Since these axes are arbitrarily placed in the plane, it is reasonable to guess that any two perpendicular unit vectors should serve just as well as a basis for R^2. In fact, since linear independence is a crucial feature of basis vectors, one might conjecture that any two nonparallel vectors would serve as a basis for R^2.

Not only is this easily proved true, but moreover the corresponding guess for R^n is also valid.

THEOREM 6.5

(a) Any n linearly independent vectors $\mathbf{v}_1, \ldots, \mathbf{v}_n$ in R^n form a basis for R^n.

(b) The coordinates $c_1, \ldots, c_n$ of a vector $\mathbf{u}$ in R^n relative to this basis are given by the formula

$$\mathbf{c} = A^{-1}\mathbf{u}, \text{ where } A = [\mathbf{v}_1| \ldots |\mathbf{v}_n] \quad.$$

■**Proof**   If $\mathbf{v}_1, \ldots, \mathbf{v}_n$ are l.i., then the $n \times n$ matrix $A = [\mathbf{v}_1| \ldots |\mathbf{v}_n]$ has rank n and is therefore invertible. Also, any representation of a vector $\mathbf{u}$ in R^n as a linear combination of $\mathbf{v}_1, \ldots, \mathbf{v}_n$, such as

$$c_1\mathbf{v}_1 + \cdots + c_n\mathbf{v}_n = \mathbf{u} \quad,$$

implies that $A\mathbf{c} = \mathbf{u}$,   or $\mathbf{c} = A^{-1}\mathbf{u}$.

That is, the choice of constants $c_1, \ldots, c_n$ in the representation of each vector $\mathbf{u}$ is always unique, and so $\mathbf{v}_1, \ldots, \mathbf{v}_n$ form a basis for R^n.   Q.E.D.

Example 9   Suppose we must show that $v_1 = (1,2)$ and $v_2 = (3,7)$ form a basis for R^2 and must find the coordinates of $u = (1,1)$ relative to this basis.

First, we compute the inverse of $A = [v_1 | v_2]$ in the usual way:

$$[A|I] = [v_1 | v_2 | I] = \begin{bmatrix} 1 & 3 & 1 & 0 \\ 2 & 7 & 0 & 1 \end{bmatrix} \rightarrow \begin{bmatrix} 1 & 3 & 1 & 0 \\ 0 & 1 & -2 & 1 \end{bmatrix}$$

$$\rightarrow \begin{bmatrix} 1 & 0 & 7 & -3 \\ 0 & 1 & -2 & 1 \end{bmatrix} = [I|A^{-1}] \quad .$$

Since A^{-1} exists, v_1, v_2 are l.i. and form a basis for R^2. The coordinates of $u = (1,1)$ are then obtained from the formula

$$c = A^{-1}u = \begin{bmatrix} 7 & -3 \\ -2 & 1 \end{bmatrix} \begin{bmatrix} 1 \\ 1 \end{bmatrix} = \begin{bmatrix} 4 \\ -1 \end{bmatrix} \quad .$$

That is, $u = 4u_1 - u_2$. (Check this.)

Note that examples 8 and 9 illustrate two different methods of finding the coordinates of a vector u in a space V, relative to a basis for V. The augmented matrix method of example 8 applies when V is any subspace of R^n. The inverse matrix method of example 9 applies only when $V = R^n$ itself.

■ **Problem**   Suppose a certain subspace V of R^3 is known to have the basis

$$v_1 = (1,2,2), \ v_2 = (3,7,6) \quad .$$

Can we determine whether $x = (2,3,0)$ is in V?

More generally, can we determine which vectors with third component zero are in V?

■ **Solution**   We know nothing about V, except that it has the basis v_1, v_2. But by direct reference to the definition of basis, we can see that x is in V if and only if x can be expressed as a linear combination of the basis elements. That is, there must be constants c_1, c_2 such that $c_1 v_1 + c_2 v_2 = x$.

We solve for the coordinates c_1, c_2 by the usual augmented matrix method:

$$[v_1 | v_2 | x] = \begin{bmatrix} 1 & 3 & 2 \\ 2 & 7 & 3 \\ 2 & 6 & 0 \end{bmatrix} \rightarrow \begin{bmatrix} 1 & 3 & 2 \\ 0 & 1 & -1 \\ 0 & 0 & -4 \end{bmatrix} \quad .$$

At this point we can see that the system is inconsistent, and so no coordinates for x exist, and x is not in V.

Observe that there is no danger of starting with a vector outside V and computing false coordinates for it. On the contrary, the fact that x is outside V can be deduced from the outcome of the calculation.

More generally, the vector $x = (a,b,0)$, which has third component zero, is in V when a and b are chosen so as to avoid inconsistency:

$$[\mathbf{v}_1|\mathbf{v}_2|\mathbf{x}] = \begin{bmatrix} 1 & 3 & a \\ 2 & 7 & b \\ 2 & 6 & 0 \end{bmatrix} \rightarrow \begin{bmatrix} 1 & 3 & a \\ 0 & 1 & b-2a \\ 0 & 0 & -2a \end{bmatrix} \cdot$$

That is, we must choose $a = 0$. Then $\mathbf{x} = (0,b,0) = b\mathbf{e}_2$. Thus the only vectors in V that have third component zero are those parallel to $\mathbf{e}_2$.

■ EXERCISES 6.2

1. Determine whether the following sets of vectors form bases for R^2. If a set forms a basis, find the coordinates of $\mathbf{v} = (8,7)$ relative to this basis.

   (a) $\mathbf{v}_1 = (1,2)$, $\mathbf{v}_2 = (3,5)$.
   (b) $\mathbf{v}_1 = (3,5)$, $\mathbf{v}_2 = (6,10)$.

2. Determine whether the following sets of vectors form bases for R^3. If a set forms a basis, find the coordinates of $\mathbf{v} = (1,2,3)$ relative to this basis.

   (a) $\mathbf{v}_1 = (-1,0,1)$, $\mathbf{v}_2 = (2,1,1)$, $\mathbf{v}_3 = (3,1,1)$.
   (b) $\mathbf{v}_1 = (1,3,-1)$, $\mathbf{v}_2 = (2,1,1)$, $\mathbf{v}_3 = (-4,3,-5)$.
   (c) $\mathbf{v}_1 = (1,0,0)$, $\mathbf{v}_2 = (1,1,0)$, $\mathbf{v}_3 = (1,1,1)$.

3. Let $A = \begin{bmatrix} 1 & 2 & -3 \\ 2 & 4 & -6 \end{bmatrix} \cdot$

   (a) Show that the generating solutions $\mathbf{u}_1$, $\mathbf{u}_2$ for the system $A\mathbf{x} = \mathbf{0}$ are l.i. and therefore form a basis for $\text{Ker}(A)$.
   (b) Determine whether the vector $\mathbf{v} = (5,2,3)$ is in $\text{Ker}(A)$, and if so find its coordinates relative to the basis $\mathbf{u}_1$, $\mathbf{u}_2$ found above.
   (c) Repeat (b) for the vector $\mathbf{v} = (4,7,8)$.

4. Let $A = \begin{bmatrix} 1 & 2 \\ 0 & 2 \\ 3 & 4 \end{bmatrix} \cdot$

   (a) Show that the two columns of A, $\mathbf{u}_1 = (1,0,3)$, $\mathbf{u}_2 = (2,2,4)$, are l.i. and therefore form a basis for Range (A).
   (b) Determine whether the vector $\mathbf{v} = (0,-4,4)$ is in Range (A), and if so find its coordinates relative to the basis $\mathbf{u}_1$, $\mathbf{u}_2$.
   (c) Repeat (b) for the vector $\mathbf{v} = (3,0,6)$.

5. Suppose that the vectors $\mathbf{v}_1, \ldots, \mathbf{v}_n$ form a basis for a vector space V. Suppose that, relative to this basis, the vectors $\mathbf{u}$ and $\mathbf{v}$ in V have coordinates $\mathbf{c} = (c_1, \ldots, c_n)$ and $\mathbf{d} = (d_1, \ldots, d_n)$ respectively.

   (a) Show that the coordinates of $t\mathbf{u}$ are $t\mathbf{c} = (tc_1, \ldots, tc_n)$.

(b) Show that the coordinates of $\mathbf{u} + \mathbf{v}$ are $\mathbf{c} + \mathbf{d} = (c_1 + d_1, \ldots, c_n + d_n)$.

(c) Let T be the map of V into R^n that sends each vector $\mathbf{v}$ to its coordinates $\mathbf{c}$, and let S be the map of R^n into V that sends each set of coordinates $\mathbf{c}$ to the vector $\mathbf{v} = c_1\mathbf{v}_1 + \cdots + c_n\mathbf{v}_n$.

   (i) Show that the maps T and S are linear.

  (ii) Show that $T \circ S$ is the identity map of R^n into R^n and that $S \circ T$ is the identity map of V into V.

Note: When a map T of V into W and a map S of W into V have properties (i) and (ii), the vector spaces V and W are said to be *isomorphic*. In particular, if a vector space has a basis with n elements, it is isomorphic to R^n.

6. Suppose vectors $\mathbf{v}_1, \ldots, \mathbf{v}_k$ span a vector space V but are linearly dependent. For example, suppose $\mathbf{v}_k$ is a linear combination of $\mathbf{v}_1, \ldots, \mathbf{v}_{k-1}$.

(a) Show that $\mathbf{v}_1, \ldots, \mathbf{v}_{k-1}$ span V.

(b) Deduce that, by successive deletion of suitably chosen elements, a subset of the original spanning set can be found that is linearly independent and still spans V.

(c) Deduce that a basis for a vector space V can be selected from among the elements of any finite set that spans V.

7. Let S be the subset of R^3 consisting of all vectors $\mathbf{x}$ that are perpendicular to $\mathbf{g} = (1,2,4)$.

(a) Show that S is a subspace of R^3.

(b) Find a basis for S.

8. Let $\mathbf{g}_1 = (1,0,3,0,4)$, $\mathbf{g}_2 = (2,1,1,3,5)$. Also, let S be the subset of R^5 consisting of all vectors $\mathbf{x}$ such that $\mathbf{g}_1 \cdot \mathbf{x} = \mathbf{g}_2 \cdot \mathbf{x} = 0$.

(a) Show that S is a subspace of R^5.

(b) Find a basis for S.

6.3 Dimension

In geometry, the dimension of a space depends on the number of coordinate axes needed to measure the location of a point. Thus the line, or R^1, has dimension 1; the plane, or R^2, has dimension 2, and so on. But the coordinate axes in R^n are themselves specified by the n standard basis vectors. Our geometric notion of dimension can therefore be transferred to the context of vector spaces by taking the number of elements in a basis as the dimension of a vector space.

This algebraic definition greatly widens the scope of the dimension concept. However, we have seen that a vector space can have a great many different bases. We must verify that all of them have the same number of elements to be sure that the dimension of a space is uniquely determined.

To this end, we begin with a detailed comparison of the coordinates of a vector in relation to different bases. We have to assume, initially, that the different bases have different numbers of elements, but the logic of coordinate transformations leads to the conclusion, in theorem 6.7, that these numbers are in fact the same. (This result may be, and often is, assumed on first reading.)

From this point, formulas are developed for the dimensions of the most important categories of vector spaces. Linkage between the dimensions of some of these spaces is to be expected. For example, the transformation given by $m \times n$ matrix A maps every vector somewhere, and the more vectors mapped to $\mathbf{0}$, the less there are to be mapped elsewhere. So the dimensions of the spaces $\text{Ker}(A)$ and $\text{Range}(A)$ should be complementary, and this is exactly the subject of the "Main Dimension Theorem."

Coordinates Relative to Different Bases

We have seen that any two nonparallel vectors form a basis for R^2 and that other spaces have many bases as well. Furthermore, we shall see in subsequent chapters that it is often necessary to work with more than one basis for a given vector space at a time. It is therefore important to know how the coordinates of a vector relative to different bases are related.

The vector $\mathbf{u} = (1,1)$ in R^2, for example, has coordinates $\mathbf{u}_1 = 1$, $\mathbf{u}_2 = 1$ relative to the standard basis $\mathbf{e}_1, \mathbf{e}_2$. But, as we saw in example 9 of the previous section, it also has coordinates $c_1 = 4$, $c_2 = -1$ relative to the basis $\mathbf{v}_1 = (1,2)$, $\mathbf{v}_2 = (3,7)$. We also constructed an invertible 2×2 matrix A, which mediates between the two sets of coordinates

$$\mathbf{c} = A^{-1}\mathbf{u} \text{ and } \mathbf{u} = A\mathbf{c} \quad .$$

Let us try to extend this observation to other cases. Suppose, for example, that $\mathbf{v}_1$, $\mathbf{v}_2$, $\mathbf{v}_3$ form a basis for a vector space V and that we know the coordinates of certain vectors $\mathbf{z}_1$, $\mathbf{z}_2$ in V, relative to this basis:

$$\mathbf{z}_1 = p_{11}\mathbf{v}_1 + p_{21}\mathbf{v}_2 + p_{31}\mathbf{v}_3 \quad ,$$

$$\mathbf{z}_2 = p_{12}\mathbf{v}_1 + p_{22}\mathbf{v}_2 + p_{32}\mathbf{v}_3 \quad .$$

Note that it is necessary to double index the constants p_{ij} since they are related both to the $\mathbf{v}_i$'s and the $\mathbf{z}_j$'s.

Suppose also that we know how to express a certain vector $\mathbf{x}$ in V as a linear combination of the $\mathbf{z}_j$'s, say,

$$\mathbf{x} = d_1\mathbf{z}_1 + d_2\mathbf{z}_2 \quad .$$

Then we can deduce the coordinates c_1, c_2, c_3 of $\mathbf{x}$ relative to the basis $\mathbf{v}_1$, $\mathbf{v}_2$, $\mathbf{v}_3$ as follows:

$$\begin{aligned} \mathbf{x} &= d_1(p_{11}\mathbf{v}_1 + p_{21}\mathbf{v}_2 + p_{31}\mathbf{v}_3) \\ &\quad + d_2(p_{12}\mathbf{v}_1 + p_{22}\mathbf{v}_2 + p_{32}\mathbf{v}_3) \\ &= \quad (p_{11}d_1 + p_{12}d_2)\mathbf{v}_1 \\ &\quad + (p_{21}d_1 + p_{22}d_2)\mathbf{v}_2 \\ &\quad + (p_{31}d_1 + p_{32}d_2)\mathbf{v}_3 \quad . \end{aligned}$$

That is, the first coordinate is $c_1 = p_{11}d_1 + p_{12}d_2$, etc., or

$$c_i = p_{i1}d_1 + p_{i2}d_2 \quad .$$

The last formula can be rewritten in matrix-vector terms as

$$\mathbf{c} = P\mathbf{d} \quad ,$$

where P is the 3×2 matrix with entries P_{ij}.

Now if $\mathbf{z}_1$, $\mathbf{z}_2$ were also a basis for V, then $\mathbf{d} = (d_1\ d_2)$ would be the coordinates of $\mathbf{x}$ relative to this basis and we could show in the same way as above that $\mathbf{d} = Q\mathbf{c}$, where Q is a 2×3 matrix.

Of course these calculations did not depend in any essential way on the fact that the two bases used for illustration had two and three elements respectively. Therefore our conclusions may be summarized in the following lemma.

Lemma 6.6 Suppose a vector space V has bases $\mathbf{v}_1, \ldots, \mathbf{v}_m$ and $\mathbf{z}_1, \ldots, \mathbf{z}_n$, so that a vector $\mathbf{x}$ in V has two sets of coordinates, $\mathbf{c}$ in R^m and $\mathbf{d}$ in R^n, relative to the two bases respectively.

Then the two sets of coordinates are related by the formulas

$$\mathbf{c} = P\mathbf{d} \text{ and } \mathbf{d} = Q\mathbf{c} \quad ,$$

where P is an $m \times n$ matrix and Q an $n \times m$ matrix.

<u>Example 1</u> The vectors $\mathbf{v}_1 = (1,2)$, $\mathbf{v}_2 = (-1,4)$, and $\mathbf{z}_1 = (7,-4)$, $\mathbf{z}_2 = (5,-2)$ form two sets of nonparallel vectors in R^2, and therefore two bases for R^2. To construct the matrix P of the above lemma, note that

$$\mathbf{z}_1 = p_{11}\mathbf{v}_1 + p_{21}\mathbf{v}_2 \text{ or } [\mathbf{v}_1|\mathbf{v}_2] \begin{bmatrix} p_{11} \\ p_{21} \end{bmatrix} = \mathbf{z}_1 \quad .$$

To solve for p_{11}, p_{21}, we reduce the augmented matrix:

$$[\mathbf{v}_1|\mathbf{v}_2|\mathbf{z}_1] = \begin{bmatrix} 1 & -1 & 7 \\ 2 & 4 & -4 \end{bmatrix} \rightarrow \begin{bmatrix} 1 & 0 & 4 \\ 0 & 1 & -3 \end{bmatrix} , \text{ or } \begin{matrix} p_{11} = & 4 \\ p_{21} = & -3 \end{matrix} \quad .$$

Also, $\quad \mathbf{z}_2 = p_{12}\mathbf{v}_1 + p_{22}\mathbf{v}_2$, or $[\mathbf{v}_1|\mathbf{v}_2] \begin{bmatrix} p_{12} \\ p_{22} \end{bmatrix} = \mathbf{z}_2$,

$$[\mathbf{v}_1|\mathbf{v}_2|\mathbf{z}_2] = \begin{bmatrix} 1 & -1 & 5 \\ 2 & 4 & -2 \end{bmatrix} \rightarrow \begin{bmatrix} 1 & 0 & 3 \\ 0 & 1 & -2 \end{bmatrix} , \text{ or } \begin{matrix} p_{12} = & 3 \\ p_{22} = & -2 \end{matrix} \quad .$$

We conclude that $\quad P = \begin{bmatrix} p_{11} & p_{12} \\ p_{21} & p_{22} \end{bmatrix} = \begin{bmatrix} 4 & 3 \\ -3 & -2 \end{bmatrix} \quad .$

Example 2 Suppose we are given that the vector $\mathbf{x}$ in R^2 has coordinates $d_1 = 6$, $d_2 = -3$ relative to the basis $\mathbf{z}_1$, $\mathbf{z}_2$ of example 1. To obtain the coordinates of $\mathbf{x}$ relative to the other basis $\mathbf{v}_1$, $\mathbf{v}_2$, we compute

$$\mathbf{c} = P\mathbf{d} = \begin{bmatrix} 4 & 3 \\ -3 & -2 \end{bmatrix} \begin{bmatrix} 6 \\ -3 \end{bmatrix} = \begin{bmatrix} 15 \\ -12 \end{bmatrix} \text{, or } \begin{array}{l} c_1 = 15 \\ c_2 = -12 \end{array}.$$

To check, we can compute $\mathbf{x}$ in two ways:

$$\mathbf{x} = d_1\mathbf{z}_1 + d_2\mathbf{z}_2 = 6(7,-4) - 3(5,-2) = (27,-18) \quad ;$$

$$\mathbf{x} = c_1\mathbf{v}_1 + c_2\mathbf{v}_2 = 15(1,2) - 12(-1,4) = (27,-18) \quad . \quad \text{Check.}$$

Dimension

Note that a single vector, say $\mathbf{v}_1$, cannot possibly form a basis for R^2. In fact, the only vectors that could be expressed in terms of such a basis would have the form $\mathbf{x} = c_1\mathbf{v}_1$ and would therefore all be parallel to $\mathbf{v}_1$.

Conversely, three or more vectors in R^2 are automatically dependent and so cannot form a basis. We conclude that all of the many bases for R^2 consist of exactly two elements.

It turns out that a comparable result holds generally.

THEOREM 6.7

Any two bases $\mathbf{v}_1, \ldots, \mathbf{v}_m$ and $\mathbf{z}_1, \ldots, \mathbf{z}_n$ for a vector space V have the same number of elements. That is, $m = n$.

■**Proof** Suppose we combine the two formulas linking the coordinates of a vector in V relative to the two bases: $\mathbf{c} = P\mathbf{d} = P(Q\mathbf{c}) = (PQ)\mathbf{c}$.

That is, $PQ = I_m = m \times m$ identity matrix.
Similarly, $QP = I_n = n \times n$ identity matrix.

But if $m < n$, then there is a nonzero vector $\mathbf{d}_0$ in the kernel of the $m \times n$ matrix P. That is, $P\mathbf{d}_0 = \mathbf{0}$. Therefore $QP\mathbf{d}_0 = \mathbf{0}$, and QP cannot be the identity matrix. So $m \geq n$. By similar arguments $n \geq m$ also. Therefore $m = n$.

COROLLARY 6.8

Coordinates $\mathbf{c}$, $\mathbf{d}$ of a vector relative to bases $\mathbf{v}_1, \ldots, \mathbf{v}_n$ and $\mathbf{z}_1, \ldots, \mathbf{z}_n$ of a vector space V are related by the formulas

$$\mathbf{c} = P\mathbf{d} \text{ and } \mathbf{d} = P^{-1}\mathbf{c},$$

where P is an invertible $n \times n$ matrix.

Example 3 For the two bases $\mathbf{v}_1, \mathbf{v}_2$ and $\mathbf{z}_1, \mathbf{z}_2$ for R^2 in the previous example 1, we
computed $P = \begin{bmatrix} 4 & 3 \\ -3 & -2 \end{bmatrix}$.

It follows from the corollary that $Q = P^{-1} = \begin{bmatrix} -2 & -3 \\ 3 & 4 \end{bmatrix}$.

In particular, the vector $\mathbf{x}$ with coordinates $\mathbf{c} = (15, -12)$ relative to basis
$\mathbf{v}_1, \mathbf{v}_2$ has the coordinates

$$\mathbf{d} = P^{-1}\mathbf{c} = \begin{bmatrix} -2 & -3 \\ 3 & 4 \end{bmatrix} \begin{bmatrix} 15 \\ -12 \end{bmatrix} = \begin{bmatrix} 6 \\ -3 \end{bmatrix} \quad \text{or} \quad \begin{cases} d_1 = 6 \\ d_2 = -3 \end{cases}$$

relative to basis $\mathbf{z}_1, \mathbf{z}_2$, which agrees with the discussion in example 2.

Suppose a vector space V has a basis with a finite number of elements,
$\mathbf{v}_1, \ldots, \mathbf{v}_n$. Then V is said to be *finite dimensional* and to have *dimension*
n. We also write dim $(V) = n$.

 The point of the previous theorem is, of course, that this definition is
consistent: there cannot be two bases with different numbers of elements,
and so a finite dimensional vector space has a *unique* dimension. The dimension is a positive integer and is an intrinsic property of the vector space.

Example 4 The vector space R^n has the standard basis $\mathbf{e}_1, \ldots, \mathbf{e}_n$ with n elements
and therefore has dimension n. This agrees with our intuitive concept of R^2
and R^3 as the two-dimensional plane and three-dimensional space.

Note that if W is a vector space and V is the subset of W consisting of just
the $\mathbf{0}$ vector, then V is a vector space simply because it satisfies the 10 properties of a vector space. It is easily checked that V can have no basis; but by
convention V is said to be finite dimensional and to have *dimension zero*.

 The importance of the dimension concept is that it allows us to measure,
in some sense, the *size* of vector spaces much less tractable than R^n. Let us
now compute basis and dimension of two important spaces.

Dimension of Ker(A)

THEOREM 6.9
Let A be an $m \times n$ matrix. Then

(a) a basis for Ker(A) is formed by the set of generating solutions
    $\mathbf{u}_1, \ldots, \mathbf{u}_k$ of the system $A\mathbf{x} = \mathbf{0}$;
(b) the dimension of the kernel, or dim $($Ker$(A))$, is n-rank (A).

■Proof   (a) We saw in section 2, example 2, that the generating solutions span the kernel. If we can also show that they are l.i., then from theorem 6.4 we can conclude that they form a basis.

Recall that the generating solutions are obtained from the parametric form of the general solution of $A\mathbf{x} = \mathbf{0}$. In example 3 of section 2, we had

$$A = \begin{bmatrix} 1 & 3 & 0 & 5 \\ 0 & 0 & 1 & 2 \\ 0 & 0 & 2 & 4 \end{bmatrix}, R = \begin{bmatrix} 1 & 3 & 0 & 5 \\ 0 & 0 & 1 & 2 \\ 0 & 0 & 0 & 0 \end{bmatrix}, \begin{cases} x_1 = -3x_2 - 5x_4 \\ x_3 = \quad\quad - 2x_4 \end{cases}.$$

The general solution is usually rewritten by giving new names to the free variables; $x_2 = c_1$, $x_4 = c_2$. Then

$$\mathbf{x} = \begin{bmatrix} -3x_2 - 5x_4 \\ x_2 \\ -2x_4 \\ x_4 \end{bmatrix} = \begin{bmatrix} -3c_1 - 5c_2 \\ c_1 \\ -2c_2 \\ c_2 \end{bmatrix} = \ldots$$

$$\ldots \quad c_1 \begin{bmatrix} -3 \\ 1 \\ 0 \\ 0 \end{bmatrix} + c_2 \begin{bmatrix} -5 \\ 0 \\ -2 \\ 1 \end{bmatrix} = c_1 \mathbf{u}_1 + c_2 \mathbf{u}_2 \quad.$$

Observe that the constants c_1, c_2 serve as the second and fourth components, respectively, of the vector $c_1 \mathbf{u}_1 + c_2 \mathbf{u}_2$. Therefore, the linear combination cannot be $\mathbf{0}$ unless c_1 and c_2 are both zero. That is, $\mathbf{u}_1$, $\mathbf{u}_2$ are l.i.

But the same thing must happen in every example, because the free variables among the components of $\mathbf{x}$ turn into the constants $c_1, \ldots, c_k$. So $c_1 \mathbf{u}_1 + \cdots + c_k \mathbf{u}_k \neq \mathbf{0}$ unless all constants are zero. That is, the generating solutions $\mathbf{u}_1, \ldots, \mathbf{u}_k$ are always l.i.

(b) Since a basis for $\mathrm{Ker}(A)$ is formed by the generating solutions, the dimension of $\mathrm{Ker}(A)$ equals the number of generating solutions. But this number was calculated, in theorem 2.2, to be n-rank (A).   Q.E.D.

Example 1   For the 3×4 matrix A used to illustrate the proof of the theorem, there were two generating solutions, and so $\dim(\mathrm{Ker}(A)) = 2$.

Dimension of Range(A)

We have seen that, although the columns of an $m \times n$ matrix A span Range(A) (see example 4 of section 4.2), they may not be linearly independent and may therefore not form a basis for the range.

Consider first the 3×4 matrix $R = \begin{bmatrix} 1 & 3 & 0 & 5 \\ 0 & 0 & 1 & 2 \\ 0 & 0 & 0 & 0 \end{bmatrix}$, which happens to be a reduced matrix of rank 2. Since the third row of R is all zeros, every linear combination of the columns $\mathbf{v}_1$, $\mathbf{v}_2$, $\mathbf{v}_3$, $\mathbf{v}_4$ of R has zero as its third component. That is, every l.c. of the columns of R is actually an l.c. of the vectors $\mathbf{e}_1$, $\mathbf{e}_2$ in R^3. That is, $\mathbf{e}_1$ and $\mathbf{e}_2$ span Range (R). Furthermore, since $\mathbf{e}_1$ and $\mathbf{e}_2$ are of course l.i., they form a basis for Range(R).

Note also that $\mathbf{e}_1$, $\mathbf{e}_2$ are the corner columns of R. So we have shown that the corner columns of R form a basis for R. The same thing happens with every reduced matrix. If there are k nonzero rows, then the corner columns are l.i. vectors $\mathbf{e}_1$, . . . , $\mathbf{e}_k$ and these form a basis for Range(R). To summarize,

Lemma 6.10 The corner columns of a reduced matrix R form a basis for Range(R), and

$\dim(\text{Range}(R)) = \text{rank}(R)$.

It turns out this result can be extended to all matrices.

THEOREM **6.11**

The corner columns of any matrix A form a basis for Range(A), and

$\dim(\text{Range}(A)) = \text{rank}(A)$.

■**Proof**   Let us compare the matrix A with its reduced form R, as in the following example:

$$A = \begin{bmatrix} 1 & 3 & 0 & 5 \\ 0 & 0 & 1 & 2 \\ 0 & 0 & 2 & 4 \end{bmatrix} = [\mathbf{u}_1|\mathbf{u}_2|\mathbf{u}_3|\mathbf{u}_4] \quad ,$$

$$R = \begin{bmatrix} 1 & 3 & 0 & 5 \\ 0 & 0 & 1 & 2 \\ 0 & 0 & 0 & 0 \end{bmatrix} = [\mathbf{v}_1|\mathbf{v}_2|\mathbf{v}_3|\mathbf{v}_4] \quad .$$

From corollary 3.5, A and R are linked by an invertible square matrix Q as follows:

$R = QA$ and $A = Q^{-1}R$.

Now, a vector $\mathbf{u}$ in Range(A) has the representation

$\mathbf{u} = A\mathbf{c} = c_1\mathbf{u}_1 + \cdots + c_n\mathbf{u}_n$,

and then the vector

$\mathbf{v} = Q\mathbf{u} = QA\mathbf{c} = R\mathbf{c} = c_1\mathbf{v}_1 + \cdots + c_n\mathbf{v}_n$

is in Range (R), and conversely.

In particular, suppose we can represent a vector $\mathbf{v}$ in terms of just the corner columns of R (columns 1 and 3 in our example) as follows:

$$\mathbf{v} = d_1\mathbf{v}_1 + 0\mathbf{v}_2 + d_3\mathbf{v}_3 + 0\mathbf{v}_4 = d_1\mathbf{v}_1 + d_3\mathbf{v}_3 \quad .$$

Then for $\mathbf{u}$ we have the representation

$$\mathbf{u} = d_1\mathbf{u}_1 + 0\mathbf{u}_2 + d_3\mathbf{u}_3 + 0\mathbf{u}_4 = d_1\mathbf{u}_1 + d_3\mathbf{u}_3$$

in terms of just the corner columns of A.

But by the previous lemma, a representation of $\mathbf{v}$ in terms of the corner columns of R exists and is unique. So a representation of $\mathbf{u}$ in terms of the corner columns of A also exists and is unique. That is, the corner columns of A form a basis for Range(A).   Q.E.D.

Example 2 A basis for the range of the matrix A used in the above proof is given by its corner columns, columns 1 and 3:

$$\mathbf{u}_1 = (1,0,0), \mathbf{u}_3 = (0,1,2) \quad ,$$

and dim (Range(A)) $= 2$.

Note that, to determine the corner columns of A, we compute its reduced form R; but to write down the basis for Range(A), we have to go back to the corner columns of the *original* matrix A.

Having completed this detailed study of the kernel and range of a matrix, we can immediately write down a fundamental theorem.

THEOREM 6.12
(*Main Dimension Theorem*)
For any $m \times n$ matrix A,

$$\dim(\text{Ker}(A)) + \dim(\text{Range}(A)) = n \quad .$$

Proof   The dimensions specified in the sum were calculated to be n-rank(A) and rank(A) respectively, and these add up to n.
                                    Q.E.D.

Example 1   For the 3×4 matrix $A = \begin{bmatrix} 1 & 3 & 0 & 5 \\ 0 & 0 & 1 & 2 \\ 0 & 0 & 2 & 4 \end{bmatrix}$,

we have now computed the dimensions of both kernel and range and have found

$$\dim(\text{Ker}(A)) + \dim(\text{Range}(A)) = 2 + 2 = 4 = n \quad ,$$

in agreement with the Main Dimension Theorem.

Example 2   Consider the matrix $A(g) = \begin{bmatrix} 1 & 0 \\ 0 & g \end{bmatrix}$,

which depends on the parameter g. The essential cases are

$$g = 0: \quad A(0) = \begin{bmatrix} 1 & 0 \\ 0 & 0 \end{bmatrix} \quad ; \quad g \neq 0 : A(g) = \begin{bmatrix} 1 & 0 \\ 0 & g \end{bmatrix} .$$

When $g = 0$, the basis for $\mathrm{Ker}\,(A(0))$ is $\mathbf{e}_2$, while the basis for $\mathrm{Range}(A(0))$ is column 1 of $A(0)$, or $\mathbf{e}_1$. When $g \neq 0$, $A(g)$ has rank $2 = n$, and so only the $\mathbf{0}$ vector is in $\mathrm{Ker}(A(g))$, and a basis for $\mathrm{Range}(A(g))$ is formed by the two corner columns of $A(g)$. Our conclusions are summarized in Table 6.1.

Parameter value	dim(Ker $(A(g))$)	dim(Range($A(g)$))	Sum
$g = 0$	1	1	2
$g \neq 0$	0	2	2

Table 6.1

Note that the *sum* of the dimensions of kernel and range is always $n = 2$, as required by the Main Dimension Theorem, although individually these dimensions vary with the parameter value.

Other Vector Spaces

Suppose $\mathbf{u}_1, \ldots, \mathbf{u}_k$ are vectors in R^n, and suppose we let the subset S of R^n consist of all linear combinations of $\mathbf{u}_1, \ldots, \mathbf{u}_k$. Then of course S is the same as the range of the matrix $A = [\mathbf{u}_1 | \ldots | \mathbf{u}_k]$, and so it is a subspace of R^n. We say that *S is the subspace of R^n generated by* $\mathbf{u}_1, \ldots, \mathbf{u}_k$.

Example 3   To find a basis for, and the dimension of, the subspace S of R^3 generated by

$$\mathbf{u}_1 = (1,1,2), \quad \mathbf{u}_2 = (2,3,4), \quad \mathbf{u}_3 = (5,7,10), \quad \mathbf{u}_4 = (0,-1,0) \quad ,$$

we reduce

$$A = [\mathbf{u}_1 | \mathbf{u}_2 | \mathbf{u}_3 | \mathbf{u}_4] = \begin{bmatrix} 1 & 2 & 5 & 0 \\ 1 & 3 & 7 & -1 \\ 2 & 4 & 10 & 0 \end{bmatrix} \rightarrow \begin{bmatrix} 1 & 2 & 5 & 0 \\ 0 & 1 & 2 & -1 \\ 0 & 0 & 0 & 0 \end{bmatrix} .$$

So a basis for $S = \mathrm{Range}(A)$ is given by the corner columns 1 and 2 of A, $\mathbf{u}_1$ and $\mathbf{u}_2$, and $\dim(S) = 2$.

We have seen that any n l.i. vectors can be used as a basis for R^n. The next result gives us similar flexibility in selecting bases for other vector spaces.

THEOREM 6.13

Suppose V is a vector space of dimension n. Then any n l.i. vectors in V form a basis for V.

■**Proof**   Let $v_1, \ldots, v_n$ be a basis for V, and suppose $z_1, \ldots, z_n$ are n l.i. vectors in V that do not form a basis for V. Then there must be a vector z_{n+1} in V that is not a linear combination of $z_1, \ldots, z_n$, and by lemma 6.1, $z_1, \ldots, z_{n+1}$ are an l.i. set.

Now let $x = d_1 z_1 + \cdots + d_{n+1} z_{n+1}$. The coordinates c of x relative to the basis $v_1, \ldots, v_n$ then have the form

$$c = Pd \quad,$$

where P is an $n \times (n+1)$ matrix, according to the discussion preceding lemma 6.6. But since P has fewer rows than columns, there must be some $d \neq 0$ such that $c = Pd = 0$. But then

$$x = d_1 z_1 + \cdots + d_{n+1} z_{n+1} = 0 v_1 + \cdots + 0 v_n = 0 \quad,$$

contradicting the independence of $z_1, \ldots, z_{n+1}$. We conclude that $z_1, \ldots, z_n$ were a basis for V after all.   Q.E.D.

Example 4 We found that $\dim(\text{Range}(A)) = 2$ for the matrix

$$A = \begin{bmatrix} 1 & 3 & 0 & 5 \\ 0 & 0 & 1 & 2 \\ 0 & 0 & 2 & 4 \end{bmatrix} \quad.$$

Since each column of A is of course in Range(A), any two l.i. columns will form a basis for Range(A). We can, for example, select columns 3 and 4 as a basis:

$$z_1 = (0,1,2), \; z_2 = (5,2,4) \quad.$$

If the dimension concept is to be useful as a measure of the "size" of a vector space, we expect that the dimension of a subspace will be less than that of the full space. Fortunately, this is true.

THEOREM **6.14**

Suppose W is a finite dimensional vector space and V is a subspace of W. Then either $V = W$ or

$$\dim(V) < \dim(W) \quad.$$

■**Proof**   Let $\dim(W) = n$.

If V contains only 0, then $\dim(V) = 0 < n$. Otherwise V contains a nonzero vector z_1, which forms an l.i. set (of one element). Either z_1 spans V, or else a second vector z_2 can be found in V such that z_1, z_2 are an l.i. set.

If we can continue this process until we have a set of n l.i. vectors $z_1, \ldots, z_n$, then, by the previous theorem, they form a basis for W and $V = W$.

Otherwise, the process stops earlier with a set of k l.i. vectors that span V, and $k < n$. These k l.i. vectors, $z_1, \ldots, z_k$, then form a basis for V, and $\dim(V) = k < n$. Q.E.D.

Example 5 All subspaces of R^n, other than R^n itself, must have dimension less than n.

■ **Problem** Let S be the subspace of R^3 generated by $u_1 = (1,2,1)$ and $u_2 = (0,1,2)$.

Let T be the subspace of R^3 generated by $v_1 = (0,4,4)$ and $v_2 = (1,1,9)$. Find all vectors that are in both S and T.

■ **Solution** If a vector z is in both S and T, then

$$z = c_1 u_1 + c_2 u_2 \text{ and } z = d_1 v_1 + d_2 v_2 \quad .$$

Subtracting these two formulas for z, we obtain

$$c_1 u_1 + c_2 u_2 + d_1(-v_1) + d_2(-v_2) = 0 \quad .$$

With $A = [u_1|u_2|-v_1|-v_2]$ and $x = (c_1,c_2,d_1,d_2)$, the interpretation of the last equation is that x is in $\mathrm{Ker}(A)$.

The numerical calculation therefore begins with the reduction of A; we quickly obtain:

$$A = \begin{bmatrix} 1 & 0 & 0 & -1 \\ 2 & 1 & -4 & -1 \\ 1 & 2 & -4 & -9 \end{bmatrix} \rightarrow \begin{bmatrix} 1 & 0 & 0 & -1 \\ 0 & 1 & 0 & -9 \\ 0 & 0 & 1 & -\frac{5}{2} \end{bmatrix} \quad \text{or} \quad \begin{array}{l} c_1 = d_2 \\ c_2 = 9d_2 \\ d_1 = \left(\frac{5}{2}\right)d_2 \end{array} \quad .$$

Or, with parameter $a = d_2$: $c_1 = a$, $c_2 = 9a$, $d_1 = \left(\frac{5}{2}\right)a$, $d_2 = a$. Then $z = c_1 u_1 + c_2 u_2 = a(1,2,1) + 9a(0,1,2) = a(1,11,19)$. We conclude that the vectors that are in both S and T are the scalar multiples of $(1,11,19)$.

This result can be checked by computing the second formula for z:

$$z = d_1 v_1 + d_2 v_2 = \left(\frac{5}{2}\right)a(0,4,4) + a(1,1,9) = a(1,11,19) \quad . \quad \text{Check.}$$

■ EXERCISES 6.3

1. For each $m \times n$ matrix A given below, compute a basis for, and the dimension of, each of the spaces $\mathrm{Ker}(A)$ and $\mathrm{Range}(A)$, and verify that the sum of their dimensions equals n. Also determine whether the given vector x is in $\mathrm{Ker}(A)$ and whether the given vector y is in $\mathrm{Range}(A)$.

(a) $A = \begin{bmatrix} 0 & 2 \\ 0 & 4 \end{bmatrix}$, $\begin{array}{l} x = (5,0) \\ y = (3,1) \end{array}$

(b) $A = \begin{bmatrix} 1 & 1 & 1 \\ 2 & 2 & 3 \end{bmatrix}$, $\begin{array}{l} x = (3,4,1) \\ y = (1,0) \end{array}$

(c) $A = \begin{bmatrix} 1 & 1 \\ 1 & 2 \\ 1 & 3 \end{bmatrix}$, $\mathbf{x} = (4,3)$
$\mathbf{y} = (2,-4,2)$

(d) $A = \begin{bmatrix} 1 & 2 & 3 \\ 2 & 4 & 6 \\ 3 & 7 & 9 \end{bmatrix}$, $\mathbf{x} = (-6,0,2)$
$\mathbf{y} = (3,6,8)$

(e) $A = \begin{bmatrix} 1 & 1 & 1 \\ 2 & 2 & 2 \\ 8 & 8 & 8 \end{bmatrix}$, $\mathbf{x} = (1,2,8)$
$\mathbf{y} = (1,2,9)$

(f) $A = \begin{bmatrix} 1 & 2 & 1 & 2 \\ 3 & 6 & 0 & 3 \\ 4 & 8 & 5 & 9 \end{bmatrix}$, $\mathbf{x} = (-6,2,-2,2)$
$\mathbf{y} = (15,-1,-3)$

2. Suppose A is a 14×17 matrix.

(a) Determine all possible values of dim(Range(A)).
(b) Determine all possible values of dim(Ker(A)).
(c) Assuming dim(Range(A)) $= 11$, compute dim(Ker(A)).

3. Repeat exercise 2, this time with A an 18×15 matrix.

4. Suppose we do not know column 4 of the 3×4 matrix A, but we do know that A has the form

$A = \begin{bmatrix} 1 & 2 & 3 & * \\ 2 & 5 & 7 & * \\ 4 & 9 & 13 & * \end{bmatrix}$,

where the stars indicate unknown entries.

(a) Determine all possible values of dim(Range(A)) and of dim(Ker(A)).
(b) Write down a nonzero vector that is in Ker(A) no matter how the fourth column of A is chosen.

5. For each of the sets of vectors $\mathbf{v}_1, \ldots, \mathbf{v}_k$ given below, find a basis for, and the dimension of, the space S they generate. Also determine whether the given vector $\mathbf{w}$ is in S and, if so, find its coordinates with respect to the basis constructed for S.

(a) $\mathbf{v}_1 = (1,0,1,3)$, $\mathbf{v}_2 = (2,1,2,5)$, $\mathbf{v}_3 = (1,2,1,1)$, $\mathbf{v}_4 = (3,1,1,4)$,
$\mathbf{w} = (4,1,6,15)$

(b) $\mathbf{v}_1 = (1,1,1,-3)$, $\mathbf{v}_3 = (1,2,1,-4)$,
$\mathbf{v}_4 = (1,-1,1,-1)$,
$\mathbf{w} = (0,3,-3,0)$

(c) $\mathbf{v}_1 = (-1,2,-2,1)$, $\mathbf{v}_2 = (1,1,-1,1)$, $\mathbf{v}_3 = (2,-1,1,0)$,
$\mathbf{v}_4 = (0,-3,0,1)$,
$\mathbf{w} = (1,1,5,-5)$

(d) $\mathbf{v}_1 = (1,0,0)$, $\mathbf{v}_2 = (1,0,1)$, $\mathbf{v}_3 = (0,0,1)$,
$\quad \mathbf{w} = (2,3,5)$

(e) $\mathbf{v}_1 = (1,1,1)$, $\mathbf{v}_2 = (1,-1,2)$, $\mathbf{v}_3 = (2,0,3)$, $\mathbf{v}_4 = (0,2,-1)$,
$\quad \mathbf{w} = (5,1,7)$

(f) $\mathbf{v}_1 = (1,1,1)$, $\mathbf{v}_2 = (2,2,3)$, $\mathbf{v}_3 = (3,4,5)$,
$\quad \mathbf{w} = (0,1,0)$

6. Suppose the subspaces S and T of R^n are generated by vectors $\mathbf{u}_1, \ldots, \mathbf{u}_h$ and $\mathbf{v}_1, \ldots, \mathbf{v}_k$ respectively. Find all vectors that are in both S and T in each of the following cases:

(a) $\mathbf{u}_1 = (1,1,-1,-1)$, $\mathbf{u}_2 = (1,-1,1,-1)$
$\quad \mathbf{v}_1 = (-1,1,1,-1)$, $\mathbf{v}_2 = (-1,3,5,-7)$.

(b) $\mathbf{u}_1 = (1,1,0,0)$, $\mathbf{u}_2 = (0,1,1,0)$, $\mathbf{u}_3 = (0,0,1,1)$
$\quad \mathbf{v}_1 = (-1,0,0,2)$, $\mathbf{v}_2 = (2,0,0,-1)$.

(c) $\mathbf{u}_1 = (1,0,4)$, $\mathbf{u}_2 = (0,1,2)$
$\quad \mathbf{v}_1 = (0,1,5)$, $\mathbf{v}_2 = (1,1,3)$.

7. Show that if the subspaces S and T of R^4 are generated by the vectors $\mathbf{u}_1$, $\mathbf{u}_2$, and $\mathbf{v}_1$, $\mathbf{v}_2$ respectively, and if the four vectors $\mathbf{u}_1$, $\mathbf{u}_2$, $\mathbf{v}_1$, $\mathbf{v}_2$ form an l.i. set, then only the $\mathbf{0}$ vector is in both S and T.

8. (a) Show that if S and T are two subspaces of vector space V, then the set $S \cap T$ of vectors that are in both S *and* T is also a subspace of V.

(b) Show that if the subspaces S and T of R^5 are generated by

$$\mathbf{u}_1 = (1,-1,0,0,0), \ \mathbf{u}_2 = (0,1,-1,0,0), \ \mathbf{u}_3 = (0,0,1,1,-2)$$
$$\mathbf{v}_1 = (1,0,0,1,-2), \ \mathbf{v}_2 = (0,2,0,2,-4),$$

then $\dim(S \cap T) = 2$.

9. (a) Suppose $\mathbf{c}$ and $\mathbf{c}'$ denote coordinates of vectors in R^2 relative to the two bases

$$\mathbf{v}_1 = (1,2), \quad \mathbf{v}_2 = (-2,1)$$
$$\mathbf{w}_1 = (14,13), \ \mathbf{w}_2 = (9,8) \quad .$$

Find the 2×2 matrix P such that $\mathbf{c} = P\mathbf{c}'$.

(b) Repeat for the case of the two bases

$$\mathbf{v}_1 = (1,3), \quad \mathbf{v}_2 = (4,5)$$
$$\mathbf{w}_1 = (6,11), \ \mathbf{w}_2 = (7,14) \quad .$$

*10. Suppose $\mathbf{c}$ and $\mathbf{c}'$ denote coordinates of vectors in R^n relative to the two bases $\mathbf{v}_1, \ldots, \mathbf{v}_n$ and $\mathbf{w}_1, \ldots, \mathbf{w}_n$. Let P denote the invertible $n \times n$ matrix such that $\mathbf{c} = P\mathbf{c}'$. Consider the $n \times 2n$ matrix

$$D = [\mathbf{v}_1| \ldots |\mathbf{v}_n|\mathbf{w}_1| \ldots |\mathbf{w}_n] \quad .$$

(a) Show that when the left block of D is reduced to the identity by row operations on D, the result is $[I|P]$.

(b) Show that when the right block of D is reduced to the identity by row operations on D, the result is $[P^{-1}|I]$.

(c) Use part (a) to compute the 3×3 matrix P linking coordinates of vectors in R^3 relative to the two bases

$$\mathbf{v}_1 = (1,0,1), \ \mathbf{v}_2 = (6,1,0), \ \mathbf{v}_3 = (5,0,4)$$
$$\mathbf{w}_1 = (12,1,5), \ \mathbf{w}_2 = (7,1,1), \ \mathbf{w}_3 = (2,0,2) \quad .$$

*11. (Alternative proof that all bases for a vector space have the same number of elements.)

Suppose $\mathbf{c}$ in R^m and $\mathbf{c}'$ in R^m are coordinates of $\mathbf{v}$ in a vector space V, relative to the two bases $\mathbf{v}_1, \ldots, \mathbf{v}_m$ and $\mathbf{z}_1, \ldots, \mathbf{z}_n$ for V. Let T, S, T', S' be the linear maps, as constructed in the exercises of section 6.2, such that

$$T(\mathbf{v}) = \mathbf{c}, \ S(\mathbf{c}) = \mathbf{v}, \ T'(\mathbf{v}) = \mathbf{c}', \ S'(\mathbf{c}') = \mathbf{v} \quad .$$

(a) Show that the maps $J = T' \circ S$ of R^n into R^n and $K = T \circ S'$ of R^n into R^m, satisfy

$$K \circ J = I = \text{identity map of } R^m \text{ into } R^m$$
$$J \circ K = I = \text{identity map of } R^n \text{ into } R^n \quad .$$

(That is, J and K constitute an isomorphism of the spaces R^m and R^n.)

(b) Deduce (as in the proof of theorem 6.7) that $m = n$.

*12. (a) Let S be the set of all linear combinations of the vectors $\mathbf{v}_1, \ldots, \mathbf{v}_k$ in a vector space V. Show that S is a subspace of V. (S is called *the subspace of V generated by* $\mathbf{v}_1, \ldots, \mathbf{v}_k$).

(b) Let $\mathbf{v}_1, \ldots, \mathbf{v}_n$ be a basis for a vector space V. Let S be the subspace of V generated by a subset of these elements, say, by $\mathbf{v}_1, \ldots, \mathbf{v}_k$. Show that $\mathbf{v}_1, \ldots, \mathbf{v}_k$ form a basis for S.

*13. Let $P(s)$ denote the vector space consisting of all polynomials in one variable s, and let $P_N(s)$ denote the subspace of $P(s)$ consisting of all polynomials in s of degree at most N.

(a) Show that the elements

$$p_0(s) = 1, \ p_1(s) = s, \ p_2(s) = s^2, \ \ldots, \ p_N(s) = s^N$$

form a basis for $P_N(s)$, and that therefore

$$\dim(P_N(s)) = N + 1 \quad .$$

(b) Deduce that $P(s)$ is not a finite dimensional vector space.

Transformations and Bases

We have been able to study linear transformations of R^n to R^m quite closely, because they have concrete representations in the form of $m \times n$ matrices. See, for example, the development of the Main Dimension Theorem in the previous section.

Our objective in this section is to show that we can achieve a comparable understanding of linear transformations between more general vector spaces, provided that these spaces are finite dimensional.

However, the matrix representation A of a given transformation T is derived in Chapter 4 by direct reference to the standard basis for R^n; in fact the jth column of A is simply $T(\mathbf{e}_j)$. Now, vector spaces other than R^n do not have a standard basis; and even in working with R^n, we often want to use a different basis as our frame of reference. We must therefore develop a concept of matrix representation that applies equally well no matter what bases are chosen for the vector spaces being transformed.

Let us proceed in stages, keeping the vector space R^n at first but selecting an arbitrary basis for it.

Mappings of R^n into R^n

Suppose, then, that $\mathbf{v}_1, \ldots, \mathbf{v}_n$ is a basis for R^n, and T a linear transformation of R^n into R^n. Suppose also that A is an $n \times n$ matrix, and that for each vector $\mathbf{x}$ in R^n the coordinates $\mathbf{c}$ of $\mathbf{x}$ and $\mathbf{d}$ of $T(\mathbf{x})$ are linked by the formula

$\mathbf{d} = A\mathbf{c}$.

Then we say that *the matrix A represents the transformation T relative to the basis $\mathbf{v}_1, \ldots, \mathbf{v}_n$ for R^n.*

Observe that this new concept of matrix representation includes the original as the special case when the basis is the standard basis $\mathbf{e}_1, \ldots, \mathbf{e}_n$. For then the coordinates $\mathbf{c}$ and $\mathbf{d}$ are simply the vectors $\mathbf{x}$ and $T(\mathbf{x})$ themselves, and so the formula $\mathbf{d} = A\mathbf{c}$ becomes $T(\mathbf{x}) = A\mathbf{x}$ as before.

In fact, once the matrix representation of T relative to the standard basis is known, its representations relative to other bases are easily derived.

THEOREM 6.15

Suppose the linear transformation T of R^n into R^n has the matrix representations A and A' relative to the bases $\mathbf{e}_1, \ldots, \mathbf{e}_n$ and $\mathbf{v}_1, \ldots, \mathbf{v}_n$ respectively.

Then $A' = P^{-1}AP$, where $P = [\mathbf{v}_1 | \ldots | \mathbf{v}_n]$.

■Proof   Say $\mathbf{x}$ and $T(\mathbf{x})$ have coordinates $\mathbf{c}$ and $\mathbf{d}$ respectively relative to the basis $\mathbf{v}_1, \ldots, \mathbf{v}_n$. That is,

$$\mathbf{x} = P\mathbf{c} \quad \text{and} \quad T(\mathbf{x}) = P\mathbf{d} \quad.$$

From the definition of A, $\qquad T(\mathbf{x}) = A\mathbf{x}$

or $\qquad\qquad\qquad\qquad\quad P\mathbf{d} = AP\mathbf{c}$

or $\qquad\qquad\qquad\qquad\quad \mathbf{d} = P^{-1}AP\mathbf{c} \quad.$

From the definition of A' $\qquad \mathbf{d} = A'\mathbf{c} \quad,$

and so by comparison, $\qquad A' = P^{-1}AP$ as required. $\qquad$ Q.E.D.

Example 1   A certain rotation transformation T has the matrix representation
$$A = \left(\tfrac{1}{5}\right) \begin{bmatrix} 3 & -4 \\ 4 & 3 \end{bmatrix} \quad \text{relative to the standard basis for } R^2.$$

To find the representation A' of T relative to the basis

$$\mathbf{v}_1 = (1,2), \ \mathbf{v}_2 = (1,3) \quad,$$

we simply compute

$$P = [\mathbf{v}_1| \ldots |\mathbf{v}_n] = \begin{bmatrix} 1 & 1 \\ 2 & 3 \end{bmatrix} \quad, \quad P^{-1} = \begin{bmatrix} 3 & -1 \\ -2 & 1 \end{bmatrix} \quad,$$

$$A' = P^{-1}AP = \begin{bmatrix} 3 & -1 \\ -2 & 1 \end{bmatrix}\left(\tfrac{1}{5}\right)\begin{bmatrix} 3 & -4 \\ 4 & 3 \end{bmatrix}\begin{bmatrix} 1 & 1 \\ 2 & 3 \end{bmatrix} = \begin{bmatrix} -5 & -8 \\ 4 & \tfrac{31}{5} \end{bmatrix} \quad.$$

Note that A' does not look like a rotation matrix, although it represents a rotation relative to the basis $\mathbf{v}_1, \mathbf{v}_2$.

Mappings of V to W

Suppose, for example, that vector spaces V and W have dimensions 2 and 3 respectively. Then a vector $\mathbf{v}$ in V has, relative to a basis $\mathbf{v}_1, \mathbf{v}_2$ for V, a set of coordinates $\mathbf{c}$ in R^2, while a vector $\mathbf{w}$ in W has, relative to a basis $\mathbf{w}_1, \mathbf{w}_2, \mathbf{w}_3$ for W, a set of coordinates $\mathbf{d}$ in R^3. Suppose also that T is a linear transformation of V into W that maps $\mathbf{v}$ to $\mathbf{w}$, or $T(\mathbf{v}) = \mathbf{w}$. To obtain a description of T in terms of the coordinates $\mathbf{c}$ and $\mathbf{d}$, write

$$T(\mathbf{v}_1) = a_{11}\mathbf{w}_1 + a_{21}\mathbf{w}_2 + a_{31}\mathbf{w}_3$$
$$T(\mathbf{v}_2) = a_{12}\mathbf{w}_1 + a_{22}\mathbf{w}_2 + a_{32}\mathbf{w}_3 \quad.$$

Then $T(\mathbf{v}) = T(c_1\mathbf{v}_1 + c_2\mathbf{v}_2)$

$\qquad\qquad = c_1\,T(\mathbf{v}_1) + c_2\,T(\mathbf{v}_2)$

$\qquad\qquad = c_1\,(a_{11}\mathbf{w}_1 + a_{22}\mathbf{w}_2 + a_{31}\mathbf{w}_3) + c_2(a_{12}\mathbf{w}_1 +$
$$a_{22}\mathbf{w}_2 + a_{32}\mathbf{w}_3)$$

$\qquad\qquad = (a_{11}c_1 + a_{12}c_2)\mathbf{w}_1 + (a_{21}c_1 + a_{22}c_2)\mathbf{w}_2 +$
$$(a_{31}c_1 + a_{32}c_2)\mathbf{w}_3 \quad.$$

That is, $\begin{cases} a_{11}c_1 + a_{12}c_2 = d_1 \\ a_{21}c_1 + a_{22}c_2 = d_2 \\ a_{31}c_1 + a_{32}c_2 = d_3 \end{cases}$ or $A\mathbf{c} = \mathbf{d}$,

where A is the 3×2 matrix with entries a_{ij} .

In other words, the coordinates of $\mathbf{v}$ and of $\mathbf{w} = T(\mathbf{v})$ are linked by a matrix transformation A of R^2 into R^3. Of course this conclusion did not depend on the selection of dimensions 2 and 3 for the spaces V and W; in fact we can summarize our calculation in a perfectly general theorem.

THEOREM 6.16

Suppose T is a linear transformation of V into W. Suppose also that V has dimension n and a basis $\mathbf{v}_1, \ldots, \mathbf{v}_n$, and W has dimension m and a basis $\mathbf{w}_1, \ldots, \mathbf{w}_m$.

Then there is an $m \times n$ matrix A such that, if

$T(\mathbf{v}) = \mathbf{w}$,

then the coordinates $\mathbf{c}$, $\mathbf{d}$ of $\mathbf{v}$, $\mathbf{w}$ relative to the given bases for V, W respectively, satisfy the equation

$A\mathbf{c} = \mathbf{d}$.

Example 2 Find a matrix representation for the mapping T of V into W, where

V is the subspace R^4 consisting of vectors of the form (a,b,b,a),

W is the subspace of R^4 consisting of vectors of the form $(p,q,r,p + q - r)$, and

$T((a,b,b,a)) = (3a, 4b, a + b, 2a + 3b)$.

First, to find a basis for V, note that a vector in V can be rewritten as

$a(1,0,0,1) + b(0,1,1,0)$

and so the linearly independent vectors

$\mathbf{v}_1 = (1,0,0,1)$ and $\mathbf{v}_2 = (0,1,1,0)$

form a basis for V. Similarly, a basis for W is formed by

$\mathbf{w}_1 = (1,0,0,1)$, $\mathbf{w}_2 = (0,1,0,1)$, and $\mathbf{w}_3 = (0,0,1,-1)$.

Furthermore,

$T(\mathbf{v}_1) = T((1,0,0,1)) = (3,0,1,2) = 3\mathbf{w}_1 + 0\mathbf{w}_2 + \mathbf{w}_3$,

$T(\mathbf{v}_2) = T((0,1,1,0)) = (0,4,1,3) = 0\mathbf{w}_1 + 4\mathbf{w}_2 + \mathbf{w}_3$.

That is, the 3×2 matrix A has the form $A = \begin{bmatrix} 3 & 0 \\ 0 & 4 \\ 1 & 1 \end{bmatrix}$.

To check this calculation, note for example that the vector $\mathbf{v}$ with coordinates $\mathbf{c} = (5,7)$, say, maps to the vector $\mathbf{w}$ with coordinates

$$\mathbf{d} = A\mathbf{c} = \begin{bmatrix} 3 & 0 \\ 0 & 4 \\ 1 & 1 \end{bmatrix} \begin{bmatrix} 5 \\ 7 \end{bmatrix} = \begin{bmatrix} 15 \\ 28 \\ 12 \end{bmatrix} .$$

That is, $\mathbf{v} = (5,7,7,5)$ maps to

$\mathbf{w} = 15\mathbf{w}_1 + 28\mathbf{w}_2 + 12\mathbf{w}_3 = (15,28,12,31)$.

Compare this with the value of $\mathbf{w}$ obtained from the definition of T:

$T((5,7,7,5)) = (15,28,12,31)$. Check.

The matrix representation A of the mapping T given by Theorem 6.16 naturally depends on the choice of bases for the spaces V and W, and thus A is called the *matrix representation of T relative to the bases* $\mathbf{v}_1, \ldots, \mathbf{v}_n$ *and* $\mathbf{w}_1, \ldots, \mathbf{w}_m$ for V and W.

Since it will be necessary to use more than one basis for each vector space in subsequent work, let us note how different matrix representations of T are related.

THEOREM 6.17

Let $\mathbf{c}$ and $\mathbf{c}'$ denote the coordinates of a vector in the n dimensional vector space V, relative to the two bases $\mathbf{v}_1, \ldots, \mathbf{v}_n$ and $\mathbf{v}'_1, \ldots, \mathbf{v}'_n$, and let P be the invertible $n \times n$ matrix such that $\mathbf{c} = P\mathbf{c}'$.

Also, let $\mathbf{d}$ and $\mathbf{d}'$ denote the coordinates of a vector in the m dimensional vector space W, relative to the two bases $\mathbf{w}_1, \ldots, \mathbf{w}_m$ and $\mathbf{w}'_1, \ldots, \mathbf{w}'_m$, and let Q be the invertible $m \times m$ matrix such that $\mathbf{d} = Q\mathbf{d}'$.

Also, suppose that the linear transformation T of V into W is represented, relative to the pair of bases $\mathbf{v}_1, \ldots, \mathbf{v}_n$ and $\mathbf{w}_1, \ldots, \mathbf{w}_m$, by the $m \times n$ matrix A; and is represented, relative to the pair of bases $\mathbf{v}'_1, \ldots, \mathbf{v}'_n$ and $\mathbf{w}'_1, \ldots, \mathbf{w}'_m$, by the $m \times n$ matrix A'.

Then $A' = Q^{-1}AP$.

■**Proof** If $\mathbf{d} = A\mathbf{c}$, then $Q\mathbf{d}' = AP\mathbf{c}'$, or

$$\mathbf{d}' = Q^{-1}AP\mathbf{c}' .$$

On comparison with the formula $\mathbf{d}' = A'\mathbf{c}'$, we see that $A' = Q^{-1}AP$.

Q.E.D.

Mappings of V to V

At this stage, we see that the theorems are becoming easier to prove, but it is going to take longer to work through examples. Perhaps, for illustrative purposes, it is best to specialize the last theorem to an interesting special case.

COROLLARY 6.18

In theorem 6.17, let $V = W$,
let basis $\mathbf{v}_1, \ldots, \mathbf{v}_n$ coincide with basis $\mathbf{w}_1, \ldots, \mathbf{w}_n$, and
let basis $\mathbf{v}_1', \ldots, \mathbf{v}_n'$ coincide with basis $\mathbf{w}_1', \ldots, \mathbf{w}_n'$, so that $P = Q$. Then

$$A' = P^{-1}AP \quad .$$

In this specialized situation, we can say that the $n \times n$ matrix A represents the transformation T of V into V relative to the basis $\mathbf{v}_1, \ldots, \mathbf{v}_n$ for V. Furthermore, when V is the space R^n, corollary 6.18 is just a restatement of theorem 6.15.

Example 3   In the first two examples of section 6.3, we considered two bases for R^2: $\mathbf{v}_1 = (1,2)$, $\mathbf{v}_2 = (-1,4)$, and $\mathbf{v}_1' = (7,-4)$, $\mathbf{v}_2' = (5, -2)$. We found that the coordinates $\mathbf{c}$, $\mathbf{c}'$ of a vector $\mathbf{v}$ relative to those two bases were linked by the formula

$$\mathbf{c} = P\mathbf{c}', \text{ where } P = \begin{bmatrix} 4 & 3 \\ -3 & -2 \end{bmatrix} \quad .$$

In the third example of section 6.3, we further calculated

$$P^{-1} = \begin{bmatrix} -2 & -3 \\ 3 & 4 \end{bmatrix} \quad .$$

Therefore, if we know that a certain linear transformation T of R^2 to R^2 is represented, relative to the basis $\mathbf{v}_1, \mathbf{v}_2$ for R^2, by the 2×2 matrix

$$A = \begin{bmatrix} 0 & 1 \\ 1 & 0 \end{bmatrix} \quad ,$$

we can deduce that, relative to the basis $\mathbf{v}_1', \mathbf{v}_2'$, T is represented by the matrix

$$A' = P^{-1}AP = \begin{bmatrix} -2 & -3 \\ 3 & 4 \end{bmatrix} \begin{bmatrix} 0 & 1 \\ 1 & 0 \end{bmatrix} \begin{bmatrix} 4 & 3 \\ -3 & -2 \end{bmatrix}$$

$$= \begin{bmatrix} -2 & -3 \\ 3 & 4 \end{bmatrix} \begin{bmatrix} -3 & -2 \\ 4 & 3 \end{bmatrix} = \begin{bmatrix} -6 & -5 \\ 7 & 6 \end{bmatrix} \quad .$$

Checking this sort of calculation can be somewhat cumbersome. In this example, however, the simplicity of A helps. We can see at a glance that $A^2 = I$. So certainly $T \circ T = I$ also. Therefore, since A' also represents T (relative to a different basis of course), $(A')^2 = I$ should hold as well. And on multiplying out $(A')^2$ we find that it does.

■ **Problem** Let $A = \begin{bmatrix} 1 & 2 \\ 3 & 4 \\ 5 & 6 \end{bmatrix}$, and $A' = \begin{bmatrix} 1 & 0 \\ 0 & 1 \\ 0 & 0 \end{bmatrix}$.

Let T be the linear transformation of R^2 into R^3 given by the formula $T(\mathbf{x}) = A\mathbf{x}$. Show that, relative to suitably chosen bases $\mathbf{v}_1, \mathbf{v}_2$ for R^2 and $\mathbf{w}_1, \mathbf{w}_2, \mathbf{w}_3$ for R^3, the transformation T is represented by the matrix A'.

■ **Solution** The coordinates $\mathbf{c}$ of $\mathbf{x}$ and $\mathbf{d}$ of $T(\mathbf{x})$, relative to the required bases for R^2 and R^3 respectively, must be linked by the formula $\mathbf{d} = A'\mathbf{c}$, or $(d_1, d_2, d_3) = (c_1, c_2, 0)$. That is,

$$T(c_1\mathbf{v}_1 + c_2\mathbf{v}_2) = d_1\mathbf{w}_1 + d_2\mathbf{w}_2 + d_3\mathbf{w}_3$$
$$= c_1\mathbf{w}_1 + c_2\mathbf{w}_2 + 0\mathbf{w}_3 \quad .$$

But $\qquad\qquad T(c_1\mathbf{v}_1 + c_2\mathbf{v}_2) = c_1 T(\mathbf{v}_1) + c_2 T(\mathbf{v}_2) \quad .$

So we need $\qquad \mathbf{w}_1 = T(\mathbf{v}_1)$ and $\mathbf{w}_2 = T(\mathbf{v}_2) \quad .$

For example, we can select $\mathbf{v}_1 = \mathbf{e}_1$ and $\mathbf{v}_2 = \mathbf{e}_2$. Then we compute

$\mathbf{w}_1 = T(\mathbf{v}_1) = A\mathbf{v}_1 = (1,3,5)$
$\mathbf{w}_2 = T(\mathbf{v}_2) = A\mathbf{v}_2 = (2,4,6) \quad .$

There is no condition on $\mathbf{w}_3$, so it can be any vector that completes a basis for R^3, say, $\mathbf{w}_3 = (1,0,0)$.

■ EXERCISES 6.4

1. Suppose the transformation T of R^2 into R^2 is represented, relative to the standard basis $\mathbf{e}_1, \mathbf{e}_2$, by the matrix $A = \begin{bmatrix} 0 & 1 \\ -1 & 0 \end{bmatrix}$. Find the matrices A', A'' representing T relative to the bases

   (a) $\mathbf{v}_1 = (1,3)$, $\mathbf{v}_2 = (4,5)$
   (b) $\mathbf{v}_1 = (6,11)$, $\mathbf{v}_2 = (7,14)$.

2. Suppose that, relative to the basis $\mathbf{v}_1 = (1,2)$, $\mathbf{v}_2 = (-2,1)$ for R^2, a transformation T is represented by the matrix $A = \begin{bmatrix} 1 & 2 \\ 2 & 4 \end{bmatrix}$. Find the matrix A' representing T relative to the basis

   $\mathbf{z}_1 = (14,13)$, $\mathbf{z}_2 = (9,8)$.

3. Let A be an $m \times n$ matrix of rank k. Let $\mathbf{v}_{k+1}, \ldots, \mathbf{v}_n$ be a basis for Ker(A), and let $\mathbf{v}_1, \ldots, \mathbf{v}_k$ be additional linearly independent vectors that complete a basis for R^n. Also let $\mathbf{w}_i = T(\mathbf{v}_i)$ for $i = 1, \ldots, k$, and let $\mathbf{w}_{k+1}, \ldots, \mathbf{w}_m$ be additional linearly independent vectors that complete a basis for R^m.

Show that relative to the bases $\mathbf{v}_1, \ldots, \mathbf{v}_n$ and $\mathbf{w}_1, \ldots, \mathbf{w}_m$ for R^n and R^m respectively, the transformation $T(\mathbf{x}) = A\mathbf{x}$ is represented by the matrix A' whose first k diagonal entries are 1's, and whose other entries are all zeros.

4. Let V be the subspace of R^3 consisting of all vectors of the form $(a,b,a+b)$. Let W be the subspace of R^3 consisting of all vectors of the form $(p,2p,q)$. Let T be the linear transformation of V into W such that

$$T(a,b,a+b) = (b,2b,a-b) \quad .$$

Find bases $\mathbf{v}_1$, $\mathbf{v}_2$ and $\mathbf{w}_1$, $\mathbf{w}_2$ for V and W respectively, and find the 2×2 matrix A that represents T relative to these bases.

5. Let $B = \begin{bmatrix} 1 & -1 & 0 \\ 4 & -3 & 1 \\ 4 & 0 & 4 \end{bmatrix}$, and let $V = \text{Range}(B)$, so that V is a subspace of R^3. Also, let T be the mapping of V into V such that, for $\mathbf{v}$ in V,

$$T(\mathbf{v}) = B\mathbf{v} \quad .$$

(a) Show that $\dim(V) = 2$, and find a basis $\mathbf{v}_1$, $\mathbf{v}_2$ for V.

(b) Find the 2×2 matrix A representing T relative to the basis just found for V.

*6. Suppose U, V, W are vector spaces with bases $\mathbf{u}_1, \ldots, \mathbf{u}_k$, $\mathbf{v}_1, \ldots, \mathbf{v}_m$, and $\mathbf{w}_1, \ldots, \mathbf{w}_n$ respectively. Suppose that, relative to these bases, the linear mapping F of U into V is represented by the $m \times k$ matrix H, and the mapping of G of V into W is represented by the $n \times m$ matrix K.

(a) Show that the composite mapping $G \circ F$ of U into W is represented by the $n \times k$ matrix KH.

(b) Suppose that $W = U$ and that the given bases for U and W are the same. Show that if $KH = I = n \times n$ identity matrix, then $G \circ F = I =$ identity mapping of U into U, and that the converse is also true.

(c) Compute the inverse A^{-1} of the 2×2 matrix A that was found to represent the transformation T of exercise 4.
Show that A^{-1} represents the transformation S given by

$$S(p,2p,q) = (p+q,p,2p+q) \quad ,$$

and that therefore $S \circ T = I$.

(d) Show that, although the 3×3 matrix B in exercise 5 does not have an inverse, the transformation T defined on the two-dimensional subspace V does have an inverse.

CHAPTER SEVEN

Metric Concepts

Introduction

The theoretical development of the main algebraic ideas encountered so far, including linear systems, matrix algebra, determinants, basis, and dimension, has not been logically dependent on the geometric concepts of length and angle. Rather, geometry has been used to interpret and illustrate the theory. Now, however, we will begin to use ideas from geometry to strengthen the algebraic theory itself.

An example of an important concept that has been developed in a purely algebraic way is that of a basis for a vector space. Bases are important because they allow the vectors in a space to be labeled by sets of coordinates. But the relation between a vector and its coordinates relative to an arbitrary basis is not obvious; in fact the coordinates have to be obtained by following a formal algebraic procedure of matrix reduction.

By contrast, the coordinates of a vector $\mathbf{x}$ in R^n relative to the standard basis $\mathbf{e}_1, \ldots, \mathbf{e}_n$ for R^n are obtained by inspection. They are simply the components of $\mathbf{x}$, which in turn are measured distances along perpendicular coordinate axes.

Bases linked to a geometric structure of perpendicular coordinate axes should therefore have a distinguished place in the further development of linear algebra. These bases, called orthonormal bases, are introduced in section 7.1. They are shown to exist in other spaces besides R^n, and a method for converting ordinary bases to orthonormal bases is presented. Computations that involve orthonormal bases prove to be far simpler than those that do not.

The behavior of lengths and angles between vectors under a linear transformation is a classical geometric problem. In fact, the transformations that test the congruence of spatially separated figures in Euclidean geometry are precisely those that preserve lengths and angles. In section 7.2, we classify the matrix transformations that have this property. They are found to have the remarkable feature that the columns of the representing matrix constitute an orthonormal basis! They are also geometrically identifiable, in R^2 at least, as rotations and reflections.

In section 7.3 we present a parallel development of metric concepts for vector spaces having complex numbers as scalars.

7.1 Orthonormal Bases

The standard basis e_1, e_2 for R^2 is especially important and easy to use. This basis consists of two unit vectors along the Cartesian coordinate axes for the plane. So the elements of the basis are perpendicular unit vectors (see Figure 7.1).

But the placement of the Cartesian axes in the plane is somewhat arbitrary; we know that any pair of perpendicular axes would serve equally well. This suggests that any two perpendicular unit vectors u_1, u_2 would form a basis for R^2 with special properties comparable to those of the standard basis.

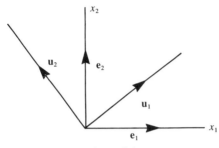

Figure 7.1

It is convenient to recast this idea in terms of the dot product, so as to be able to work in R^n and translate geometric concepts into algebra efficiently.

Orthonormal Sets

We say that vectors u, v in R^n are *orthogonal* if $u \cdot v = 0$. In R^2 and R^3, this simply means that the vectors are perpendicular.

An *orthonormal set* in R^n consists of unit vectors $u_1, \ldots, u_k$ that are orthogonal to each other. That is,

$$|u_i| = 1, \text{ for } i = 1, \ldots, k$$

and $\quad u_i \cdot u_j = 0, \text{ for } i \neq j \quad .$

Example 1 The vectors $v_1 = (3,4)$, $v_2 = (-8,6)$ are orthogonal, since $v_1 \cdot v_2 = (3,4) \cdot (-8,6) = -24 + 24 = 0$. However they do not form an orthonormal set, since they are not unit vectors:

$$|v_1| = (3^2 + 4^2)^{\frac{1}{2}} = 5, |v_2| = (8^2 + 6^2)^{\frac{1}{2}} = 10 \quad .$$

Example 2 Suppose we divide the orthogonal vectors v_1, v_2 by their lengths:

$$\frac{v_1}{|v_1|} = \left(\tfrac{1}{5}\right)(3,4), \quad \frac{v_2}{|v_2|} = \left(\tfrac{1}{10}\right)(-8,6) \quad .$$

Then we obtain unit vectors that are, of course, still orthogonal. That is, we obtain an orthonormal set

$$u_1 = \left(\tfrac{3}{5}, \tfrac{4}{5}\right), \quad u_2 = \left(-\tfrac{4}{5}, \tfrac{3}{5}\right) \quad .$$

Example 3 The standard basis vectors $e_1, \ldots, e_n$ for R^n evidently form an orthonormal set.

The next theorem shows how easy it is to isolate the distinctive properties of orthonormal sets.

THEOREM 7.1

The vectors $u_1, \ldots, u_k$ of an orthonormal set in R^n have the following properties:

(a) They form a basis for the subspace S for R^n that they generate (and are said to form an *orthonormal basis* for S). In particular, when $k = n$, they form an orthonormal basis for R^n.

(b) The coordinates $c_1, \ldots, c_k$ of a vector u relative to the orthonormal basis $u_1, \ldots, u_k$ are

$$c_i = u \cdot u_i \quad .$$

■**Proof** (a) Suppose $c_1 u_1 + c_2 u_2 + \cdots + c_k u_k = 0$.

If we take the dot product of both sides of this equation with u_1, we obtain

$$c_1(u_1 \cdot u_1) + c_2(u_2 \cdot u_1) + \cdots + c_k(u_k \cdot u_1) = 0 \cdot u_1 \quad ,$$

or $\qquad\qquad c_1(1) + c_2(0) + \cdots + c_k(0) \qquad = 0 \quad ,$

or $\qquad\qquad\qquad\qquad c_1 = 0 \quad .$

Similarly, $c_2, \ldots, c_k$ are all zero, and so $u_1, \ldots, u_k$ are l.i. Also, since $u_1, \ldots, u_k$ span S and are l.i., they form a basis for S. Finally, if $k = n$, we have n linearly independent vectors in R^n, which by theorem 6.5 form a basis for R^n.

(b) Given that $c_1 u_1 + \cdots + c_k u_k = u$, we again take the dot product of the equation with u_1 and this time obtain $c_1 = u \cdot u_1$. Similarly we obtain $c_i = u \cdot u_i$. Q.E.D.

Example 4 The coordinates of the vector $\mathbf{u} = (2,1)$ relative to the orthonormal basis for R^2,

$$\mathbf{u}_1 = \left(\tfrac{3}{5}, \tfrac{4}{5}\right), \mathbf{u}_2 = \left(-\tfrac{4}{5}, \tfrac{3}{5}\right) \quad,$$

are

$$c_1 = \mathbf{u}\cdot\mathbf{u}_1 = (2,1)\cdot\left(\tfrac{3}{5}, \tfrac{4}{5}\right) = 2 \quad,$$

$$c_2 = \mathbf{u}\cdot\mathbf{u}_2 = (2,1)\cdot\left(-\tfrac{4}{5}, \tfrac{3}{5}\right) = -1 \quad.$$

That is,

$$\mathbf{u} = 2\mathbf{u}_1 - \mathbf{u}_2 \quad.$$

Completion of Orthonormal Basis for R^n

Example 5 Given the orthonormal set in R^3

$$\mathbf{u}_1 = \left(\tfrac{1}{3}\right)(1,2,2), \mathbf{u}_2 = \left(\tfrac{1}{3}\right)(2,1,-2) \quad,$$

suppose we want to find a third vector $\mathbf{u}_3$ that completes an orthonormal basis for R^3.

First observe that any vector $\mathbf{x}$ perpendicular to both $\mathbf{u}_1$ and $\mathbf{u}_2$ satisfies the conditions

$$\mathbf{u}_1\cdot\mathbf{x} = 0 \quad \text{or} \quad \left(\tfrac{1}{3}\right)(x_1 + 2x_2 + 2x_3) = 0$$

and

$$\mathbf{u}_2\cdot\mathbf{x} = 0 \quad \text{or} \quad \left(\tfrac{1}{3}\right)(2x_1 + x_2 - 2x_3) = 0 \quad.$$

After canceling the scalar factors of $\tfrac{1}{3}$ from these equations, we solve the linear system for $\mathbf{x}$ in the usual way:

$$\begin{bmatrix} 1 & 2 & 2 \\ 2 & 1 & -2 \end{bmatrix} \rightarrow \begin{bmatrix} 1 & 2 & 2 \\ 0 & -3 & -6 \end{bmatrix} \rightarrow \begin{bmatrix} 1 & 2 & 2 \\ 0 & 1 & 2 \end{bmatrix} \rightarrow \begin{bmatrix} 1 & 0 & -2 \\ 0 & 1 & 2 \end{bmatrix} \quad,$$

or $x_1 = 2x_3$, $x_2 = -2x_3$, and $\mathbf{x} = a(2,-2,1)$. Now we select $a = \tfrac{1}{3}$ so as to obtain a *unit* vector perpendicular to $\mathbf{u}_1$ and $\mathbf{u}_2$, that is, $\mathbf{u}_3 = \left(\tfrac{1}{3}\right)(2,-2,1)$.

The procedure for completing an orthonormal basis for R^n works just as well when there is more than one basis element to be found, as the next example shows.

Example 6 Given $\mathbf{u}_1 = \left(\tfrac{1}{\sqrt{3}}\right)(1,1,1)$, we start to complete an orthonormal basis for R^3 by solving $\mathbf{u}_1\cdot\mathbf{x} = 0$, or $x_1 + x_2 + x_3 = 0$ and obtain

$$\mathbf{x} = a_1(-1,1,0) + a_2(-1,0,1) \quad.$$

We can select the solution $\mathbf{x} = (-1,1,0)$ and divide it by its length to obtain $\mathbf{u}_2 = \left(\tfrac{1}{\sqrt{2}}\right)(-1,1,0)$. To obtain the third basis element $\mathbf{u}_3$, we go on to solve the system

$$\mathbf{u}_1\cdot\mathbf{x} = 0 \qquad \text{or} \qquad x_1 + x_2 + x_3 = 0 \quad,$$

$$\mathbf{u}_2\cdot\mathbf{x} = 0 \qquad \text{or} \qquad -x_1 + x_2 \qquad = 0$$

and find $\mathbf{x} = a(-1,-1,2)$. Therefore, $\mathbf{u}_3 = \left(\tfrac{1}{\sqrt{6}}\right)(-1,-1,2)$.

Construction of Orthonormal Basis for Subspace S of R^n

Whenever $\mathbf{v}_1$ and $\mathbf{v}_2$ are nonparallel vectors in R^n, they generate a subspace S of dimension two. However, unless $\mathbf{v}_1$ and $\mathbf{v}_2$ happen to be orthogonal, they cannot be adjusted in length to form an orthonormal basis for S.

We do obtain one unit vector in S by defining

$$\mathbf{u}_1 = \frac{\mathbf{v}_1}{|\mathbf{v}_1|} \quad .$$

Although we could find all the vectors $\mathbf{x}$ in R^n orthogonal to $\mathbf{u}_1$ by solving $\mathbf{u}_1 \cdot \mathbf{x} = 0$, it would not be easy to determine which ones were in S. Instead we reason as follows. If $\mathbf{u}_1$, $\mathbf{u}_2$ are an orthonormal basis for S, then by theorem 7.1, $\mathbf{v}_2$ is the sum of vectors parallel to $\mathbf{u}_1$ and $\mathbf{u}_2$:

$$\mathbf{v}_2 = c_1\mathbf{u}_1 + c_2\mathbf{u}_2$$
$$= (\mathbf{v}_2 \cdot \mathbf{u}_1)\mathbf{u}_1 + (\mathbf{v}_2 \cdot \mathbf{u}_2)\mathbf{u}_2 \quad .$$

Now, since we know $\mathbf{v}_2$ and $\mathbf{u}_1$, we know the part, $(\mathbf{v}_2 \cdot \mathbf{u}_1)\mathbf{u}_1$, of $\mathbf{v}_2$ that is parallel to $\mathbf{u}_1$. By subtracting this, we get the part of $\mathbf{v}_2$ parallel to $\mathbf{u}_2$:

$$(\mathbf{v}_2 \cdot \mathbf{u}_2)\mathbf{u}_2 = \mathbf{v}_2 - (\mathbf{v}_2 \cdot \mathbf{u}_1)\mathbf{u}_1 \quad .$$

Or, on solving for $\mathbf{u}_2$ itself,

$$\mathbf{u}_2 = k(\mathbf{v}_2 - (\mathbf{v}_2 \cdot \mathbf{u}_1)\mathbf{u}_1) \quad .$$

Of course the constant k must be chosen so that $\mathbf{u}_2$ is a unit vector. Let us summarize this construction in a theorem.

THEOREM 7.2

An orthonormal basis for the two-dimensional subspace S of R^n generated by the nonparallel vectors $\mathbf{v}_1$, $\mathbf{v}_2$ is

$$\mathbf{u}_1 = \frac{\mathbf{v}_1}{\text{(length numerator)}} \quad ,$$

$$\mathbf{u}_2 = \frac{(\mathbf{v}_2 - (\mathbf{v}_2 \cdot \mathbf{u}_1)\mathbf{u}_1)}{\text{(length numerator)}} \quad .$$

Example 7   To find an orthonormal basis for the subspace S of R^3 generated by $\mathbf{v}_1 = (1, -2, 2)$, $\mathbf{v}_2 = (4, -5, 2)$, we first compute

$$|\mathbf{v}_1| = 3, \ \mathbf{u}_1 = \frac{\mathbf{v}_1}{|\mathbf{v}_1|} = \left(\tfrac{1}{3}\right)(1, -2, 2) \quad .$$

Then $\mathbf{v}_2 \cdot \mathbf{u}_1 = (4,-5,2) \cdot \left(\frac{1}{3}\right)(1,-2,2) = \left(\frac{1}{3}\right)(18) = 6$,

$\qquad \mathbf{v}_2 - (\mathbf{v}_2 \cdot \mathbf{u}_1)\mathbf{u}_1 = (4,-5,2) - 6\left(\left(\frac{1}{3}\right)(1,-2,2)\right) = (2,-1,-2)$,

and $\mathbf{u}_2 = \dfrac{(2,-1,-2)}{\text{(length numerator)}} = \left(\frac{1}{3}\right)(2,-1,-2)$.

The idea underpinning theorem 7.2 extends further; for example, if $\mathbf{u}_1$, $\mathbf{u}_2$, $\mathbf{u}_3$ are an orthonormal basis for a subspace S of R^n, then, by theorem 7.1, any vector $\mathbf{v}$ in S is the sum of three vectors parallel to $\mathbf{u}_1$, $\mathbf{u}_2$, $\mathbf{u}_3$ respectively:

$\qquad \mathbf{v} = (\mathbf{v} \cdot \mathbf{u}_1)\mathbf{u}_1 + (\mathbf{v} \cdot \mathbf{u}_2)\mathbf{u}_2 + (\mathbf{v} \cdot \mathbf{u}_3)\mathbf{u}_3$.

Now, if we know $\mathbf{v}$, $\mathbf{u}_1$, $\mathbf{u}_2$, then we can compute the parts of $\mathbf{v}$ parallel to $\mathbf{u}_1$ and $\mathbf{u}_2$ and so we can deduce the part of $\mathbf{v}$ parallel to $\mathbf{u}_3$:

$\qquad (\mathbf{v} \cdot \mathbf{u}_3)\mathbf{u}_3 = \mathbf{v} - (\mathbf{v} \cdot \mathbf{u}_1)\mathbf{u}_1 - (\mathbf{v} \cdot \mathbf{u}_2)\mathbf{u}_2$.

Then we divide the right side by its length to obtain a unit vector $\mathbf{u}_3$.

Example 8   Let S be the subspace of R^4 spanned by

$\qquad \mathbf{v}_1 = (1,1,1,1),\ \mathbf{v}_2 = (2,0,2,0),\ \mathbf{v}_3 = (5,9,-3,1)$.

The steps in constructing an orthonormal basis for S are as follows:

Step 1. $|\mathbf{v}_1| = 2,\ \mathbf{u}_1 = \mathbf{v}_1/|\mathbf{v}_1| = \left(\frac{1}{2}\right))(1,1,1,1)$.

Step 2. $\mathbf{v}_2 \cdot \mathbf{u}_1 = (2,0,2,0) \cdot \left(\frac{1}{2}\right)(1,1,1,1) = 2$

$\qquad\quad \mathbf{v}_2 - (\mathbf{v}_2 \cdot \mathbf{u}_1)\,\mathbf{u}_1 = (2,0,2,0) - 2\left(\frac{1}{2}\right)(1,1,1,1)$
$\qquad\qquad\qquad\qquad\qquad = (1,-1,1,-1)$,

and $\mathbf{u}_2 = \dfrac{(1,-1,1,-1)}{\text{(length numerator)}} = \left(\frac{1}{2}\right)(1,-1,1,-1)$.

Step 3. $\mathbf{v}_3 \cdot \mathbf{u}_1 = (5,9,-3,1) \cdot \left(\frac{1}{2}\right)(1,1,1,1) = 6$,
$\qquad\quad \mathbf{v}_3 \cdot \mathbf{u}_2 = (5,9,-3,1) \cdot \left(\frac{1}{2}\right)(1,-1,1,-1) = -4$.

Therefore, $\mathbf{v}_3 - (\mathbf{v}_3 \cdot \mathbf{u}_1)\mathbf{u}_1 - (\mathbf{v}_3 \cdot \mathbf{u}_2)\mathbf{u}_2$
$\qquad\qquad = (5,9,-3,1) - 6\left(\frac{1}{2}\right)(1,1,1,1) - (-4)\left(\frac{1}{2}\right)(1,-1,1,-1)$
$\qquad\qquad = (4,4,-4,-4)$,

and $\mathbf{u}_3 = \dfrac{(4,4,-4,-4)}{\text{(length numerator)}}$
$\qquad\quad = (4,4,-4,-4)/8 = \left(\frac{1}{2}\right)(1,1,-1,-1)$.

By repeating the arguments given above, we can easily verify that the orthogonalisation process illustrated in the last few examples works for all subspaces of R^n.

Suppose the l.i. vectors $\mathbf{v}_1, \ldots, \mathbf{v}_k$ in R^n generate the k dimensional subspace S of R^n.

Then the elements $\mathbf{u}_1, \ldots, \mathbf{u}_k$ of an orthonormal basis can be constructed in turn by the following *Gram-Schmidt* process:

$$\mathbf{u}_1 = \frac{\mathbf{v}_1}{\text{(length numerator)}}$$

$$\mathbf{u}_2 = \frac{(\mathbf{v}_2 - (\mathbf{v}_2 \cdot \mathbf{u}_1)\,\mathbf{u}_1)}{\text{(length numerator)}}$$

$$\cdots\cdots\cdots\cdots\cdots\cdots\cdots\cdots\cdots\cdots\cdots\cdots$$

$$\mathbf{u}_k = \frac{(\mathbf{v}_k - (\mathbf{v}_k \cdot \mathbf{u}_1)\mathbf{u}_1 - \ldots - (\mathbf{v}_k \cdot \mathbf{u}_{k-1})\mathbf{u}_{k-1})}{\text{(length numerator)}} \quad .$$

We refrain from presenting further examples of the process with subspaces S of large dimension. The calculation, although always similar in form to the examples already done, can become numerically unmanageable even when the given generators $\mathbf{v}_1, \ldots, \mathbf{v}_k$ have integer components. The reason is that the process of dividing a vector by its length to obtain a unit vector, called *normalization* of the vector, usually introduces square roots, which accumulate as the calculation progresses.

■ **Problem**   Find an orthonormal basis $\mathbf{u}_1, \mathbf{u}_2, \mathbf{u}_3$ for R^3 such that $\mathbf{u}_1$ is parallel to $(2,2,1)$ and $\mathbf{u}_2$ is orthogonal to $\mathbf{e}_2$.

■ **Solution**   Since the direction of $\mathbf{u}_1$ is given,

$$\mathbf{u}_1 = \frac{(2,2,1)}{\text{(length numerator)}}$$

or $\qquad\qquad \mathbf{u}_1 = \left(\tfrac{1}{3}\right)(2,2,1) \quad .$

Now if a vector $\mathbf{x} = (a,b,c)$ is orthogonal to $\mathbf{e}_2$, then

$$\mathbf{x} \cdot \mathbf{e}_2 = (a,b,c) \cdot (0,1,0) = b = 0 \quad .$$

So $\mathbf{u}_2$ has the form $\mathbf{u}_2 = (a,0,c)$. But $\mathbf{u}_2$ is orthogonal to $\mathbf{u}_1$ as well,

and therefore $\quad \mathbf{u}_2 \cdot \mathbf{u}_1 = (a,0,c) \cdot (2,2,1) = 2a+c = 0 \;$ or $\; c = -2a \quad .$

That is, $\qquad\qquad \mathbf{u}_2 = (a,0,-2a) = a(1,0,-2) \quad .$

We can choose $\qquad \mathbf{u}_2 = \frac{(1,0,-2)}{\text{(length numerator)}}$

$$= \left(\tfrac{1}{\sqrt{5}}\right)(1,0,-2) \quad .$$

Now if a vector $\mathbf{x} = (x_1, x_2, x_3)$ is orthogonal to $\mathbf{u}_2$ and $\mathbf{u}_1$, then

$$-2x_3 = 0 \quad,$$

and $\quad 2x_1 + \quad 2x_2 + x_3 = 0 \quad,$

which quickly reduces to $\mathbf{x} = c(4, -5, 2)$.

Therefore $\qquad \mathbf{u}_3 = \dfrac{(4, -5, 2)}{(\text{length numerator})}$

$$= \tfrac{1}{3\sqrt{5}}(4, -5, 2) \quad .$$

■ EXERCISES 7.1

1. For each of the following orthonormal sets of vectors $\mathbf{u}_1, \mathbf{u}_2, \ldots$, complete an orthonormal basis $\mathbf{u}_1, \ldots, \mathbf{u}_n$ for R^n, Also, find the coordinates $c_1, \ldots, c_n$ of the given vector $\mathbf{w}$ relative to this orthonormal basis.

(a) $\mathbf{u}_1 = \left(\tfrac{1}{\sqrt{10}}\right))(3,1)$, $\mathbf{w} = (1,4)$

(b) $\mathbf{u}_1 = \left(\tfrac{1}{13}\right)(5,12)$, $\mathbf{w} = (3,7)$

(c) $\mathbf{u}_1 = \left(\tfrac{1}{9}\right)(1,8,-4)$, $\mathbf{u}_2 = \left(\tfrac{1}{9}\right)(8,1,4)$, $\mathbf{w} = (23,22,3)$

(d) $\mathbf{u}_1 = \left(\tfrac{1}{3}\right)(1,-2,2)$, $\mathbf{u}_2 = \left(\tfrac{1}{15}\right)(2,11,10)$, $\mathbf{w} = (2,1,0)$

(e) $\mathbf{u}_1 = \left(\tfrac{1}{\sqrt{2}}\right)(1,1,0,0)$, $\mathbf{u}_2 = \left(\tfrac{1}{\sqrt{2}}\right)(0,0,1,1)$,
$\quad \mathbf{u}_3 = \left(\tfrac{1}{2}\right)(-1,1,1,-1)$, $\mathbf{w} = (-3,3,-1,1)$

2. Each of the following sets of vectors $\mathbf{v}_1, \mathbf{v}_2, \ldots$ generates a subspace of R^n. Find an orthonormal basis $\mathbf{u}_1, \mathbf{u}_2, \ldots$ for S, such that $\mathbf{u}_1$ is parallel to $\mathbf{v}_1$. Also, determine whether the given vector $\mathbf{w}$ is in S, and if so, determine its coordinates relative to the orthonormal basis.

(a) $\mathbf{v}_1 = (0,1,1)$, $\mathbf{v}_2 = (1,2,3)$, $\mathbf{w} = (3,-1,2)$

(b) $\mathbf{v}_1 = (2,3,6)$, $\mathbf{v}_2 = (-1,9,4)$, $\mathbf{w} = (3,8,-5)$

(c) $\mathbf{v}_1 = (2,6,9)$, $\mathbf{v}_2 = (18,32,15)$, $\mathbf{w} = (15,1,-4)$

(d) $\mathbf{v}_1 = (1,1,1,1)$, $\mathbf{v}_2 = (4,4,-2,-2)$, $\mathbf{v}_3 = (6,0,2,-4)$,
$\quad \mathbf{w} = (6,4,3,1)$

3. Let $A = \begin{bmatrix} -2 & 1 & 0 & 2 \\ 1 & -1 & 1 & -2 \\ 0 & 1 & -2 & 2 \end{bmatrix}$. Find an orthonormal basis:

(a) for Range(A).

(b) for Ker(A).

4. Suppose $\mathbf{u}_1, \ldots, \mathbf{u}_k$ is an orthonormal basis for a subspace S of R^n. Suppose $\mathbf{w}$ is in S, and has coordinates $\mathbf{c} = (c_1, \ldots, c_k)$ relative to the above basis.

(a) Show that $|\mathbf{w}| = (c_1^2 + \cdots + c_k^2)^{\frac{1}{2}} = |\mathbf{c}|$.

(b) Assuming that $\mathbf{z}$ is another vector in S with coordinates $\mathbf{d} = (d_1, \ldots, d_k)$, show that $\mathbf{w} \cdot \mathbf{z} = \mathbf{c} \cdot \mathbf{d}$.

(c) Deduce that the angle between $\mathbf{w}$ and $\mathbf{z}$ in R^n is the same as the angle between $\mathbf{c}$ and $\mathbf{d}$ in R^k.

5. (a) Find an orthonormal basis $\mathbf{u}_1$, $\mathbf{u}_2$, $\mathbf{u}_3$ for R^3 such that
   (i) $\mathbf{u}_1$ is parallel to the vector (4,4,7), and
   (ii) $\mathbf{u}_2$ is parallel to the x_1x_2 coordinate plane.
   (b) How many such bases are there?

*6. Describe how the Gram-Schmidt process could be modified to cover the case when the generators $\mathbf{v}_1, \ldots, \mathbf{v}_k$ are not known to be linearly independent.

7.2 Orthogonal Transformations

In Chapter 4 we introduced a few important types of geometric transformations and used matrix methods to study them. Now we are ready to use another strategy: we will specify a significant geometric condition on transformations and use matrix algebra to classify the transformations that have the stated property.

The first step is to express the geometric concept of orthogonal vectors in matrix terms. In fact, a column vector $\mathbf{u}$ in R^n is an $n \times 1$ matrix, and so its transpose $\mathbf{u}^t$ is a $1 \times n$ matrix. Therefore the product $\mathbf{u}^t\mathbf{v}$ makes sense and is a 1×1 matrix, in effect a scalar:

$$\mathbf{u}^t\mathbf{v} = [u_1 \ldots u_n] \begin{bmatrix} v_1 \\ \cdot \\ \cdot \\ \cdot \\ v_n \end{bmatrix} = u_1 v_1 + \cdots + u_n v_n \quad .$$

That is, $$\mathbf{u}^t\mathbf{v} = \mathbf{u} \cdot \mathbf{v} \quad .$$

Therefore the vectors $\mathbf{u}$, $\mathbf{v}$ in R^n are orthogonal if $\mathbf{u}^t\mathbf{v} = 0$.

Example 1 Let $\mathbf{x}$, $\mathbf{y}$ be vectors in R^n, and let C and D be $n \times n$ matrices. Then the dot product $(C\mathbf{x}) \cdot (D\mathbf{y})$ of vectors in R^n can be rewritten as a product of four matrices as follows:

$$(C\mathbf{u}) \cdot (D\mathbf{v}) = (C\mathbf{u})^t D\mathbf{v} = \mathbf{u}^t C^t D\mathbf{v} \quad .$$

Note that at the last step the law for the transpose of a product from Chapter 3 was applied.

Now we can proceed to transformations.

A linear transformation T of R^n into R^n is said to be *orthogonal* if it preserves the length of each vector; that is,

$$|T(\mathbf{u})| = |\mathbf{u}| \quad \text{for all } \mathbf{u} \text{ in } R^n \ .$$

Example 2   First, the length of a vector $\mathbf{u}$ in R^2 remains the same when it is rotated through an angle θ, and so the rotation transformation R_θ is an orthogonal transformation of R^2. On similar geometric grounds, a rotation about an axis in R^3 is always an orthogonal transformation of R^3.

Suppose we let the geometric transformation T be represented by the $n \times n$ matrix A, that is, $T(\mathbf{u}) = A\mathbf{u}$. Then we can re-express the condition for the orthogonality of T as an algebraic condition on the matrix A. In fact, the condition that all vectors $\mathbf{u}$ satisfy $|T(\mathbf{u})| = |\mathbf{u}|$ can be rewritten as $|A\mathbf{u}| = |\mathbf{u}|$, or equivalently, $|A\mathbf{u}|^2 = |\mathbf{u}|^2$, or $(A\mathbf{u}) \cdot (A\mathbf{u}) = \mathbf{u} \cdot \mathbf{u}$, or, on converting the dot products to matrix products,

$$\mathbf{u}^t A^t A \mathbf{u} = \mathbf{u}^t \mathbf{u} \quad \text{for all } \mathbf{u} \text{ in } R^n \ .$$

If, for example, it happens that $A^t A = I$, then the condition is certainly satisfied and T is orthogonal. Interestingly, the converse also holds.

THEOREM 7.3.

Suppose the linear transformation T of R^n into R^n is represented by the $n \times n$ matrix A. Then the transformation T is orthogonal if and only if
$A^t A = I$.

■**Proof**   Let $B = A^t A$, and suppose T is orthogonal, so that

$$\mathbf{u}^t B \mathbf{u} = \mathbf{u}^t \mathbf{u} \quad \text{for all } \mathbf{u} \text{ in } R^n \ .$$

We want to deduce that $B = I$. First, to show that the diagonal entries of B satisfy $b_{ii} = 1$, let $\mathbf{u} = \mathbf{e}_1$.

Then, using the 2×2 case for illustration, we get

$$\begin{bmatrix} 1 & 0 \end{bmatrix} \begin{bmatrix} b_{11} & b_{12} \\ b_{21} & b_{22} \end{bmatrix} \begin{bmatrix} 1 \\ 0 \end{bmatrix} = \begin{bmatrix} 1 & 0 \end{bmatrix} \begin{bmatrix} 1 \\ 0 \end{bmatrix} \ , \quad \text{or} \quad b_{11} = 1 \ ,$$

and similarly $b_{ii} = 1$ for all i.

To show that the other entries of B satisfy $b_{ij} = 0$, let $\mathbf{u} = \mathbf{e}_1 + \mathbf{e}_2$. Then

$$\begin{bmatrix} 1 & 1 \end{bmatrix} \begin{bmatrix} b_{11} & b_{12} \\ b_{21} & b_{22} \end{bmatrix} \begin{bmatrix} 1 \\ 1 \end{bmatrix} = \begin{bmatrix} 1 & 1 \end{bmatrix} \begin{bmatrix} 1 \\ 1 \end{bmatrix} \ ,$$

or $b_{11} + b_{12} + b_{21} + b_{22} = 2$,

or $b_{12} + b_{21} = 0$, or $b_{21} = -b_{12}$, and similarly $b_{ji} = -b_{ij}$ for $j \neq i$.

However, from the way in which B is defined, we also have

$$B^t = (A^tA)^t = A^t(A^t)^t = A^tA = B \quad ,$$

which means that $b_{ji} = b_{ij}$ for all i, j. It follows that $b_{ij} = 0$ for $j \neq i$. Therefore, $B = I$.   Q.E.D.

Example 3   Since we know, on geometric grounds, that rotations are orthogonal transformations, their representing matrices should satisfy $A^tA = I$. For example,

$$(R_{60°})^t R_{60°} = \begin{bmatrix} \frac{1}{2} & \frac{\sqrt{3}}{2} \\ \frac{\sqrt{3}}{2} & \frac{1}{2} \end{bmatrix} \begin{bmatrix} \frac{1}{2} & -\frac{\sqrt{3}}{2} \\ \frac{\sqrt{3}}{2} & \frac{1}{2} \end{bmatrix} = \begin{bmatrix} 1 & 0 \\ 0 & 1 \end{bmatrix} = I \quad . \quad \text{Check}$$

An $n \times n$ matrix A that satisfies $A^tA = I$ is said to be an *orthogonal matrix*. Then the previous theorem can be restated as follows:

In R^n, the transformation T is orthogonal (preserves lengths) if and only if the representing matrix A is orthogonal ($A^tA = I$).

But matrix orthogonality reduces, in turn, to vector concepts, as the next theorem shows.

THEOREM 7.4

An $n \times n$ matrix A is orthogonal if and only if its columns form an orthonormal set of vectors.

■**Proof**   Suppose the columns of A are $\mathbf{u}_1, \ldots, \mathbf{u}_n$. Then

$$\begin{aligned} (A^tA)_{ij} &= (\text{row } i \text{ of } A^t) \cdot (\text{column } j \text{ of } A) \\ &= (\text{column } i \text{ of } A) \cdot (\text{column } j \text{ of } A) \\ &= \mathbf{u}_i \cdot \mathbf{u}_j \quad . \end{aligned}$$

Now the orthogonality condition for A means, in particular, that

$$(A^tA)_{ii} = I_{ii} = 1, \text{ or } \mathbf{u}_i \cdot \mathbf{u}_i = 1, \text{ or } |\mathbf{u}_i| = 1 \quad ,$$

and also that for $j \neq i$,

$$(A^tA)_{ij} = I_{ij} = 0, \text{ or } \mathbf{u}_i \cdot \mathbf{u}_j = 0 \quad .$$

But these conditions on the columns $\mathbf{u}_1, \ldots, \mathbf{u}_n$ are precisely the conditions that they must satisfy to form an orthonormal set.   Q.E.D.

Example 4   The vectors $\mathbf{u}_1 = (0,1)$, $\mathbf{u}_2 = (1,0)$ clearly form an orthonormal set, and so the matrix

$$A = [\mathbf{u}_1 | \mathbf{u}_2] = \begin{bmatrix} 0 & 1 \\ 1 & 0 \end{bmatrix}$$

must, by theorem 7.4, be orthogonal and represent an orthogonal transformation. In fact, this matrix represents reflection across the line $y = x$, as we may remember from Chapter 4. More generally, we recognize from geometry that a vector and its mirror image have the same length, and so we realize at this point that all reflections must form a second category of orthogonal transformations.

What further orthogonal transformations, besides rotations and reflections, can there be in R^2? Let us use theorem 7.4 to classify them. If A is a 2×2 orthogonal matrix, then column 1 of A, or $\mathbf{u}_1$, must be a unit vector in R^2. Suppose $\mathbf{u}_1$ is placed at angle ω counterclockwise from the positive x_1-axis. Then, by trigonometry,

$$\mathbf{u}_1 = (\cos \omega, \sin \omega) \quad .$$

Now column 2 of A, or $\mathbf{u}_2$, is also a unit vector in the plane and is perpendicular to $\mathbf{u}_1$. So it is 90° away from $\mathbf{u}_1$ and is therefore placed at angle $\omega + 90°$ or $\omega - 90°$. That is,

$$\text{either} \quad \mathbf{u}_2 = [\cos(\omega + 90°), \sin(\omega + 90°)] = (-\sin \omega, \cos \omega) \quad ,$$

$$\text{or} \quad \mathbf{u}_2 = [\cos(\omega - 90°), \sin (\omega - 90°)] = (\sin \omega, -\cos \omega) \quad .$$

Therefore,

$$A = [\mathbf{u}_1 | \mathbf{u}_2] = \begin{bmatrix} \cos \omega & -\sin \omega \\ \sin \omega & \cos \omega \end{bmatrix} \quad \text{or} \quad \begin{bmatrix} \cos \omega & \sin \omega \\ \sin \omega & -\cos \omega \end{bmatrix} \quad .$$

But the first matrix is of course the rotation R_θ, where $\theta = \omega$ (see section 4.1), and the second is the reflection F_α, where $\alpha = \frac{\omega}{2}$ (see section 4.2). We have therefore proved the following theorem.

THEOREM 7.5

The orthogonal transformations in R^2 are the rotations R_θ and the reflections F_α.

The concept of orthogonal transformation has been presented in terms of length rather than angle. But it is easy to see geometrically that rotations and reflections both preserve the angles between pairs of vectors $\mathbf{u}$, $\mathbf{v}$ in R^2. To determine whether an orthogonal transformation T in R^3 must also preserve angles, suppose

$$\mathbf{x} = T(\mathbf{u}) = A\mathbf{u}, \mathbf{y} = T(\mathbf{v}) = A\mathbf{v} \quad .$$

Then $|\mathbf{x}| = |T(\mathbf{u})| = |\mathbf{u}|$, and similarly $|\mathbf{y}| = |\mathbf{v}|$. Also, let β represent the angle between $\mathbf{x}$ and $\mathbf{y}$, and γ the angle between $\mathbf{u}$ and $\mathbf{v}$. Now from the formula for the angle between vectors in section 2.1,

$$\cos \beta = \frac{(\mathbf{x} \cdot \mathbf{y})}{(|\mathbf{x}||\mathbf{y}|)} = \frac{((A\mathbf{u}) \cdot (A\mathbf{v}))}{(|\mathbf{u}||\mathbf{v}|)}$$

$$= \frac{(\mathbf{u}^t A^t A\mathbf{v})}{(|\mathbf{u}||\mathbf{v}|)} = \frac{(\mathbf{u}^t \mathbf{v})}{(|\mathbf{u}||\mathbf{v}|)}$$

$$= \frac{(\mathbf{u} \cdot \mathbf{v})}{(|\mathbf{u}||\mathbf{v}|)} = \cos \gamma$$

That is, $\beta = \gamma$, and so angles between vectors in R^3 are preserved.

Furthermore, if, as suggested in section 2.1, the above formula for angles between vectors is used to *define* angles in R^n, we would have obtained the same result in R^n.

THEOREM 7.6

Orthogonal transformations in R^n preserve angles between vectors. In particular, an orthogonal transformation maps every pair of orthogonal vectors to another pair of orthogonal vectors.

■ **Problem**   Find all orthogonal matrices of the form

$$A = \begin{bmatrix} * & \frac{12}{13} \\ * & * \end{bmatrix} \quad ,$$

where the stars denote unknown entries.

■ **Solution**   The columns of $A = [\mathbf{u}_1|\mathbf{u}_2]$ are unit vectors. In particular, the second column $\mathbf{u}_2 = \left(\frac{12}{13}, *\right) = \left(\frac{12}{13}, b\right)$ satisfies

$$\left(\tfrac{12}{13}\right)^2 + b^2 = 1 \quad \text{or} \quad b = \pm\tfrac{5}{13} \quad .$$

Suppose, for example, that

$$\mathbf{u}_2 = \left(\tfrac{12}{13}, \tfrac{5}{13}\right) = \left(\tfrac{1}{13}\right)(12, 5) \quad .$$

Then, since $\mathbf{u}_1$ is a unit vector orthogonal to $(12, 5)$, therefore,

$$\mathbf{u}_1 = \pm\left(\tfrac{1}{13}\right)(5, -12) \quad .$$

That is, $A = \left(\tfrac{1}{13}\right)\begin{bmatrix} 5 & 12 \\ -12 & 5 \end{bmatrix}$ or $\left(\tfrac{1}{13}\right)\begin{bmatrix} -5 & 12 \\ 12 & 5 \end{bmatrix}$.

Similarly, if we choose $\mathbf{u}_2 = \left(\tfrac{12}{13}, -\tfrac{5}{13}\right) = \left(\tfrac{1}{13}\right)(12, -5)$,

then $\mathbf{u}_1 = \pm\left(\tfrac{1}{13}\right)(5, 12)$,

and $A = \left(\tfrac{1}{13}\right)\begin{bmatrix} 5 & 12 \\ 12 & -5 \end{bmatrix}$,   or $\left(\tfrac{1}{13}\right)\begin{bmatrix} -5 & 12 \\ -12 & -5 \end{bmatrix}$.

We conclude that there are exactly four matrices of the required form.

■ EXERCISES 7.2

1. Find a 2×2 orthogonal matrix whose first column is the vector
$$\mathbf{u}_1 = \left(\tfrac{1}{13}\right)(5,12).$$

2. Find a 3×3 orthogonal matrix whose first and third columns are, respectively, the vectors
$$\mathbf{u}_1 = \left(\tfrac{1}{13}\right)(3,4,12), \ \mathbf{u}_3 = \left(\tfrac{1}{13}\right)(4,-12,3) \quad .$$

3. Find all orthogonal 2×2 matrices of the form $\begin{bmatrix} \tfrac{1}{2} & * \\ * & * \end{bmatrix}$.

4. Find all orthogonal matrices of the forms

   (a) $\begin{bmatrix} * & * & 1 \\ * & * & * \\ 1 & * & * \\ * & * & 1 \end{bmatrix}$ (b) $\begin{bmatrix} * & 1 & * & * \\ * & * & * & \tfrac{3}{5} \\ * & * & 1 & * \\ * & * & * & * \end{bmatrix}$.

   (b)

5. (a) Show that, with the notations R_θ and F_α for rotations and reflections in R^2 introduced in Chapter 4,
   $$R_\theta F_\alpha = F_{\left(\tfrac{\theta}{2}\right)+\alpha} \quad .$$

   (b) Show that the composite of two orthogonal transformations in R^n is another orthogonal transformation.

   (c) Show that the product of two $n \times n$ orthogonal matrices is another $n \times n$ orthogonal matrix.

   (d) Deduce from (b) and (c) that the product $R_\theta F_\alpha$ in (a) had to be a reflection or rotation.

   (e) Verify that Table 7.1, which shows the products of orthogonal transformations in R^2, is correct.

		Right Factor	
		R_ω	F_α
Left	R_θ	$R_{\theta + \omega}$	$F_{\left(\tfrac{\theta}{2}\right)+\alpha}$
Factor	F_β	$F_{\beta-\left(\tfrac{\omega}{2}\right)}$	$R_{2\beta - 2\alpha}$

Table 7.1

6. (a) Find all 2×2 orthogonal transformations that map $(1,-1)$ to $(1,1)$.

   (b) Find all 2×2 orthogonal transformations that map $(1,1)$ to $(1,0)$.

7. Suppose A is an $n \times n$ orthogonal matrix. Show that

   (a) $A^{-1} = A^t$.
   (b) $AA^t = I$.
   (c) A^t is also an orthogonal matrix.
   (d) the n *rows* of A form an orthonormal set.

8. Suppose A and B are $n \times n$ matrices. Show that, if A is orthogonal and if $A^tBA = I$, then $B = I$.

9. (a) Show that every $n \times n$ orthogonal matrix A satisfies $\det A = \pm 1$.
   (b) Deduce that every orthogonal transformation in R^2 (or R^3) preserves areas (or volumes). (See section 5.3 for geometric background.)
   (c) Is it true that a linear transformation in R^2 (or R^3) that preserves areas (or volumes) must be orthogonal?

10. Show that every permutation matrix is orthogonal. (See exercises 3.3 for discussion of permutation matrices.)

11. Give algebraic and geometric reasons why projections, and shearing transformations other than the identity, cannot be orthogonal.

12. (Orthocomplementary subspaces.) Let V be a subspace of R^n. Let W be the subset of R^n consisting of those vectors that are orthogonal to every vector in V. That is,

    $\mathbf{y}$ is in W if and only if $\mathbf{y} \cdot \mathbf{x} = 0$ for all $\mathbf{x}$ in V .

    Then W is called the *orthocomplement* of V.

    (a) Show that the orthocomplement is always a subspace of R^n.
    (b) Let $\mathbf{v}_1, \ldots, \mathbf{v}_h$ be an orthonormal basis for V, and $\mathbf{w}_1, \ldots, \mathbf{w}_k$ an orthonormal basis for the orthocomplement W of V. Show that $\mathbf{v}_1, \ldots, \mathbf{v}_h, \mathbf{w}_1, \ldots, \mathbf{w}_k$ forms an orthonormal basis for R^n.
    (c) Deduce that if W is the orthocomplement of V, then V is the orthocomplement of W. (The spaces V and W are said to be *orthocomplementary*.)
    (d) Show that orthocomplementary subspaces V and W of R^n satisfy the condition

    $\dim(V) + \dim(W) = n$.

13. Let A be an $m \times n$ matrix.

    (a) Suppose $\mathbf{x}$ is in R^n, and $\mathbf{w}$ is in R^m. Show that

    $(A^t\mathbf{w}) \cdot \mathbf{x} = \mathbf{w} \cdot (A\mathbf{x})$.

(b) Show that **x** is orthogonal to every vector of the form $A^t\mathbf{v}$, if and only if **x** is in $\text{Ker}(A)$.

(c) Deduce that the space $\text{Ker}(A)$ is the orthocomplement of the space Range (A^t), that is, that $\text{Ker}(A)$ and Range (A^t) are orthocomplementary subspaces of R^n.

(d) Show that $\text{Ker}(A^t)$ and Range (A) are orthocomplementary subspaces of R^m.

14. Let V be the subspace of R^n spanned by the vectors $\mathbf{u}_1, \ldots, \mathbf{u}_h$.

Let $A = [\mathbf{u}_1|\ldots|\mathbf{u}_h]$.

Show that the space V and its orthocomplement W are the spaces Range(A) and $\text{Ker}(A^t)$ respectively.

15. Suppose the subspace V of R^4 is spanned by the vectors

$\mathbf{u}_1 = (1,3,0,2)$, $\mathbf{u}_2 = (2,7,1,1)$, $\mathbf{u}_3 = (1,4,1,-1)$.

Find bases for, and the dimensions of, the space V and its orthocomplement W.

∎*7.3 The Complex Case

In this short section we extend the notions of orthonormal basis, orthogonal transformation, and orthogonal matrix to the case where the complex numbers form our number system.

This material may be skipped on first reading; indeed for many applications of linear algebra, the use of complex numbers is never necessary. Yet in such interesting topics as the representation of electron spin in physics, complex vectors and matrices play an essential role.

The reader may wish to refer to the brief review of complex numbers given in Appendix A.

Complex Orthogonal Vectors

The vector space C^n consists of all n-tuples with complex components; thus $\mathbf{w} = (3+4i,12i)$ is a vector in C^2. Every vector **w** in C^n can be expressed in the form

$\mathbf{w} = \mathbf{p} + i\mathbf{q}$,

where **p** and **q** are vectors in R^n. For example,

$\mathbf{w} = (3+4i,12i) = (3,0) + i(4,12) = \mathbf{p} + i\mathbf{q}$,

where $\mathbf{p} = (3,0)$, $\mathbf{q} = (4,12)$.

The *complex conjugate* of the vector $\mathbf{w} = \mathbf{p} + i\mathbf{q}$ is the vector $\mathbf{w}^* = \mathbf{p} - i\mathbf{q}$. In our example,

$$\mathbf{w}^* = (3,0) - i(4,12) = (3-4i, -12i) \quad .$$

Note that, for every complex vector $\mathbf{w}$,

$$\begin{aligned}
\mathbf{w}^* \cdot \mathbf{w} &= (\mathbf{p}+i\mathbf{q}) \cdot (\mathbf{p}-i\mathbf{q}) \\
&= \mathbf{p} \cdot \mathbf{p} + i\mathbf{q} \cdot \mathbf{p} - i\mathbf{p} \cdot \mathbf{q} + \mathbf{q} \cdot \mathbf{q} \\
&= |\mathbf{p}|^2 + |\mathbf{q}|^2 > 0 \quad \text{unless } \mathbf{w} = \mathbf{0} \quad,
\end{aligned}$$

and the *length* of a complex vector $\mathbf{w}$ is defined to be the quantity

$$|\mathbf{w}| = (\mathbf{w}^* \cdot \mathbf{w})^{\frac{1}{2}} \quad .$$

In our example,

$$|\mathbf{w}|^2 = |\mathbf{p}|^2 + |\mathbf{q}|^2 = 3^2 + (4^2 + 12^2) = 169 \quad,$$

or $|\mathbf{w}| = 13$.

Complex vectors $\mathbf{u}, \mathbf{v}$ are said to be *orthogonal* if $\mathbf{u}^* \cdot \mathbf{v} = 0$. For example, $\mathbf{u} = (1, 2i)$ and $\mathbf{v} = (2i, 1)$ are orthogonal because $\mathbf{u}^* = (1, -2i)$ and so

$$\mathbf{u}^* \cdot \mathbf{v} = (1, -2i) \cdot (2i, 1) = 2i - 2i = 0 \quad .$$

Complex vectors $\mathbf{u}_1, \ldots, \mathbf{u}_k$ in C^n are said to form an *orthonormal set* if

$$|\mathbf{u}_i| = 1 \quad, \quad \text{for } i = 1, \ldots, k$$

and $\quad \mathbf{u}_i \cdot \mathbf{u}_j \neq 0 \quad$, for $i \neq j$.

For example, $\mathbf{u}_1 = \left(\frac{1}{5}\right)(3, 4i)$ and $\mathbf{u}_2 = \left(\frac{1}{5}\right)(4, -3i)$ are easily seen to form an orthonormal set in C^2.

The role played by complex orthonormal sets in C^n is closely parallel to that of real orthonormal sets in R^n. In fact, the formula for coordinates relative to a complex orthonormal basis is derived in just the same way as in the real case. Let us simply restate the result in a theorem.

THEOREM 7.1 (Complex Context)

The vectors $\mathbf{u}_1, \ldots, \mathbf{u}_k$ of an orthonormal set in C^n have the following properties:

(i) They form a basis for the subspace S of C^n, which they generate (and are said to form an *orthonormal basis* for S). In particular, when $k = n$, they form an orthonormal basis for C^n.

(ii) The coordinates $c_1, \ldots, c_k$ of a vector $\mathbf{u}$ relative to the orthonormal basis $\mathbf{u}_1, \ldots, \mathbf{u}_k$ are

$$c_i = \mathbf{u} \cdot \mathbf{u}_i^* \quad .$$

Example 1   Relative to the orthonormal basis $\mathbf{u}_1$, $\mathbf{u}_2$ for C^2 just cited, the coordinates of the vector $\mathbf{u} = (5,10i)$ can be computed as follows:

$$c_1 = \mathbf{u} \cdot \mathbf{u}_1^* = \left(\tfrac{1}{5}\right)(5,10i) \cdot (3,-4i) = \left(\tfrac{1}{5}\right)(15+40) = 11$$

$$c_2 = \mathbf{u} \cdot \mathbf{u}_2^* = \left(\tfrac{1}{5}\right)(5,10i) \cdot (4,\ 3i) = \left(\tfrac{1}{5}\right)(20-30) = -2 \quad.$$

To check our result, we compute as follows:

$$c_1\mathbf{u}_1 + c_2\mathbf{u}_2 = 11\left(\tfrac{1}{5}\right)(3,4i) - 2\left(\tfrac{1}{5}\right)(4,-3i)$$
$$= (5,10i) = \mathbf{u}, \text{ as expected.}$$

Unitary Transformations

A linear transformation T of C^n into C^n is said to be *unitary* if it preserves the length of each vector; that is:

$$|T(\mathbf{u})| = |\mathbf{u}| \quad \text{for all } \mathbf{u} \text{ in } C^n \quad.$$

As in the real context, we want to translate this definition into matrix terms. We therefore let A be the complex $n \times n$ matrix that represents the transformation T. Then the condition for T to be unitary is that for all $\mathbf{u}$ in C^n,

$$|A\mathbf{u}| = |\mathbf{u}|$$
or $$|A\mathbf{u}|^2 = |\mathbf{u}|^2$$

or $$(A\mathbf{u})^* \cdot (A\mathbf{u}) = \mathbf{u}^* \cdot \mathbf{u} \quad.$$

Now, the complex conjugate of a product is taken factor by factor (see Appendix A), and this is so whether the factors are scalars, vectors, or matrices. Thus for the matrix-vector product $A\mathbf{u}$, we have $(A\mathbf{u})^* = A^*\mathbf{u}^*$. The condition for T to be unitary can therefore be rewritten

$$(A^*\mathbf{u}^*) \cdot (A\mathbf{u}) = \mathbf{u}^* \cdot \mathbf{u} \quad,$$
or $$(\mathbf{u}^*)^t (A^*)^t A\mathbf{u} = (\mathbf{u}^*)^t \mathbf{u} \quad \text{for all } \mathbf{u} \text{ in } C^n \quad.$$

The matrix $(A^*)^t$ often occurs in conjunction with the complex matrix A; it is called the *transpose conjugate* of A and is denoted by a superscript dagger; i.e., we write $A^\dagger$ for $(A^*)^t$.

Basic properties of the transpose conjugate follow immediately from those of the complex conjugate and the transpose itself; it is easily checked, for example, that

$$(AB)^\dagger = B^\dagger A^\dagger \quad \text{and} \quad (A^\dagger)^\dagger = A \quad.$$

Our earlier theorem 7.3, which described orthogonal transformations, carries over naturally to the complex case.

THEOREM 7.3 (Complex Context)

Suppose the linear transformation T of C^n into C^n is represented by the complex $n \times n$ matrix A.

Then T is unitary if and only if

$$A^\dagger A = I \ .$$

■**Proof**   Let $B = A^\dagger A$. The condition for T to be unitary is then

$$\mathbf{u}^\dagger B \mathbf{u} = \mathbf{u}^\dagger \mathbf{u} \qquad \text{for all } \mathbf{u} \text{ in } C^n \ .$$

Take the case $n = 2$ for simplicity. Just as in the real context, we can deduce that

(i) $b_{11} = 1$,   by using $\mathbf{u} = \mathbf{e}_1$

(ii) $b_{22} = 1$,   by using $\mathbf{u} = \mathbf{e}_2$

(iii) $b_{12} + b_{21} = 0$,   by using $\mathbf{u} = \mathbf{e}_1 + \mathbf{e}_2 = (1,1)$

(iv) $b_{12} - b_{21} = 0$,   by using $\mathbf{u} = \mathbf{e}_1 + i\mathbf{e}_2 = (1,i)$.

Combining (iii) and (iv), we conclude that $b_{12} = b_{21} = 0$. Thus $B = I$ as required.   Q.E.D.

We say that a complex $n \times n$ matrix is *unitary* if $A^\dagger A = I$. With this terminology, the above theorem can be reworded to say that a linear transformation is unitary (preserves lengths) if and only if its representing matrix is unitary.

Finally, unitary matrices, like orthogonal matrices, are characterized by their column structure. The proof of the following theorem is exactly like the one previously given for the real case and is left as an exercise.

THEOREM 7.4 (Complex Context)

An $n \times n$ matrix is unitary if and only if its columns form an orthonormal set.

Example 2   Let $A = \begin{bmatrix} \left(\dfrac{4+3i}{13}\right) & -\dfrac{12i}{13} \\ \dfrac{12}{13} & \left(\dfrac{3+4i}{13}\right) \end{bmatrix}$

Then the columns of A are easily seen to be of unit length; and

$$\mathbf{v}_1^* \cdot \mathbf{v}_2 = \left(\tfrac{1}{13}\right)(4-3i,12) \cdot \left(\tfrac{1}{13}\right)(-12i,3+4i)$$
$$= \left(\tfrac{1}{13}\right)^2((4-3i)(-12i) + 12(3+4i))$$
$$= \left(\tfrac{1}{13}\right)^2(0) = 0 \quad.$$

Therefore A is a unitary matrix.

■ **Problem** Classify the 2×2 unitary matrices.

■ **Solution** A 2×2 matrix $A = [\mathbf{v}_1|\mathbf{v}_2]$ is unitary if its columns $\mathbf{v}_1, \mathbf{v}_2$ form an orthonormal set of complex vectors in C^2. But orthonormal sets come in bunches. Suppose, in fact, that α and β are complex numbers of absolute value 1, and let

$$\mathbf{w}_1 = \alpha\mathbf{v}_1 \quad, \quad \mathbf{w}_2 = \beta\mathbf{v}_2 \quad.$$

Then $\mathbf{w}_1$ and $\mathbf{w}_2$ also have unit length, and

$$\mathbf{w}_1^* \cdot \mathbf{w}_1 = (\alpha\mathbf{v}_1)^* \cdot (\beta\mathbf{v}_2)$$
$$= (\alpha^*\beta)(\mathbf{v}_1^* \cdot \mathbf{v}_2)$$
$$= (\alpha^*\beta)(0) = 0 \quad.$$

Thus $\mathbf{w}_1, \mathbf{w}_2$ also form an orthonormal set.

Clearly, in classifying orthonormal sets in C^2, or equivalently, 2×2 unitary matrices, we need only one representative matrix from each "bunch." One way to select a representative matrix is to require that the first component of $\mathbf{v}_1$ be real and positive and that the second component of $\mathbf{v}_2$ be real and negative. That is,

$$\mathbf{v}_1 = (a,c) \quad, \qquad \text{where } a \geq 0$$
$$\mathbf{v}_2 = (b,d) \quad, \qquad \text{where } d \leq 0 \quad.$$

Then
$$0 = \mathbf{v}_1^* \cdot \mathbf{v}_2 = (a,c^*) \cdot (b,d) = ab + c^*d \quad,$$

or
$$c^* = -(a/d)b$$

or
$$c = -(a/d)b^* \quad.$$

Thus
$$\mathbf{v}_1 = (a, -(a/d)b^*) = -(a/d)(-d, b^*) \quad.$$

But since
$$1 = |\mathbf{v}_1| = (a/d)^2(d^2 + b^2)$$

and
$$1 = |\mathbf{v}_2| = d^2 + b^2 \quad,$$

therefore
$$a/d = -1 \text{ or } d = -a \quad,$$

and in consequence
$$c = -(-1)b^* = b^* \quad.$$

Accordingly, the representative unitary matrix is

$$A = [\mathbf{v}_1|\mathbf{v}_2] = \begin{bmatrix} a & b \\ b^* & -a \end{bmatrix} \quad, \quad \text{where } a \geq 0 \text{ and } |a|^2 + |b|^2 = 1.$$

Also, the general 2×2 unitary matrix has the form

$$A = \begin{bmatrix} \alpha a & \beta b \\ \alpha b^* & -\beta a \end{bmatrix} = \begin{bmatrix} a & b \\ b^* & -a \end{bmatrix} \begin{bmatrix} \alpha & 0 \\ 0 & \beta \end{bmatrix} \cdots$$

$$\cdots \quad, \quad \text{with } a, b \text{ as above}$$
$$\text{and } |\alpha| = |\beta| = 1 \quad.$$

An example of this factorization of a unitary matrix is

$$A = \begin{bmatrix} -i/3 & 2(i-1)/3 \\ 2(1-i)/3 & 1/3 \end{bmatrix} = \begin{bmatrix} 1/3 & 2(1-i)/3 \\ 2(i+i)/3 & -1/3 \end{bmatrix} \begin{bmatrix} -i & 0 \\ 0 & -1 \end{bmatrix}.$$

In this case the parameters are $a = \frac{1}{3}$, $b = 2(1-i)$, $\alpha = -i$, $\beta = -1$.

■ EXERCISES 7.3

1. (a) Verify that $\mathbf{u}_1 = (\frac{1}{13})(5,12i)$, $\mathbf{u}_2 = (\frac{1}{13})(12i,5)$ form an orthonormal basis for C^2.

   (b) Find the coordinates of $\mathbf{u} = (5i,1)$ relative to this basis.

2. (a) Show that the vectors $\mathbf{v}_1 = (2+i,2)$, $\mathbf{v}_2 = (2i,-1-2i)$ are orthogonal but do not form an orthonormal set.

   (b) Divide $\mathbf{v}_1$ and $\mathbf{v}_2$ by their lengths and thereby obtain an orthonormal basis $\mathbf{u}_1$, $\mathbf{u}_2$ for C^2.

   (c) Find the coordinates of the vector $\mathbf{u} = (3,9i)$ relative to this orthonormal basis.

3. The Gram-Schmidt process works in C^n as well as R^n. The formulas that convert k linearly independent vectors $\mathbf{v}_1, \ldots, \mathbf{v}_k$ into an orthonormal basis $\mathbf{u}_1, \ldots, \mathbf{u}_k$ for a subspace S of dimension k are

$$\mathbf{u}_1 = \frac{\mathbf{v}_1}{\text{(length numerator)}}$$

$$\mathbf{u}_1 = \frac{(\mathbf{v}_2 - (\mathbf{v}_2 \cdot \mathbf{u}_1^*)\mathbf{u}_1)}{\text{(length numerator)}}$$

$$\ldots\ldots\ldots\ldots\ldots\ldots\ldots\ldots\ldots\ldots\ldots\ldots\ldots\ldots\ldots\ldots\ldots$$

$$\mathbf{u}_k = \frac{(\mathbf{v}_k - (\mathbf{v}_k \cdot \mathbf{u}_1^*)\mathbf{u}_1 - \cdots - (\mathbf{v}_k \cdot \mathbf{u}_{k-1}^*)\mathbf{u}_{k-1})}{\text{(length numerator)}}.$$

   (a) Obtain an orthonormal basis for the subspace of C^3 spanned by the vectors $\mathbf{v}_1 = (1,2,2i)$, $\mathbf{v}_2 = (-3,6,0)$.

   (b) Verify that the complex Gram-Schmidt formulas given above do, in fact, produce an orthonormal set $\mathbf{u}_1, \ldots, \mathbf{u}_k$.

4. Find a unitary matrix of the form $U = \begin{bmatrix} \frac{3}{5} & * \\ * & * \end{bmatrix}$ that satisfies

   $\det U = i$.

5. (a) Show that if U is unitary, then so are U^t, U^*, and U^{-1}.

   (b) Show that the product of unitary matrices is again unitary.

6. Show that a 2×2 unitary matrix A satisfies $A^\dagger = A$ if and only if it has the form

$$A = H(p,q,r) = \begin{bmatrix} r & p+iq \\ p-iq & -r \end{bmatrix} \quad,$$

where p,q,r are real and $p^2 + q^2 + r^2 = 1$.

*7. Let $\mathbf{p}$ denote the vector (p,q,r) in R^3, and let $H(\mathbf{p})$ denote the matrix $H(p,q,r)$ of the preceding exercise.

(a) Show that if U is a unitary matrix, then

$$UH(\mathbf{p})U^\dagger = H(\mathbf{p}'), \quad \text{where } |\mathbf{p}'| = |\mathbf{p}| \quad .$$

(b) Fix a unitary matrix U. Show that the map from $\mathbf{p}$ to $\mathbf{p}'$, determined by part (a), is a linear map of R^3 into R^3 that preserves lengths, in other words, that

$$\mathbf{p}' = B\mathbf{p} \quad, \quad \text{where } B \text{ is a } 3\times3 \text{ orthogonal matrix.}$$

(c) Suppose the notation $B = f(U)$ is introduced for the 3×3 orthogonal matrix B obtained in part (b) from the 2×2 unitary matrix U. Show that

$$f(U_1 U_2) = f(U_1)f(U_2) \quad \text{and} \quad f(U^{-1}) = f(U)^{-1}.$$

(d) Suppose we write the general 2×2 unitary matrix in the form

$$U = \begin{bmatrix} \alpha a & \beta b \\ \alpha b^* & -\beta a \end{bmatrix} \quad.$$

Show that, if $UH(0,0,1)U^\dagger = H(0,0,1)$, then $a = \pm1$ and $b = 0$. Show also that if, in addition, $UH(1,0,0)U^\dagger = H(1,0,0)$, then $\alpha = -\beta$.

(e) Deduce that, if $UH(\mathbf{p})U^\dagger = H(\mathbf{p})$ for all $\mathbf{p}$, in other words, if $f(U) = I_3$, then $U = \alpha I_2$.

(f) The 2×2 unitary matrices U satisfying $\det U = 1$ are collectively denoted $SU(2)$. Show that, if U is in $SU(2)$ and $f(U) = I_3$, then $U = \pm I_2$.

(g) Deduce that, if U_1 and U_2 are in $SU(2)$ and $f(U_1) = f(U_2)$, then $U_1 = \pm U_2$. That is, deduce that the mapping f of $SU(2)$ into the 3×3 orthogonal matrices is a two-to-one mapping.

CHAPTER EIGHT

Eigenvectors and Applications

Introduction

Many applications of the matrix product require us to compute vector quantities of the form $A^m\mathbf{v}$, as in the fleet predictor and market share predictor formulas of section 3.2. The computation of high powers of a given numerical matrix can be slow, however, and it can be quite difficult to find an algebraic expression for the general power A^m in terms of m. The eigenvector method to be developed in this chapter allows us to overcome these problems in many cases and also provides great insight into the structure of matrices and the transformations they represent.

The basic idea is very simple. Suppose the matrix A maps a vector $\mathbf{v}$ to a scalar multiple λ of itself; that is, $A\mathbf{v} = \lambda\mathbf{v}$. Then repeated application of A to $\mathbf{v}$ only produces additional factors of λ, so that $A^m\mathbf{v} = \lambda^m\mathbf{v}$. In this case the form of the vector quantities $A^m\mathbf{v}$ is transparent. The vectors $\mathbf{v}$ that fit this description are called "eigenvectors" of A, and the only trouble is that, for a typical matrix A, there are not very many of them.

It turns out, however, that all we really need is that an $n \times n$ matrix A have n linearly independent eigenvectors. Since these form a basis for R^n, all other vectors can be described in terms of them, and likewise the action of A on any vector can be described in terms of its action on the eigenvectors.

We begin section 8.1 by showing how easily $A^m\mathbf{v}$ is computed once we have the necessary quota of eigenvectors. Then we turn our attention to the fundamental problem of locating these eigenvectors. The method is straightforward and computational. We present a special sequence of examples that illustrate the various eigenvalue patterns that can occur and that serve as a data base for testing the concepts developed through the rest of the chapter.

When only diagonal matrices are involved, matrix algebra is simplified dramatically, separating into several distinct copies of ordinary, scalar algebra. This can be traced to the fact that the standard basis vectors are always eigenvectors of a diagonal matrix. In section 8.2 we show that comparable simplicity in calculation can be achieved whenever an $n \times n$ matrix A has enough eigenvectors to form a basis for R^n. In fact, the linear transformation represented by the matrix A can be given an alternative description in terms of a diagonal matrix. Matrices enjoying this useful property are said to be "diagonalizable"; the rest of the chapter is devoted to their study.

Thus in section 8.3, we work out the implications of the predictor formulas in various practical concepts, while in section 8.4 we consider processes that evolve continuously in time rather than, for example, in year-by-year steps. The continuous processes are modeled by certain systems of differential equations that our eigenvector methodology deals with easily.

Finally we discuss matrices that have the special algebraic property of being unchanged by transposition. Called "symmetric matrices," they share the surprising feature of always having an orthonormal eigenvector basis, and, in consequence, always being diagonalizable in a way that preserves their geometric significance as transformations. Their theory is summarized in the important "Principal Axes Theorem," whose proof, however, is fairly long and may well be skipped on first reading. Then, in section 8.6, we present a superb geometric application to the classification of quadratic curves and surfaces. It should be noted that the analysis of quadratic forms involved here has extensive applications in mathematics, statistics, and elsewhere.

Sometimes, in order to diagonalize a matrix, complex numbers must be admitted as scalars. Specific attention is given to this case, especially in a subsection of section 8.5 under the heading "Hermitian Matrices." The discussion leans on the complex methods introduced in the starred section 7.3.

There are, as well, some matrices that simply cannot be diagonalized, no matter what number system is used. For them, the concept of "Jordan form" is developed in Chapter 9; it successfully substitutes for the "diagonal form" applicable to the matrices emphasized in this chapter.

8.1 Eigenvectors

This section is divided into three parts. First we introduce the eigenvector concept and explain how the use of eigenvectors facilitates computation. Next we show how to locate the eigenvectors of any given matrix. The illustrative sequence of examples 4 through 9 confirms that the list of eigenvectors can assume a variety of patterns, and that a "shortage" of eigenvectors sometimes occurs. Finally we give some attention to general theory and, in particular, derive conditions under which an adequate complement of eigenvectors is assured.

Suppose that the $n \times n$ matrix A and the vector $\mathbf{v} \neq \mathbf{0}$ in R^n satisfy the *eigenvector equation*

$$A\mathbf{v} = \lambda\mathbf{v} \quad .$$

Then the scalar λ is said to be an *eigenvalue* of the matrix A, and the vector $\mathbf{v}$ is said to be an *eigenvector of A corresponding to the eigenvalue* λ.

Alternative names for eigenvector occur in the literature; they include "characteristic vector" and "proper vector." Similarly, an eigenvalue may be called a "characteristic value" or "proper value."

Example 1 Let $A = \begin{bmatrix} -1 & -6 \\ 1 & 4 \end{bmatrix}$, $\mathbf{v} = \begin{bmatrix} -2 \\ 1 \end{bmatrix}$.

Then $A\mathbf{v} = \begin{bmatrix} -1 & -6 \\ 1 & 4 \end{bmatrix} \begin{bmatrix} -2 \\ 1 \end{bmatrix} = \begin{bmatrix} -4 \\ 2 \end{bmatrix} = 2 \begin{bmatrix} -2 \\ 1 \end{bmatrix} = 2\mathbf{v}$.

Therefore $(-2,1)$ is an eigenvector of the matrix A corresponding to the eigenvalue 2.

As a first application of the eigenvector concept, consider the following lemma and its immediate consequences.

Lemma 8.1 If $A\mathbf{v} = \lambda\mathbf{v}$, then $A^m\mathbf{v} = \lambda^m\mathbf{v}$.

■**Proof** $A^2\mathbf{v} = A(A\mathbf{v}) = A(\lambda\mathbf{v}) = \lambda A\mathbf{v} = \lambda(\lambda\mathbf{v}) = \lambda^2\mathbf{v}$.
Similarly, when A is applied to $A^2\mathbf{v}$, a third factor of λ will be produced, and after m applications of A we will have m factors of λ.   Q.E.D.

Example 2 With A and $\mathbf{v}$ as in example 1, we saw that $\mathbf{v}$ was an eigenvector of A corresponding to eigenvalue 2. It follows that $A^m\mathbf{v} = 2^m\mathbf{v}$.

Admittedly, our shortcut for computing $A^m\mathbf{v}$, as described so far, applies only when $\mathbf{v}$ happens to be an eigenvector of A, which is often not the case. However, when the $n \times n$ matrix A has enough eigenvectors to form a basis for R^n, this limitation effectively disappears.

Lemma 8.2 Suppose that $\mathbf{v}_1, \ldots, \mathbf{v}_n$ are eigenvectors of the $n \times n$ matrix A, corresponding to the eigenvalues $\lambda_1, \ldots, \lambda_n$ respectively. Suppose also that $\mathbf{v}_1, \ldots, \mathbf{v}_n$ form a basis for R^n and that the coordinates of a given vector $\mathbf{v}$ relative to this basis are $c_1, \ldots, c_n$. Then

$A^m\mathbf{v} = c_1\lambda_1^m\mathbf{v}_1 + \ldots + c_n\lambda_n^m\mathbf{v}_n$.

■**Proof** $A\mathbf{v} = A(c_1\mathbf{v}_1 + \ldots + c_n\mathbf{v}_n)$
$= c_1(A\mathbf{v}_1) + \ldots + c_n(A\mathbf{v}_n)$
$= c_1\lambda_1\mathbf{v}_1 + \ldots + c_n\lambda_n\mathbf{v}_n$.

Similarly, application of A m times produces the mth powers of the eigenvalues $\lambda_1, \ldots, \lambda_n$, as required.   Q.E.D.

Example 3 Let $A =$

$\begin{bmatrix} -1 & -6 \\ 1 & 4 \end{bmatrix}$, $\mathbf{v}_1 = \begin{bmatrix} 3 \\ -1 \end{bmatrix}$, $\mathbf{v}_2 = \begin{bmatrix} -2 \\ 1 \end{bmatrix}$, $\mathbf{v} = \begin{bmatrix} 6 \\ -1 \end{bmatrix}$.

It happens that $\mathbf{v}_1$ and $\mathbf{v}_2$ are both eigenvectors of A, corresponding to the eigenvalues $\lambda_1 = 1$, $\lambda_2 = 2$ respectively. From this information, an algebraic formula for $A^m\mathbf{v}$ can be obtained as follows. First, we find the coordinates of $\mathbf{v}$ relative to the basis $\mathbf{v}_1$, $\mathbf{v}_2$ for R^2:

$$[\mathbf{v}_1|\mathbf{v}_2|\mathbf{v}] = \begin{bmatrix} 3 & -2 & | & 6 \\ -1 & 1 & | & -1 \end{bmatrix} \rightarrow \begin{bmatrix} 1 & 0 & | & 4 \\ 0 & 1 & | & 3 \end{bmatrix} \quad , \quad \text{or}$$

$$\mathbf{v} = 4\mathbf{v}_1 + 3\mathbf{v}_2 \quad .$$

Then, by lemma 8.2, we conclude that

$$A^m\mathbf{v} = 4 \cdot 1^m\mathbf{v}_1 + 3 \cdot 2^m\mathbf{v}_2$$

$$= 4 \begin{bmatrix} 3 \\ -1 \end{bmatrix} + 3 \cdot 2^m \begin{bmatrix} -2 \\ 1 \end{bmatrix} = \begin{bmatrix} 12 - 6 \cdot 2^m \\ -4 + 3 \cdot 2^m \end{bmatrix} \quad .$$

Computation of Eigenvectors

Our next objective is to develop methods of determining how many eigenvectors a given matrix has and a procedure for computing them.

Lemma 8.3 The scalar λ is an eigenvalue of the $n \times n$ matrix A if and only if λ satisfies the *characteristic equation*

$$\det(\lambda I - A) = 0 \quad .$$

■Proof The eigenvector equation, $A\mathbf{v} = \lambda\mathbf{v}$, can be rewritten as
$$\lambda I\mathbf{v} = A\mathbf{v}$$
or $\quad (\lambda I - A)\mathbf{v} = \mathbf{0} \quad .$

But this equation has a nonzero solution, $\mathbf{v}$, if and only if rank $(\lambda I - A) < n$, or equivalently, $\det(\lambda I - A) = 0$. 	Q.E.D.

Example 4 Let $A = \begin{bmatrix} 0 & 1 \\ -2 & -3 \end{bmatrix}$. Then

$$\lambda I - A = \begin{bmatrix} \lambda & 0 \\ 0 & \lambda \end{bmatrix} - \begin{bmatrix} 0 & 1 \\ -2 & -3 \end{bmatrix} = \begin{bmatrix} \lambda & -1 \\ 2 & \lambda+3 \end{bmatrix} \quad , \quad \text{and}$$

$$\det(\lambda I - A) = (\lambda)(\lambda + 3) - (-1)(2) = \lambda^2 + 3\lambda + 2 =$$
$$(\lambda + 1)(\lambda + 2) \quad .$$

So $\det(\lambda I - A) = 0$ when $\lambda = -1$ or $\lambda = -2$.

That is, the eigenvalues of A are $\lambda_1 = -1$, $\lambda_2 = -2$.

Also, an eigenvector $\mathbf{v}$ of A corresponding to the eigenvalue λ satisfies $(\lambda I - A)\mathbf{v} = \mathbf{0}$; in other words, $\mathbf{v}$ is in the kernel of the matrix

$$\lambda I - A = \begin{bmatrix} \lambda & -1 \\ 2 & \lambda+3 \end{bmatrix} \quad .$$

CASE $\lambda_1 = -1$: $\lambda I - A = \begin{bmatrix} -1 & -1 \\ 2 & 2 \end{bmatrix}$, which reduces to $\begin{bmatrix} 1 & 1 \\ 0 & 0 \end{bmatrix}$.

The general solution is

$$x_1 = -x_2, \text{ or } \mathbf{x} = \begin{bmatrix} x_1 \\ x_2 \end{bmatrix} = \begin{bmatrix} -x_2 \\ x_2 \end{bmatrix} = x_2 \begin{bmatrix} -1 \\ 1 \end{bmatrix} \quad .$$

Therefore $\mathbf{v}_1 = (-1, 1)$ is an eigenvector of A corresponding to the eigenvalue $\lambda_1 = -1$, and any other eigenvector corresponding to the same eigenvalue is a scalar multiple of $\mathbf{v}_1$.

CASE $\quad \lambda_2 = -2: \quad \lambda I - A = \begin{bmatrix} -2 & -1 \\ 2 & 1 \end{bmatrix} \quad .$

Again, we can reduce this matrix, but note that it *is* a rank 1, 2×2 matrix, and so we *know* that the generating solution is $\pm (-b, a)$, where (a, b) is the first row of the matrix. Therefore $\mathbf{v}_2 = (1, -2)$ is an eigenvector corresponding to eigenvalue $\lambda_2 = -2$.

We conclude that the 2×2 matrix A has enough linearly independent eigenvectors, $\mathbf{v}_1 = (-1, 1)$, $\mathbf{v}_2 = (1, -2)$, to form a basis for R^2.

Some information about the eigenvalues of the general 2×2 matrix can be obtained by direct calculation:

$$\text{If } A = \begin{bmatrix} a & b \\ c & d \end{bmatrix} \quad , \quad \text{then } \lambda I - A = \begin{bmatrix} \lambda - a & -b \\ -c & \lambda - d \end{bmatrix} \quad , \quad \text{and}$$

$\det(\lambda I - A) = (\lambda - a)(\lambda - d) - (-b)(-c) = \lambda^2 - (a+d)\lambda + ad - bc$. That is, the characteristic equation of a 2×2 matrix is always a quadratic in λ, and so a 2×2 matrix always has at most two eigenvalues.

Example 5 Let $A = \begin{bmatrix} 0 & -1 \\ 1 & 0 \end{bmatrix} \quad .$ Then $\lambda I - A = \begin{bmatrix} \lambda & 1 \\ -1 & \lambda \end{bmatrix} \quad ,$

and $\det(\lambda I - A) = (\lambda)(\lambda) - (1)(-1) = \lambda^2 + 1 \quad .$

Therefore, the characteristic equation is $\lambda^2 + 1 = 0$, which has no real solutions. Therefore, assuming that, as usual, we are using the real numbers as our scalars, we must conclude that A has no eigenvalues and, in consequence, no eigenvectors.

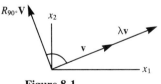

Figure 8.1

There is a geometric interpretation of this result. The matrix A represents rotation by $90°$, and so A rotates *every* vector $\mathbf{v}$ by $90°$. But for $\mathbf{v}$ to be an eigenvector of A, the equation $A\mathbf{v} = \lambda \mathbf{v}$ would have to hold; in other words $A\mathbf{v}$ and $\mathbf{v}$ would have to be parallel. These conditions on $\mathbf{v}$ are inconsistent! (See Figure 8.1.)

Example 5a With A as in example 5, but admitting complex numbers as our scalars, the characteristic equation $\lambda^2 + 1 = 0$ does have roots $\lambda_1 = i$, $\lambda_2 = -i$. The corresponding eigenvectors can then be computed just as in example 4.

CASE $\lambda_1 = i$: $\lambda I - A = \begin{bmatrix} i & 1 \\ -1 & i \end{bmatrix}$; $v_1 = \begin{bmatrix} 1 \\ -i \end{bmatrix}$

CASE $\lambda_2 = -i$: $\lambda I - A = \begin{bmatrix} -i & 1 \\ -1 & -i \end{bmatrix}$; $v_2 = \begin{bmatrix} 1 \\ i \end{bmatrix}$.

Example 6 Let $A = \begin{bmatrix} 2 & -2 & 2 \\ 0 & 1 & 1 \\ -4 & 8 & 3 \end{bmatrix}$. Then $\lambda I - A = \begin{bmatrix} \lambda-2 & 2 & -2 \\ 0 & \lambda-1 & -1 \\ 4 & -8 & \lambda-3 \end{bmatrix}$.

It is convenient to expand $\det(\lambda I - A)$ down column 1:

$$\det(\lambda I - A) = (\lambda - 2)\det\begin{bmatrix} \lambda-1 & -1 \\ -8 & \lambda-3 \end{bmatrix} - 0 + 4\det\begin{bmatrix} 2 & -2 \\ \lambda-1 & -1 \end{bmatrix}$$
$$= (\lambda - 2)((\lambda - 1)(\lambda - 3) - (-1)(-8)) + 4(2(-1) - (-2)(\lambda - 1))$$
$$= (\lambda - 2)(\lambda^2 - 4\lambda - 5) + 8(\lambda - 2)$$
$$= (\lambda - 2)(\lambda^2 - 4\lambda + 3)$$
$$= (\lambda - 2)(\lambda - 1)(\lambda - 3) .$$

That is, the eigenvalues are $\lambda_1 = 1$, $\lambda_2 = 2$, $\lambda_3 = 3$.

Now the eigenvectors corresponding to each eigenvalue in turn can be computed.

CASE $\lambda_1 = 1$: $\lambda I - A = \begin{bmatrix} -1 & 2 & -2 \\ 0 & 0 & -1 \\ 4 & -8 & -2 \end{bmatrix} \rightarrow \begin{bmatrix} -1 & 2 & -2 \\ 0 & 0 & -1 \\ 0 & 0 & -10 \end{bmatrix} \rightarrow$

$\begin{bmatrix} -1 & 2 & 0 \\ 0 & 0 & -1 \\ 0 & 0 & 0 \end{bmatrix} \rightarrow \begin{bmatrix} 1 & -2 & 0 \\ 0 & 0 & 1 \\ 0 & 0 & 0 \end{bmatrix}$. Thus the general solution is

$\begin{cases} x_1 = 2x_2 \\ x_3 = 0 \end{cases}$, or $x = \begin{bmatrix} x_1 \\ x_2 \\ x_3 \end{bmatrix} = \begin{bmatrix} 2x_2 \\ x_2 \\ 0 \end{bmatrix} = x_2\begin{bmatrix} 2 \\ 1 \\ 0 \end{bmatrix}$.

Therefore, $\mathbf{v}_1 = (2,1,0)$ is an eigenvector corresponding to eigenvalue $\lambda_1 = 1$.

CASE $\lambda_2 = 2$: $\lambda I - A = \begin{bmatrix} 0 & 2 & -2 \\ 0 & 1 & -1 \\ 4 & -8 & -1 \end{bmatrix} \rightarrow \begin{bmatrix} 4 & -8 & -1 \\ 0 & 1 & -1 \\ 0 & 2 & -2 \end{bmatrix} \rightarrow$

$\begin{bmatrix} 4 & 0 & -9 \\ 0 & 1 & -1 \\ 0 & 0 & 0 \end{bmatrix} \rightarrow \begin{bmatrix} 1 & 0 & -\frac{9}{4} \\ 0 & 1 & -1 \\ 0 & 0 & -0 \end{bmatrix}$. The general solution is

$\begin{cases} x_1 = \left(\frac{9}{4}\right)x_3 \\ x_2 = x_3 \end{cases}$, or $\mathbf{x} = \begin{bmatrix} x_1 \\ x_2 \\ x_3 \end{bmatrix} = \begin{bmatrix} \left(\frac{9}{4}\right)x_3 \\ x_3 \\ x_3 \end{bmatrix} = x_3\begin{bmatrix} \frac{9}{4} \\ 1 \\ 1 \end{bmatrix}$.

Choosing the free variable $x_3 = 4$ to clear fractions, we obtain $\mathbf{v}_2 = (9,4,4)$ as an eigenvector corresponding to eigenvalue $\lambda_2 = 2$.

CASE $\lambda_3 = 3$: $\lambda I - A = \begin{bmatrix} 1 & 2 & -2 \\ 0 & 2 & -1 \\ 4 & -8 & 0 \end{bmatrix} \rightarrow \begin{bmatrix} 1 & 2 & -2 \\ 0 & 2 & -1 \\ 0 & -16 & 8 \end{bmatrix} \rightarrow$

$\begin{bmatrix} 1 & 0 & 1 \\ 0 & 2 & -1 \\ 0 & 0 & 0 \end{bmatrix} \rightarrow \begin{bmatrix} 1 & 0 & -1 \\ 0 & 1 & -\frac{1}{2} \\ 0 & 0 & 0 \end{bmatrix}$. Thus the general solution is

$\begin{cases} x_3 = x_3 \\ x_2 = \left(\frac{1}{2}\right)x_3 \end{cases}$, or $\mathbf{x} = \begin{bmatrix} x_1 \\ x_2 \\ x_3 \end{bmatrix} = \begin{bmatrix} x_3 \\ \left(\frac{1}{2}\right)x_3 \\ x_3 \end{bmatrix} = x_3 \begin{bmatrix} 1 \\ \frac{1}{2} \\ 1 \end{bmatrix}$.

Choosing $x_3 = 2$ to clear fractions, we obtain $\mathbf{v}_3 = (2,1,2)$ as an eigenvector corresponding to the eigenvalue $\lambda_2 = 3$.

It is easily verified that the eigenvectors $\mathbf{v}_1, \mathbf{v}_2, \mathbf{v}_3$ just computed are linearly independent and so form a basis for R^3.

The numerical computation of eigenvalues and eigenvectors becomes lengthier as the size of the matrix increases, but the work can easily be checked along the way. In fact, if we compute a false eigenvalue λ_0, so that $\det(\lambda_0 I - A) \neq 0$, then we will discover the mistake when $\lambda_0 I - A$ reduces to the identity instead of yielding an eigenvector. Furthermore, each eigenvector $\mathbf{v}$ can be verified by multiplying out the product $A\mathbf{v}$, which should equal $\lambda\mathbf{v}$. Thus to check on eigenvector $\mathbf{v}_1$ of example 6, we compute:

$$A\mathbf{v}_1 = \begin{bmatrix} 2 & -2 & 2 \\ 0 & 1 & 1 \\ -4 & 8 & 3 \end{bmatrix} \begin{bmatrix} 2 \\ 1 \\ 0 \end{bmatrix} = \begin{bmatrix} 2 \\ 1 \\ 0 \end{bmatrix} = 1\mathbf{v}_1 = \lambda_1\mathbf{v}_1. \text{ Check.}$$

Triangular Matrices

There are a few special types of matrices whose eigenvalues can be obtained without computation.

Example 7 Consider the upper triangular matrix

$$A = \begin{bmatrix} 2 & 3 & 8 \\ 0 & 3 & 4 \\ 0 & 0 & 4 \end{bmatrix} . \quad \text{Then } \lambda I - A = \begin{bmatrix} \lambda-2 & -3 & -8 \\ 0 & \lambda-3 & -4 \\ 0 & 0 & \lambda-4 \end{bmatrix} .$$

Thus $\lambda I - A$ is also upper triangular, and so its determinant is the product of the diagonal entries

$$\det(\lambda I - A) = (\lambda - 2)(\lambda - 3)(\lambda - 4) .$$

Therefore the eigenvalues of A are 2, 3, and 4. That is, the eigenvalues are simply the diagonal entries of A. It is easy to see that this argument applies to every triangular matrix.

Lemma 8.4 The eigenvalues of a triangular $n \times n$ matrix A are its diagonal entries $a_{11}, \ldots, a_{nn}$.

Even for a triangular matrix, as in example 7, there is no corresponding shortcut for determining the eigen*vectors*. They must be computed for each eigenvalue in turn, just as in the previous examples.

The Characteristic Polynomial and Its Roots

We have seen that, for a 2×2 matrix A, $\det(\lambda I - A)$ is a polynomial of degree 2, and examples 6 and 7 indicate that for a 3×3 matrix it is a polynomial of degree 3. In fact it is easily shown (see exercises) that, for an $n \times n$ matrix, $p(\lambda) = \det(\lambda I - A)$ is a polynomial of degree n, and it is called the *characteristic polynomial of A*. The eigenvalues of A can therefore be described as the roots of the characteristic polynomial. In the next example, we address the problem of locating these roots.

Example 8 Let $A = \begin{bmatrix} 3 & 2 & 4 \\ 2 & 0 & 2 \\ 4 & 2 & 3 \end{bmatrix}$. Then $\lambda I - A = \begin{bmatrix} \lambda - 3 & -2 & -4 \\ -2 & \lambda & -4 \\ -4 & -2 & \lambda - 3 \end{bmatrix}$.

On expansion down column one, we obtain

$$
\begin{aligned}
p(\lambda) &= \det(\lambda I - A) \\
&= (\lambda - 3)\det \begin{bmatrix} \lambda & -2 \\ -2 & \lambda - 3 \end{bmatrix} - (-2)\det \begin{bmatrix} -2 & -4 \\ -2 & \lambda - 3 \end{bmatrix} \cdots \\
&\qquad\qquad \cdots + (-4)\det \begin{bmatrix} -2 & -4 \\ \lambda & -2 \end{bmatrix} \\
&= (\lambda - 3)(\lambda^2 - 3\lambda - 4) + 2(-2\lambda - 2) - 4(4\lambda + 4) \\
&= \lambda^3 - 6\lambda^2 - 15\lambda - 8 \quad .
\end{aligned}
$$

Often the roots of a cubic are complicated numbers that must be approximated by numerical techniques. But since in this case the coefficients of $p(\lambda)$ are integers, there is a reasonable chance that one or more roots may be integers. Let us check on this. First, $p(0) = -8$, $p(1) = -28$, $p(2) = -54$, etc. Evidently $p(\lambda)$ is decreasing as λ increases, and so no root is in sight. Next, try negative integers. Since $p(-1) = 0$, $\lambda = -1$ is a root.

Now, once a root, $\lambda = a$, of a polynomial $p(\lambda)$ has been found, the *remainder theorem* of elementary algebra says that $\lambda - a$ is a factor of $p(\lambda)$. To find the other factor of $p(\lambda)$, we use long division, sometimes called synthetic division.

$$(\lambda - a))\overline{p(\lambda)} \text{ takes the form } (\lambda + 1))\overline{\lambda^3 - 6\lambda^2 - 15\lambda - 8}$$

with the long division showing:

$$\begin{array}{r} \lambda^2 - 7\lambda - 8 \\ \hline (\lambda + 1)\overline{)\ \lambda^3 - 6\lambda^2 - 15\lambda - 8} \\ \lambda^3 + \ \lambda^2 \\ \hline -7\lambda^2 - 15\lambda \\ -7\lambda^2 - \ 7\lambda \\ \hline -8\lambda - 8 \\ -8\lambda - 8 \\ \hline 0 \end{array}$$

That is, $p(\lambda) = (\lambda + 1)(\lambda^2 - 7\lambda - 8) = (\lambda + 1)(\lambda + 1)(\lambda - 8)$,

or $\quad\quad p(\lambda) = (\lambda + 1)^2(\lambda - 8)$.

The eigenvalues of A are therefore $\lambda_1 = -1$, $\lambda_2 = 8$. Calculation of the eigenvectors proceeds according to method.

CASE $\quad \lambda_1 = -1: \quad \lambda I - A =$

$$\begin{bmatrix} -4 & -2 & -4 \\ -2 & -1 & -2 \\ -4 & -2 & -4 \end{bmatrix} \rightarrow \begin{bmatrix} -4 & -2 & -4 \\ 0 & 0 & 0 \\ 0 & 0 & 0 \end{bmatrix} \rightarrow \begin{bmatrix} 1 & \frac{1}{2} & 1 \\ 0 & 0 & 0 \\ 0 & 0 & 0 \end{bmatrix} .$$

The general solution is $x_1 = -\frac{1}{2}x_2 - x_3$, or

$$\mathbf{x} = \begin{bmatrix} x_1 \\ x_2 \\ x_3 \end{bmatrix} = \begin{bmatrix} -\frac{1}{2}x_2 - x_3 \\ x_2 \\ x_3 \end{bmatrix} = x_2 \begin{bmatrix} -\frac{1}{2} \\ 1 \\ 0 \end{bmatrix} + x_3 \begin{bmatrix} -1 \\ 0 \\ 1 \end{bmatrix} .$$

That is, there are two linearly independent eigenvectors corresponding to the eigenvalue $\lambda_1 = -1$:

$$\mathbf{v}_1 = (-1, 2, 0) \text{ and } \mathbf{v}_2 = (-1, 0, 1) .$$

It is an easy exercise to check that $\mathbf{v}_3 = (2, 1, 2)$ is an eigenvector corresponding to the eigenvalue $\lambda_2 = 8$. Furthermore, it is easily checked that $\mathbf{v}_1, \mathbf{v}_2, \mathbf{v}_3$ are an l.i. set and therefore a basis for R^3. So in this example the 3×3 matrix A has enough eigen*vectors* to form a basis for R^3, even though A has fewer than the maximum possible number, n, of eigen*values*.

The final example in this series shows that even when all eigenvalues of a matrix are real numbers, there may be a shortage of eigenvectors.

Example 9 Let $A = \begin{bmatrix} 1 & 2 \\ 0 & 1 \end{bmatrix}$. Then $\lambda I - A = \begin{bmatrix} \lambda-1 & -2 \\ 0 & \lambda-1 \end{bmatrix}$,

and $\det(\lambda I - A) = (\lambda - 1)^2 - (-2)(0) = (\lambda - 1)^2$. So the only eigenvalue of A is $\lambda_1 = 1$. The corresponding eigenvectors are in the kernel of

$$\lambda_1 I - A = \begin{bmatrix} 0 & -2 \\ 0 & 0 \end{bmatrix} \rightarrow \begin{bmatrix} 0 & \boxed{1} \\ 0 & 0 \end{bmatrix} .$$

Thus all the eigenvectors are multiples of $\mathbf{v}_1 = (1,0)$, and these are not enough to form a basis for R^2.

The following basic theorem on eigenvector structure seems reasonable in view of the results of the examples we have just worked through.

Condition for Eigenvector Basis

THEOREM 8.5

Suppose that the characteristic polynomial, $p(\lambda) = \det(\lambda I - A)$, of $n \times n$ matrix A has n distinct roots, $\lambda_1, \ldots, \lambda_n$.

Then A has n linearly independent eigenvectors.

■**Proof** Corresponding to each of the n distinct eigenvalues, $\lambda_1, \ldots,$ λ_n, there is at least one eigenvector; so we have n eigenvectors $\mathbf{v}_1, \ldots, \mathbf{v}_n$ corresponding to eigenvalues $\lambda_1, \ldots, \lambda_n$ respectively. We need only prove that these eigenvectors are l.i.

Note first that, since $A\mathbf{v}_j = \lambda_j \mathbf{v}_j$,

therefore $(\lambda_1 I - A)\mathbf{v}_j = \lambda_1 \mathbf{v}_j - \lambda_j \mathbf{v}_j = (\lambda_1 - \lambda_j)\mathbf{v}_j$.

Suppose now that there are constants $c_1, \ldots, c_n$, not all zero, such that $c_1\mathbf{v}_1 + \ldots + c_n\mathbf{v}_n = \mathbf{0}$. For example, say $c_n \neq 0$. On applying the matrix $\lambda_1 I - A$ to this equation, we obtain

$$c_1(\lambda_1 I - A)\mathbf{v}_1 + c_2(\lambda_1 I - A)\mathbf{v}_2 + \ldots + c_n(\lambda_1 I - A)\mathbf{v}_n = \mathbf{0}$$
or $c_1(\lambda_1 - \lambda_1)\mathbf{v}_1 + c_2(\lambda_1 - \lambda_2)\mathbf{v}_2 + \ldots + c_n(\lambda_1 - \lambda_n)\mathbf{v}_n = \mathbf{0}$
or $d_2\mathbf{v}_2 + \ldots + \qquad\qquad d_n\mathbf{v}_n = \mathbf{0}$,

where $d_j = c_j(\lambda_1 - \lambda_j)$. In particular, d_n is the product of the numbers $c_n \neq 0$ and $\lambda_1 - \lambda_n \neq 0$, and so $d_n \neq 0$.

Similarly, we eliminate the term $d_2\mathbf{v}_2$ by applying $\lambda_2 I - A$ to the new equation, etc., finally reaching an equation of the form $k_n\mathbf{v}_n = \mathbf{0}$ with $k_n \neq 0$, which is impossible since the eigenvector $\mathbf{v}_n$ is nonzero. We conclude that no nonzero linear combination of $\mathbf{v}_1, \ldots, \mathbf{v}_n$ adds to $\mathbf{0}$, and so they are l.i.   Q.E.D.

In particular, the eigenvectors of the matrices in examples 4 and 6 of this section were shown by direct calculation to form bases for R^n. This information could have been deduced from theorem 8.5 as soon as we had determined that these matrices had n distinct eigenvalues.

It is useful to realize that theorem 8.5 can be thought of as two separate theorems: one for the context in which real numbers are used as scalars, and a second for the context of complex numbers as scalars.

THEOREM 8.5 (Real Context)

Let A be an $n \times n$ matrix with real entries, and suppose A has n distinct real eigenvalues. Then A has n linearly independent eigenvectors with components (which therefore form a basis for R^n).

THEOREM 8.5 (Complex Context)

Let A be an $n \times n$ matrix with complex entries, and suppose A has n distinct complex eigenvalues. Then A has n linearly independent eigenvectors with complex components (which therefore form a basis for C^n).

In particular, the 2×2 matrix A of example 5a was found to have two distinct complex eigenvalues $\lambda_1 = i$, $\lambda_2 = -i$. The existence of two linearly independent complex eigenvalues could then have been deduced from theorem 8.5 (complex context). By contrast, this matrix has no real eigenvalues, and so theorem 8.5 (real context) says nothing about it.

When an $n \times n$ matrix has fewer than n distinct eigenvalues, direct calculation of the eigenvectors is necessary to determine whether there are enough of them to form a basis. Two such cases were examples 8 and 9 above. Calculation showed that in example 8 there was an eigenvector basis but that in example 9 there was not.

Eigenspaces

The existence of the eigenvector basis in example 8, despite the shortage of eigenvalues, resulted from the presence of more than one l.i. eigenvector corresponding to the eigenvalue $\lambda_1 = -1$. The following terminology is useful in discussing this type of situation.

Suppose λ is an eigenvalue of the $n \times n$ matrix A. Then the set of all vectors $\mathbf{v}$ satisfying $(\lambda I - A)\mathbf{v} = \mathbf{0}$ is called the *eigenspace of A corresponding to the eigenvalue* λ, and denoted E_λ.

Clearly the eigenspace E_λ consists of all eigenvectors corresponding to eigenvalue λ, plus the $\mathbf{0}$ vector. And since the set E_λ is the kernel of the matrix $\lambda I - A$, it is always a vector space.

In particular, the eigenspaces of the matrix of example 8 are

E_{-1}, with basis $\mathbf{v}_1 = (-1,2,0)$, $\mathbf{v}_2 = (-1,0,1)$
E_8, with basis $\mathbf{v}_3 = (2,1,2)$.

Note that $\dim(E_{-1}) = 2$, while $\dim(E_8) = 1$.

A generalization of theorem 8.5 using eigenspace dimensions is given in exercise 11.

■ **Problem** (a) Suppose the $n \times n$ matrices A and B commute and that $\mathbf{x}$ is an eigenvector of A. Show that the vectors $B\mathbf{x}$, $B^2\mathbf{x}$, $\ldots$, $B^k\mathbf{x}$, $\ldots$, are, when nonzero, further eigenvectors of A and correspond to the same eigenvalue as $\mathbf{x}$ itself.

(b) Show that the matrices

$$A = \begin{bmatrix} 5 & -4 & -2 \\ -4 & 5 & -2 \\ -2 & -2 & 8 \end{bmatrix}, \quad B = \begin{bmatrix} 1 & -2 & 2 \\ -2 & 4 & -4 \\ 1 & -4 & 4 \end{bmatrix}$$

commute and that $\mathbf{x} = (1,1,-4)$ is an eigenvector of A.

(c) How many linearly independent eigenvectors of A can be derived from the given eigenvector $\mathbf{x}$ by the method of part (a)?

■ **Solution** (a) By hypothesis, $A\mathbf{x} = \lambda\mathbf{x}$; also $AB = BA$, and therefore $AB^k = B^k A$. It follows that

$$A(B^k\mathbf{x}) = (AB^k)\mathbf{x} = (B^k A)\mathbf{x}$$
$$= B^k(A\mathbf{x}) = B^k(\lambda\mathbf{x}) = \lambda(B^k\mathbf{x}) \quad ;$$

in other words $B^k\mathbf{x}$ is also an eigenvector of A and corresponds to the eigenvalue λ as required.

(b) For the given numerical matrices A and B and vector $\mathbf{x}$, it is easily verified that $AB = BA$ and that $A\mathbf{x} = 9\mathbf{x}$.

(c) By straightforward computation,

$$B\mathbf{x} = 9(-1,2,-2)$$

and $\qquad B^2\mathbf{x} = 81(-1,2,-2)$.

Similarly, $\qquad B^k\mathbf{x} = 9^k(-1,2,-2)$.

We conclude that the method yields just one extra l.i. eigenvector of A, because all the eigenvectors generated are multiples of $\mathbf{y} = (-1,2,-2)$.

■ EXERCISES 8.1

1. Let $A = \begin{bmatrix} 4 & 3 \\ 7 & 8 \end{bmatrix}$, $\mathbf{v}_1 = \begin{bmatrix} 1 \\ -1 \end{bmatrix}$, $\mathbf{v}_2 = \begin{bmatrix} 3 \\ 7 \end{bmatrix}$.

(a) Compute $A\mathbf{v}_1$ and $A\mathbf{v}_2$.

(b) Deduce that $\mathbf{v}_1$ and $\mathbf{v}_2$ are eigenvectors of A and find the eigenvalue to which each of them corresponds.

(c) Using part (b), determine the vector $A^{100}\mathbf{v}_1$.

(d) Using part (b), construct a formula for $A^m\mathbf{v}_2$.

2. Let $A = \begin{bmatrix} -11 & -20 \\ 6 & 11 \end{bmatrix}$, $\mathbf{v}_1 = \begin{bmatrix} 5 \\ -3 \end{bmatrix}$, . . .

$$. . . \quad \mathbf{v}_2 = \begin{bmatrix} 2 \\ -1 \end{bmatrix} , \quad \mathbf{v} = \begin{bmatrix} 7 \\ -5 \end{bmatrix} .$$

(a) Show that $\mathbf{v}_1$ and $\mathbf{v}_2$ are eigenvectors of A and find the eigenvalue to which each of them corresponds.

(b) Show that $\mathbf{v}_1$ and $\mathbf{v}_2$ form a basis for R^2 and find the coordinates c_1, c_2 of $\mathbf{v}$ relative to this basis.

(c) Using lemma 8.2, construct a formula for $A^m\mathbf{v}$.

(d) Compute $A^{1001}\mathbf{v}$.

3. Let $A = \begin{bmatrix} 1 & 2 \\ 4 & 3 \end{bmatrix}$, $\mathbf{v} = \begin{bmatrix} 4 \\ -1 \end{bmatrix}$.

(a) Find the characteristic equation of A.

(b) Deduce the eigenvalues and eigenvectors of A.

(c) Construct a formula for $A^m\mathbf{v}$.

4. Compute the eigenvectors for the matrix of example 7 of this section and verify that a basis for R^3 can be constructed from them.

5. For which choices of angle of rotation θ does the rotation matrix R_θ have real eigenvalues? (Geometric reasoning, like that used in the discussion of example 5 of this section, can be used to eliminate most possible values of θ.)

In the next group of exercises, compute all eigenvalues and eigenvectors of each matrix and determine in each case whether the matrix has enough eigenvectors to form a basis for R^n.

6. (a) $\begin{bmatrix} 2 & 2 \\ 2 & -1 \end{bmatrix}$ (b) $\begin{bmatrix} 14 & 16 \\ -9 & -10 \end{bmatrix}$ (c) $\begin{bmatrix} 3 & 2 \\ -5 & 1 \end{bmatrix}$

   (d) $\begin{bmatrix} 3 & 8 \\ 10 & 14 \end{bmatrix}$ (e) $\begin{bmatrix} -2 & 5 \\ 4 & 6 \end{bmatrix}$ (f) $\begin{bmatrix} 5 & 4 \\ -1 & 1 \end{bmatrix}$

7. (a) $\begin{bmatrix} 2 & 1 & 0 \\ 6 & 1 & -1 \\ 0 & 0 & 1 \end{bmatrix}$ (b) $\begin{bmatrix} 3 & 10 & 10 \\ 0 & -2 & -5 \\ 0 & 1 & 2 \end{bmatrix}$ (c) $\begin{bmatrix} 0 & 0 & -1 \\ 1 & 1 & 1 \\ 2 & 0 & 3 \end{bmatrix}$

   (d) $\begin{bmatrix} 2 & 0 & 6 \\ 0 & 3 & 1 \\ 0 & 0 & 3 \end{bmatrix}$ (e) $\begin{bmatrix} 2 & -1 & 0 \\ 2 & 0 & 1 \\ 2 & -2 & -2 \end{bmatrix}$ (f) $\begin{bmatrix} -1 & 1 & 3 \\ 1 & 3 & 1 \\ 3 & 1 & -1 \end{bmatrix}$

   (g) $\begin{bmatrix} 1 & 0 & 0 \\ -2 & 3 & 0 \\ 8 & -4 & 7 \end{bmatrix}$ (h) $\begin{bmatrix} 2 & -6 & -6 \\ -1 & 1 & 2 \\ 3 & -6 & -7 \end{bmatrix}$ (i) $\begin{bmatrix} 2 & 3 & -2 \\ 2 & 3 & 0 \\ 6 & -6 & 7 \end{bmatrix}$

8. (a) $\begin{bmatrix} 0 & -1 & -1 & 0 \\ -1 & 0 & 0 & -1 \\ -1 & 0 & 0 & -1 \\ 0 & -1 & -1 & 0 \end{bmatrix}$ (b) $\begin{bmatrix} 0 & 0 & 1 & 2 \\ 0 & 0 & 2 & 1 \\ 1 & 2 & 0 & 0 \\ 2 & 1 & 0 & 0 \end{bmatrix}$

9. (a) Verify that $\mathbf{x} = (0,1,1)$ is an eigenvector of the matrix
$$A = \begin{bmatrix} 3 & 2 & -2 \\ 2 & 0 & 4 \\ -2 & 4 & 0 \end{bmatrix}, \quad \text{and that}$$
$$B = \begin{bmatrix} 2 & 2 & -8 \\ 2 & -7 & 10 \\ -8 & 10 & -4 \end{bmatrix} \quad \text{commutes with } A.$$

   (b) Use the method outlined in the problem immediately preceding these exercises to obtain further eigenvectors of A. How many l.i. eigenvectors, besides $\mathbf{x}$, are obtained?

10. Suppose that each entry of the square matrix B is linear in x. That is,

   $b_{ij} = p_{ij}x + q_{ij}$, where p_{ij}, q_{ij} are constants.

   (a) Show that when B is 2×2, det B is a polynomial in x of degree ≤ 2.

   (b) Deduce, by Laplace expansion, that when B is 3×3, det B is a polynomial in x of degree ≤ 3.

   (c) Deduce, by Laplace expansion, that when B is $n \times n$, det B is a polynomial in x of degree $\leq n$.

11. Show that for an $n \times n$ matrix A, the characteristic polynomial $\det(\lambda I - A)$ is always a polynomial in λ of degree n. (Use the method of Laplace expansion and the result of the preceding exercise.)

*12. Suppose that the $n \times n$ matrix A has eigenvalues $\lambda_1, \ldots, \lambda_k$; and let $E_1, \ldots, E_k$ denote the corresponding eigenspaces. Show that, if

$$\sum_{i=1}^{k} \dim (E_i) = n \quad,$$

then a basis for R^n can be formed from the eigenvectors of A. (Modify the proof of theorem 8.5.)

*13. Show that, although an eigenvector of an $n \times n$ matrix A is necessarily an eigenvector of A^2 as well, the converse is not true. That is, show that an eigenvector of A^2 is not necessarily an eigenvector of A.

8.2 Diagonalization

Introduction

The method developed in the previous section for computing $A^m \mathbf{v}$ can be modified rather easily into a technique for computing powers A^m of the $n \times n$ matrix A itself.

Observe first that the product of 2×2 diagonal matrices takes the form

$$\begin{bmatrix} a & 0 \\ 0 & b \end{bmatrix} \begin{bmatrix} c & 0 \\ 0 & d \end{bmatrix} = \begin{bmatrix} ac & 0 \\ 0 & bd \end{bmatrix} \ .$$

That is, the matrix multiplication simplifies to a scalar multiplication for each diagonal entry. Similarly,

$$\begin{bmatrix} a & 0 \\ 0 & b \end{bmatrix}^m = \begin{bmatrix} a^m & 0 \\ 0 & b^m \end{bmatrix} \ ,$$

and the same clearly applies to $n \times n$ diagonal matrices.

The indicated strategy for computing A^m is therefore to link A algebraically to a diagonal matrix D.

Diagonalization

In fact, suppose $\mathbf{v}_1, \ldots , \mathbf{v}_n$ are eigenvectors of A corresponding to the eigenvalues $\lambda_1, \ldots , \lambda_n$. Let $P = [\mathbf{v}_1 | \ldots | \mathbf{v}_n]$.

Then $\quad P\mathbf{c} = c_1 \mathbf{v} + \cdots + c_n \mathbf{v}_n$

and $\quad AP\mathbf{c} = \lambda_1 c_1 \mathbf{v}_1 + \cdots + \lambda_n c_n \mathbf{v}_n \ .$

Also, let $D = \begin{bmatrix} \lambda_1 & & \\ & \ddots & \\ & & \lambda_n \end{bmatrix} \ ,$

so that $D\mathbf{c} = \begin{bmatrix} \lambda_1 & & \\ & \ddots & \\ & & \lambda_n \end{bmatrix} \begin{bmatrix} c_1 \\ \vdots \\ c_n \end{bmatrix} = \begin{bmatrix} \lambda_1 c_1 \\ \vdots \\ \lambda_n c_n \end{bmatrix} \ .$

Then $PD\mathbf{c} = \lambda_1 c_1 \mathbf{v}_1 + \cdots + \lambda_n c_n \mathbf{v}_n \ .$

Therefore, $\quad AP\mathbf{c} = PD\mathbf{c} \quad$ for all $\mathbf{c}$ in $R^n \ .$

And so $\quad AP = PD \ .$

The link just established between the square matrix A and the diagonal matrix D is strengthened under the conditions of the next theorem.

THEOREM 8.6

Suppose $\mathbf{v}_1, \ldots, \mathbf{v}_n$ are eigenvectors of the $n \times n$ matrix A, corresponding to eigenvalues $\lambda_1, \ldots, \lambda_n$ respectively, and that these eigenvectors form a basis for R^n.

Let $P = [\mathbf{v}_1|\ldots|\mathbf{v}_n]$, $D = \begin{bmatrix} \lambda_1 & & \\ & \cdot & \\ & & \cdot \\ & & & \lambda_n \end{bmatrix}$. Then

(a) $A = PDP^{-1}$ and the matrix A is said to be *diagonalizable*.

(b) $A^m = PD^mP^{-1}$ for each integer $m \geq 0$.

(c) $A^m = PD^mP^{-1}$ for every integer m, provided that none of the eigenvalues $\lambda_1, \ldots, \lambda_n$ is zero.

■**Proof**   (a) Since $\mathbf{v}_1, \ldots, \mathbf{v}_n$ are a basis for R^n, they are l.i., and so P is invertible. Therefore the equation $AP = PD$ derived above can be multiplied on the right by P^{-1} to produce the stated formula.

(b) $A^2 = PDP^{-1}PDP^{-1} = PDDP^{-1} = PD^2P^{-1}$, and similarly for the mth power.

(c) If the diagonal entries $\lambda_1, \ldots, \lambda_n$ are nonzero, then D is invertible, and:

$$A^{-1} = (PDP^{-1})^{-1} = (P^{-1})^{-1}D^{-1}P^{-1} = PD^{-1}P^{-1} \quad,$$

and similarly for negative integer powers.   Q.E.D.

Example 1   Let $A = \begin{bmatrix} 0 & 1 \\ -2 & -3 \end{bmatrix}$. To obtain a formula for A^m, recall from example 4 of the previous section that A has eigenvectors $\mathbf{v}_1 = (-1,1)$, $\mathbf{v}_2 = (1,-2)$, corresponding to eigenvalues $\lambda_1 = -1$, $\lambda_2 = -2$, and that these eigenvectors form a basis for R^2. Therefore theorem 8.6 applies:

Write $P = [\mathbf{v}_1|\mathbf{v}_2] = \begin{bmatrix} -1 & 1 \\ 1 & -2 \end{bmatrix}$ and compute

$$P^{-1} = \begin{bmatrix} -2 & -1 \\ -1 & -1 \end{bmatrix} \quad.$$

Write $D = \begin{bmatrix} \lambda_1 & \\ & \lambda_2 \end{bmatrix} = \begin{bmatrix} -1 & 0 \\ 0 & -2 \end{bmatrix}$ and compute:

$$D^m = \begin{bmatrix} (-1)^m & 0 \\ 0 & (-2)^m \end{bmatrix} = (-1)^m \begin{bmatrix} 1 & 0 \\ 0 & 2^m \end{bmatrix} \quad.$$

Then $\quad A^m = PD^mP^{-1} = \begin{bmatrix} -1 & 1 \\ 1 & -2 \end{bmatrix} (-1)^m \begin{bmatrix} 1 & 0 \\ 0 & 2^m \end{bmatrix} \cdots$

$$\cdots \begin{bmatrix} -2 & -1 \\ -1 & -1 \end{bmatrix}$$

$$= (-1)^m \begin{bmatrix} -1 & 1 \\ 1 & -2 \end{bmatrix} \begin{bmatrix} -2 & -1 \\ -2^m & -2^m \end{bmatrix}$$

$$= (-1)^m \begin{bmatrix} 2 - 2^m & 1 - 2^m \\ -2 + 2^{m+1} & -1 + 2^{m+1} \end{bmatrix} \quad.$$

In particular, with $m = 10$, and $m = -2$, we obtain

$$A^{10} = \begin{bmatrix} -1{,}022 & -1{,}023 \\ 2{,}046 & 2{,}047 \end{bmatrix} \quad \text{and} \quad A^{-2} = \begin{bmatrix} \frac{7}{4} & \frac{3}{4} \\ -\frac{3}{2} & -\frac{1}{2} \end{bmatrix} \quad.$$

Geometric Interpretation of Diagonalization

Suppose A is an $n \times n$ diagonalizable matrix, as in theorem 8.6. Then, as noted above, the coordinates of the vectors $\mathbf{x}$ and $A\mathbf{x}$, relative to the eigenvector basis $\mathbf{v}_1, \ldots, \mathbf{v}_n$, will be $\mathbf{c}$ and $D\mathbf{c}$ respectively. For this reason, we say that the *diagonal matrix D represents the matrix transformation A, relative to the eigenvector basis.*

Moreover, according to theorem 6.5 the coordinates $\mathbf{c}$ of the vector $\mathbf{x}$, relative to the eigenvector basis, are expressed through the matrix $P = [\mathbf{v}_1|\ldots|\mathbf{v}_n]$ by the formula

$$\mathbf{c} = P^{-1}\mathbf{x} \quad \text{or, equivalently,} \quad \mathbf{x} = P\mathbf{c} \quad.$$

For this reason, we say that P^{-1} *implements the change from old coordinates* (relative to standard basis) *to new coordinates* (relative to eigenvector basis). Similarly, of course, P is said to implement the change from new coordinates back to old.

Finally the formula $A = PDP^{-1}$ means that the matrix transformation A can be applied to each vector $\mathbf{x}$ in three stages; first P^{-1} is applied, then D, then P.

Step 1   *Change to new coordinates* by applying P^{-1} to $\mathbf{x}$, obtaining $c = P^{-1}\mathbf{x}$.

Step 2   *Implement transformation* by applying the representing matrix D to $\mathbf{c}$, obtaining the new coordinates $D\mathbf{c}$ of the transformed vector.

Step 3   *Return to old coordinates* by applying P to $D\mathbf{c}$, thereby recovering the transformed vector $A\mathbf{x}$ itself.

Example 2   Consider the matrix transformation $A = \frac{1}{25} \begin{bmatrix} 7 & 24 \\ 24 & -7 \end{bmatrix} \quad.$

Then the characteristic equation is

$$\det(\lambda I - A) = \det \begin{bmatrix} \lambda - \frac{7}{25} & -\frac{24}{25} \\ -\frac{24}{25} & \lambda + \frac{7}{25} \end{bmatrix} = \lambda^2 - 1 = 0 \quad.$$

Therefore the eigenvalues are $\lambda_1 = 1$, $\lambda_2 = -1$.

The corresponding eigenvectors are easily seen to be $(4,3)$ and $(-3,4)$, or, if we wish to select unit eigenvectors,

$$\mathbf{v}_1 = \left(\tfrac{1}{5}\right)(4,3), \ \mathbf{v}_2 = \left(\tfrac{1}{5}\right)(-3,4) \ .$$

The c_1 and c_2 coordinate axes can then be drawn in the directions of the new basis vectors $\mathbf{v}_1$, $\mathbf{v}_2$, which happen to be perpendicular (see Figure 8.2.).

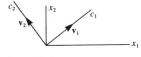

Figure 8.2

Now, relative to the eigenvector basis $\mathbf{v}_1$, $\mathbf{v}_2$, the matrix transformation A is represented by the diagonal matrix

$$D = \begin{bmatrix} \lambda_1 & 0 \\ 0 & \lambda_2 \end{bmatrix} = \begin{bmatrix} 1 & 0 \\ 0 & -1 \end{bmatrix} \ .$$

Therefore $D\mathbf{c} = \begin{bmatrix} 1 & 0 \\ 0 & -1 \end{bmatrix} \begin{bmatrix} c_1 \\ c_2 \end{bmatrix} = \begin{bmatrix} c_1 \\ -c_2 \end{bmatrix} \ .$

That is, in terms of the new coordinates c_1, c_2 of a vector, the effect of the transformation is to reverse the sign of the c_2 coordinate. That is, the transformation is reflection across the c_1-axis. That is, the transformation is reflection across the line ℓ through the origin in direction $\mathbf{u} = \mathbf{v}_1 = \left(\tfrac{1}{5}\right)(4,3)$. (This is, of course, the line $3x - 4y = 0$.)

To check our conclusion, we can verify that the reflection matrix $T_{\mathbf{u}}$ constructed in section 4.2 is indeed the same as the above matrix A:

$$T_{\mathbf{u}} = 2P_{\mathbf{u}} - I = 2 \begin{bmatrix} u_1^2 & u_1 u_2 \\ u_1 u_2 & u_2^2 \end{bmatrix} - I$$

$$= 2 \begin{bmatrix} \frac{16}{25} & \frac{12}{25} \\ \frac{12}{25} & \frac{9}{25} \end{bmatrix} - \begin{bmatrix} 1 & 0 \\ 0 & 1 \end{bmatrix}$$

$$= \begin{bmatrix} \frac{7}{25} & \frac{24}{25} \\ \frac{24}{25} & -\frac{7}{25} \end{bmatrix} = A \ . \qquad \text{Check.}$$

■ **Problem** When an invertible matrix P can be found such that $P^{-1}AP$ and $P^{-1}BP$ are both diagonal, the matrices A and B are said to be *simultaneously diagonalizable*.

Show that if the $n\times n$ matrix A has n distinct eigenvalues, and if B commutes with A, then A and B are simultaneously diagonalizable.

■ **Solution** In view of theorems 8.5 and 8.6, the matrix A is certainly diagonalizable; hence we have a formula

$$A = PDP^{-1}, \text{ or equivalently, } D = P^{-1}AP \ .$$

Now let $G = P^{-1}BP$. Then G and D commute, since

$$GD = (P^{-1}BP)(P^{-1}AP) = P^{-1}BAP$$
$$= P^{-1}ABP = (P^{-1}AP)(P^{-1}BP) = DG \ .$$

But $GD = \begin{bmatrix} g_{11} \cdots g_{1n} \\ \cdot \quad \cdot \\ \cdot \quad \cdot \\ \cdot \quad \cdot \\ g_{n1} \cdots g_{nn} \end{bmatrix} \begin{bmatrix} \lambda_1 \\ \quad \cdot \\ \quad \quad \cdot \\ \quad \quad \quad \lambda_n \end{bmatrix} = \begin{bmatrix} \lambda_1 g_{11} \cdots \lambda_n g_{1n} \\ \cdot \quad \quad \cdot \\ \cdot \quad \quad \cdot \\ \cdot \quad \quad \cdot \\ \lambda_1 g_{n1} \cdots \lambda_n g_{nn} \end{bmatrix}$.

In other words, $(GD)_{ij} = \lambda_j g_{ij}$.

Similarly, $(DG)_{ij} = \lambda_i g_{ij}$,

so we must have $\lambda_j g_{ij} = \lambda_i g_{ij}$ or $(\lambda_j - \lambda_i)g_{ij} = 0$ for all i,j .

But since $\lambda_j - \lambda_i \neq 0$ when $i \neq j$, therefore $g_{ij} = 0$ when $i \neq j$. That is, G is a diagonal matrix as required.

■ EXERCISES 8.2

1. Let $A = \begin{bmatrix} -3 & 1 \\ -4 & 2 \end{bmatrix}$.

   (a) Diagonalize A (that is, find a diagonal matrix D and an invertible matrix P such that $A = PDP^{-1}$).
   (b) Deduce the formula
   $$A^m = \tfrac{1}{3}\begin{bmatrix} -1 & 1 \\ -4 & 4 \end{bmatrix} + \tfrac{(-2)^m}{3}\begin{bmatrix} 4 & -1 \\ 4 & -1 \end{bmatrix} .$$
   (c) Deduce A^{10}.

2. By means of the diagonalization process, obtain a formula for A^m for each of the following square matrices A:

   (a) $\begin{bmatrix} 4 & 3 \\ 7 & 8 \end{bmatrix}$ (b) $\begin{bmatrix} -11 & -20 \\ 6 & 11 \end{bmatrix}$ (c) $\begin{bmatrix} 1 & 2 \\ 4 & 3 \end{bmatrix}$.

3. Diagonalize the matrix A and deduce a geometric interpretation of the transformation it represents in the cases:

   (a) $A = \begin{bmatrix} 0 & 1 \\ 1 & 0 \end{bmatrix}$ (b) $A = \begin{bmatrix} 5 & -2 \\ -2 & 2 \end{bmatrix}$.

4. Let A be a square matrix and let h and k be constants. Show that

   (a) If $\mathbf{v}$ is an eigenvector of A corresponding to the eigenvalue λ, then $\mathbf{v}$ is also an eigenvector of $hA + kI$, corresponding to the eigenvalue $h\lambda + k$.
   (b) If A is diagonalizable, so is $hA + kI$.

5. Prove the following converse of theorem 8.6:

   Suppose $A = PDP^{-1}$, where $P = [v_1|\ldots|v_n]$ is an invertible $n \times n$ matrix and D is the diagonal matrix $D = \begin{bmatrix} \lambda_1 & & \\ & \cdot & \\ & & \cdot \\ & & & \lambda_n \end{bmatrix}$.

   Then $v_1, \ldots, v_n$ are eigenvectors of A corresponding to the eigenvalues $\lambda_1, \ldots, \lambda_n$ respectively.

6. Show that if the $n \times n$ matrices A and B are simultaneously diagonalizable, then A and B commute.

7. For each of the following pairs of matrices, find a simultaneous diagonalization or show that none exists.

   (a) $A = \begin{bmatrix} 3 & 4 \\ 4 & -3 \end{bmatrix}$, $B = \begin{bmatrix} 4 & 2 \\ 2 & 1 \end{bmatrix}$

   (b) $A = \begin{bmatrix} 0 & 1 \\ 1 & 0 \end{bmatrix}$, $B = \begin{bmatrix} 3 & 4 \\ 4 & -3 \end{bmatrix}$

8. Suppose the $n \times n$ matrices A and B commute and that A has the diagonalization $A = PDP^{-1}$, where $D = \begin{bmatrix} 1 & 0 & 0 \\ 0 & 1 & 0 \\ 0 & 0 & 2 \end{bmatrix}$.

   Show that A and B may not be simultaneously diagonalizable.

9. The $n \times n$ matrices A and B are said to be *similar* if there is an invertible matrix Q such that $A = QBQ^{-1}$. (For example, a diagonalizable matrix A is one that is similar to a diagonal matrix D.)

   (a) Show that similar matrices A, B have the same characteristic polynomial and therefore the same eigenvalues.

   (b) Let E_λ, F_λ be the eigenspaces of the similar matrices A, B corresponding to the eigenvalue λ. Show that $\dim(E_\lambda) = \dim(F_\lambda)$.

   (c) Show that if A is diagonalizable, so is any matrix similar to A.

   (d) Show that if A and B are similar, say $A = QBQ^{-1}$, then $A^m = QB^mQ^{-1}$ for positive integers m, and also for negative integers when A is invertible.

   (e) By reference to section 6.4, show that similar matrices represent the same transformation of R^n, relative to different bases for R^n.

*10. Suppose an $n \times n$ matrix has k l.i. eigenvectors $v_1, \ldots, v_k$ corresponding to the eigenvalue $\lambda = \lambda_1$. Show that its characteristic polynomial, $p(\lambda)$, is divisible by $(\lambda - \lambda_1)^k$.

8.3 Applications to Discrete Processes

Introduction

In earlier chapters we obtained predictor formulas for various discrete processes. By directly computing the powers of their governing matrices, we were able to see how these processes would evolve a few steps into the future from a given initial state.

But the eigenvector method that we now have at our disposal allows us, in most cases, to derive a formula, $f(k)$, for the predictor $A^k \mathbf{x}_n$ in terms of the step parameter k. Using this powerful instrument, we can expect to infer the long-range behavior of the processes under study. Rather than take up again the processes considered in previous chapters, however, we leave them to the exercises. Instead we introduce new categories of illustrative material.

Linear models for the fluctuations of biological populations are presented. It is important to realize that population dynamics should not, in general, be modeled by linear systems because resource limitations always put upper limits on population sizes that cannot readily be incorporated in linear theory. Yet the fluctuations of populations about their normal, equilibrium levels can reasonably be described in linear terms, and that is the type of population process we discuss.

All of the practical models discussed in this book are of course prototype models: the simplest meaningful models of a category of processes. In real life, most important processes are influenced by a large number of variables and require vast amounts of time and space for adequate analysis. Prototype models are simply the foundation from which more comprehensive studies can develop. Our population models, in particular, involve only pairs of species and do not reflect the true complexity of ecosystems. They can nevertheless provide insight into the sources of population fluctuation in nature.

We also discuss, briefly in the narrative and at length in the exercises, prototype stochastic processes. These describe the cumulative effects of statistically regulated random events. Some of them were encountered, without due acknowledgement, in earlier chapters, and they are, in any event, central to modern scientific thought.

Example 1   Suppose that, in an agricultural delta, populations of owls and mice normally exist in a predator-prey relationship. Since predators are usually much less numerous than their prey, suppose that the owl population is measured in thousands of birds, the mouse population in millions of animals. Suppose that under normal conditions there are K (thousands) of owls and L (millions) of mice. Suppose, however, that once in many years extreme winter weather conditions reduce the owl population drastically. Our objective is to construct a simple, prototype model describing how the owl and mouse populations might gradually be restored to their normal equilibrium.

Let x_n and y_n denote the *deviations*, during year n, of the owl and mouse populations from their normal levels. (That is, the *actual* populations during year n are $K + x_n$ thousand owls and $L + y_n$ million mice.)

Suppose the *change* in the owl population between year n and year $n + 1$, or $x_{n+1} - x_n$, is given by the formula

$$x_{n+1} - x_n = -\left(\tfrac{1}{10}\right)x_n + \left(\tfrac{2}{10}\right)y_n \quad .$$

The terms $-\left(\tfrac{1}{10}\right)x_n$ and $\left(\tfrac{2}{10}\right)y_n$, which determine the change in the owl population, may be interpreted as follows:

1. A low owl population in year n (i.e. $x_n < 0$) means that there is more food per owl and results in a positive increment, $\left(-\tfrac{1}{10}\right)x_n$, in the owl population next year. Conversely, a high owl population in year n (i.e. $x_n > 0$) causes a decline in the owl population next year.

2. A high mouse population in year n (i.e. $y_n > 0$) also means more food per owl, and so also results in a positive increment, $\left(\tfrac{2}{10}\right)y_n$, in the owl population next year. Conversely, a shortage of mice causes a decline in the owl population next year.

Suppose also that the change in the mouse population between year n and year $n + 1$, or $y_{n+1} - y_n$, is given by the formula

$$y_{n+1} - y_n = -\left(\tfrac{1}{10}\right)x_n - \left(\tfrac{4}{10}\right)y_n \quad .$$

This time, both terms on the right side have minus signs, since low owl and mouse populations both cause increases in the mouse population next year.

The changes in the populations from year to year are then governed by the equations

$$x_{n+1} - x_n = -\left(\tfrac{1}{10}\right)x_n + \left(\tfrac{2}{10}\right)y_n$$
$$y_{n+1} - y_n = -\left(\tfrac{1}{10}\right)x_n - \left(\tfrac{4}{10}\right)y_n \quad ,$$

or

$$x_{n+1} = \left(\tfrac{9}{10}\right)x_n + \left(\tfrac{2}{10}\right)y_n$$
$$y_{n+1} = -\left(\tfrac{1}{10}\right)x_n + \left(\tfrac{6}{10}\right)y_n \quad ,$$

or, with vector notation $\mathbf{x}_n = (x_n, y_n)$,

$$\mathbf{x}_{n+1} = A\mathbf{x}_n, \text{ where } A = \left(\tfrac{1}{10}\right)\begin{bmatrix} 9 & 2 \\ -1 & 6 \end{bmatrix} \quad .$$

Evidently the population vector can be predicted k years ahead with the following *population predictor* formula (compare the predictor formulas of section 3.2):

$$\mathbf{x}_{n+k} = A^k \mathbf{x}_n \quad .$$

Now A has eigenvalues $\lambda_1 = \tfrac{7}{10}$, $\lambda_2 = \tfrac{8}{10}$, with corresponding eigenvectors $\mathbf{v}_1 = (1, -1)$, $\mathbf{v}_2 = (2, -1)$. Therefore A can be diagonalized, with

$$D = \begin{bmatrix} \tfrac{7}{10} & 0 \\ 0 & \tfrac{8}{10} \end{bmatrix} \quad , \quad P = \begin{bmatrix} 1 & 2 \\ -1 & -1 \end{bmatrix} \quad ,$$

and $A^m = PD^mP^{-1} = \begin{bmatrix} 1 & 2 \\ -1 & -1 \end{bmatrix} \begin{bmatrix} \left(\frac{7}{10}\right)^m & 0 \\ 0 & \left(\frac{8}{10}\right)^m \end{bmatrix} \begin{bmatrix} -1 & -2 \\ 1 & 1 \end{bmatrix}$

$$= \left(\tfrac{7}{10}\right)^m \begin{bmatrix} -1 & -2 \\ 1 & 2 \end{bmatrix} + \left(\tfrac{8}{10}\right)^m \begin{bmatrix} 2 & 2 \\ -1 & -1 \end{bmatrix}.$$

Suppose, for example, that as a result of winter conditions the owl population in year 0 is only half its normal level K; in other words its deviation from normal is $x_0 = \left(-\frac{1}{2}\right)K$; while the mouse population is at its normal level, or $y_0 = 0$. Then the population vector is initially $\mathbf{x}_0 = (-K/2)(1,0)$, and in year k is

$$\mathbf{x}_k = A^k\mathbf{x}_0 = \left(\tfrac{K}{2}\right) \begin{bmatrix} \left(\frac{7}{10}\right)^k - 2\left(\frac{8}{10}\right)^k \\ \left(\frac{8}{10}\right)^k - \left(\frac{7}{10}\right)^k \end{bmatrix}.$$

From this formula we can read off the pattern of population change: the owl population gradually increases back to normal, while the mouse population increases for the first few years and then gradually falls back to normal. The numerical population data (with $K = 1$ for convenience) for the first few years is given by Table 8.1.

		Year						
		0	**1**	**2**	**3**	**4**	**5**	**10**
Population	Owls	−.500	−.450	−.395	−.341	−.290	−.244	−.093
Deviation	Mice	0	.050	.075	.085	.085	.080	.040

Table 8.1

Example 2 (Prototype stochastic process) Suppose that three jars, labeled A, B, and C, initially contain masses x_0, y_0, z_0 of sand. Suppose that the following procedure is used to redistribute the sand among the jars. First, the sand is removed from each jar. Then

Of the sand originally in jar A, $\frac{1}{3}$ is placed in jar B and $\frac{2}{3}$ is placed in jar C;
Of the sand originally in jar B, $\frac{1}{3}$ is placed in jar A, and $\frac{2}{3}$ is returned to jar B;
Of the sand originally in jar C, $\frac{2}{3}$ is placed in jar A, and $\frac{1}{3}$ is placed in jar B.

The problem is to find the quantities of sand the jars contain after the redistribution has been repeated m times and to decide whether these quantities approach fixed levels as m increases.

Let the components of the vector $\mathbf{x}_m = (x_m, y_m, z_m)$ denote the masses of sand in jars A, B, C after m repetitions of the redistribution procedure. Then

$$x_{n+1} = \left(\tfrac{1}{3}\right)y_n + \left(\tfrac{2}{3}\right)z_n$$
$$y_{n+1} = \left(\tfrac{1}{3}\right)x_n + \left(\tfrac{2}{3}\right)y_n + \left(\tfrac{1}{3}\right)z_n$$
$$z_{n+1} = \left(\tfrac{2}{3}\right)x_n,$$

or $\quad \mathbf{x}_{n+1} = A\mathbf{x}_n$, where $A = \frac{1}{3}\begin{bmatrix} 0 & 1 & 2 \\ 1 & 2 & 1 \\ 2 & 0 & 0 \end{bmatrix}$.

Note that the matrix $3A$ has integer components; so its eigenvalues are easily computed to be 3, 1, and -2. Therefore, the eigenvalues of A are $\lambda_1 = 1$, $\lambda_2 = \frac{1}{3}$, and $\lambda_3 = -\frac{2}{3}$, and the corresponding eigenvectors are $\mathbf{v}_1 = (3,5,2)$, $\mathbf{v}_2 = (1,-3,2)$, $\mathbf{v}_3 = (1,0,-1)$.

Then $\quad \mathbf{x}_m = A^m \mathbf{x}_0 = PD^m P^{-1} \mathbf{x}_0$,

or $\quad \mathbf{x}_m = \begin{bmatrix} 3 & 1 & 1 \\ 5 & -3 & 0 \\ 2 & 2 & -1 \end{bmatrix} \begin{bmatrix} 1 & 0 & 0 \\ 0 & (\frac{1}{3})^m & 0 \\ 0 & 0 & (-\frac{2}{3})^m \end{bmatrix} \frac{1}{30} \cdots$

$$\cdots \begin{bmatrix} 3 & 3 & 3 \\ 5 & -5 & 5 \\ 16 & -4 & -14 \end{bmatrix} \mathbf{x}_0 \quad .$$

Also, as m increases, the entries $(\frac{1}{3})^m$ and $(-\frac{2}{3})^m$ approach 0, and so $\mathbf{x}_m$ approaches the vector

$$\begin{bmatrix} 3 & 1 & 1 \\ 5 & -3 & 0 \\ 2 & 2 & -1 \end{bmatrix} \begin{bmatrix} 1 & 0 & 0 \\ 0 & 0 & 0 \\ 0 & 0 & 0 \end{bmatrix} \frac{1}{30} \begin{bmatrix} 3 & 3 & 3 \\ 5 & -5 & 5 \\ 16 & -4 & -14 \end{bmatrix} \mathbf{x}_0$$

$$= \frac{1}{10} \begin{bmatrix} 3 & 3 & 3 \\ 5 & 5 & 5 \\ 2 & 2 & 2 \end{bmatrix} \begin{bmatrix} x_0 \\ y_0 \\ z_0 \end{bmatrix} = \begin{bmatrix} (\frac{3}{10})(x_0 + y_0 + z_0) \\ (\frac{3}{10})(x_0 + y_0 + z_0) \\ (\frac{2}{10})(x_0 + y_0 + z_0) \end{bmatrix} \quad .$$

That is, no matter how the total mass $x_0 + y_0 + z_0$ of sand was originally distributed among the three jars A, B, C, the proportions of the sand they contain approach the levels 30%, 50%, 20% respectively through repetition of the redistribution procedure.

Stochastic Matrices

The last example described the movement of a certain population among various locations, in this case a population of sand grains moving among three jars. Examples in the same category occurred in section 3.2: for instance, a population of breakfast cereal consumers moving among several competing cereal brands.

In such problems, the population vector $\mathbf{x}_n$ must always be a *positive vector*; that is, all components must be positive or zero. Similarly, the *transition matrix A* that links successive states of the population vector through the formula $\mathbf{x}_{m+1} = A\mathbf{x}_m$ must be a *positive matrix*; that is, all entries must be positive or zero.

Observe, too, that the entries a_{1j}, a_{2j}, a_{3j} in column j of the sand problem transition matrix are the fractions of the sand in jar j that are transferred to jars 1, 2, 3 respectively. Since all the sand in jar j must go somewhere, the sum of these fractions $a_{1j} + a_{2j} + a_{3j}$ must add to 1. (Check that each column of the sand transition matrix adds to 1.)

We conclude that the transition matrices for the sand and cereal processes fall into the category of *stochastic matrices*, which are the positive $n \times n$ matrices in which the sum of the entries in each column is 1.

Example 3   The transition matrix for the cereal problem (example 2, section 3.2), $A = \begin{bmatrix} \frac{8}{10} & \frac{3}{10} \\ \frac{2}{10} & \frac{7}{10} \end{bmatrix}$, is stochastic.

Processes represented by stochastic transition matrices form a significant part of the vast category of "Markov processes" in general. These processes describe everything from the progress of gambling games to the movement of electrons. Although the technical definition of a Markov process is complex, the essential characteristic is that the present state of the process ($\mathbf{x}_m$) can be used to predict its future states ($\mathbf{x}_{m+1}, \mathbf{x}_{m+2}, \ldots$) and that the earlier history of the process ($\mathbf{x}_{m-1}, \mathbf{x}_{m-2}, \ldots$) provides no further information about the future. A Markov process is "forgetful."

■ **Problem**   Classify the 2×2 matrices A such that A and A^{-1} are both stochastic.

■ **Solution**   Consider the representations

$$A = \begin{bmatrix} a & b \\ c & d \end{bmatrix} \quad , \quad A^{-1} = (\det A)^{-1} \begin{bmatrix} d & -b \\ -c & a \end{bmatrix} .$$

If A is stochastic, then a,b,c,d are all positive or zero. Also, the entries of A^{-1} are, except for a common factor $(\det A)^{-1}$, the numbers $a,d,-b,-c$; and so if A^{-1} is stochastic these four numbers cannot have mixed signs. This means that if A and A^{-1} are both stochastic, then

either   (i) b and c are both zero

or   (ii) a and d are both zero.

In case (i), $A = \begin{bmatrix} a & 0 \\ 0 & d \end{bmatrix} = \begin{bmatrix} 1 & 0 \\ 0 & 1 \end{bmatrix}$,

since the entries in each column must add to 1, and similarly,

in case (ii), $A = \begin{bmatrix} 0 & b \\ c & 0 \end{bmatrix} = \begin{bmatrix} 0 & 1 \\ 1 & 0 \end{bmatrix}$.

Observe that in both solutions there is no "mixing" of the two components of a vector $\mathbf{v}$ in the stochastic process; they are either (i) unchanged from step to step, or (ii) exchanged at each step. In fact, the two solutions are exactly the two 2×2 permutation matrices (section 3.3 exercises), and it can be proved that the $n \times n$ permutation matrices are the only solutions to our problem in dimension n.

■ EXERCISES 8.3

1. Suppose that a forest contains populations of two species, A and B, of finches, which forage for various kinds of insects, nuts, and berries. Suppose that when these populations are displaced from their normal levels by x_0, y_0 thousands of birds respectively, the population deviations x_n, y_n in subsequent years are governed by the equations

$$x_{n+1} - x_n = \frac{(-6x_n - 2y_n)}{10} \quad ,$$

$$y_{n+1} - y_n = \frac{(-3x_n - 5y_n)}{10} \quad .$$

(The small negative terms $-2y_n$, $-3x_n$ on the right sides of the equations occur because the two species are competing for some of the same foods, and therefore they inhibit each other's population growth. The self-limiting effect that each species has on its own population growth is larger, as expressed in the larger negative terms $-6x_n$, $-5y_n$.) Obtain a formula for the population displacements in year m and describe the way in which the populations gradually return to their usual levels, assuming initial displacements $x_0 = 1$, $y_0 = 6$ (thousands of birds).

2. Suppose that the finch populations of the previous exercise were governed by somewhat different equations:

$$x_{n+1} - x_n = \frac{(-2x_n - 2y_n)}{10} \quad ,$$

$$y_{n+1} - y_n = \frac{(-3x_n - y_n)}{10} \quad .$$

Show that, if the initial population displacements were $x_0 = 2$, $y_0 = -3$, then in subsequent years there would be progressively larger fluctuations in the populations.

3. In example 2 of section 3.2, a model for the market shares of two cereal brands was constructed. Show that the transition matrix in this model is diagonalizable, obtain a formula for the market share vector in year m, and show that the market shares always approach a 60% to 40% split, regardless of their initial proportions.

4. In exercise 4 of section 3.2, a model for three competing cereal brands was constructed. Show that the 3×3 transition matrix in this model is stochastic and has three real eigenvalues: $\lambda_1 = 1$, and λ_2, λ_3 satisfying $|\lambda| < 1$. Deduce, as in example 2 of this section, that the market shares approach fixed levels as time goes on, and determine these market shares.

5. In example 3 of section 3.2, a model for the state of a taxi fleet was constructed. Show that the 3×3 transition matrix of this model is stochastic, with one real eigenvalue $\lambda_1 = 1$, and two complex eigenvalues λ_2, λ_3 satisfying $|\lambda| < 1$. Deduce, by using lemma 8.2, that the state of the fleet eventually stabilizes and determine the ultimate age distribution of the cars in the fleet. (*Note*: use the fact that for complex numbers λ, $|\lambda^m| = |\lambda|^m$. There is no need to compute the complex eigenvectors.)

6. Suppose that in the owl-mouse population model of example 1 of this section, the intensity of predation increased so that the model equations were changed to

$$x_{n+1} - x_n = -\left(\tfrac{1}{10}\right)x_n + \left(\tfrac{3}{10}\right)y_n \quad,$$
$$y_{n+1} - y_n = -\left(\tfrac{2}{10}\right)x_n - \left(\tfrac{4}{10}\right)y_n \quad.$$

Show that the transition matrix of the model would then have complex eigenvalues λ_1, λ_2 satisfying $|\lambda| < 1$. Deduce, from lemma 8.2, that the owl and mouse populations would still, after being disrupted in some way, return to their normal equilibrium.

7. Show that if the $n \times n$ matrix A has n l.i. eigenvectors corresponding to eigenvalues $\lambda_1, \ldots, \lambda_n$, then $A^m = \lambda_1^m M_1 + \ldots \lambda_n^m M_n$, where $M_1, \ldots, M_n$ are $n \times n$ numerical matrices of rank 1 that each have the eigenvalues 1 and 0. (See method of example 2 of this section.)

8. (a) Show, by finding an example, that a stochastic matrix A may not be invertible.

   (b) Suppose the stochastic matrix A links the vectors $\mathbf{x}_0, \mathbf{x}_1, \mathbf{x}_2, \ldots$ through the formula $\mathbf{x}_{m+1} = A\mathbf{x}_m$. Suppose also that $\mathbf{x}_p$ is known; say $\mathbf{x}_p = \left(\tfrac{2}{3}, \tfrac{1}{3}\right)$. Show that although $\mathbf{x}_{p+1}, \mathbf{x}_{p+2}, \ldots$ can be deduced from $\mathbf{x}_p$, it may not be possible to deduce $\mathbf{x}_{p-1}, \mathbf{x}_{p-2}, \ldots, \mathbf{x}_0$ from $\mathbf{x}_p$.

9. Show that if A is a stochastic matrix, then for each positive integer k, A^k is also a stochastic matrix.

10. Let $\mathbf{1}$ denote the vector $(1, 1, \ldots, 1)$.

    (a) Show that if A is a stochastic matrix, then $A^t \mathbf{1} = \mathbf{1}$, so that $\lambda = 1$ is an eigenvalue of A^t.

    (b) Show that for any square matrix B,

    $$\det(\lambda I - B) = \det(\lambda I - B^t).$$

    (c) Deduce that $\lambda = 1$ is an eigenvalue of every stochastic matrix.

*11. Let Σx_i denote the sum of the components of a vector $\mathbf{x}$.

    (a) Show that if $A\mathbf{x} = \lambda\mathbf{x}$, then $\Sigma|(A\mathbf{x})_i| = |\lambda|\Sigma|x_i|$.

    (b) Show that if A is a stochastic matrix and $\mathbf{x}$ a positive vector, then $\Sigma(A\mathbf{x})_i = \Sigma x_i$.

    (c) Deduce that a positive eigenvector of a stochastic matrix always corresponds to the eigenvalue $\lambda = 1$.

*12. Given a vector $\mathbf{x}$, let $\mathbf{x}^+$ denote the vector with components $x_i^+ = |x_i|$. For example, if $\mathbf{x} = (1, 3+4i, -2)$, then $\mathbf{x}^+ = (1,5,2)$. Also, let A be a stochastic matrix.

    (a) Show that

$$\Sigma|(A\mathbf{x})_i| \leq \Sigma(A\mathbf{x}^+)_i = \Sigma x_i^+ = \Sigma|x_i| \quad .$$

    (b) Using part (a) of the previous question, show that if λ is an eigenvalue of a stochastic matrix, then $|\lambda| \leq 1$.

*13. The stochastic matrix A is said to be *strictly positive* if all its entries are strictly positive, i.e. satisfy $a_{ij} > 0$. Using the steps indicated below, prove the following theorem concerning strictly positive stochastic matrices.

    Theorem Let A be a strictly positive stochastic matrix. Then:

    (i) All eigenvalues of A satisfy $\lambda = 1$ or $|\lambda| < 1$.

    (ii) The eigenspace E_1 satisfies $\dim(E_1) = 1$.

    (iii) The eigenspace E_1 contains (and is therefore spanned by) a positive eigenvector.

    Steps to the proof:

    (a) Suppose $A\mathbf{x} = \lambda\mathbf{x}$, with $|\lambda| = 1$. Deduce from the previous question that:

$$\Sigma|(A\mathbf{x})_i| = \Sigma(A\mathbf{x}^+)_i = \Sigma|x_i| \quad .$$

    (b) Deduce that $\mathbf{x} = c\mathbf{x}^+$.

    (c) Deduce that $\lambda = 1$ (completing the proof of part (i)).

    (d) Deduce from (b) that the eigenspace E_1 consists of positive eigenvectors and their scalar multiples.

    (e) Let $\mathbf{x}_1, \mathbf{x}_2$ be positive eigenvectors in E_1. Show that if t is a real scalar, then $\mathbf{x}_1 - t\mathbf{x}_2$ is also in E_1, and deduce that it cannot have both positive and negative components. Deduce further that for some choice of t, $\mathbf{x}_1 = t\mathbf{x}_2$, and conclude that $\mathbf{x}_1, \mathbf{x}_2$ are l.d.

14. (a) Show that $A_1 = \begin{bmatrix} 1 & 0 \\ 0 & 1 \end{bmatrix}$ and $A_2 = \begin{bmatrix} 0 & 1 \\ 1 & 0 \end{bmatrix}$ are stochastic matrices that are not strictly positive.

    (b) Show that for the stochastic matrix A_1, $\dim(E_1) > 1$.

(c) Show that for the stochastic matrix A_2, $\lambda = -1$ is an eigenvalue.

*15. Show that every 2×2 stochastic matrix is diagonalizable, but that there are 3×3 stochastic matrices that are not diagonalizable (even with complex scalars).

Note: Exercises 9.3 contain further discussion of stochastic matrices.

8.4 Applications to Systems of Differential Equations

Introduction

We have discussed such processes as brand competition in consumer markets as though they proceeded in discrete steps, say once-a-year realignments of brand preference. Of course this was just a deliberate simplifying assumption about a process that surely takes place continuously and from day to day. It would therefore be more precise to talk about *rates* of shifts in brand preferences. Now rates of change are usually expressed mathematically as derivatives of functions, in this case the functions expressing the market shares of the competing brands. A continuous formulation of the brand competition process will therefore be expressed in terms of equations containing derivatives of unknown functions or, as they are usually called, *differential equations*.

The systems of differential equations that we consider can be thrown into the matrix-vector framework with very little effort, and eigenvector theory proves ideally suited to facilitate the solution of these systems. The choice of a continuous or discrete model for a process can therefore be determined by the intrinsic character of the process itself, since both types of model can be analyzed with equal ease and by largely identical methods.

Our main illustrative example in this section is from radioactive decay theory. We also propose continuous models for some of the processes discussed earlier. Many problems in the physical sciences are formulated in terms of "second order" differential equations. In the problem at the end of the section, we show how such equations can be converted to the systems formulation that we have emphasized.

Further discussion of differential equations is given in section 9.3.

Systems of Differential Equations

Recall that a radioactive isotope has the property that it decays into other kinds of atoms at a rate proportional to the amount, $x(t)$, present at time t.

Now the rate of change of a quantity $x(t)$ is given by its derivative $\dfrac{d}{dt}x(t)$; therefore the radioactive decay law says that $\dfrac{d}{dt}x(t) = -kx(t)$, where k is a proportionality constant expressing the decay rate.

Suppose we know the amount, x_0, of the substance present initially and also the decay rate k. Then $x(t)$, the amount present at time t, must satisfy the following *initial value problem* (IVP), consisting of

a *differential equation* (DE), $\dfrac{d}{dt}x(t) = -kx(t)$,

and an *initial condition*, $x(0) = x_0$.

It is proved in elementary calculus that the unique solution of this IVP is

$$x(t) = x_0 e^{-kt} \ ,$$

which is the familiar exponential decay formula. Consider now a compound decay problem.

Example 1   Suppose substance A undergoes radioactive decay (with decay rate k) to a substance B. Suppose B also undergoes radioactive decay (with decay rate h) to a third substance C. If the initial amounts of substances A and B are 5 gm and 7 gm respectively, and the decay rates are $k = 3, h = 2$, how much of each substance is present at time $t > 0$?

Let $x(t)$, $y(t)$ represent the amounts of substances A, B respectively present at time t. Then

$$\frac{dx}{dt} = -kx \quad \text{(A decays, no replenishment), and}$$

$$\frac{dy}{dt} = kx - hy \quad \text{(B decays, but also } gains \text{ from A).}$$

Note that the kx term in the second differential equations carries a plus sign because this is the rate at which B is gaining material from A.

This system of two differential equations can be rewritten in matrix-vector notation as

$$\begin{bmatrix} \dfrac{dx}{dt} \\ \dfrac{dy}{dt} \end{bmatrix} = \begin{bmatrix} -k & 0 \\ k & -h \end{bmatrix} \begin{bmatrix} x \\ y \end{bmatrix} \ .$$

On substituting the given numerical data, we obtain the IVP:

$$\begin{bmatrix} \dfrac{dx}{dt} \\ \dfrac{dy}{dt} \end{bmatrix} = \begin{bmatrix} -3 & 0 \\ 3 & -2 \end{bmatrix} \begin{bmatrix} x \\ y \end{bmatrix} \quad \text{(system of DE's)}$$

$$\begin{bmatrix} x(0) \\ y(0) \end{bmatrix} = \begin{bmatrix} 5 \\ 7 \end{bmatrix} \quad \text{(initial conditions) .}$$

To streamline the notation, let $\dfrac{d\mathbf{x}}{dt}$ denote $\begin{bmatrix} \dfrac{dx}{dt} \\ \dfrac{dy}{dt} \end{bmatrix}$. In other words, vector differentiation is, by definition, to be done *componentwise*.

Then the IVP can be rewritten as $\dfrac{d\mathbf{x}}{dt} = A\mathbf{x}$,

$$\mathbf{x}(0) = \mathbf{x}_0$$

where $A = \begin{bmatrix} -3 & 0 \\ 3 & -2 \end{bmatrix}$, $\mathbf{x}_0 = \begin{bmatrix} 5 \\ 7 \end{bmatrix}$.

As is customary in calculus, we try to integrate the differential equation first and then consider the initial conditions. But the unknown function $\mathbf{x}(t)$ is now a vector, and so any exponential solution would have to take the form of a constant *vector* $\mathbf{v}$ times an exponential factor.

That is, $\mathbf{x}(t) = e^{\lambda t}\mathbf{v}$. That such solutions do exist is shown by the following lemma.

Lemma 8.7 The vector function $\mathbf{x}(t) = e^{\lambda t}\mathbf{v}$ is a solution of the linear system of differential equations $\dfrac{d\mathbf{x}}{dt} = A\mathbf{x}$ if and only if $\mathbf{v}$ is an eigenvector of A corresponding to the eigenvalue λ.

■**Proof** Observe that $A(e^{\lambda t}\mathbf{v}) = e^{\lambda t}A\mathbf{v}$, while, since vectors are differentiated componentwise,

$$\frac{d}{dt}(e^{\lambda t}\mathbf{v}) = \frac{d}{dt}(e^{\lambda t}v_1, \ldots, e^{\lambda t}v_n)$$
$$= (\lambda e^{\lambda t}v_1, \ldots, \lambda e^{\lambda t}v_n)$$
$$= \lambda e^{\lambda t}\mathbf{v}.$$

Therefore $\mathbf{x}(t) = e^{\lambda t}\mathbf{v}$ satisfies the DE if and only if

$$e^{\lambda t}A\mathbf{v} = \lambda e^{\lambda t}\mathbf{v},$$

or $\quad A\mathbf{v} = \lambda\mathbf{v}$. $\qquad\qquad$ Q.E.D.

In the case of example 1, we note that A, a lower triangular matrix, has eigenvalues $\lambda_1 = -3$, $\lambda_2 = -2$ and corresponding eigenvectors $\mathbf{v}_1 = (1, -3)$, $\mathbf{v}_2 = (0, 1)$. Therefore

$$\mathbf{x}_1(t) = e^{-3t}(1, -3), \quad \mathbf{x}_2(t) = e^{-2t}(0, 1)$$

are two solutions of the DE.

In order to satisfy the initial condition $\mathbf{x}(0) = \mathbf{x}_0$, we must first enlarge the set of solutions of the DE by means of the next theorem.

THEOREM 8.8

Suppose $v_1, \ldots, v_n$ are eigenvectors of the $n \times n$ matrix A corresponding to eigenvalues $\lambda_1, \ldots, \lambda_n$ respectively and that these eigenvectors form a basis for R^n. Then every initial value problem

$$\frac{d\mathbf{x}}{dt} = A\mathbf{x}$$

$$\mathbf{x}(0) = \mathbf{x}_0$$

has a solution of the form

$$\mathbf{x}(t) = c_1 e^{\lambda_1 t} \mathbf{v}_1 + \cdots + c_n e^{\lambda_n t} \mathbf{v}_n \quad .$$

■ **Proof** First, it is easily verified (see exercises) that a linear combination of solutions of a DE, $\dfrac{d\mathbf{x}}{dt} = A\mathbf{x}$, is also a solution. Therefore the given function $\mathbf{x}(t)$, which is a linear combination of the solutions found in the previous lemma, is also a solution. Also, since $e^0 = 1$, the initial value of this solution is simply

$$\mathbf{x}(0) = c_1 \mathbf{v}_1 + \cdots + c_n \mathbf{v}_n$$
$$= P\mathbf{c}, \text{ where } P = [\mathbf{v}_1 | \cdots | \mathbf{v}_n] \quad .$$

But since the eigenvectors form a basis for R^n, the matrix P is invertible. So by choosing $\mathbf{c} = P^{-1}\mathbf{x}_0$, we obtain $\mathbf{x}(0) = P(P^{-1}\mathbf{x}_0) = \mathbf{x}_0$ as required.

Q.E.D.

To complete the analysis of example 1, note that the solutions given by theorem 8.8 are

$$\mathbf{x}(t) = c_1 e^{-3t} \begin{bmatrix} 1 \\ -3 \end{bmatrix} + c_2 e^{-2t} \begin{bmatrix} 0 \\ 1 \end{bmatrix} \quad .$$

The initial data $\mathbf{x}_0 = (5,7)$ then implies that

$$c_1 \begin{bmatrix} 1 \\ -3 \end{bmatrix} + c_2 \begin{bmatrix} 0 \\ 1 \end{bmatrix} = \begin{bmatrix} 5 \\ 7 \end{bmatrix} \quad , \quad \text{or} \quad \begin{cases} c_1 = 5 \\ -3c_1 + c_2 = 7 \end{cases} ,$$

or $c_1 = 5$, $c_2 = 22$. A solution of the IVP is therefore

$$\mathbf{x}(t) = 5e^{-3t} \begin{bmatrix} 1 \\ -3 \end{bmatrix} + 22e^{-2t} \begin{bmatrix} 0 \\ 1 \end{bmatrix} = \begin{bmatrix} 5e^{-3t} \\ -15e^{-3t} + 22e^{-2t} \end{bmatrix} \quad .$$

The first component shows the simple exponential decay of the amount of substance A, while the formula for the remnant of substance B is, not surprisingly, more complicated.

The more comprehensive discussions of linear systems in texts on differential equations include proofs that the above initial value problems always have unique solutions. In particular, the solution we have obtained for the radioactive decay problem is indeed the unique solution, as intuition would suggest. See also section 9.3 for further development of the theory.

Some of the processes modeled earlier in terms of a transition matrix A and discrete time steps $n = 0,1,2, \ldots$, through the formula $\mathbf{x}_{n+1} = A\mathbf{x}_n$, can also be described as continuous processes. For example, the shifting of consumers among competing brands can plausibly be regarded as a continuous process rather than as an annual realignment of brand choice.

Example 2   Suppose, for example, that consumers of brand X shift to brand Y at the steady rate of 1% per week, while consumers of brand Y shift to brand X at the faster rate of 2% per week. Then, with time t measured in weeks, the market shares $x(t)$ and $y(t)$ of brands X and Y respectively would satisfy

$$\frac{dx}{dt} = -\tfrac{1}{100}\, x + \tfrac{2}{100}\, y$$

$$\frac{dy}{dt} = \tfrac{1}{100}\, x - \tfrac{2}{100}\, y \quad,$$

or $\dfrac{d\mathbf{x}}{dt} = A\mathbf{x}$, where $A = \left(\tfrac{1}{100}\right) \begin{bmatrix} -1 & 2 \\ 1 & -2 \end{bmatrix}$.

The eigenvectors of A are easily computed to be $\lambda_1 = 0$, $\lambda_2 = -3$, and the corresponding eigenvectors $\mathbf{v}_1 = (2,1)$, $\mathbf{v}_2 = (1,-1)$. Therefore,

$$\mathbf{x}(t) = c_1 e^{0t} \begin{bmatrix} 2 \\ 1 \end{bmatrix} + c_2 e^{-3t} \begin{bmatrix} 1 \\ -1 \end{bmatrix} = c_1 \begin{bmatrix} 2 \\ 1 \end{bmatrix} + c_2 e^{-3t} \begin{bmatrix} 1 \\ -1 \end{bmatrix} .$$

For example, if at time $t = 0$, brand X enters a market monopolized by brand Y, then the initial shares are $\mathbf{x}_0 = (0,1)$. The initial condition on $\mathbf{x}(t)$ is then

$$\mathbf{x}(0) = c_1 \begin{bmatrix} 2 \\ 1 \end{bmatrix} + c_2 \begin{bmatrix} 1 \\ -1 \end{bmatrix} = \begin{bmatrix} 0 \\ 1 \end{bmatrix} \quad, \quad \text{or} \begin{cases} 2c_1 + c_2 = 0 \\ c_1 - c_2 = 1 \end{cases} ,$$

or $c_1 = \tfrac{1}{3}$, $c_2 = -\tfrac{2}{3}$. The shares of the brands at time t are therefore the components of the vector

$$\mathbf{x}(t) = \tfrac{1}{3} \begin{bmatrix} 2 \\ 1 \end{bmatrix} - \tfrac{2}{3} e^{-3t} \begin{bmatrix} 1 \\ -1 \end{bmatrix} .$$

So as time goes on, the market shares of the two brands X and Y approach the fixed proportions $\tfrac{2}{3}, \tfrac{1}{3}$ respectively.

Note that, as time passes, the solution $\mathbf{x}(t)$ always approaches $c_1(2,1)$ because of the exponential decay factor in the other term of the solution formula. So no matter what the initial market shares of the two brands had been, they would ultimately approach the same fixed proportions $\tfrac{2}{3}, \tfrac{1}{3}$.

■ **Problem**   An "mth order" differential equation is one containing an unknown function $x(t)$ and its derivatives up to and including the mth derivative. Show that the second order differential equation

$$L(x) = a_2 \frac{d^2x}{dt^2} + a_1 \frac{dx}{dt} + a_0 x = 0$$

can be rewritten as a first order system $\dfrac{dx}{dt} = A\mathbf{x}$ and that the eigenvalues of the 2×2 matrix A are the roots of the polynomial $a_2\lambda^2 + a_1\lambda + a_0$ (which is called the characteristic polynomial of the differential equation $L(x) = 0$).

■ **Solution** The "second derivative" symbol $\dfrac{d^2x}{dt^2}$ means simply that the the function $x(t)$ is differentiated twice. So if we give its first derivative a name by writing $\dfrac{dx}{dt} = y(t)$, then $\dfrac{d^2x}{dt^2}$ is the same as $\dfrac{dy}{dt}$. Also, the given second order DE, $L(x) = 0$, can then be rewritten as $a_2\dfrac{dy}{dt} + a_1 y + a_0 x = 0$. We now have two first order DE's containing x and y, which we can rewrite in a regular pattern as

$$\frac{dx}{dt} = y$$

$$\frac{dy}{dt} = -\left(\frac{a_0}{a_2}\right) x - \left(\frac{a_1}{a_2}\right) y \quad .$$

But this system has the familiar form $\dfrac{d\mathbf{x}}{dt} = A\mathbf{x}$,

with
$$\mathbf{x}(t) = \begin{bmatrix} x(t)^* \\ y(t) \end{bmatrix} \quad \text{and} \quad A = \begin{bmatrix} 0 & 1 \\ -\dfrac{a_0}{a_2} & -\dfrac{a_1}{a_2} \end{bmatrix} \quad .$$

Also, $\det(\lambda I - A) = \lambda^2 + \left(\dfrac{a_1}{a_2}\right)\lambda + \dfrac{a_0}{a_2}$

$$= \left(\frac{a_2\lambda^2 + a_1\lambda + a_0}{a_0}\right) \quad ;$$

therefore the eigenvalues of A are the roots of the characteristic polynomial of the given second order DE, as required.

■ EXERCISES 8.4

1. Solve the following initial value problems.

(a) $\dfrac{dx}{dt} = x + 6y$

$\dfrac{dy}{dt} = -x + 6y$

$\mathbf{x}(0) = (7,3)$

(b) $\dfrac{dx}{dt} = -2x + 5y$

$\dfrac{dy}{dt} = 4x + 6y$

$\mathbf{x}(0) = (-2,8)$

(c) $\dfrac{dx_1}{dt} = \qquad x_2$

$\dfrac{dx_2}{dt} = x_1 + x_2 - x_3$

$\dfrac{dx_3}{dt} = \qquad - x_2$

$\mathbf{x}(0) = (0,1,2)$

(d) $\dfrac{dx_1}{dt} = \qquad\qquad - x_3$

$\dfrac{dx_2}{dt} = x_1 + x_2 + \quad x_3$

$\dfrac{dx_3}{dt} = 2x_1 \qquad + 3x_3$

$\mathbf{x}(0) = (2,2,-2)$

2. Suppose substance A is radioactive, with two decay modes: to substance B with decay rate $p = 2$ and also to substance C with decay rate $q = 3$. Suppose that substance B is also radioactive, decaying to substance C at decay rate $r = 4$. Assuming that at time $t = 0$ there is 50 gm of substance A and none of B and C, how much of substance C is there at time $t > 0$?

3. Suppose brands X, Y, and Z of a consumer staple struggle for shares of a fixed market but that brand X gains twice as fast from brands Y and Z as they gain from X or from each other. Assuming that at time $t = 0$, X and Y enter a market monopolized by brand Z, and that X gains from Z at the rate of 1% of Z's market share per week, find the market shares of the three brands after 25 weeks.

4. For animal species living in nonseasonal environments, continuous-time population models are sometimes realistic. For example, the owl-mouse population processes in the text and exercises of section 8.3 might be cast in the following versions:

(a) $\dfrac{d\mathbf{x}}{dt} = r \begin{bmatrix} -1 & 2 \\ -1 & -4 \end{bmatrix} \mathbf{x}$ (b) $\dfrac{d\mathbf{x}}{dt} = r \begin{bmatrix} -1 & 3 \\ -2 & -4 \end{bmatrix} \mathbf{x}$.

The parameter r in each model must be chosen to suit the rate at which the overall population process occurs.

Show that in each case the population displacements represented by the components of $\mathbf{x}(t)$ gradually approach zero, just as they do in the discrete time models of section 8.3.

Note: In case (b), complex eigenvalues occur; $e^{\lambda t}$ is then simplified by use of the identity

$$e^{p+iq} = e^p(\cos q + i \sin q) \quad .$$

5. (a) Show that if $\mathbf{z}(t)$ and $\mathbf{w}(t)$ are vectors, and c_1 and c_2 are scalar constants, then

$$\frac{d}{dt}[c_1\mathbf{z}(t) + c_2\mathbf{w}(t)] = c_1\frac{d\mathbf{z}}{dt} + c_2\frac{d\mathbf{w}}{dt} \quad .$$

  (b) Show that if $\mathbf{z}(t)$ and $\mathbf{w}(t)$ are both solutions of the linear system $\frac{d\mathbf{x}}{dt} = A\mathbf{x}$, then so is any linear combination, $c_1\mathbf{z}(t) + c_2\mathbf{w}(t)$, of them.

6. (a) Show that if M is a constant (numerical) matrix, then

$$\frac{d}{dt}[M\mathbf{x}(t)] = M\frac{d\mathbf{x}}{dt} \quad .$$

   (Compute the components of $M\mathbf{x}(t)$, then differentiate component-wise.)

  (b) Suppose $A = PDP^{-1}$, where $P = [\mathbf{v}_1|\ldots|\mathbf{v}_n]$, and the diagonal matrix D has diagonal entries $\lambda_1, \ldots, \lambda_n$. Also, let $\mathbf{y}(t) = P^{-1}\mathbf{x}(t)$.
   Show that the system $\frac{d\mathbf{x}}{dt} = A\mathbf{x}$ can be rewritten as $\frac{d\mathbf{y}}{dt} = D\mathbf{y}$, or

$$\frac{dy_1}{dt} = \lambda_1 y_1$$

$$\cdots \cdots \cdots$$

$$\frac{dy_n}{dt} = \lambda_n y_n \quad .$$

  Deduce that in the new $\mathbf{y}$-coordinates, each differential equation in the system can be solved separately and that each solution takes the form

$$y_1(t) = y_1(0)e^{\lambda_1 t}, \ldots, y_n(t) = y_n(0)e^{\lambda_n t} \quad .$$

  Note: The above procedure is called *uncoupling* the system $\frac{d\mathbf{x}}{dt} = A\mathbf{x}$. It simplifies the system into n separate, scalar DE's. However, it is often difficult to see the significance of the solution in the new coordinates. For example, in the consumer brand models, each new coordinate $y_j(t)$ would represent a certain linear combination of the market shares of various brands!

7. (a) Using the method outlined in the problem at the end of this section, convert the second-order DE

$$L(x) = 2\frac{d^2x}{dt^2} + 6\frac{dx}{dt} - 8x = 0$$

  to a first-order system.

(b) Show that the initial conditions

$$x(0) = b_0 \quad, \qquad (0) = b_1$$

correspond to the initial conditions $\mathbf{x}(0) = (b_0, b_1)$ for the first order system.

(c) In particular, solve the "initial-value problem"

$$L(x) = 0$$
$$x(0) = 7 \quad, \qquad \frac{dx'}{dt}(0) = -3 \quad.$$

8.5 Symmetric Matrices

Introduction

An $n \times n$ matrix A is said to be *symmetric* if $A^t = A$. Since the ijth entry of A^t is the same as the jith entry of A, we can say that A is symmetric if $a_{ij} = a_{ji}$ for all i and j.

Example 1 $A = \begin{bmatrix} 1 & 2 \\ 2 & 3 \end{bmatrix}$ is symmetric, since $A^t = \begin{bmatrix} 1 & 2 \\ 2 & 3 \end{bmatrix} = A$,

or equivalently, because $a_{21} = a_{12} = 2$.

Example 2 $A = \begin{bmatrix} 1 & 2 & 3 \\ 2 & 5 & 6 \\ 3 & 7 & 8 \end{bmatrix}$ is not symmetric, since $A^t = \begin{bmatrix} 1 & 2 & 3 \\ 2 & 5 & 7 \\ 3 & 6 & 8 \end{bmatrix} \neq A$,

or, equivalently, because $a_{32} = 7$ while $a_{23} = 6$.

Our main objective in this section is to prove that every symmetric matrix has a set of eigenvectors that form an orthonormal basis for R^n. This quite surprising and extremely useful result is called the Principal Axes Theorem; it is also an important special case of one of the central theorems in mathematics, the Spectral Theorem.

Note, however, that the Principal Axes Theorem is primarily of theoretical rather than computational significance: the actual eigenvalues and eigenvectors of a symmetric matrix are computed by the usual methods; the theorem simply predicts the essential features of the result of the computation. So the proof can be skipped on first reading.

Theory of Symmetric Matrices

We begin by noting some immediate consequences of symmetry.

Lemma 8.9 (a) An $n \times n$ matrix A is symmetric if and only if, for all vectors $\mathbf{u}$, $\mathbf{v}$ in R^n,

$$\mathbf{u} \cdot A\mathbf{v} = \mathbf{v} \cdot A\mathbf{u} \quad .$$

(b) If vectors $\mathbf{u}$, $\mathbf{v}$ in R^n are eigenvectors of the symmetric matrix A and correspond to different eigenvalues, then $\mathbf{u}$, $\mathbf{v}$ are orthogonal.

■**Proof**   (a) For any square matrix A,

$$\mathbf{u} \cdot A\mathbf{v} = (A\mathbf{v}) \cdot \mathbf{u} = (A\mathbf{v})^t\mathbf{u} = \mathbf{v}^t A^t \mathbf{u} = \mathbf{v} \cdot A^t\mathbf{u} \quad .$$

Thus if $A^t = A$, then $\mathbf{u} \cdot A\mathbf{v} = \mathbf{v} \cdot A\mathbf{u}$.

Conversely, if the last identity holds, then on substituting $\mathbf{u} = \mathbf{e}_i$ and $\mathbf{v} = \mathbf{e}_j$, we obtain $a_{ij} = a_{ji}$ and therefore A is symmetric.

(b) Suppose $A\mathbf{u} = \lambda_1\mathbf{u}$ and $A\mathbf{v} = \lambda_2\mathbf{v}$, and A is symmetric. Then on substitution in the identity in part (a), we obtain

$$\mathbf{u} \cdot (\lambda_2\mathbf{v}) = \mathbf{v} \cdot (\lambda_1\mathbf{u}) \text{ or } (\lambda_2 - \lambda_1)\mathbf{v} \cdot \mathbf{u} = 0 \quad .$$

But since, by assumption, $\lambda_2 - \lambda_1 \neq 0$,
therefore $\mathbf{v} \cdot \mathbf{u} = 0$, and $\mathbf{u}$, $\mathbf{v}$ are orthogonal.   Q.E.D.

So far in this text, we have proved theorems about real vectors, matrices, etc., without using complex numbers. The next result, however, is so much easier to prove with a very minimal use of complex numbers than with real methods only that we make an exception.

Recall that if $w = p + iq$ is a complex number, then $w^* = p - iq$ is called the *complex conjugate* of w. Also $w^*w = (p - iq)(p + iq) = p^2 + q^2 > 0$ unless $w = 0$. Thus if $w = 3 + 4i$, then $w^* = 3 - 4i$, and $w^*w = 9 + 16 = 25$.

Now, a complex vector, such as $\mathbf{w} = (3 + 4i, -2i)$, can be expressed in the form $\mathbf{w} = \mathbf{p} + i\mathbf{q}$, with $\mathbf{p}$ and $\mathbf{q}$ real vectors. In our example, $\mathbf{w} = (3,0) + i(4,-2)$. The *complex conjugate* of $\mathbf{w}$ is the vector $\mathbf{w}^* = \mathbf{p} - i\mathbf{q}$. In our example, $\mathbf{w}^* = (3,0) - i(4,-2) = (3 - 4i, 2i)$.

There is also a positive dot product:

$$\mathbf{w}^* \cdot \mathbf{w} = (\mathbf{p} - i\mathbf{q}) \cdot (\mathbf{p} + i\mathbf{q}) = \mathbf{p} \cdot \mathbf{p} + \mathbf{q} \cdot \mathbf{q}$$
$$= |\mathbf{p}|^2 + |\mathbf{q}|^2 > 0 \text{ unless } \mathbf{w} = \mathbf{0} \quad .$$

Finally, $\mathbf{w}^{**} = \mathbf{w}$, and $\mathbf{w}^* = \mathbf{w}$ when the vector $\mathbf{w}$ is real.

THEOREM **8.10**

All eigenvalues of a symmetric matrix A are real, and the number of real eigenvalues is at least 1.

■Proof   If $A\mathbf{w} = \lambda\mathbf{w}$, with eigenvector $\mathbf{w} \neq 0$, then

$$\mathbf{w}^* \cdot A\mathbf{w} = \lambda\mathbf{w}^* \cdot \mathbf{w} .$$

But $\mathbf{w}^* \cdot \mathbf{w}$ is a positive real number. Also,

$$\begin{aligned}
\mathbf{w}^* \cdot A\mathbf{w} &= (\mathbf{p} - i\mathbf{q}) \cdot [A(\mathbf{p} + i\mathbf{q})] \\
&= (\mathbf{p} - i\mathbf{q}) \cdot (A\mathbf{p} + iA\mathbf{q}) \\
&= \mathbf{p} \cdot A\mathbf{p} + i\mathbf{p} \cdot A\mathbf{q} - i\mathbf{q} \cdot A\mathbf{p} + \mathbf{q} \cdot A\mathbf{q} .
\end{aligned}$$

The two terms containing i cancel, by the previous lemma, and so $\mathbf{w}^* \cdot A\mathbf{w}$ is also a real number.

The eigenvalue λ is therefore a quotient of real numbers and must be a real number.

Finally, every square matrix has at least one eigenvalue (see exercises).

Q.E.D.

Example 3   In section 8.2, the symmetric matrix $A = \frac{1}{25}\begin{bmatrix} 7 & 24 \\ 24 & -7 \end{bmatrix}$ was found to have eigenvectors $\mathbf{v}_1 = \left(\frac{4}{5}, \frac{3}{5}\right)$, $\mathbf{v}_2 = \left(-\frac{3}{5}, \frac{4}{5}\right)$ corresponding to the distinct eigenvalues $\lambda_1 = 1$, $\lambda_2 = -1$. The dot product $\mathbf{v}_1 \cdot \mathbf{v}_2$ is zero, as lemma 8.9 (b) predicts, and the eigenvalues are real, as theorem 8.10 predicts.

In fact, $\mathbf{v}_1$ and $\mathbf{v}_2$ clearly form an orthonormal basis for R^2, and in consequence the diagonalization of A took the form

$$A = PDP^{-1} = PDP^t ,$$

where $P = [\mathbf{v}_1|\mathbf{v}_2] = \begin{bmatrix} \frac{4}{5} & -\frac{3}{5} \\ \frac{3}{5} & \frac{4}{5} \end{bmatrix}$ = orthogonal matrix,

and $D = \begin{bmatrix} \lambda_1 & 0 \\ 0 & \lambda_2 \end{bmatrix} = \begin{bmatrix} 1 & 0 \\ 0 & -1 \end{bmatrix}$ = diagonal matrix.

The same kind of diagonalization holds for symmetric matrices in general.

THEOREM 8.11

(Principal Axes Theorem)

Suppose A is a symmetric $n \times n$ matrix. Then A has a set of eigenvectors that form an orthonormal basis for R^n, and A has an *orthogonal diagonalization*,

$$A = PDP^{-1} = PDP^t ,$$

where P is orthogonal and D diagonal.

In fact, if $\mathbf{v}_1, \ldots, \mathbf{v}_n$ are an orthonormal set of eigenvectors of A and correspond to eigenvalues $\lambda_1, \ldots, \lambda_n$, then one choice of P and D is

$$P = [\mathbf{v}_1|\ldots|\mathbf{v}_n], \quad D = \begin{bmatrix} \lambda_1 & & \\ & \ddots & \\ & & \lambda_n \end{bmatrix} .$$

■**Proof**   *Step 1: Partial diagonalization.* By the previous theorem, A has at least one real eigenvalue λ_1 and corresponding real eigenvector $\mathbf{v}_1$. Suppose that we can find an orthonormal set of k eigenvectors $\mathbf{v}_1, \ldots, \mathbf{v}_k$ of A, corresponding to eigenvalues $\lambda_1, \ldots, \lambda_k$. Then the orthonormal set $\mathbf{v}_1, \ldots, \mathbf{v}_k$ can be extended to an orthonormal basis $\mathbf{v}_1, \ldots, \mathbf{v}_k, \mathbf{w}_{k+1}, \ldots, \mathbf{w}_n$ for R^n by the methods of section 7.1. Of course the extra elements $\mathbf{w}_{k+1}, \ldots, \mathbf{w}_n$ may not be eigenvectors of A.

However, $Q = [\mathbf{v}_1|\ldots|\mathbf{v}_k|\mathbf{w}_{k+1}|\ldots|\mathbf{w}_n]$ is an orthogonal matrix, and for $1 \leq j \leq k$, we have

$$Q\mathbf{e}_j = \mathbf{v}_j \quad \text{and} \quad \mathbf{e}_j = Q^{-1}\mathbf{v}_j \quad .$$

Let $B = Q^{-1}AQ = Q^tAQ$. Then B is symmetric, since

$$B^t = (Q^tAQ)^t = Q^tA^tQ^{tt} = Q^tAQ = B \quad .$$

Also, $\mathbf{e}_j$ is an eigenvector of B, for $1 \leq j \leq k$, since

$$B\mathbf{e}_j = Q^{-1}AQ\mathbf{e}_j = Q^{-1}A\mathbf{v}_j = Q^{-1}\lambda_j\mathbf{v}_j = \lambda_j\mathbf{e}_j \quad .$$

But the equation $B\mathbf{e}_j = \lambda_j\mathbf{e}_j$ means that $\lambda_j\mathbf{e}_j$ is column j of B; and since B is symmetric, $\lambda_j\mathbf{e}_j$ is also row j of B. That is, B has the form

$$B = \begin{bmatrix} \lambda_1 & & & & 0 \\ & \cdot & & & \cdot \\ & & \cdot & & \cdot \\ & & & \cdot & \cdot \\ & & & \lambda_k & 0 \\ \hline 0\cdots 0 & & & & B_0 \end{bmatrix} \quad \text{or} \quad B = \begin{bmatrix} D & O \\ \hline 0^t & B_0 \end{bmatrix} \quad ,$$

where D is a $k \times k$ diagonal matrix, and B_0 an $(n-k) \times (n-k)$ square matrix. In fact, since B is symmetric, it is easily deduced that B_0 is also symmetric.

Step 2: Finding an extra eigenvector. Since B_0 is symmetric, it has at least one real eigenvalue λ_0, and corresponding unit eigenvector $\mathbf{u}_0$ in R^{n-k}.

Now let $\mathbf{u}$ be the column vector $\begin{bmatrix} \mathbf{0} \\ \mathbf{u}_0 \end{bmatrix}$, where $\mathbf{0}$ is the k-component zero vector. Then $\mathbf{u}$ is in R^n, and:

$$B\mathbf{u} = \begin{bmatrix} D & O \\ O & B_0 \end{bmatrix} \begin{bmatrix} \mathbf{0} \\ \mathbf{u}_0 \end{bmatrix} = \begin{bmatrix} \mathbf{0} \\ B_0\mathbf{u}_0 \end{bmatrix} = \cdots$$

$$\cdots \begin{bmatrix} \mathbf{0} \\ \lambda_0\mathbf{u}_0 \end{bmatrix} = \lambda_0\mathbf{u} \quad .$$

That is, $\mathbf{u}$ is a new unit eigenvector of B which is orthogonal to the previously known eigenvectors $\mathbf{e}_1, \ldots, \mathbf{e}_k$.

Finally, we convert our results back into terms of A. Let $\mathbf{v}_{k+1} = Q\mathbf{u}$. Then, since $\mathbf{e}_1, \ldots, \mathbf{e}_k, \mathbf{u}$ are an orthonormal set and Q an orthogonal matrix; therefore $Q\mathbf{e}_1, \ldots, Q\mathbf{e}_k, Q\mathbf{u}$, in other words $\mathbf{v}_1, \ldots, \mathbf{v}_{k+1}$, are also an orthonormal set. Of course, $\mathbf{v}_{k+1}$ is also an eigenvector of A:

$$A\mathbf{v}_{k+1} = (QBQ^{-1})(Q\mathbf{u}) = QB\mathbf{u} = Q(\lambda_0\mathbf{u}) = \lambda_0\mathbf{v}_{k+1} \quad .$$

We conclude that an orthonormal set of eigenvectors of A can be enlarged, one element at a time, until an orthonormal basis for R^n is found.   Q.E.D.

Note on Computation

As illustrated by the example preceding the formal statement of the Principal Axes Theorem, the actual orthogonal matrix P and diagonal matrix D follow automatically from the computation of eigenvalues and eigenvectors. The only situation needing special note is the presence of an eigenspace of higher dimension.

Example 1   The symmetric matrix $A = \begin{bmatrix} 3 & 2 & 4 \\ 2 & 0 & 2 \\ 4 & 2 & 3 \end{bmatrix}$ was found, in example 8

of section 8.1, to have eigenvalues $\lambda_1 = -1$, $\lambda_2 = 8$, and corresponding eigenspaces

E_{-1}, spanned by $\mathbf{v}_1 = (-1,2,0)$, $\mathbf{v}_2 = (-1,1,0)$,
E_8, spanned by $\mathbf{v}_3 = (2,1,2)$.

Then the basis $\mathbf{v}_1$, $\mathbf{v}_2$ for E_{-1} is converted to an orthonormal basis by using the Gram-Schmidt process of section 7.1. The result is $\mathbf{u}_1 = \left(\frac{1}{\sqrt{5}}\right)(-1,2,0)$, $\mathbf{u}_2 = \left(\frac{1}{3\sqrt{5}}\right)(-4,-2,5)$. Of course the orthonormal basis for E_8 is simply

$\mathbf{u}_3 = \left(\frac{1}{3}\right)(2,1,2)$.

Then we have the required orthonormal eigenvector basis.

Hermitian Matrices

Let us resume the discussion of linear algebra in the complex context, using the method introduced in section *7.3. In particular, recall that the *transpose conjugate* of a complex matrix A is the matrix $A^\dagger = (A*)^t$. For example,

if $B = \begin{bmatrix} 12 & 4+3i \\ 4+3i & 12i \end{bmatrix}$,   then $B^\dagger = \begin{bmatrix} 12 & 4-3i \\ 4-3i & -12i \end{bmatrix}$,

while if $C = \begin{bmatrix} 12 & 4+3i \\ 4-3i & -12 \end{bmatrix}$,   then $C^\dagger = \begin{bmatrix} 12 & 4+3i \\ 4-3i & -12 \end{bmatrix}$.

We say that a square matrix A is *Hermitian* if $A^\dagger = A$. Thus matrix C above is Hermitian, but matrix B is not.

It is easily seen that a square matrix A is Hermitian if and only if

$a_{ij} = a*_{ji}$ for all indices i, j .

That is, the diagonal entries of a Hermitian matrix are real and the off-diagonal entries come in complex conjugate pairs.

Our immediate purpose is to show that Hermitian and symmetric matrices share the same crucial properties.

THEOREM 8.9/8.10 (Complex Context)

(a) The $n \times n$ matrix A is Hermitian if and only if for all $\mathbf{u}, \mathbf{v}$ in C^n,

$$\mathbf{u}^\dagger A \mathbf{v} = (\mathbf{v}^\dagger A \mathbf{u})^\dagger \quad .$$

(b) If A is Hermitian, then eigenvectors of A corresponding to different eigenvalues are orthogonal.

(c) If A is Hermitian, then all eigenvalues of A are real and the number of real eigenvalues is at least 1.

■**Proof**   (a) $(\mathbf{v}^\dagger A \mathbf{u})^\dagger = \mathbf{u}^\dagger A^\dagger \mathbf{v}^{\dagger\dagger} = \mathbf{u}^\dagger A^\dagger \mathbf{v}$.

Then if $A^\dagger = A$, the required identity holds. Conversely, if it holds, then on taking $\mathbf{u} = \mathbf{e}_i$, $\mathbf{v} = \mathbf{e}_j$, we obtain $a_{ij} = a_{ji}^*$ and so A is Hermitian.

(b) Let $A\mathbf{u} = \lambda_1 \mathbf{u}$, and $A\mathbf{v} = \lambda_2 \mathbf{v}$, with $\lambda_1 \neq \lambda_2$. On substitution in the identity of part (a), we obtain

$$\mathbf{u}^\dagger \lambda_2 \mathbf{v} = (\mathbf{v}^\dagger \lambda_1 \mathbf{u})^\dagger$$

or $\qquad \lambda_2 \mathbf{u}^\dagger \mathbf{v} = \lambda_1 \mathbf{u}^\dagger \mathbf{v}$

or $\qquad (\lambda_2 - \lambda_1) \mathbf{u}^\dagger \mathbf{v} = 0$.

But since $\lambda_2 - \lambda_1 \neq 0$, therefore $\mathbf{u}^\dagger \mathbf{v} = \mathbf{u}^* \cdot \mathbf{v} = 0$. That is, $\mathbf{u}$ and $\mathbf{v}$ are orthogonal.

(c) Let λ be an eigenvalue of A, so that $A\mathbf{u} = \lambda \mathbf{u}$ for some $\mathbf{u} \neq \mathbf{0}$, and let $\mathbf{v} = \mathbf{u}$. On substitution in the identity of part (a), we obtain

$\mathbf{u}^\dagger \lambda \mathbf{u} = (\mathbf{u}^\dagger \lambda \mathbf{u})^\dagger$

$\lambda \mathbf{u}^\dagger \mathbf{u} = \lambda^\dagger \mathbf{u}^\dagger \mathbf{u}$.

But since $\mathbf{u}^\dagger \mathbf{u} = |\mathbf{u}|^2 > 0$, therefore $\lambda = \lambda^\dagger$ or $\lambda = \lambda^*$, and λ is real.

Q.E.D.

THEOREM 8.11 (Complex Context)

Suppose A is an $n \times n$ Hermitian matrix.

Then A has a set of eigenvectors that form an orthonormal basis for C^n, and A has a *unitary diagonalization*

$$A = UDU^{-1} = UDU^\dagger \quad ,$$

where U is unitary and D is diagonal.

In fact, we can choose the columns of U to be the orthonormal eigenvectors and the diagonal entries of D to be the corresponding eigenvalues.

Note that it was easier to prove that the eigenvalues are real this time because we allowed ourselves the use of the method of the starred section 7.3.

With the preliminary results established, the proof of the Principal Axes Theorem for Hermitian matrices now proceeds in a manner exactly parallel to the earlier demonstration for symmetric matrices.

■ Proof   Exercise.

Example 2   To diagonalize the Hermitian matrix $A = \begin{bmatrix} -3 & 2+2i \\ 2-2i & 4 \end{bmatrix}$,

we first compute det $(A-\lambda I) = \lambda^2 - \lambda - 20 = (\lambda+4)(\lambda-5)$.

For the eigenvalue $\lambda_1 = -4$, $A+4I = \begin{bmatrix} 1 & 2+2i \\ 2-2i & 8 \end{bmatrix}$;

so, by inspection, an eigenvector is $\mathbf{v}_1 = (2+2i, -1)$.

For the eigenvalue $\lambda_2 = 5$, $A-5I = \begin{bmatrix} -8 & 2+2i \\ 2-2i & -1 \end{bmatrix}$,

and a corresponding eigenvector is $\mathbf{v}_2 = (1, 2-2i)$.

Unit eigenvectors are then easily computed as

$$\mathbf{u}_1 = \left(\tfrac{1}{3}\right)(2+2i, -1), \quad \mathbf{u}_2 = \left(\tfrac{1}{3}\right)(1, 2-2i) \ .$$

Thus a unitary diagonalization of A is $A = UPU^\dagger$, where

$$U = \begin{bmatrix} \mathbf{u}_1 \mid \mathbf{u}_2 \end{bmatrix} = \left(\tfrac{1}{3}\right)\begin{bmatrix} 2+2i & 1 \\ -1 & 2-2i \end{bmatrix},$$

$$D = \begin{bmatrix} \lambda_1 & 0 \\ 0 & \lambda_2 \end{bmatrix} = \begin{bmatrix} -4 & 0 \\ 0 & 5 \end{bmatrix}.$$

■ EXERCISES 8.5

1. For each of the matrices listed below, find an orthogonal matrix P and a diagonal matrix D that constitute an orthogonal diagonalization.

   (a) $\begin{bmatrix} 3 & 1 \\ 1 & 3 \end{bmatrix}$ (b) $\begin{bmatrix} -3 & 4 \\ 4 & 3 \end{bmatrix}$ (c) $\begin{bmatrix} 6 & -2 \\ -2 & 9 \end{bmatrix}$

   (d) $\begin{bmatrix} 4 & 2 & 2 \\ 2 & 3 & 0 \\ 2 & 0 & 5 \end{bmatrix}$ (e) $\begin{bmatrix} 0 & 1 & 1 \\ 1 & 0 & 1 \\ 1 & 1 & 0 \end{bmatrix}$ (f) $\begin{bmatrix} 1 & 2 & 0 \\ 2 & 3 & -2 \\ 0 & -2 & 1 \end{bmatrix}$

2. In the example in the text following the Principal Axes Theorem, apply the Gram-Schmidt process to the vectors $\mathbf{v}_1$, $\mathbf{v}_2$ spanning E_{-1} to obtain the elements $\mathbf{u}_1$, $\mathbf{u}_2$ of the orthonormal eigenvector basis.

3. Show that, if a symmetric matrix A has only one eigenvalue $\lambda = a$, then $A = aI$.

4. Suppose A is a symmetric matrix. Show that A is also orthogonal if and only if it has no eigenvalues other than $\lambda = \pm 1$.

5. Show that, in sigma notation,

$$\mathbf{u} \cdot A\mathbf{v} = \sum_{i=1}^{n} \sum_{j=1}^{n} a_{ij} u_i v_j \quad ,$$

and deduce lemma 8.9(a).

6. The Fundamental Theorem of Algebra states that every polynomial, other than a constant, has at least one (complex) root. Deduce, by reference to the characteristic polynomial, that every square matrix has at least one (complex) eigenvalue.

*7. (Classification of 3×3 orthogonal matrices) A corollary of the Fundamental Theorem of Algebra is that every polynomial with real coefficients and of odd degree has at least one real root.

   (a) Deduce that if A is a 3×3 orthogonal matrix, then either 1 or -1 is an eigenvalue of A.

   (b) Deduce, by partial diagonalization of A, that relative to a suitable orthonormal basis for R^3, the orthogonal transformation A is represented by the orthogonal matrix

$$B = \left[\begin{array}{c|cc} \pm 1 & 0 & 0 \\ \hline 0 & a & b \\ 0 & c & d \end{array} \right] \quad , \quad \text{where} \quad B_0 = \begin{bmatrix} a & b \\ c & d \end{bmatrix} \quad \text{is}$$

   a 2×2 orthogonal matrix.

   (c) Deduce that every orthogonal transformation of R^3 can be classified as one of the following:

      (i) A rotation about an axis.
      (ii) A reflection across a plane.
      (iii) The composite of a reflection across a plane, followed by a rotation about an axis perpendicular to that plane.

*8. (Alternative proof of Principal Axes Theorem) Suppose A is a symmetric, $n \times n$ matrix with an orthonormal set of eigenvectors $\mathbf{v}_1, \ldots, \mathbf{v}_k$. Let W be the subspace of R^n consisting of those vectors that are perpendicular to all of $\mathbf{v}_1, \ldots, \mathbf{v}_k$.

   (a) Using lemma 8.9(a), show that if $\mathbf{w}$ is in W, then $A\mathbf{w}$ is also in W.

   (b) Deduce that there is a linear transformation T of W into W such that $T(\mathbf{w}) = A\mathbf{w}$.

(c) Using a matrix representation of T relative to a basis for W (see section *6.4), show that there is a unit vector $\mathbf{w}_0$ in W and a scalar λ_0 such that $T(\mathbf{w}_0) = \lambda_0 \mathbf{w}_0$.

(d) Deduce that $\mathbf{v}_{k+1} = \mathbf{w}_0$ extends the orthonormal set of eigenvalues of A.

(e) Complete proof of theorem as in text.

9. Obtain a unitary matrix U and a diagonal matrix D that constitute a unitary diagonalization, for each of the following matrices

(a) $\begin{bmatrix} 0 & i \\ -i & 0 \end{bmatrix}$ (b) $\begin{bmatrix} 12 & 3+4i \\ 3-4i & -12 \end{bmatrix}$

(c) $\begin{bmatrix} 2 & 1+2i \\ 1-2i & -2 \end{bmatrix}$ $\begin{bmatrix} 0 & i & i \\ -i & 0 & -1 \\ -i & -1 & 0 \end{bmatrix}$

10. Suppose A is a Hermitian matrix. Show that A is also unitary if and only if it has no eigenvalues other than $\lambda = \pm 1$.

*11. Show that every unitary matrix can be diagonalized (adapt the method indicated for 3×3 orthogonal matrices in exercise 7).

12. Let $H = \begin{bmatrix} r & p+iq \\ p-iq & -r \end{bmatrix}$.

Show that the eigenvalues of H are $\pm|\mathbf{p}|$, where $\mathbf{p} = (p,q,r)$.

8.6 Quadratic Forms

Introduction

A *quadratic form* in two variables x, y is a function of the form

$$Q(x,y) = px^2 + 2qxy + ry^2 .$$

Some of the most familiar plane curves have equations of the form $Q(x,y) = k^2$, with k a constant.

Example 1 $Q(x,y) = x^2 + y^2 = a^2$
represents a *circle* of radius a.

Example 2 $Q(x,y) = (x/a)^2 + (y/b)^2 = 1$
represents an *ellipse* with x-intercepts $\pm a$, and y-intercepts $\pm b$.

Example 3 $\quad Q(x,y) = (x/a)^2 - (y/b)^2 = 1$

represents a *hyperbola* with x-intercepts $\pm a$, and asymptotes $y/x - \pm b/a$.

However, when $Q(x,y)$ contains an xy term (or *cross term*) it is less easy to identify the curve that the equation $Q(x,y) = k^2$ represents. Our strategy will be to rewrite the quadratic form in matrix terms and then apply diagonalization theory to eliminate cross terms.

Diagonalization

First, we rewrite $Q(x,y)$ as a matrix product:

$$
\begin{aligned}
Q(x,y) &= px^2 + 2qxy + ry^2 \\
&= x(px + qy) + y(qx + ry) \\
&= [x \quad y] \begin{bmatrix} px + qy \\ qx + ry \end{bmatrix} \\
&= [x \quad y] \begin{bmatrix} p & q \\ q & r \end{bmatrix} \begin{bmatrix} x \\ y \end{bmatrix} \quad .
\end{aligned}
$$

Or $\quad Q(x,y) = \mathbf{x}^t A \mathbf{x}$, where $\mathbf{x} = \begin{bmatrix} x \\ y \end{bmatrix}$, $A = \begin{bmatrix} p & q \\ q & r \end{bmatrix}$.

Note that the coefficients p, r of x^2, y^2 respectively in $Q(x,y)$ become the diagonal entries of A, while the coefficient $2q$ of the cross term xy is split equally between the two off-diagonal entries of A. Thus A is a *symmetric* matrix.

Next, recall that a symmetric matrix A has the orthogonal diagonalization

$$ A = PDP^{-1} = PDP^t $$

and that the new coordinates $\mathbf{x}' = (x',y')$ relative to the eigenvector basis satisfy

$$ \mathbf{x}' = P^{-1}\mathbf{x} = P^t\mathbf{x} \quad . $$

This allows us to rewrite the quadratic form as

$$
\begin{aligned}
Q(x,y) &= \mathbf{x}^t PDP^t\mathbf{x} \\
&= (P^t\mathbf{x})^t D(P^t\mathbf{x})
\end{aligned}
$$

or $\quad Q(x',y') = (\mathbf{x}')^t D\mathbf{x}' = [x' \ y'] \begin{bmatrix} \lambda_1 & 0 \\ 0 & \lambda_2 \end{bmatrix} \begin{bmatrix} x' \\ y' \end{bmatrix}$

or $\quad Q(x',y') = \lambda_1(x')^2 + \lambda_2(y')^2 \quad .$

That is, we have *diagonalized* the quadratic form by eliminating the cross terms.

Example 4 $\quad Q(x,y) = 5x^2 + 8xy + 5y^2 = 9$.

In this case, $p = 5$, $r = 5$, and $2q = 8$ or $q = 4$.

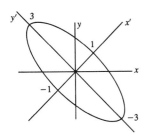

Figure 8.3

Therefore $\quad Q(x,y) = \mathbf{x}^{\,t} \begin{bmatrix} 5 & 4 \\ 4 & 5 \end{bmatrix} \mathbf{x}$.

It is easily checked that the eigenvalues are $\lambda_1 = 9$, $\lambda_2 = 1$, and corresponding unit eigenvectors $\mathbf{v}_1 = \left(\frac{1}{\sqrt{2}}\right)(1,1)$, $\mathbf{v}_2 = \left(\frac{1}{\sqrt{2}}\right)(-1,1)$. The eigenvectors $\mathbf{v}_1$ and $\mathbf{v}_2$ give the directions of the new x' and y' coordinate axes, and in new coordinates:

$$Q(x',y') = 9(x')^2 + 1(y')^2 = 9 \quad .$$

Evidently the curve represented is the ellipse:

$$(x'/1)^2 + (y'/3)^2 = 1 \quad . \text{(See Figure 8.3)}.$$

Example 5 $\quad Q(x,y) = 11x^2 - 24xy + 4y^2 = 20$.

In this case, $p = 11$, $r = 4$, $2q = -24$ or $q = -12$, and we easily obtain the symmetric matrix $A = \begin{bmatrix} 11 & -12 \\ -12 & 4 \end{bmatrix}$, eigenvalues $\lambda_1 = 20$, $\lambda_2 = -5$, and the diagonalized quadratic form $Q(x',y') = 20(x')^2 - 5(y')^2 = 20$.

So the curve represented is the hyperbola

$$(x'/1)^2 - (y'/2)^2 = 1 \text{ (see Figure 8.4)}.$$

To find the directions of the new x',y' coordinate axes we would compute the eigenvectors; they are

$$\mathbf{v}_1 = \left(\tfrac{4}{5}, -\tfrac{3}{5}\right), \mathbf{v}_2 = \left(\tfrac{3}{5}, \tfrac{4}{5}\right) \quad .$$

But to see the *shape* of the curve this is not necessary; we can draw the x' and y' axes as the conventional horizontal and vertical lines; in other words, we disregard the original x,y coordinate system completely.

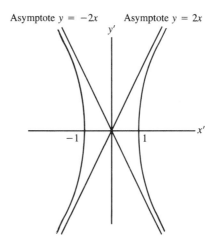

Figure 8.4

Higher Dimensions

Our diagonalization method works without change for quadratic forms in three or more variables.

Example 6 $Q(x,y,z) = 3x^2 + 3z^2 + 4xy + 8xz + 4yz$

$$= [x\ y\ z] \begin{bmatrix} 3 & 2 & 4 \\ 2 & 0 & 2 \\ 4 & 2 & 3 \end{bmatrix} \begin{bmatrix} x \\ y \\ z \end{bmatrix} = \mathbf{x}^t A \mathbf{x}\ .$$

The correspondence between terms in Q and entries in A is shown in Table 8.2.

Squared Term in Q	Diagonal Entry in A
$3x^2$	$a_{11} = 3$
$0y^2$	$a_{22} = 0$
$3z^2$	$a_{33} = 3$

Cross Term in Q	Off-Diagonal Entry in A
$4xy$	$a_{12} = a_{21} = \frac{4}{2} = 2$
$8xz$	$a_{13} = a_{31} = \frac{8}{2} = 4$
$4yz$	$a_{23} = a_{32} = \frac{4}{2} = 2$

Table 8.2

At the end of section 8.5, we studied this symmetric matrix, and found the orthogonal diagonalization:

$$D = \begin{bmatrix} -1 & & \\ & -1 & \\ & & 8 \end{bmatrix}\ ,\quad P = \text{orthogonal matrix with complicated entries.}$$

In the new coordinates $\mathbf{x}' = P^{-1}\mathbf{x}$, the quadratic form becomes

$$Q(x',y',z') = -(x')^2 - (y')^2 + 8(z')^2\ .$$

For example, to sketch the surface $Q(x,y,z) = 8$ in *new* coordinates, we solve the equation

$$Q(x',y',z') = -(x')^2 - (y')^2 + 8(z')^2 = 8\ ,$$

obtaining $z' = \pm \sqrt{1 + \left(\dfrac{(x')^2 + (y')^2}{8} \right)}.$

Thus the z' intercept, where $x' = y' = 0$, is $z' = \pm 1$.

As we move away from z' axis, $(x')^2 + (y')^2$ gradually increases, and so z' either increases from 1 or decreases from -1.

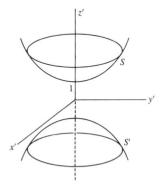

Hyperboloid of two sheets

Figure 8.5

So the surface has one bowl-shaped part S, entirely above the $x'y'$ plane, and a second part S', the mirror image of S, below the $x'y'$ plane. This type of surface is called a "hyperboloid of two sheets" (see Figure 8.5.).

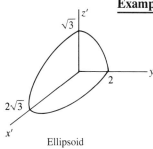

Ellipsoid

Figure 8.6

Example 7 Determine the shape of the surface

$$Q(x,y,z) = 3x^2 + 2y^2 + 3z^2 - 2xy - 2yz = 12 \quad .$$

In this case, $Q = \mathbf{x}^t A \mathbf{x}$, where

$$A = \begin{bmatrix} 3 & -1 & 0 \\ -1 & 2 & -1 \\ 0 & -1 & 3 \end{bmatrix} \quad .$$

It is easily verified that A has the eigenvalues $\lambda_1 = 1$, $\lambda_2 = 3$, $\lambda_3 = 4$; so in the new coordinates the surface has the equation

$$(x')^2 + 3(y')^2 + 4(z')^2 = 12 \quad .$$

Equivalently, the surface can be represented as

$$\left(\frac{x'}{2\sqrt{3}}\right)^2 + \left(\frac{y'}{2}\right)^2 + \left(\frac{z'}{\sqrt{3}}\right)^2 = 1 \quad ,$$

which is an ellipsoid with intercepts on the new coordinate axes at $x' = \pm 2\sqrt{3}$, $y' = \pm 2$, $z' = \pm\sqrt{3}$. Figure 8.6 shows the part of the surface in the first octant.

Summary

It is a direct corollary of the Principal Axes Theorem that quadratic forms can be diagonalized.

COROLLARY 8.12

(Diagonalization of quadratic forms)
Let A be an $n \times n$ symmetric matrix, and let $Q(\mathbf{x}) = \mathbf{x}^t A \mathbf{x}$ be a quadratic form in n variables.
Let $D = \begin{bmatrix} \lambda_1 & & \\ & \cdot & \\ & & \cdot \lambda_n \end{bmatrix}$, $P = [\mathbf{v}_1 | \ldots | \mathbf{v}_n]$ be an orthogonal diagonalization of A. Then, in coordinates $\mathbf{x}' = P^{-1}\mathbf{x}$, Q has the diagonal form

$$Q(\mathbf{x}') = (\mathbf{x}')^t D \mathbf{x}' = \lambda_1(x_1')^2 + \ldots + \lambda_n(x_n')^2 \quad .$$

A quadratic form $Q(\mathbf{x}) = \mathbf{x}^t A \mathbf{x}$ is said to be *nonsingular* if all eigenvalues of the symmetric matrix A are nonzero. (All examples discussed in this section have been nonsingular.) We have developed a classification of those quadratic curves and surfaces that correspond to nonsingular quadratic forms.

THEOREM **8.13**

(Classification of quadratic curves and surfaces)

(a) Suppose $Q(x,y)$ is a nonsingular quadratic form in two variables. Then the plane curve $Q(x,y) = 1$ can only be an ellipse or hyperbola.

(b) Suppose $Q(x,y,z)$ is a nonsingular quadratic form in three variables. Then the surface $Q(x,y,z) = 1$ is geometrically congruent to one of the following surfaces.

(i) $\left(\dfrac{x}{a}\right)^2 + \left(\dfrac{y}{b}\right)^2 + \left(\dfrac{z}{c}\right)^2 = 1$ (ellipsoid)

(ii) $\left(\dfrac{x}{a}\right)^2 + \left(\dfrac{y}{b}\right)^2 - \left(\dfrac{z}{c}\right)^2 = 1$ (hyperboloid of one sheet)

(iii) $-\left(\dfrac{x}{a}\right)^2 - \left(\dfrac{y}{b}\right)^2 - \left(\dfrac{z}{c}\right)^2 = 1$ (hyperboloid of two sheets)

■**Proof**   (a) In view of the diagonal form

$$Q(x',y') = \lambda_1(x')^2 + \lambda_2(y')^2 = 1 \quad ,$$

the signs of the eigenvalues λ_1, λ_2 determine the classification: ellipse when both positive, hyperbola when of opposite sign.

(b) As in the reasoning for part (a), three positive eigenvalues imply case (i); two plus and one minus, case (ii); one plus and two minus, case (iii).   Q.E.D.

Note that we have sketched an example of all surfaces in theorem 8.13 except the hyperboloid of one sheet. The form of the latter surface, as specified in part b(ii) of the theorem, is shown in Figure 8.7.

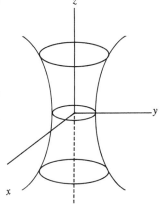

Hyperboloid of one sheet

Figure 8.7

■ **Problem**   Suppose the quadratic form $Q(x,y) = 2x^2 + 4xy - y^2$ is modified by the introduction of a parameter k, to

$$Q_k(x,y) = 2x^2 + 4xy - y^2 + k(x^2 + y^2) \quad .$$

For which values of the parameter k is the curve $Q_k(x,y) = 1$

(a) an ellipse   (b) a hyperbola?

■ **Solution** $Q(x,y) = \mathbf{x}'A\mathbf{x}$,   where $A = \begin{bmatrix} 2 & 2 \\ 2 & -1 \end{bmatrix}$.

Therefore $\begin{aligned} Q_k(x,y) &= \mathbf{x}'A\mathbf{x} + k\mathbf{x}'\mathbf{x} \\ &= \mathbf{x}^t(A + kI)\mathbf{x} \quad . \end{aligned}$

Now, the eigenvalues of A are easily seen to be $\lambda_1 = -2$, $\lambda_2 = 3$.

But if $A\mathbf{v} = \lambda\mathbf{v}$, then $(A+kI)\mathbf{v} = (\lambda+k)\mathbf{v}$. That is, the matrix $A+kI$ has the same eigenvectors as A, but the corresponding eigenvalues shift to

$$\lambda_1 = k-2, \quad \lambda_2 = k+3 \quad .$$

Therefore the curve $Q_k(x,y) = 1$ is:

(a) an ellipse, if both eigenvalues are positive, that is if $k > 2$.

(b) a hyperbola, if the two eigenvalues have opposite signs, that is, if $2 > k > -3$.

■ EXERCISES 8.6

1. For each of the following quadratic forms, $Q(\mathbf{x})$,

   (i) find a symmetric matrix A such that $Q(\mathbf{x}) = \mathbf{x}^t A\mathbf{x}$,

   (ii) diagonalize the quadratic form,

   (iii) classify the curve or surface represented by the equation $Q(\mathbf{x}) = 1$.

   (*Note*: some of the symmetric matrices occurring here were studied in the exercises of section 8.5.)

   (a) $Q(x,y) = 3x^2 + 2xy + 3y^2$

   (b) $Q(x,y) = -3x^2 + 8xy + 3y^2$

   (c) $Q(x,y) = 6x^2 - 4xy + 9y^2$

   (d) $Q(x,y) = 7x^2 + 48xy - 7y^2$

   (e) $Q(x,y) = 2xy$

   (f) $Q(x,y,z) = 4x^2 + 3y^2 + 5z^2 + 4xy + 4xz$

   (g) $Q(x,y,z) = 2xy + 2yz + 2zx$

   (h) $Q(x,y,z) = x^2 + 3y^2 + z^2 + 4xy - 4yz$

2. Classify the curve or surface $Q(\mathbf{x}) = -1$, for quadratic forms (b), (g), and (h) of exercise 1.

3. (a) Suppose $Q(x,y)$ is a quadratic form in two variables, and the curve $Q(x,y) = 1$ is a hyperbola. Show that the graph $Q(x,y) = 0$ is a pair of nonparallel straight lines.

   (b) Suppose $Q(x,y,z)$ is a quadratic form in three variables, and the surface $Q(x,y,z) = 1$ is a hyperboloid. Show that the graph $Q(x,y,z) = 0$ is a cone.

*4. Classify the singular quadratic forms in two and three variables.

5. Describe how the classification of the curve

$$2x^2 - 6xy - 6y^2 + k(x^2 + y^2) = 1$$

varies with the value of the parameter k.

6. Describe how the classification of the surface

$$z^2 - x^2 + 4xy + 4yz + k(x^2 + y^2 + z^2) = 1$$

varies with the value of the parameter k.

7. A quadratic form $Q(\mathbf{x})$ is said to be *positive definite* if $Q(\mathbf{x}) > 0$ for all $\mathbf{x} \neq 0$.

Show that the equation form $Q(\mathbf{x}) = \mathbf{x}^t A \mathbf{x}$ is positive definite if and only if every eigenvalue of the symmetric matrix A is positive.

8. Suppose the quadratic form $Q(\mathbf{x}) = \lambda_1(x_1)^2 + \ldots + \lambda_n(x_n)^2$ is positive definite, and let G be the $n \times n$ diagonal matrix whose nonzero entries are

$$g_{ii} = (\lambda_i)^{-\frac{1}{2}} \ .$$

Show that, in terms of the new coordinates $\mathbf{x}' = G^{-1}\mathbf{x}$,

$$Q(\mathbf{x}') = (\mathbf{x}')^t I \mathbf{x}' = (x_1')^2 + \ldots + (x_n')^2 \ .$$

*9. Suppose $Q(\mathbf{x})$ and $R(\mathbf{x})$ are quadratic forms in n variables, $\mathbf{x}$, and that new coordinates $\mathbf{x}' = K^{-1}\mathbf{x}$ can be found such that Q and R assume the diagonal forms

$$Q(\mathbf{x}') = \mu_1(x_1')^2 + \ldots + \mu_n(x_n')^2$$
$$R(\mathbf{x}') = \nu_1(x_1')^2 + \ldots + \nu_n(x_n')^2 \ .$$

Then Q and R are said to be *simultaneously diagonalizable*.

Show that if one of Q and R is positive definite, then Q and R are simultaneously diagonalizable. (*Note:* the matrix K that implements the coordinate change will not, in most cases, be orthogonal.)

CHAPTER NINE

Jordan Form

Introduction

One method of gaining insight into the structure of a linear transformation is to calculate its matrix representation relative to a strategically chosen basis, From Chapter 8, we know that if A is a matrix transformation of C^n into C^n. and A has enough eigenvectors to form a basis for C^n, then relative to the eigenvector basis, the mapping A is represented by a diagonal matrix D.

But diagonalizable matrices do not display some of the more interesting features of matrix algebra. For example, if the diagonal matrix D has diagonal entries $\lambda_1, \ldots, \lambda_n$, then D^k has diagonal entries $\lambda_1^k, \ldots, \lambda_n^k$. So D^k is the zero matrix only if D itself is the zero matrix, and similarly $A^k = O$ only if $A = O$.

By contrast, the nonzero matrix $A = \begin{bmatrix} 2 & -1 \\ 4 & -2 \end{bmatrix}$ does satisfy $A^2 = \mathbf{0}$,

and so is necessarily nondiagonalizable. In fact, it is easily checked that the sole eigenvalue of A is $\lambda = 0$ and that there is just one linearly independent eigenvector, $\mathbf{v} = (1,2)$. The sole eigenspace, E_0, therefore has dimension 1, and there are vectors in C^2 outside this eigenspace.

To see how A acts on an arbitrary vector $\mathbf{x}$, compute:

$$A\mathbf{x} = \begin{bmatrix} 2 & -1 \\ 4 & -2 \end{bmatrix} \begin{bmatrix} x_1 \\ x_2 \end{bmatrix} = \begin{bmatrix} 2x_1 - x_2 \\ 4x_1 - 2x_2 \end{bmatrix} = (2x_1 - x_2) \begin{bmatrix} 1 \\ 2 \end{bmatrix} .$$

That is, $A\mathbf{x}$ is always a scalar multiple of $\mathbf{v}$, and so $A\mathbf{x}$ is always in the eigenspace E_0. In consequence, A maps $A\mathbf{x}$ to $\mathbf{0}$, and $A^2\mathbf{x}$ is always $\mathbf{0}$.

In this example, therefore, the eigenspace E_0 is the core of a larger space, C^2, where vectors act like eigenvectors under *repeated* application of the mapping A.

Similarly, we might expect any eigenspace E_λ to be the core of a larger space, G_λ, where vectors act like eigenvectors under repeated application of the mapping $A - \lambda I$. The spaces G_λ are called "generalized eigenspaces," and we will show that they play the same role for nondiagonalized matrices as the ordinary eigenspaces E_λ otherwise do.

In section 9.1, we form "strings" of vectors linked by successive applications of the mapping $A - \lambda I$ and use them to form bases for generalized eigenspaces. We then show, in section 9.2, that within a generalized eigenspace and relative to a string basis, a matrix A has a relatively simple matrix representation called a "Jordan block." Then we assemble the Jordan blocks into a "Jordan form" matrix that represents the full mapping A relative to a string basis for C^n.

A Jordan block has λ's on the diagonal, 1's immediately above the diagonal, and zeros elsewhere. For example, a 3×3 Jordan block looks like this:

$$\begin{bmatrix} \lambda & 1 & 0 \\ 0 & \lambda & 1 \\ 0 & 0 & \lambda \end{bmatrix} \quad .$$

Computations with matrices in Jordan form are not quite as simple as those with diagonal matrices. But we demonstrate in section 9.3 that such essential operations as computation of powers of a matrix and solution of systems of differential equations in the form $\dfrac{d\mathbf{x}}{dt} = A\mathbf{x}$ can be implemented and fall into usable, standardized patterns.

9.1 Strings

To begin, let us recall the place of eigenspaces in the theory of diagonalization. Every square $n \times n$ matrix A has one or more (possibly complex) eigenvalues, say $\lambda_1, \ldots, \lambda_k$; and corresponding eigenspaces $E_{\lambda_1}, \ldots, E_{\lambda_k}$. The dimension of the eigenspace E_λ, written $\dim(E_\lambda)$, gives the number of linearly independent eigenvectors in E_λ. When these dimensions add up to n, so that

$$\dim(E_\lambda) + \cdots + \dim(E_{\lambda_k}) = n \quad ,$$

then A has enough eigenvectors to form a basis for C^n, and A is diagonalizable.

Conversely, when A is not diagonalizable, there are some vectors in C^n that are outside the span of the eigenvectors. In this case the action of the matrix transformation A on vectors has entirely new, and at this stage unforeseeable, features.

We may reasonably want to keep the familiar eigenvalues and eigenspaces at the center of the structural theory of nondiagonalizable matrices. For this to be possible, at least one of the eigenspaces E_λ must have additional vectors, besides its own elements, "under its sway." There must be, in effect, a larger subspace G_λ containing E_λ whose vectors are mapped by A according to some pattern related to the eigenvalue λ.

In order to make this intuitive notion more precise, let us start with the simplest possible context in which nondiagonalizable matrices occur—that of 2×2 matrices with only one eigenvalue.

Example 1   The upper triangular matrix $A = \begin{bmatrix} 0 & 1 \\ 0 & 0 \end{bmatrix}$ has only the eigenvalue $\lambda = 0$, and corresponding eigenvector $\mathbf{v}_1 = (1,0)$. Thus the eigenspace E_0 satisfies $\dim(E_0) = 1$. But to obtain the structure of the matrix transformation A, we must describe how A acts on a set of basis vectors for C^2. We must therefore reach outside the span of the sole eigenspace E_0 and select a second basis vector, say $\mathbf{v}_2 = (0,1)$. Then $A\mathbf{v}_2 = (1,0) = \mathbf{v}_1$.

We conclude that A acts on the basis $\mathbf{v}_1$, $\mathbf{v}_2$ in the following way:

$$\begin{cases} A\mathbf{v}_2 = \mathbf{v}_1 \\ A\mathbf{v}_1 = \mathbf{0} \end{cases} .$$

Using arrows to indicate the effect of the mapping A, we can combine the above equations in the schematic diagram

$$\mathbf{v}_2 \rightarrow \mathbf{v}_1 \rightarrow \mathbf{0} \quad .$$

This example shows that vectors outside *any* eigenspace may, by repeated application of the transformation A, be mapped first into an eigenspace, say E_0, and then, like an element of E_0, to $\mathbf{0}$. The "larger space" G_0 whose elements are mapped in a pattern related to the eigenvalue $\lambda = 0$ is therefore the space spanned by $\mathbf{v}_1$ and $\mathbf{v}_2$ and which satisfies $\dim(G_0) = 2 > \dim(E_0)$.

To extend this tentative notion of the space G_λ to the case of an eigenvalue $\lambda \neq 0$, observe that the contents of the eigenspace E_λ are mapped to $\mathbf{0}$, not by A itself, but by the matrix $A - \lambda I$. The space G_λ should therefore consist of those vectors that are eventually mapped to $\mathbf{0}$ by repeated application $A - \lambda I$. We use the terminology described below.

The *generalized eigenspace* G_λ corresponding to an eigenvalue λ of an $n \times n$ matrix A, consists of all vectors $\mathbf{v}$ such that, for some positive integer k,

$$(A - \lambda I)^k \mathbf{v} = \mathbf{0} \quad .$$

Also, every nonzero vector in G_λ is called a *generalized eigenvector of A corresponding to the eigenvalue* λ.

It is easily checked (see exercises) that a generalized eigenspace is always a subspace of C^n.

Example 2   To find the generalized eigenspaces of the matrix

$$A = \begin{bmatrix} 3 & 1 & 0 \\ 0 & 3 & 0 \\ 0 & 0 & 2 \end{bmatrix} ,$$

observe that the eigenvalues are $\lambda = 3, 2$. Corresponding to $\lambda = 3$, the first few powers of $A - 3I$ are

$$A - 3I = \begin{bmatrix} 0 & 1 & 0 \\ 0 & 0 & 0 \\ 0 & 0 & -1 \end{bmatrix} , \ (A-3I)^2 = \begin{bmatrix} 0 & 0 & 0 \\ 0 & 0 & 0 \\ 0 & 0 & 1 \end{bmatrix} , \ldots$$

$$\ldots (A-3I)^3 = \begin{bmatrix} 0 & 0 & 0 \\ 0 & 0 & 0 \\ 0 & 0 & -1 \end{bmatrix} .$$

In fact, $(A-3I)^k = \pm (A-3I)^2$ for $k \geq 2$.

By inspection, we see that the eigenspace $E_3 = \operatorname{Ker}(A - 3I)$ is spanned by the vector $(1,0,0) = e_1$. Also, $\operatorname{Ker}((A - 3I)^2)$ is clearly spanned by e_1 and e_2, and for $k > 2$, $\operatorname{Ker}((A - 3I)^2)$ is of course the same as for $k = 2$.

Thus the generalized eigenspace corresponding to $\lambda = 3$ is

$$G_3 = \operatorname{Ker}((A - 3I)^2) ,$$

and $\quad \dim G_3 = 2 > \dim E_3 = 1$.

For the eigenvalue $\lambda = 2$, we have

$$(A-2I) = \begin{bmatrix} 1 & 1 & 0 \\ 0 & 1 & 0 \\ 0 & 0 & 0 \end{bmatrix} , \ (A-2I)^2 = \begin{bmatrix} 1 & 2 & 0 \\ 0 & 1 & 0 \\ 0 & 0 & 0 \end{bmatrix} , \ldots$$

$$\ldots (A-2I)^k = \begin{bmatrix} 1 & * & 0 \\ 0 & 1 & 0 \\ 0 & 0 & 0 \end{bmatrix} .$$

The eigenspace $E_2 = \operatorname{Ker}(A - 2I)$ is spanned by e_3, as are the kernels of all the matrices $(A - 2I)^k$. So in this case, the generalized eigenspace is $G_2 = E_2$.

Note that within the generalized eigenspaces, $A - \lambda I$ acts in the following patterns:

within G_3: $\quad e_2 \to e_1 \to \mathbf{0}$
within G_2: $\quad\quad\ e_3 \to \mathbf{0}$.

Example 3 The matrix $A = \begin{bmatrix} 4 & 1 & 0 \\ 0 & 4 & 1 \\ 0 & 0 & 4 \end{bmatrix}$ has only the eigenvalue $\lambda = 4$.

To find the generalized eigenspace G_4, we compute

$$A - 4I = \begin{bmatrix} 0 & 1 & 0 \\ 0 & 0 & 1 \\ 0 & 0 & 0 \end{bmatrix} , \ldots$$

$$\ldots (A - 4I)^2 = \begin{bmatrix} 0 & 0 & 1 \\ 0 & 0 & 0 \\ 0 & 0 & 0 \end{bmatrix} , \ (A - 4I)^3 = O.$$

The eigenspace $E_4 = \operatorname{Ker}(A - 4I)$ is spanned by e_1. However, $\operatorname{Ker}((A - 4I)^2)$ is spanned by e_1 and e_2, while $\operatorname{Ker}((A - 4I)^3) = C^3$.

Thus $G_4 = C^3$.

In this example, we conclude that

$$\dim G_4 = 3 > \dim E_4 = 1$$

and that within the generalized eigenspace G_4, $A - 4I$ acts according to the pattern

$$\mathbf{e}_3 \rightarrow \mathbf{e}_2 \rightarrow \mathbf{e}_1 \rightarrow \mathbf{0} \quad .$$

These examples suggest that the generalized eigenvectors of a square matrix A can be used to form a basis for C^n when the ordinary eigenvectors do not suffice. Furthermore, the elements of such a basis seem to fall into linked sequences. These perceptions are formalized in the following definitions.

Suppose that A is a square matrix and that the nonzero vectors $\mathbf{v}_1, \ldots, \mathbf{v}_m$ satisfy the conditions

$$\begin{cases} (A - \lambda I)\mathbf{v}_m & = \mathbf{v}_{m-1} \\ (A - \lambda I)\mathbf{v}_{m-1} = \mathbf{v}_{m-2} \\ \cdots\cdots\cdots\cdots\cdots\cdots\cdots\cdots \\ (A - \lambda I)\mathbf{v}_2 & = \mathbf{v}_1 \\ (A - \lambda I)\mathbf{v}_1 & = \mathbf{0} \quad . \end{cases}$$

Then $\mathbf{v}_1, \ldots, \mathbf{v}_m$ are said to form a *string* of generalized eigenvectors, corresponding to the eigenvalue λ of A. Also, the action of the matrix $A - \lambda I$ on the string is symbolically denoted

$$\mathbf{v}_m \rightarrow \mathbf{v}_{m-1} \rightarrow \cdots \rightarrow \mathbf{v}_1 \rightarrow \mathbf{0} \quad ,$$

and the string is said to be *of length m*.

A *string basis* for a generalized eigenspace G_λ is a collection of strings whose elements, taken together, form a basis for G_λ.

Often each generalized eigenspace of a matrix is spanned by the elements of a single string. This is true of all the matrices discussed so far. But the next example shows that more than one string is sometimes necessary.

Example 4   The matrix $A = \begin{bmatrix} 4 & 1 & 0 \\ 0 & 4 & 0 \\ 0 & 1 & 4 \end{bmatrix}$ has only one eigenvalue, $\lambda = 4$.

The powers of $A - \lambda I$ are

$$A - \lambda I = \begin{bmatrix} 0 & 1 & 0 \\ 0 & 0 & 0 \\ 0 & 1 & 0 \end{bmatrix} \quad , (A - \lambda I)^2 = O \quad .$$

The sole eigenspace $E_\lambda = \text{Ker}(A - \lambda I)$ is spanned by $\mathbf{e}_1$, $\mathbf{e}_3$. However, every vector in C^3 is mapped to $\mathbf{0}$ by $(A - \lambda I)^2$, and so the generalized eigenspace is:

$$G_\lambda = \text{Ker}((A - \lambda I)^2) = C^3 \quad .$$

Thus the longest string in G_λ will have length 2 and will assume the form

$$\mathbf{v}_2 \to \mathbf{v}_1 \to \mathbf{0} \quad .$$

To find such a string, select a vector that is in G_λ but not E_λ, say $\mathbf{v}_2 = \mathbf{e}_2$. Then compute

$$\mathbf{v}_1 = (A - \lambda I)\mathbf{v}_2 = \begin{bmatrix} 0 & 1 & 0 \\ 0 & 0 & 0 \\ 0 & 1 & 0 \end{bmatrix} \begin{bmatrix} 0 \\ 1 \\ 0 \end{bmatrix} = \begin{bmatrix} 1 \\ 0 \\ 1 \end{bmatrix} = \mathbf{e}_1 + \mathbf{e}_3 \quad .$$

That is, a string of length 2 is

$$\mathbf{e}_2 \to \mathbf{e}_1 + \mathbf{e}_3 \to \mathbf{0} \quad .$$

A further string, of length 1, is formed by any eigenvector, say,

$$\mathbf{e}_1 \to \mathbf{0} \quad .$$

The three vectors appearing in these two strings, $\mathbf{e}_2$, $\mathbf{e}_1 + \mathbf{e}_3$, $\mathbf{e}_1$, form a basis for G_λ.

This example suggests that a generalized eigenspace G_λ may contain many strings of varied lengths. We will have to work carefully to collect all the strings necessary to form a string basis. We begin our theoretical development with a lemma that clarifies the structure of G_λ and limits the lengths of the strings it contains.

Lemma 9.1 Suppose G_λ is the generalized eigenspace corresponding to an eigenvalue λ of an $n \times n$ matrix A. Then

(a) The elements of any string G_λ form a linearly independent set.

(b) The maximum length of strings in G_λ is an integer m satisfying $m \le n$, and

$$G_\lambda = \mathrm{Ker}((A - \lambda I)^m) \quad .$$

■Proof (a) Let $\mathbf{v}_k \to \cdots \to \mathbf{v}_1 \to \mathbf{0}$ be a string in G_λ, and suppose $c_1 \mathbf{v}_1 + \cdots + c_k \mathbf{v}_k = \mathbf{0}$. Application of the matrix $(A - \lambda I)^{k-1}$ to this equation yields

$$c_1 \mathbf{0} + \cdots + c_{k-1} \mathbf{0} + c_k \mathbf{v}_1 = \mathbf{0} \quad ,$$

or simply $c_k \mathbf{v}_1 = \mathbf{0}$. But since $\mathbf{v}_1 \ne \mathbf{0}$, c_k must be zero. In turn, application of $(A - \lambda I)^{k-2}$ to the original equation implies that $c_{k-1} = \mathbf{0}$, etc. Thus all the constants $c_1, \ldots, c_k$ must be zero, and the string $\mathbf{v}_1, \ldots, \mathbf{v}_k$ form an l.i. set.

(b) If G_λ contained a string of length greater than n, its elements would form a set of more than n l.i. vectors in C^n, which is impossible. So $m \le n$. Also, let $\mathbf{x}$ be any vector in G_λ, and consider the string

$$\mathbf{x} \to (A - \lambda I)\mathbf{x} \to \cdots \to (A - \lambda I)^m \mathbf{x} \quad .$$

Because the maximum string length is m, the last element must be $\mathbf{0}$. That is $(A - \lambda I)^m \mathbf{x} = \mathbf{0}$ for all $\mathbf{x}$ in G_λ, and so $G_\lambda = \mathrm{Ker}((A - \lambda I)^m)$ as required.

Q.E.D.

Given a string $\mathbf{v}_k \to \cdots \to \mathbf{v}_1 \to \mathbf{0}$ in a generalized eigenspace G_λ, its *terminal element* $\mathbf{v}_1$ is necessarily in the eigenspace E_λ. Thus if a collection of strings form a basis for G_λ, their terminal elements will form an l.i. set within E_λ.

Therefore, in the course of developing a string basis for G_λ, we might reasonably expect to construct a string basis for E_λ from the terminal elements of the strings we select. However, example 4 above shows that randomly selected basis elements for E_λ will not necessarily serve as the terminal elements of the longer strings needed to form a string basis.

In fact, the simple basis $\mathbf{e}_1$, $\mathbf{e}_3$ might easily have been chosen for the eigenspace E_λ of example 4. This would have blocked the completion of the string basis with a string of length 2, which necessarily terminates with the vector $\mathbf{e}_1 + \mathbf{e}_3$, or a scalar multiple thereof.

The following lemma tells us how to develop a basis for E_λ that will accommodate the terminal elements of the long strings.

Lemma 9.2 Suppose A is a square matrix with a generalized eigenspace G_λ in which maximum string length is m.

Let F_j be the subset of G_λ that contains

(i) the terminal element of each string of length j in G_λ
(ii) the zero vector.

Then (a) Each F_j, for $1 \leq j \leq m$, is a subspace of the eigenspace E_λ.

(b) $F_1 = E_\lambda$, and F_j contains F_{j+1}, for $1 \leq j \leq m-1$.

■**Proof**   (a) Note that if $\mathbf{v}_j \to \cdots \to \mathbf{v}_1 \to \mathbf{0}$
and $\mathbf{w}_j \to \cdots \to \mathbf{w}_1 \to \mathbf{0}$

are strings of length j, then so are

$\mathbf{v}_j + \mathbf{w}_j \to \cdots \to \mathbf{v}_1 + \mathbf{w}_1 \to \mathbf{0}$, unless $\mathbf{v}_1 + \mathbf{w}_1 = \mathbf{0}$,

and $t\mathbf{v}_j \to \cdots \to t\mathbf{v}_1 \to \mathbf{0}$, unless $t\mathbf{v}_1 = \mathbf{0}$.

That is, starting with elements $\mathbf{v}_1$, $\mathbf{w}_1$ in F_j, their sums and products $\mathbf{v}_1 + \mathbf{w}_1$ and $t\mathbf{v}_1$ are still in F_j either because they are still terminal elements of strings of length j, or because they are $\mathbf{0}$. Therefore F_j does have the "no escape" properties of a subspace.

(b) If $\mathbf{v}_1$ is the terminal element of a string

$\mathbf{v}_{j+1} \to \mathbf{v}_j \to \cdots \to \mathbf{v}_1 \to \mathbf{0}$,

it is also the terminal element of the shortened string

$\mathbf{v}_j \to \cdots \to \mathbf{v}_1 \to \mathbf{0}$.

Therefore F_j contains F_{j+1} as stated.

Also, every eigenvector $\mathbf{v}$ forms a string of length 1, $\mathbf{v} \to \mathbf{0}$, and so $E_\lambda = F_1$.

Q.E.D.

The eigenspace E_λ is now ready to sprout a collection of strings, which, as we will prove, form a basis for the larger, generalized eigenspace G_λ.

To start, let the dimensions of the subspaces just constructed be

$$p = \dim F_m, \, q = \dim F_{m-1}, \ldots, r = \dim F_1 = \dim E_\lambda \quad .$$

Since the subspaces are nested, we have $p \le q \le \ldots \le r$.

Then let $\mathbf{z}_1, \ldots, \mathbf{z}_p$ be a basis for F_m, and let $s_1, \ldots, s_p$ be strings of length m with terminal elements $\mathbf{z}_1, \ldots, \mathbf{z}_p$ respectively. These strings may be represented in the form

string s_1: $\mathbf{a}_1 \to \mathbf{b}_1 \to \cdots \to \mathbf{y}_1 \to \mathbf{z}_1 \to \mathbf{0}$

· ·

string s_p: $\mathbf{a}_p \to \mathbf{b}_p \to \cdots \to \mathbf{y}_p \to \mathbf{z}_p \to \mathbf{0}$.

Note that we have given letter names to the elements of each string, from **a** at the top to **z** at the terminal step. The subscripts on the string elements serve to identify the strings by number.

Next, extend the basis for F_m to a basis for the larger subspace F_{m-1} by adjoining additional basis elements $\mathbf{z}_{p+1}, \ldots, \mathbf{z}_q$. Also, let $s_{p+1}, \ldots, s_q$ be strings of length $m-1$ with terminal elements $\mathbf{z}_{p+1}, \ldots, \mathbf{z}_q$ respectively.

Continue adding strings terminating in the successive subspaces $F_{m-2}, \ldots, F_1$ in term. The last strings are those of length 1; the last one, for example, being $\mathbf{z}_r \to \mathbf{0}$. We now have the following collection of strings:

Strings of length m	$\left\{\vphantom{\begin{array}{c}a\\b\end{array}}\right.$	string s_1 : $\mathbf{a}_1 \to \mathbf{b}_1 \to \cdots \to \mathbf{y}_1 \to \mathbf{z}_1 \to \mathbf{0}$
		· ·
		string s_p : $\mathbf{a}_p \to \mathbf{b}_p \to \cdots \to \mathbf{y}_p \to \mathbf{z}_p \to \mathbf{0}$
Strings of length $m-1$	$\left\{\vphantom{\begin{array}{c}a\\b\end{array}}\right.$	string s_{p+1} : $\mathbf{b}_{p+1} \to \cdots \to \mathbf{y}_{p+1} \to \mathbf{z}_{p+1} \to \mathbf{0}$
		· ·
		string s_q : $\to \mathbf{b}_q \to \cdots \to \mathbf{y}_q \to \mathbf{z}_q \to \mathbf{0}$

· ·

Strings of length 1 $\left\{\vphantom{\begin{array}{c}a\\b\end{array}}\right.$ ·

$\mathbf{z}_r \to \mathbf{0}$

$\underbrace{}$ $\underbrace{}$ $\qquad$ $\underbrace{}$ $\underbrace{}$

string level m $\quad$ string level $m-1$ $\quad$ string level 2 $\quad$ string level 1

With the notations used in the table, the string elements $\mathbf{a}_1, \ldots, \mathbf{a}_p$ are all m steps from $\mathbf{0}$ and may be said to be at string level m. Similarly the elements $\mathbf{b}_1, \ldots, \mathbf{b}_q$ are at string level $m-1$, and finally the elements $\mathbf{z}_1, \ldots, \mathbf{z}_r$ are at string level 1, adjacent to $\mathbf{0}$.

We now show that this collection of strings provides a basis for the generalized eigenspace G_λ. This constitutes the key theoretical result for the chapter.

THEOREM 9.3

Every generalized eigenspace has a string basis.

■**Proof** Let G_λ be a generalized eigenspace. To show that a string basis for G_λ is formed by the strings $s_1, \ldots, s_r$ constructed above, we first establish that their collected elements are l.i.

Step 1. Suppose, then, that a linear combination of them adds to $\mathbf{0}$,

$$c_1\mathbf{a}_1 + \cdots + c_p\mathbf{a}_p + c_{p+1}\mathbf{b}_1 + \cdots + c_k\mathbf{z}_r = \mathbf{0} \quad ,$$

and apply the matrix $(A - \lambda I)^{m-1}$ to this equation. The matrix maps all the vectors in the equation to $\mathbf{0}$, except for the $\mathbf{a}_i$'s, which are at the top of strings of length m and are mapped to the corresponding $\mathbf{z}_i$'s. We therefore obtain

$$c_1\mathbf{z}_1 + \cdots + c_p\mathbf{z}_p = \mathbf{0} \quad .$$

But since $\mathbf{z}_p, \ldots, \mathbf{z}_p$, as basis elements for F_m, are l.i., the constants $c_1, \ldots, c_p$ are all zero. We then proceed to apply lower powers, $(A - \lambda I)^{m-2}$ etc., to the original equation and eventually conclude that all the constants are zero. So the string elements are l.i. as required.

Step 2. *It must also be shown that the collected string elements span G_λ.* Suppose, then, that $\mathbf{a}$ is any element of G_λ outside their span; and suppose, for instance, that $\mathbf{a}$ is at the top of a string of length m. Then $(A - \lambda I)^{m-1}\mathbf{a} = \mathbf{z}$, where $\mathbf{z}$ is in F_m. Relative to the basis we constructed for F_m, $\mathbf{z}$ has a representation

$$\begin{aligned}\mathbf{z} &= c_1\mathbf{z}_1 + \cdots + c_p\mathbf{z}_p \\ &= (A - \lambda I)^{m-1}(c_1\mathbf{a}_1 + \cdots + c_p\mathbf{a}_p) \quad .\end{aligned}$$

Subtracting the two formulas we have for $\mathbf{z}$, we obtain

$$(A - \lambda I)^{m-1}((\mathbf{a} - (c_1\mathbf{a}_1 + \cdots + c_p\mathbf{a}_p)) = \mathbf{0} \quad .$$

That is, the vector $\mathbf{b} = \mathbf{a} - (c_1\mathbf{a}_1 + \cdots + c_p\mathbf{a}_p)$ is at the top of a shorter string than $\mathbf{a}$. Also, since $\mathbf{a}$ is outside the span of the string elements, so is $\mathbf{b}$.

Repetition of the argument produces vectors outside the span at successively lower levels, until finally the $\mathbf{0}$ vector, at level 0, is reached, a contradiction. Q.E.D.

■ **Problem** Show that each generalized eigenspace of the $n \times n$ matrix

$$A = \begin{bmatrix} 0 & 1 & 0 & . & . & . & 0 \\ 0 & 0 & 1 & . & . & . & 0 \\ . & . & . & . & & & . \\ . & . & . & & . & & . \\ . & . & . & & & . & . \\ . & . & & & & . & 1 \\ -c_0 & -c_1 & . & . & . & . & -c_{n-1} \end{bmatrix}$$

has a basis consisting of a single string. (This result is of theoretical importance in the theory of differential equations; see exercises of section 9.3.)

■ **Solution**   Suppose G_λ is a generalized eigenspace of A. Then each string, s_i, in a string basis for G_λ terminates with an element z_i that is in the ordinary eigenspace E_λ. Now, if there are k strings in the string basis for G_λ, then their terminal elements $z_1, \ldots, z_k$ form k l.i. elements in E_λ, and $\dim(E_\lambda)$ is at least k.

But $E_\lambda = \text{Ker}(A - \lambda I)$, and so from the Main Dimension Theorem (section 6.3)

$$\dim E_\lambda = n - \dim[(\text{Range}(A - \lambda I)] \quad .$$

Also, $A - \lambda I = \begin{bmatrix} -\lambda & 1 & 0 & . & . & . & 0 \\ 0 & -\lambda & 1 & . & . & . & 0 \\ . & . & . & & & & \\ . & . & . & & & . & \\ . & . & & . & & & \\ . & . & & & . & & 1 \\ -c_0 & -c_1 & . & . & . & . & -\lambda - c_{n-1} \end{bmatrix}$;

therefore by inspection columns 2 through n of $A - \lambda I$ form an l.i. set, and $\dim\{\text{Range}(A - \lambda I)\}$ is at least $n - 1$. Thus $\dim(E_\lambda)$ can only be 1 and there is only one string in the string basis for G_λ, as required.

■ EXERCISES 9.1

1. For the matrix $A = \begin{bmatrix} 0 & 0 & 1 & 1 \\ 0 & 0 & 0 & 0 \\ 0 & 0 & 0 & 0 \\ 0 & 0 & 0 & 0 \end{bmatrix}$,

   (a) determine all eigenvalues of A,
   (b) find the dimension of each generalized eigenspace G_λ, and each ordinary eigenspace E_λ, of A, and
   (c) find the number of strings of each length in a string basis for each generalized eigenspace G_λ.

2. Repeat exercise 1 for the following matrices.

   (a) $\begin{bmatrix} 14 & 16 \\ -9 & -10 \end{bmatrix}$ (b) $\begin{bmatrix} 3 & 2 \\ -5 & 1 \end{bmatrix}$ (c) $\begin{bmatrix} 5 & 4 \\ -1 & 1 \end{bmatrix}$

   (d) $\begin{bmatrix} 4 & 0 & 2 \\ 0 & 4 & 1 \\ 0 & 0 & 4 \end{bmatrix}$ (e) $\begin{bmatrix} 2 & 0 & 6 \\ 0 & 3 & 1 \\ 0 & 0 & 3 \end{bmatrix}$ (f) $\begin{bmatrix} 2 & -1 & 0 \\ 2 & 0 & 1 \\ 0 & -2 & -2 \end{bmatrix}$

3. Suppose $\mathbf{u}$ is in $\text{Ker}((A - \lambda I)^j)$ and $\mathbf{v}$ is in $\text{Ker}((A - \lambda I)^k)$. Let h be the larger of j, k.

   (a) Show that $\mathbf{u} + \mathbf{v}$ is in $\text{Ker}((A - \lambda I)^h)$.
   (b) Deduce that a generalized eigenspace G_λ is always a subspace of C^n.

4. (a) Show that if a string basis for a generalized eigenspace includes a string of length p or greater, then

$$\dim\{\mathrm{Ker}((A - \lambda I)^p)\} > \dim\{\mathrm{Ker}((A - \lambda I)^{p-1})\} \quad .$$

(b) Show that if maximum string length in a string basis for a generalized eigenspace is m, then

$$\dim\{\mathrm{Ker}((A - \lambda I)^{m+1})\} = \dim\{\mathrm{Ker}(A - \lambda I)^m)\} >$$
$$\dim\{\mathrm{Ker}((A - \lambda I)^{m-1})\} \quad .$$

5. Let λ be an eigenvalue of the square matrix A, and let

$$d_p = \dim\{\mathrm{Ker}(A - \lambda I)^p\} \quad .$$

Show that in a string basis for the generalized eigenspace G_λ,

(a) The number of strings of length p or greater is $d_p - d_{p-1}$.

(b) The number of strings of length exactly p is

$$h_p = 2d_p - d_{p+1} - d_{p-1} \quad .$$

6. Let $A = \begin{bmatrix} 0 & 0 & 1 & 0 \\ 0 & 0 & 0 & 1 \\ 0 & 0 & 0 & 0 \\ 0 & 0 & 0 & 0 \end{bmatrix}$.

Using the method of the previous exercise, show that the string basis for the generalized eigenspace G_0 of A has the pattern

$$\begin{cases} s_1: \mathbf{a}_1 \to \mathbf{b}_1 \to \mathbf{0} \\ s_2: \mathbf{a}_2 \to \mathbf{b}_2 \to \mathbf{0} \end{cases} \quad .$$

7. For each generalized eigenspace of each of the following matrices, determine the pattern of a string basis.

(a) $\begin{bmatrix} 0 & 1 & 0 & 0 \\ 0 & 0 & 0 & 0 \\ 1 & 0 & 0 & 0 \\ 0 & 0 & 0 & 0 \end{bmatrix}$ (b) $\begin{bmatrix} 0 & 0 & 0 & 1 \\ 0 & 2 & 0 & 0 \\ 0 & 1 & 2 & 0 \\ 0 & 0 & 0 & 0 \end{bmatrix}$

8. (Conjugate strings) Let A be a real $n \times n$ matrix.

(a) Show that $(A - \lambda I)^k \mathbf{v} = \mathbf{0}$ if and only if $(A - \lambda^* I)^k \mathbf{v}^* = \mathbf{0}$, where λ^*, $\mathbf{v}^*$ are the complex conjugates of λ, $\mathbf{v}$ respectively.

(b) Deduce that, if λ is a complex eigenvalue of A, so is λ^*.

(c) Suppose $s: \mathbf{v}_p \to \cdots \to \mathbf{v}_1 \to \mathbf{0}$ is a string in the generalized eigenspace G_λ of A. Show that $s^*: \mathbf{v}_p^* \to \cdots \to \mathbf{v}_1^* \to \mathbf{0}$ is a string in G_{λ^*}.

(d) Show that, if a collection of strings $s_1, \ldots, s_r$ forms a basis for G_λ, then the collection $s_1^*, \ldots, s_r^*$ forms a basis for G_{λ^*}.

9.2 Jordan Form

Introduction

A diagonal matrix D, with diagonal entries $d_{jj} = \lambda_j$, maps a set of coordinates $(c_1, \ldots, c_n)$ to $(\lambda_1 c_1, \ldots, \lambda_n c_n)$. It can therefore be regarded as an assembly of n distinct one-dimensional transformations of the form $c_j \rightarrow \lambda_j c_j$.

The transformation represented by a nondiagonalizable matrix A is also assembled from smaller, simpler transformations, but not exclusively one-dimensional ones. In fact, we start with the subspace of S spanned by the elements of a single string $\mathbf{v}_p \rightarrow \cdots \rightarrow \mathbf{v}_1 \rightarrow \mathbf{0}$ within a generalized eigenspace G_λ. Because A acts simply on the string elements, it is easy to construct a $p \times p$ matrix (denoted $J(\lambda;p)$) that represents the mapping A within the subspace S.

But because G_λ has a string basis, the blocks $J(\lambda;p)$ of various sizes can be assembled along the diagonal of a larger matrix (denoted $J(\lambda;\mathbf{p})$) that represents the mapping A within G_λ. Of course this tactic must be repeated; the larger blocks $J(\lambda;\mathbf{p})$ must be assembled along the diagonal of a still larger matrix (denoted J) that represents the entire mapping A. But the second step is more difficult; to justify it we have to establish that the string bases for the generalized eigenspaces G_λ combine into a basis for C^n. This is the essential theoretical context of this section.

With the construction of the Jordan form J of a square matrix A, we complete a classification of matrix transformations of C^n into C^n. They are all assembled from the mapping structure embodied in a string. The Jordan form J, although necessarily more complicated than diagonal form, proves quite satisfactory for purposes of calculation. Its properties are developed through the applications in the next section.

Single Strings

Suppose that $\mathbf{v}_p \rightarrow \cdots \rightarrow \mathbf{v}_1 \rightarrow \mathbf{0}$ is a string within a generalized eigenspace G_λ of a square matrix A. Then the action of the transformation A on the string elements takes the form

$$\left\{ \begin{array}{l} (A - \lambda I)\mathbf{v}_p = \mathbf{v}_{p-1} \\ \cdots\cdots\cdots\cdots\cdots \\ (A - \lambda I)\mathbf{v}_2 = \mathbf{v}_1 \\ (A - \lambda I)\mathbf{v}_2 = \mathbf{0} \end{array} \right.$$

or

$$\left\{ \begin{array}{l} A\mathbf{v}_p = \lambda\mathbf{v}_p + \mathbf{v}_{p-1} \\ \cdots\cdots\cdots\cdots\cdots \\ A\mathbf{v}_2 = \lambda\mathbf{v}_2 + \mathbf{v}_1 \\ A\mathbf{v}_1 = \lambda\mathbf{v}_1 \quad . \end{array} \right.$$

Now let S be the subspace of G_λ spanned by the string elements $v_1, \ldots, v_p$. Any vector x in S has a representation of the form

$$x = c_1 v_1 + \cdots + c_p v_p$$

and

$$\begin{aligned} Ax &= c_1(Av_1) + \cdots + c_p(Av_p) \\ &= c_1(\lambda v_1) + c_2(\lambda v_2 + v_1) + \cdots + c_p(\lambda v_p + v_{p-1}) \\ &= (\lambda c_1 + c_2)v_1 + \cdots + (\lambda c_{p-1} + c_p)v_{p-1} + \lambda c_p v_p \quad. \end{aligned}$$

Thus Ax is also in S, and relative to the basis $v_1, \ldots, v_p$, its coordinates are $\lambda c_1 + c_2, \ldots, \lambda c_{p-1} + c_p, \lambda c_p$. If we use c and c' to denote the coordinates of x and Ax respectively relative to the basis, we have

$$\begin{bmatrix} c_1' \\ \cdot \\ \cdot \\ \cdot \\ c_p' \end{bmatrix} = \begin{bmatrix} \lambda & 1 & & & \\ & \lambda & 1 & & \\ & & \cdot & \cdot & \\ & & & \cdot & 1 \\ & & & & \lambda \end{bmatrix} \begin{bmatrix} c_1 \\ \cdot \\ \cdot \\ \cdot \\ c_p \end{bmatrix},$$

or $\quad c' = Jc, \quad$ where $J = \begin{bmatrix} \lambda & 1 & & & \\ & \lambda & 1 & & \\ & & \cdot & \cdot & \\ & & & \cdot & 1 \\ & & & & \lambda \end{bmatrix} \quad.$

Observe that the matrix J has a rather simple structure; it has λ's on the diagonal, 1's immediately above the diagonal, and, of course, zeros elsewhere. A matrix of this sort is fully determined as soon as we know its dimensions, $p \times p$, and the value of its diagonal entries, λ. We call such a matrix a *Jordan block* and denote it $J(\lambda; p)$.

Example 1 Let $A = \begin{bmatrix} -3 & 9 \\ -4 & 9 \end{bmatrix} \quad.$

Then $\det(A - \lambda I) = \det \begin{bmatrix} -3-\lambda & 9 \\ -4 & 9-\lambda \end{bmatrix} = \lambda^2 - 6\lambda + 9 = (\lambda - 3)^2 \quad.$

The sole eigenvalue of A is therefore $\lambda = 3$, and

$$A - 3I = \begin{bmatrix} -6 & 9 \\ -4 & 6 \end{bmatrix}, \quad \text{while } (A - 3I)^2 = \begin{bmatrix} 0 & 0 \\ 0 & 0 \end{bmatrix} \quad.$$

Thus $\operatorname{Ker}((A - 3I)^2) = C^2$, and so the generalized eigenspace G_3 is all of C^2.

To find a string, let $v_2 = (1,0)$. Then $(A - 3I)v_2 = (-6, -4) = v_1$; and we have $v_2 \to v_1 \to 0$. The string elements v_1, v_2 span a space S of dimension 2 (which is necessarily C^2). Relative to the basis v_1, v_2 for C^2, therefore, the action of A on coordinates is given by the Jordan block matrix

$$J = \begin{bmatrix} 3 & 1 \\ 0 & 3 \end{bmatrix}, \quad \text{or, in streamlined notation, } J(3;2).$$

Example 2 Let $A = \begin{bmatrix} 6 & 1 & 0 \\ 0 & 5 & 1 \\ -1 & -1 & 4 \end{bmatrix}$.

Then $\det(A - \lambda I = \begin{bmatrix} 6-\lambda & 1 & 0 \\ 0 & 5-\lambda & 1 \\ -1 & -1 & 4-\lambda \end{bmatrix}$

$$= (6-\lambda) \det \begin{bmatrix} 5-\lambda & 1 \\ -1 & 4-\lambda \end{bmatrix} - 1 \det \begin{bmatrix} 1 & 0 \\ 5-\lambda & 1 \end{bmatrix}$$

$$= (6-\lambda)(\lambda^2 - 9\lambda + 21) - 1$$

$$= -\lambda^3 + 15\lambda^2 - 75\lambda + 125 = -(\lambda - 5)^3.$$

The sole eigenvalue is $\lambda = 5$, and

$$A - 5I = \begin{bmatrix} 1 & 1 & 0 \\ 0 & 0 & 1 \\ -1 & -1 & -1 \end{bmatrix} ,$$

$$(A - 5I)^2 = \begin{bmatrix} 1 & 1 & 1 \\ -1 & -1 & -1 \\ 0 & 0 & 0 \end{bmatrix} , (A - 5I)^3 = O .$$

So the generalized eigenspace G_5 equals all of C^3.

Starting with, say $\mathbf{v}_3 = \mathbf{e}_1 = (1,0,0)$, we obtain

$$\mathbf{v}_2 = (A - 5I)\mathbf{v}_3 = (1, 0, -1)$$
$$\mathbf{v}_1 = (A - 5I)\mathbf{v}_2 = (1, -1, 0)$$
$$(A - 5I)\mathbf{v}_1 = \mathbf{0} .$$

Thus we have a string of length 3. The Jordan block corresponding to this string is, of course,

$$J = J(5;3) = \begin{bmatrix} 5 & 1 & 0 \\ 0 & 5 & 1 \\ 0 & 0 & 5 \end{bmatrix} .$$

Multiple Strings

Of course more than one string often occurs in a generalized eigenspace G_λ. In order to describe the action of A on all the vectors in G_λ, we have to combine the information carried by the several strings. The next example illustrates the procedure.

Example 3 Let $A = \begin{bmatrix} 8 & 0 & 0 & 0 \\ 0 & 8 & 0 & 3 \\ 4 & 0 & 8 & 0 \\ 0 & 0 & 0 & 8 \end{bmatrix}$.

Then it is easily seen that $\det(A - \lambda I) = (8 - \lambda)^4$, and so the sole eigenvalue is $\lambda = 8$. Also,

$$A - 8I = \begin{bmatrix} 0 & 0 & 0 & 0 \\ 0 & 0 & 0 & 3 \\ 4 & 0 & 0 & 0 \\ 0 & 0 & 0 & 0 \end{bmatrix} \quad , \quad \text{and } (A - 8I)^2 = O \quad .$$

Thus strings have maximum length 2, and the generalized eigenspace G_8 is all of C_4. It is easy to find strings of length 2; for example:

Let $\qquad$ $\mathbf{a}_1 = \mathbf{e}_1$.

Then $\qquad$ $\mathbf{b}_1 = (A - 8I)\mathbf{a}_1 = 4\mathbf{e}_3$,

while $\qquad$ $(A - 8I)\mathbf{b}_1 = \mathbf{0}$.

Also, let $\qquad$ $\mathbf{a}_2 = \mathbf{e}_4$.

Then $\qquad$ $\mathbf{b}_2 = (A - 8I)\mathbf{a}_2 = 3\mathbf{e}_2$

while $\qquad$ $(A - 8I)\mathbf{b}_2 = \mathbf{0}$.

We conclude that the strings spanning the generalized eigenspace assume the form

string s_1: $\mathbf{a}_1 \rightarrow \mathbf{b}_1 \rightarrow \mathbf{0}$

string s_2: $\mathbf{a}_2 \rightarrow \mathbf{b}_2 \rightarrow \mathbf{0}$.

We already know how A acts on a vector such as $\mathbf{x} = c_1\mathbf{a}_1 + c_2\mathbf{b}_1$, which is within the span of a single string s_1. In fact $A\mathbf{x} = c_1'\mathbf{a}_1 + c_2'\mathbf{b}_1$, where

$$\begin{bmatrix} c_1' \\ c_2' \end{bmatrix} = J(8;2) \begin{bmatrix} c_1 \\ c_2 \end{bmatrix}$$

$$= \begin{bmatrix} 8 & 1 \\ 0 & 8 \end{bmatrix} \begin{bmatrix} c_1 \\ c_2 \end{bmatrix} = \begin{bmatrix} 8c_1 + c_2 \\ 8c_2 \end{bmatrix} \quad .$$

Similarly, $\mathbf{x} = c_3\mathbf{a}_2 + c_4\mathbf{b}_2$ is mapped to $A\mathbf{x} = c_3'\mathbf{a}_2 + c_4'\mathbf{b}_2$, where

$$\begin{bmatrix} c_3' \\ c_4' \end{bmatrix} = J(8;2) \begin{bmatrix} c_3 \\ c_4 \end{bmatrix} = \begin{bmatrix} 8c_3 + c_4 \\ 8c_4 \end{bmatrix} \quad .$$

It is easily deduced that an arbitrary vector in G_λ, say

$\qquad$ $\mathbf{x} = c_1\mathbf{a}_1 + c_2\mathbf{b}_1 + c_3\mathbf{a}_2 + c_4\mathbf{b}_2$, is mapped to

$A\mathbf{x} = c_1'\mathbf{a}_1 + c_2'\mathbf{b}_1 + c_3'\mathbf{a}_2 + c_4'\mathbf{b}_2$, where

$$\begin{bmatrix} c_1' \\ c_2' \\ c_3' \\ c_4' \end{bmatrix} = \begin{bmatrix} 8c_1 + c_2 \\ 8c_2 \\ 8c_3 + c_4 \\ 8c_4 \end{bmatrix} = \begin{bmatrix} 8 & 1 & 0 & 0 \\ 0 & 8 & 0 & 0 \\ 0 & 0 & 8 & 1 \\ 0 & 0 & 0 & 8 \end{bmatrix} \begin{bmatrix} c_1 \\ c_2 \\ c_3 \\ c_4 \end{bmatrix} \quad ,$$

or, $\quad \mathbf{c}' = J\mathbf{c}$, where $J = \begin{bmatrix} 8 & 1 & 0 & 0 \\ 0 & 8 & 0 & 0 \\ \hline 0 & 0 & 8 & 1 \\ 0 & 0 & 0 & 8 \end{bmatrix}$.

Observe that the Jordan blocks corresponding to the individual strings are placed along the diagonal of J, which otherwise consists of zeros. We can therefore express J in terms of 2×2 matrices:

$$J = \left[\begin{array}{c|c} J(8;2) & O \\ \hline O & J(8;2) \end{array}\right] \quad .$$

The method of example 3 clearly applies to any generalized eigenspace, no matter how many strings are present in its string basis. Here is a summary of the construction.

Lemma 9.4 Suppose strings $s_1, \ldots, s_k$, of lengths $p_1, \ldots, p_k$ respectively, form a string basis for a generalized eigenspace G_λ of a square matrix A.

Then the coordinates of vectors in G_λ, relative to the string basis, are transformed by the mapping A according to the formula

$$\mathbf{c}' = J\,\mathbf{c}, \text{ where } J = \left[\begin{array}{c|c|c|c} J(\lambda;p_1) & & & \\ \hline & J(\lambda;p_2) & & \\ \hline & & \ddots & \\ \hline & & & J(\lambda;p_k) \end{array}\right] \quad .$$

That is, the Jordan blocks $J(\lambda;p_1), \ldots, J(\lambda;p_k)$ are placed along the diagonal of J, which otherwise consists only of zeros.

As a streamlined notation for J, we write $J = J(\lambda;p_1, \ldots, p_k)$, or $J = J(\lambda;\mathbf{p})$, where $\mathbf{p} = (p_1, \ldots, p_k)$. We also refer to J as a *Jordan block for the eigenspace G_λ*.

Example 4 In the previous section (example 4), we found a string basis for an eigenspace G_4 consisting of strings

$$s_1 : \mathbf{a}_1 \to \mathbf{b}_1 \to \mathbf{0}$$
$$s_2 : \qquad \mathbf{b}_2 \to \mathbf{0} \quad .$$

The Jordan blocks corresponding to strings s_1, s_2 respectively are

$$J(4;2) = \begin{bmatrix} 4 & 1 \\ 0 & 4 \end{bmatrix} \quad , \quad J(4;1) = [4] \quad .$$

The Jordan block for the generalized eigenspace G_4 is therefore

$$J(4;2,1) = \left[\begin{array}{c|c} J(4;2) & \\ \hline & J(4;1) \end{array}\right] = \left[\begin{array}{cc|c} 4 & 1 & 0 \\ 0 & 4 & 0 \\ \hline 0 & 0 & 4 \end{array}\right] \quad .$$

Jordan Form

Our examples suggest that for matrices with just one eigenvalue, a string basis for G_λ will always serve as a basis for C^n. As for the more typical case when a square matrix A has several eigenvalues, we might reasonably expect to obtain a basis for C^n by collecting the elements of string bases for all of the generalized eigenspaces. Then we could place the Jordan blocks for all of the generalized eigenspaces along the diagonal of an $n \times n$ matrix J, which would represent the transformation A relative to the string basis so formed.

Let us first illustrate the construction of J with a numerical example, and then prove that the method always works.

Example 5 Let $A = \begin{bmatrix} 3 & 0 & 0 \\ 4 & 6 & 6 \\ -3 & -1 & 1 \end{bmatrix}$. To find the eigenvalues, compute

$$\det(A - \lambda I) = \det \begin{bmatrix} 3-\lambda & 0 & 0 \\ 4 & 6-\lambda & 6 \\ -3 & -1 & 1-\lambda \end{bmatrix} \begin{array}{l} = (3-\lambda)((6-\lambda)(1-\lambda) + 6) \\ = (3-\lambda)(\lambda^2 - 7\lambda + 12) \\ = -\ (\lambda-3)^2(\lambda-4) \end{array} .$$

Corresponding to the eigenvalue $\lambda_1 = 3$, compute

$$A - 3I = \begin{bmatrix} 0 & 0 & 0 \\ 4 & 3 & 6 \\ -3 & -1 & -2 \end{bmatrix} ,$$

$$(A-3I)^2 = \begin{bmatrix} 0 & 0 & 0 \\ -6 & 3 & 6 \\ 2 & -1 & -2 \end{bmatrix} = (A-3I)^3 .$$

So strings in the generalized eigenspace G_3 have maximum length 2. A basis for $G_3 = \mathrm{Ker}((A-3I)^2)$ is easily computed by reducing $(A-3I)^2$ to $\begin{bmatrix} 1 & -\frac{1}{2} & 1 \\ 0 & 0 & 0 \\ 0 & 0 & 0 \end{bmatrix}$: the basis elements are $\mathbf{v}_1 = (1,2,0)$, $\mathbf{v}_2(-1,0,1)$, and dim $G_3 = 2$.

We then obtain a string $\mathbf{a}_1 \to \mathbf{b}_1 \to \mathbf{0}$ by

choosing $\mathbf{a}_1 = \mathbf{v}_1 = (1,2,0)$, and computing

$$\mathbf{b}_1 = (A - 3I)\, \mathbf{a}_1 = (0,10,-5)$$

and checking that $(A-3I)\mathbf{b}_1 = \mathbf{0}$.

The l.i. string elements $\mathbf{a}_1$, $\mathbf{b}_1$ necessarily form a basis for the two-dimensional space G_3. Accordingly, the Jordan block for the space G_3 is

$$J(3;2) = \begin{bmatrix} 3 & 1 \\ 0 & 3 \end{bmatrix} .$$

Similarly, it is easy to check that the generalized eigenspace G_4 has dimension 1 and is spanned by the eigenvector $\mathbf{a}_2 = (0, -3, 1)$. So a string basis for G_4 is simply $s_2 : \mathbf{a}_2 \to \mathbf{0}$, and the corresponding Jordan block is $J(4;1) = [4]$.

The collected elements of the string bases for G_3 and G_4 are

$$\mathbf{a}_1 = (1,2,0), \ \mathbf{b}_1 = (0,10,-5), \ \mathbf{a}_2 = (0,-3,1) .$$

It is easily checked that these are l.i. and so form a basis for C^3, as expected. Relative to this string basis for C^3, the matrix transformation A is represented by

$$J = \left[\begin{array}{c|c} J(3;2) & \\ \hline & J(4;1) \end{array}\right] = \left[\begin{array}{cc|c} 3 & 1 & 0 \\ 0 & 3 & 0 \\ \hline 0 & 0 & 4 \end{array}\right] \quad .$$

To show that the above construction of J is always successful, we must prove that the collected string elements always form a basis for C^n.

Lemma 9.5 (Existence of string basis for C^n)

Suppose that, for each eigenvalue λ of the $n \times n$ matrix A, a string basis is selected for the corresponding generalized eigenspace G_λ. Then the collection of all the elements in all these string bases forms a basis for C^n (and is called a *string basis for C^n*).

■**Proof**   *Step 1* (linear independence)

Within a generalized eigenspace G_λ, the mapping A is represented by a Jordan block $J(\lambda;\mathbf{p})$. Similarly $A - rI$ is represented, within G_λ, by $J(\lambda;\mathbf{p}) - rI$. But the sole eigenvalue of $J(\lambda;\mathbf{p})$ is λ; so the sole eigenvalue of $J(\lambda;\mathbf{p}) - rI$ is $\lambda - r$. Therefore, if $\lambda - r \neq 0$, $A - rI$ maps any nonzero vector in G_λ to another nonzero vector in G_λ. Let m denote maximum string length in G_λ, and let $\mathbf{w}$ be any nonzero vector in G_λ.

Then $(A - \lambda I)^m \mathbf{w} = \mathbf{0}$

and $(A - rI)^m \mathbf{w} \neq \mathbf{0}$ for $r \neq \lambda$.

Suppose now that a nonzero linear combination of the collected string basis elements adds to $\mathbf{0}$. If we take all the nonzero terms in the sum that are from the same generalized eigenspace and combine them into a single vector $\mathbf{w}$, we obtain an equation of the form

$\mathbf{w}_1 + \cdots + \mathbf{w}_k = \mathbf{0}$.

Each term on the left-hand side of this equation is nonzero and is in a different generalized eigenspace. Suppose that $\mathbf{w}_1$ is in G_λ, and apply $(A - \lambda I)^m$ to the equation. Then a new equation

$\mathbf{0} + \mathbf{w}_2' + \cdots + \mathbf{w}_k' = \mathbf{0}$

is obtained, where $\mathbf{w}_i' = (A - \lambda I)^m \mathbf{w}_i \neq \mathbf{0}$ for $i = 2, \ldots, k$. We can repeat this procedure to eliminate one term at a time until we have a single nonzero vector $\mathbf{w}_k''$ equal to $\mathbf{0}$, a contradiction. So the collected string basis elements are l.i., as required.

Step 2 (spanning C^n).

If the collected string basis elements span a subspace S smaller than C^n then additional l.i. vectors $\mathbf{v}_1, \ldots, \mathbf{v}_q$ can be found to complete a basis for C^n. Let us relabel the collected string basis elements $\mathbf{v}_{q+1}, \ldots, \mathbf{v}_n$.

Then the coordinates of a vector $\mathbf{x}$ in S have the form $\mathbf{c} = (0, \ldots, 0, c_{q+1}, \ldots, c_n)$; and since $A\mathbf{x}$ is in S also, it has coordinates $\mathbf{c}' = (0, \ldots, 0, c'_{q+1}, \ldots, c'_n)$. Therefore the matrix M representing the mapping A relative to this basis contains a $q \times (n-q)$ block of zeros:

$$M = \left[\begin{array}{c|c} P & O \\ \hline Q & R \end{array}\right] \quad,$$

where P is $q \times q$, Q is $(n-q) \times q$, and R is $(n-q) \times (n-q)$. But the square matrix P must have at least one eigenvalue λ and a corresponding eigenvector $\mathbf{w}_0 = (w_1, \ldots, w_q)$. Then, writing $w = (w_1, \ldots, w_q, 0, \ldots, 0)$, we have:

$$(M - \lambda I)\mathbf{w} = \mathbf{y} \quad, \qquad \text{where } \mathbf{y} \text{ is in } S \quad.$$

If $\mathbf{y}$ happens to be $\mathbf{0}$, then $\mathbf{w}$ is a new eigenvector of M, outside of S, contradicting the definition of S.

If $\mathbf{y}$ is not $\mathbf{0}$, we write $\mathbf{y} = \mathbf{y}_1 + \mathbf{y}_2$, where $\mathbf{y}_1$ is in G_λ, and $\mathbf{y}_2$ is in the subspace S_2 of S spanned by the other generalized eigenspaces of A. (Of course, if λ is not an eigenvalue of A, there is no G_λ, and the $\mathbf{y}_1$ summand can be deleted!) Then $\text{Range}(R - \lambda I) = S_2$, and so we can find a vector $\mathbf{z}$ in S such that:

$$(M - \lambda I)\mathbf{z} = \mathbf{y}_2 \quad.$$

Now, with $\mathbf{u} = \mathbf{w} - \mathbf{z}$, we have:

$$(M - \lambda I)\mathbf{u} = \mathbf{y} - \mathbf{y}_2 = \mathbf{y}_1 \quad.$$

That is, $(M - \lambda I)\mathbf{u}$ is in G_λ. So by definition of generalized eigenspace, $\mathbf{u}$ is also in G_λ. But then $\mathbf{w} = \mathbf{u} + \mathbf{z}$ is in S, contradiction.   Q.E.D.

As an immediate consequence of this lemma, we can state our main theoretical conclusion for this chapter.

THEOREM 9.6

(Jordan Form) Let A be a square, $n \times n$ matrix. Then the mapping A is represented (relative to a string basis for C^n) by the matrix

$$J = \left[\begin{array}{c|c|c} J(\lambda_1, \mathbf{p}_1) & & \\ \hline & \ddots & \\ \hline & & J(\lambda_k, \mathbf{p}_k) \end{array}\right] \quad,$$

in which the Jordan blocks corresponding to the generalized eigenspaces of A are placed along the diagonal and zeros are placed elsewhere in J.

Note that when several strings of various lengths occur in a string basis, the dimension n is necessarily rather high; in fact, it is too high for any but artificially simple illustrative matrices A to be converted to Jordan form.

However, to write out examples of matrices J in Jordan form is simple; it just takes space!

Example 6 A typical matrix in Jordan form is

$$J = \begin{bmatrix} 3 & 1 & & & & \\ & 3 & 1 & & & \\ & & 3 & & & \\ & & & 3 & & \\ & & & & 8 & 1 \\ & & & & & 8 \\ & & & & & & 5 \end{bmatrix}.$$

Since a Jordan form matrix is necessarily upper triangular, its eigenvalues can be read off the diagonal. They are $\lambda = 3, 8, 5$. The Jordan blocks corresponding to the generalized eigenspaces are evidently

$J(3;3,1)$ for $\lambda = 3$, $J(8;2)$ for $\lambda = 8$, $J(5;1)$ for $\lambda = 5$.

■ EXERCISES 9.2

1. Write down the Jordan form of each of the matrices in exercises 1, 2, 6, and 7 of section 9.1.

2. Write down the Jordan form of each of the following matrices.

(a) $\begin{bmatrix} 0 & 0 & 1 & 1 \\ 0 & 0 & 1 & 1 \\ 0 & 0 & 1 & 1 \\ 0 & 0 & 0 & 0 \end{bmatrix}$ (b) $\begin{bmatrix} 3 & 5 & 0 & 0 \\ 0 & 3 & 6 & 0 \\ 0 & 0 & 4 & 7 \\ 0 & 0 & 0 & 4 \end{bmatrix}$ (c) $\begin{bmatrix} 3 & 0 & 0 & 0 \\ 0 & 3 & 5 & 0 \\ 0 & 0 & 4 & 6 \\ 0 & 0 & 0 & 4 \end{bmatrix}$

3. Suppose a square matrix has eigenvalues $\lambda = 4, 7$, and the string bases for its generalized eigenspaces have the following patterns:

for G_4, string basis $\begin{cases} \mathbf{a}_1 \rightarrow \mathbf{b}_1 \rightarrow \mathbf{c}_1 \rightarrow \mathbf{0} \quad , \\ \qquad \mathbf{b}_2 \rightarrow \mathbf{c}_2 \rightarrow \mathbf{0} \end{cases}$

for G_7, string basis $\begin{cases} \mathbf{a}_3 \rightarrow \mathbf{b}_3 \rightarrow \mathbf{0} \quad . \\ \qquad \mathbf{b}_4 \rightarrow \mathbf{0} \end{cases}$

Write down the Jordan form of the matrix.

4. Suppose that a square matrix has two eigenvalues $\lambda = 2, 5$, and that the dimensions $d_p = \dim\{\mathrm{Ker}\,(A - \lambda I)^p\}$ are as follows:

for $\lambda = 2$: $d_1 = 2$, $d_2 = 4$, $d_p = 5$ for $p \geq 3$, and
for $\lambda = 5$: $d_1 = 1$, $d_p = 2$ for $p \geq 2$.

Write down the Jordan form of the matrix.

5. Using the properties of the dimension numbers d_p, h_p developed in exercises 9.1, show that the Jordan form of a given square matrix A is unique, except for the ordering of the Jordan blocks along the diagonal.
Note. Because of its essential uniqueness the Jordan form of a matrix is also called the *Jordan canonical form* of the matrix.

*6. Suppose the real $n \times n$ matrix A has at least one complex eigenvalue, $\lambda = a + ib$, and let

$$ s \quad : \mathbf{v}_p \to \cdots \to \mathbf{v}_1 \to \mathbf{0} $$

and $\quad s^* : \mathbf{v}_p^* \to \cdots \to \mathbf{v}_1^* \to \mathbf{0}$

be conjugate strings in G_λ, G_λ^* respectively. (See exercises 9.1.)
For $j = 1, \ldots, p$, let

$$ \mathbf{g}_j = \left(\tfrac{1}{2}\right)(\mathbf{v}_j + \mathbf{v}_j^*), \text{ and } \mathbf{h}_j = \left(\tfrac{1}{2i}\right)(\mathbf{v}_j - \mathbf{v}_j^*) \quad . $$

(a) Show that the real vectors $\mathbf{g}_1, \mathbf{h}_1, \ldots, \mathbf{g}_p, \mathbf{h}_p$ form a basis for the subspace S of C^n spanned by the elements of the two strings s and s^*.

(b) Show that $\quad A\mathbf{g}_j = a\mathbf{g}_j - b\mathbf{h}_j + \mathbf{g}_{j-1}$
$$ A\mathbf{h}_j = b\mathbf{g}_j + a\mathbf{h}_j + \mathbf{h}_{j-1} \quad . $$

(c) Deduce that, within the subspace S and relative to the basis $\mathbf{g}_1, \mathbf{h}_1, \ldots, \mathbf{g}_p, \mathbf{h}_p$, the mapping is represented by the $2p \times 2p$ matrix

$$ K(\lambda;p) = \begin{bmatrix} Y_\lambda & I_2 & & \\ & \ddots & \ddots & \\ & & \ddots & I_2 \\ & & & Y_\lambda \end{bmatrix} $$

which contains p identical 2×2 blocks of the form

$$ Y_\lambda = \begin{bmatrix} a & b \\ -b & a \end{bmatrix} $$

along the diagonal, 2×2 identity blocks immediately above the diagonal blocks, and zeros elsewhere.

(d) Suppose that within the generalized eigenspace G_λ, the mapping A is represented by the Jordan block $J(\lambda;\mathbf{p}) = J(\lambda;p_1, \ldots, p_k)$. Let S_λ be the subspace of C^n spanned by the combined elements of G_λ and G_λ^*. Show that within S_λ the mapping A is represented by the matrix $K(\lambda;\mathbf{p}) = K(\lambda;p_1, \ldots, p_k)$ formed by placing the blocks $K(\lambda;p_1), \ldots, K(\lambda;p_k)$ along the diagonal.

(e) Deduce the following *theorem* (canonical representation of a real, $n \times n$ matrix). Suppose A is a real $n \times n$ matrix. Suppose that in the Jordan form J of A, the two blocks $J(\lambda;\mathbf{p})$ and $J(\lambda^*;\mathbf{p})$ corresponding to each pair of conjugate eigenvectors λ, λ^* are replaced by the single real block $K(\lambda;\mathbf{p})$. Then the resulting real matrix K represents the mapping A, relative to a suitable basis for R^n.

7. Let $A = \begin{bmatrix} 0 & 0 & 1 & 1 \\ 0 & 0 & -1 & 0 \\ 0 & 1 & 0 & 0 \\ -1 & 1 & 0 & 0 \end{bmatrix}$, $K = \left[\begin{array}{cc|cc} 0 & 1 & 1 & 0 \\ -1 & 0 & 0 & 1 \\ \hline 0 & 0 & 0 & 1 \\ 0 & 0 & -1 & 0 \end{array} \right]$.

Show that K is the real canonical representation of A.

8. Suppose the Jordan canonical form of a real matrix A is

$$J = \left[\begin{array}{cc|cc|c|c} 1+2i & 1 & 0 & 0 & 0 & 0 \\ 0 & 1+2i & 0 & 0 & 0 & 0 \\ \hline 0 & 0 & 1-2i & 1 & 0 & 0 \\ 0 & 0 & 0 & 1-2i & 0 & 0 \\ \hline 0 & 0 & 0 & 0 & i & 0 \\ \hline 0 & 0 & 0 & 0 & 0 & -i \end{array} \right].$$

Find the real canonical form, K, of A.

9.3 Applications of Jordan Form

Introduction

A great many problems in matrix algebra, in systems of differential equations, in the theory of stochastic processes, etc., contain a square matrix A to which calculations must be applied. Many of these calculations were done in Chapter 8 on the assumption of diagonalizability. Now, through the use of Jordan form, they can be applied to any square matrix whatsoever. Of course the solutions to the problems contain new features arising from the distinct properties of matrices that are not diagonalizable.

Powers of a Square Matrix

Let N, or $N(p)$, denote the $p \times p$ matrix that has 1's immediately above the diagonal and zeros elsewhere. For example,

$$N(2) = \begin{bmatrix} 0 & 1 \\ 0 & 0 \end{bmatrix},$$

$$N(4) = \begin{bmatrix} 0 & 1 & 0 & 0 \\ 0 & 0 & 1 & 0 \\ 0 & 0 & 0 & 1 \\ 0 & 0 & 0 & 0 \end{bmatrix}.$$

It is easily checked that

$$N(4)^2 = \begin{bmatrix} 0 & 0 & 1 & 0 \\ 0 & 0 & 0 & 1 \\ 0 & 0 & 0 & 0 \\ 0 & 0 & 0 & 0 \end{bmatrix} , \quad N(4)^3 = \begin{bmatrix} 0 & 0 & 0 & 1 \\ 0 & 0 & 0 & 0 \\ 0 & 0 & 0 & 0 \\ 0 & 0 & 0 & 0 \end{bmatrix} , \quad N(4)^4 = O \ ;$$

a similar pattern holds for any value of the dimension p. In fact, those entries of N^k that are k steps above the diagonal are 1's and all the other entries are zeros.

The Jordan block $J(\lambda;p)$ equals $\lambda I + N(p)$, or, in abbreviated notation, $J = \lambda I + N$. We have now expressed $J(\lambda;p)$ as the sum of two matrices, λI and $N(p)$, which commute and which are easily raised to the mth power. This enables us to develop a formula for J^m rather easily. In fact, it is most efficient to define and compute general polynomials in J instead of just powers of J.

Given a polynomial $f(x)$, the corresponding polynomial in the square matrix A is obtained by substituting A for x and inserting I into the constant term:

if $\quad f(x) = c_r x^r + \cdots + c_1 x + c_0 \ ,$
then $\quad f(A) = c_r A^r + \cdots + c_1 A + c_0 I \ .$

It is readily checked that, as a special case of this formula, $f(\lambda I) = f(\lambda)I$.

More generally, if A and B are two square matrices that commute, and $f(x,y)$ is a polynomial in two variables, then $f(A,B)$ is obtained by substituting A and B for x and y respectively and I for the constant term. For example,

if $\quad f(x,y) = (x + y)^2 = x^2 + 2xy + y^2 \ ,$
then $\quad f(A,B) = (A + B)^2 = A^2 + 2AB + B^2 \ .$

The consistency of this definition depends, of course, on the commutativity assumption $AB = BA$; otherwise the expansion of $(A + B)^2$ would contain the two unequal terms AB and BA instead of simply $2AB$.

Suppose now that $f(x)$ is a polynomial of degree r. Then the Taylor expansion from calculus gives us the following polynomial identity:

$$f(x + y) = f(x) + f'(x)y + \frac{f''(x)}{2!}y^2 + \cdots + \frac{f^{(r)}(x)}{r!} y^r \ ,$$

where $f', f'', \ldots, f^{(r)}$ represent successive derivatives of f. On substituting λI for x and N for y, we obtain

$$f(\lambda I + N) = f(\lambda I) + f'(\lambda I)N + \frac{f''(\lambda I)N^2}{2!} + \cdots + \frac{f^{(r)}(\lambda I)}{r!}N^r \ ,$$

or $\quad f((J(\lambda;p))) = f(\lambda)I + f'(\lambda)N + \frac{f''(\lambda)}{2!} N^2 + \cdots + \frac{f^{(r)}(\lambda)}{r!}N^r \ ,$

or $\quad f((J(\lambda;p))) = \begin{bmatrix} f(\lambda) & f'(\lambda) & \dfrac{f''(\lambda)}{2!} \\ & f(\lambda) & f'(\lambda) & \dfrac{f''(\lambda)}{2!} & \cdots \\ & & \ddots & \ddots & \ddots \\ & & & \ddots & \ddots & \dfrac{f''(\lambda)}{2!} \\ & & & & \ddots & f'(\lambda) \\ & & & & & f(\lambda) \end{bmatrix}$.

Example 1 To compute $J(\lambda;4)^2$ by this method, put $f(\lambda) = \lambda^2$. Then

$$f'(\lambda) = 2\lambda, f''(\lambda) = 2, f'''(\lambda) = 0. \text{ Also, } \frac{f''(\lambda)}{2!} = \frac{2}{2} = 1 \quad .$$

Therefore,

$$J(\lambda;4)^2 = \begin{bmatrix} \lambda^2 & 2\lambda & 1 & 0 \\ 0 & \lambda^2 & 2\lambda & 1 \\ 0 & 0 & \lambda^2 & 2\lambda \\ 0 & 0 & 0 & \lambda^2 \end{bmatrix} \quad .$$

Example 2 To compute $J(\lambda;3)^{10}$, let $f(\lambda) = \lambda^{10}$.

Then $f'(\lambda) = 10\lambda^9$; $\frac{f''(\lambda)}{2!} = 45\lambda^8$, and

$$J(\lambda;3)^{10} = \begin{bmatrix} \lambda^{10} & 10\lambda^9 & 45\lambda^8 \\ 0 & \lambda^{10} & 10\lambda^9 \\ 0 & 0 & \lambda^{10} \end{bmatrix} \quad .$$

Example 3 To compute $J(\lambda;2)^r$, let $f(\lambda) = \lambda^r$.
Then $f'(\lambda) = r\lambda^{r-1}$, and

$$J(\lambda;2) = \begin{bmatrix} \lambda^r & r\lambda^{r-1} \\ 0 & \lambda^r \end{bmatrix} \quad .$$

The formula we have developed for $f((J(\lambda;p))$ actually allows us to compute $f(A)$ for any square matrix A. First, observe that whenever a matrix A is constructed from square diagonal blocks $D_1, \ldots, D_k$ and otherwise contains only zeros, then A^r is likewise constructed from the blocks $(D_1)^r, \ldots, (D_k)^r$. That is,

if $A = \begin{bmatrix} D_1 & & \\ & \ddots & \\ & & D_k \end{bmatrix}$, then $A^r = \begin{bmatrix} (D_1)^r & & \\ & \ddots & \\ & & (D_k)^r \end{bmatrix}$.

So the computation of $f(J)$ for a matrix J in Jordan form amounts to the computation of $f((J(\lambda;p))$ for the Jordan blocks $J(\lambda;p)$ contained in J.

Example 4   To compute J^4, where $J = \begin{bmatrix} 2 & 1 & 0 \\ 0 & 2 & 0 \\ 0 & 0 & 3 \end{bmatrix}$, observe that the Jordan

blocks in J are $J(2;2)$ and $J(3;1)$. Now as in example 3,

$$J(2;2)^4 = \begin{bmatrix} 2^4 & 4 \cdot 2^{4-1} \\ 0 & 2^4 \end{bmatrix} = \begin{bmatrix} 16 & 32 \\ 0 & 16 \end{bmatrix} ,$$

while $J(3;1)^4 = [3^4] = [81]$.

Therefore $J^4 = \begin{bmatrix} 16 & 32 & 0 \\ 0 & 16 & 0 \\ 0 & 0 & 81 \end{bmatrix}$.

Now let A be an arbitrary $n \times n$ square matrix, with string basis $\mathbf{v}_1, \ldots, \mathbf{v}_n$ relative to which the mapping A is represented by the Jordan matrix J. Then, with $P = [\mathbf{v}_1 | \ldots | \mathbf{v}_n]$, we have, according to the theory of basis change in section 6.4, the formulas

$$A = PJP^{-1}$$
and $$A^r = PJ^rP^{-1} .$$

It follows easily that if $f(x)$ is any polynomial,

$$f(A) = Pf(J)P^{-1} .$$

Example 5   Let $A = \begin{bmatrix} 1 & 1 \\ -1 & 3 \end{bmatrix}$. To compute A^r, first obtain string basis and Jordan form for A:

$$\det(A - \lambda I) = \det \begin{bmatrix} 1-\lambda & 1 \\ -1 & 3-\lambda \end{bmatrix} = \lambda^2 - 4\lambda + 4 = (\lambda - 2)^2.$$

Also, since $A - 2I = \begin{bmatrix} -1 & 1 \\ -1 & 1 \end{bmatrix} \rightarrow \begin{bmatrix} 1 & -1 \\ 0 & 0 \end{bmatrix}$, the sole eigenvector is $(1,1)$. There must therefore be a string of length 2; say $\mathbf{a}_1 = (1,0)$; then $\mathbf{b}_1 = (A - 2I)\mathbf{a}_1 = (-1, -1)$.

Then the Jordan form is $J(2;2) = \begin{bmatrix} 2 & 1 \\ 0 & 2 \end{bmatrix}$, while

$$P = [\mathbf{b}_1 | \mathbf{a}_1] = \begin{bmatrix} -1 & 1 \\ -1 & 0 \end{bmatrix} , \text{ and } P^{-1} = \begin{bmatrix} 0 & -1 \\ 1 & -1 \end{bmatrix} .$$

So $A^r = \begin{bmatrix} -1 & 1 \\ -1 & 0 \end{bmatrix} \begin{bmatrix} 2 & 1 \\ 0 & 2 \end{bmatrix}^r \begin{bmatrix} 0 & -1 \\ 1 & -1 \end{bmatrix}$

$= \begin{bmatrix} -1 & 1 \\ -1 & 0 \end{bmatrix} \begin{bmatrix} 2^r & r2^{r-1} \\ 0 & 2^r \end{bmatrix} \begin{bmatrix} 0 & -1 \\ 1 & -1 \end{bmatrix} = 2^{r-1} \begin{bmatrix} 2-r & r \\ -r & 2+r \end{bmatrix}$.

■ Problem Given a square matrix A, find the polynomial $f(x)$ of lowest degree such that $f(A) = O$.

■ Solution Let J be the Jordan form of A. Since $f(A) = Pf(J)P^{-1}$, $f(A) = O$ if an only if $f(J) = O$. Also, if $J(\lambda;p)$ is a Jordan block of J, then $f(J(\lambda;p))$ is a Jordan block of $f(J)$. We must therefore find a polynomial $f(x)$ such that, for every Jordan block $J(\lambda;p)$ of J, $f(J(\lambda;p)) = O$ holds.

But we derived a formula for $f(J(\lambda;p))$, and it equals the zero matrix only where $f(\lambda), f'(\lambda), \ldots, f^{(p-1)}(\lambda)$ are all zero. Thus $f(x)$ and its first $p-1$ derivatives must vanish at $x = \lambda$; in other words, $(x-\lambda)^p$ must be a factor of f.

Let $\lambda_1, \ldots, \lambda_k$ be the eigenvalues of A, and $m_1, \ldots, m_k$ the maximum lengths of strings in the generalized eigenspaces $G_{\lambda 1}, \ldots, G_{\lambda k}$. Then the maximum size of the Jordan blocks $J(\lambda_i;p)$ corresponding to eigenvalue λ_i is $p = m_i$. We therefore choose

$$f(x) = (x - \lambda_1)^{m_1} \ldots (x - \lambda_k)^{m_k}.$$

Then $f(J(\lambda;p)) = \mathbf{0}$ for every Jordan block in J, and so $f(J)$, and $f(A)$, are the zero matrix.

Note: The function $f(x)$ of lowest degree satisfying $f(A) = O$ is called the *minimal polynomial* of the square matrix A. From the method of construction, it is clear that for an $n \times n$ matrix, the degree of the minimal polynomial is at most n.

Once we have computed the Jordan form of a matrix A, we can write down its minimal polynomial by inspection. Thus the sole Jordan block in the Jordan form of the immediately preceding 2×2 matrix A is $J(2;2)$, and so the minimal polynomial of A is simply $f(x) = (x-2)^2$.

For the 7×7 matrix that is shown immediately after theorem 9.6, the minimal polynomial is

$$f(x) = (x-3)^3(x-8)^2(x-5) \quad,$$

which is of degree 6, less than the dimension $n = 7$ of the matrix.

Linear Systems of Differential Equations

For a diagonalizable $n \times n$ matrix A, a solution of the *initial value problem* (*IVP*),

$$\begin{cases} \dfrac{d\mathbf{x}}{dt} = A\mathbf{x} \\ \mathbf{x}(0) = \mathbf{x}_0 \quad, \end{cases}$$

was obtained in section 8.4. In fact, from the eigenvalues $\lambda_1, \ldots, \lambda_n$ of A and the corresponding eigenvectors $\mathbf{v}_1, \ldots, \mathbf{v}_n$ that form a basis for R^n, we can construct the matrix $P = [\mathbf{v}_1| \ldots |\mathbf{v}_m]$ used before, and also the diagonal matrix

$$U(t) = \begin{bmatrix} e^{\lambda_1 t} & & & \\ & \cdot & & \\ & & \cdot & \\ & & & \cdot \\ & & & & e^{\lambda_n t} \end{bmatrix},$$

which satisfies $U(0) = I$.

Then $PU(t) = [e^{\lambda_1 t}\mathbf{v}_1| \ldots e^{\lambda_n t}\mathbf{v}_n]$

and $PU(t)\mathbf{c} = c_1 e^{\lambda_1 t}\mathbf{v}_1 + \cdots + c_n e^{\lambda_n t}\mathbf{v}_n$.

That is, parametrized solutions to the system are given by the formula

$$\mathbf{x}(t) = PU(t)\mathbf{c} \quad .$$

The initial condition demands that

$$\mathbf{x}_0 = \mathbf{x}(0) = PU(0)\mathbf{c} = PI\,\mathbf{c} = P\,\mathbf{c}, \text{ or } \mathbf{c} = P^{-1}\mathbf{x}_0 \quad .$$

The solution of the *IVP* can therefore be written

$$\mathbf{x}(t) = PU(t)P^{-1}\mathbf{x}_0 \quad ,$$

or, in terms of the *propagator* $P(t) = PU(t)P^{-1}$,

$$\mathbf{x}(t) = P(t)\mathbf{x}_0 \quad .$$

Using the Jordan form, however, we can solve the *IVP* without having to assume that A is diagonalizable. We begin by rewriting the problem in terms of the Jordan matrix.

The given matrix A and its Jordan form J are linked by the formulas

$$A = PJP^{-1} \text{ and } J = P^{-1}AP \quad ,$$

where the columns of P are string basis elements $\mathbf{v}_1, \ldots, \mathbf{v}_n$ instead of ordinary eigenvectors. Now introduce new coordinates $\mathbf{y}$ through the formulas

$$\mathbf{y} = P^{-1}\mathbf{x} \text{ and } \mathbf{x} = P\mathbf{y} \quad .$$

Then $\dfrac{d\mathbf{y}}{dt} = \dfrac{d}{dt}P^{-1}\mathbf{x} = P^{-1}\dfrac{d\mathbf{x}}{dt} = P^{-1}A\mathbf{x} = P^{-1}AP\mathbf{y}$ or $\dfrac{d\mathbf{y}}{dt} = J\mathbf{y}.$

For example, if $J = \begin{bmatrix} 2 & 1 & 0 \\ 0 & 2 & 0 \\ \hline 0 & 0 & 3 \end{bmatrix}$, the system in terms of the $\mathbf{y}$

variables assumes the form

$$\begin{cases} \begin{cases} y_1' = 2y_1 + y_2 \\ y_2' = 2y_2 \\ y_3' = 3y_3 \end{cases} \end{cases} \quad .$$

(Here primes are again used to denote derivatives of scalar functions.) Observe that this system can be separated into smaller subsystems: the first two equations form a system containing only y_1, y_2; and the third equation forms a system containing only y_3. This corresponds to the placement of 2×2 and 1×1 Jordan blocks along the diagonal of J. In effect, we can deal with one Jordan block $J(\lambda;p)$ at a time.

For convenience, suppose that $J(\lambda;p)$ is the first Jordan block in J, to which correspond the variables $y_1, \ldots, y_p$. These p variables form a vector in R^p, which we denote

$$\mathbf{w} = (w_1, \ldots, w_p) = (y_1, \ldots, y_p) \quad .$$

Then $\mathbf{w}$ satisfies the system

$$\frac{d\mathbf{w}}{dt} = J(\lambda;p)\mathbf{w} = (\lambda I + N)\mathbf{w} = \lambda\mathbf{w} + N\mathbf{w} \quad ,$$

or $\quad \dfrac{d\mathbf{w}}{dt} - \lambda\mathbf{w} = N\mathbf{w},$

or $\quad e^{-\lambda t}\dfrac{d\mathbf{w}}{dt} - \lambda e^{-\lambda t}\mathbf{w} = N(e^{-\lambda t}\mathbf{w}) \quad ,$

or $\quad \dfrac{d}{dt}(e^{-\lambda t}\mathbf{w}) = N(e^{-\lambda t}\mathbf{w}) \quad .$

With the further change of variables $\mathbf{z} = e^{-\lambda t}\mathbf{w}$, we have

$$\frac{d\mathbf{z}}{dt} = N\mathbf{z} \quad ,$$

or
$$\begin{bmatrix} z_1' \\ \cdot \\ \cdot \\ \cdot \\ z_p' \end{bmatrix} = \begin{bmatrix} 0 & & 1 & \\ & \cdot & & \cdot \\ & & \cdot & & 1 \\ & & & \cdot \\ & & & & 0 \end{bmatrix} \begin{bmatrix} z_1 \\ \cdot \\ \cdot \\ \cdot \\ z_p \end{bmatrix} \quad , \quad \text{or} \quad \begin{cases} z_1' = z_2 \\ z_2' = z_3 \\ \cdots\cdots \\ z_{p-1}' = z_p \\ z_p' = 0 \end{cases} .$$

This system has a simple interpretation: once z_1 is chosen, say $z_1 = q(t)$, then z_2, z_3, etc. are successive derivatives of $q(t)$. But finally $z_p' = 0$, or $q^{(p)}(t) = 0$. That is, $q(t)$ must be chosen so that its pth derivative is zero, and we know from calculus that this means $q(t)$ is a polynomial of degree not exceeding p.

For example, we can write down the following list of solutions $\mathbf{z}(t)$:

choose $q(t) = 1$, obtain $\mathbf{z}_1(t) = (1, 0, \ldots, 0)$;

" $\quad q(t) = t$, " $\quad \mathbf{z}_2(t) = (t, 1, \ldots, 0)$;

" $\quad q(t) = \dfrac{t^2}{2!}$, " $\quad \mathbf{z}_3(t) = (\dfrac{t^2}{2!}, t, 1, \ldots, 0)$

$\cdots\cdots\cdots\cdots\cdots\cdots\cdots\cdots\cdots\cdots\cdots\cdots\cdots\cdots\cdots\cdots\cdots$

$\quad q^{(t)} = \dfrac{t^{(p-1)}}{(p-1)!}$, " $\quad z_p(t) = \left(\dfrac{t^{p-1}}{(p-1)!}, \ldots, 1\right)$.

Further solutions are then obtained as linear combinations of the ones listed above:

$$\mathbf{z}(t) = c_1\mathbf{z}_1(t) + \cdots + c_p\mathbf{z}_p(t) \quad ,$$

or, with $U(t;p) = [\mathbf{z}_1(t)| \ldots |\mathbf{z}_p(t)] = \begin{bmatrix} 1 & t & \ldots & \frac{t^{p-1}}{(p-1)!} \\ 0 & 1 & t & \cdot \\ & & \cdot & \cdot & \cdot \\ & & & \cdot & \cdot \\ & & & & t \\ & & & & 1 \end{bmatrix}$,

$$\mathbf{z}(t) = U(t;p)\mathbf{c} \quad .$$

In terms of the $\mathbf{w}$ variables, the solutions are

$$\mathbf{w}(t) = e^{\lambda t}U(t;p)\mathbf{c} \quad ,$$

or, writing $U(t;\lambda;p) = e^{\lambda t}U(t;p)$,

$$\mathbf{w}(t) = U(t;\lambda;p)\mathbf{c} \quad .$$

Now form the matrix $U(t)$ by placing the blocks $U(t;\lambda;p)$ along its diagonal, to obtain solutions of the system $\frac{d\mathbf{y}}{dt} = J\mathbf{y}$ in the form

$$\mathbf{y}(t) = U(t)\mathbf{c} \quad .$$

Note that $U(t)$ assumes a special form at $t = 0$. In fact, its diagonal blocks $U(t;p)$ satisfy $U(0;p) = I_p$ by inspection. So $U(0) = I_n$ likewise.

Finally, we return to the original $\mathbf{x}$-variables, given by $\mathbf{x} = P\mathbf{y}$, and conclude that the solutions are

$$\mathbf{x}(t) = PU(t)\mathbf{c} \quad .$$

The initial condition then assumes the form

$$\mathbf{x}_0 = \mathbf{x}(0) = PU(0)\mathbf{c} = PI\mathbf{c} = P\mathbf{c}, \quad \text{or} \quad \mathbf{c} = P^{-1}\mathbf{x}_0 \quad .$$

Therefore the solution of the initial value problem is

$$\mathbf{x}(t) = PU(t)P^{-1}\mathbf{x}_0 \quad .$$

In terms of the *propagator matrix*

$$P(t) = PU(t)P^{-1} \quad ,$$

we can write the solution of the *IVP* as

$$\mathbf{x}(t) = P(t)\mathbf{x}_0 \quad .$$

Example 1 To solve the initial value problem

$$\begin{cases} x_1' = x_1 + x_2 \\ x_2' = -x_1 + 3x_2 \\ x_1(0) = 1 \\ x_2(0) = 4 \quad , \end{cases}$$

we rewrite the system in the form

$$\begin{cases} \dfrac{d\mathbf{x}}{dt} = A\mathbf{x} \ , & \text{where } A = \begin{bmatrix} 1 & 1 \\ -1 & 3 \end{bmatrix} \ . \\ \mathbf{x}(0) = (1,4) \end{cases}$$

In an earlier example in this section, we calculated the Jordan form of A as $J = \begin{bmatrix} 2 & 1 \\ 0 & 2 \end{bmatrix}$, relative to the string basis consisting of the single string $\mathbf{v}_2 \to \mathbf{v}_1 \to \mathbf{0}$, where $\mathbf{v}_1 = (-1,-1)$, $\mathbf{v}_2 = (1,0)$.

To construct the propagator $P(t)$, we write down

$$U(t,p) = U(t;2) = \begin{bmatrix} 1 & t \\ 0 & 1 \end{bmatrix} \ ,$$

$$U(t;\lambda;p) = U(t;2;2) = e^{2t} \begin{bmatrix} 1 & t \\ 0 & 1 \end{bmatrix} \ ,$$

$$P = [\mathbf{v}_1|\mathbf{v}_2] = \begin{bmatrix} -1 & 1 \\ -1 & 0 \end{bmatrix} \ , \quad P^{-1} = \begin{bmatrix} 0 & -1 \\ 1 & -1 \end{bmatrix} \ ,$$

$$P(t) = PU(t)P^{-1} = \begin{bmatrix} -1 & 1 \\ -1 & 0 \end{bmatrix} e^{2t} \begin{bmatrix} 1 & t \\ 0 & 1 \end{bmatrix} \cdots$$

$$\cdots \begin{bmatrix} 0 & -1 \\ 1 & -1 \end{bmatrix} = e^{2t} \begin{bmatrix} 1-t & t \\ -t & 1+t \end{bmatrix} \ .$$

Then the solution of the *IVP* is

$$\mathbf{x}(t) = P(t)\mathbf{x}_0 = e^{2t} \begin{bmatrix} 1-t & t \\ -t & 1+t \end{bmatrix} \begin{bmatrix} 1 \\ 4 \end{bmatrix} = e^{2t} \begin{bmatrix} 1+3t \\ 4+3t \end{bmatrix} \ .$$

■ EXERCISES 9.3

1. For each of the following matrices A, compute the powers A^2, A^3, and A^r.

   (a) $\begin{bmatrix} 1 & 1 \\ 0 & 1 \end{bmatrix}$ (b) $\begin{bmatrix} 3 & 1 \\ 0 & 3 \end{bmatrix}$ (c) $\begin{bmatrix} 14 & 16 \\ -9 & -10 \end{bmatrix}$

   (d) $\begin{bmatrix} 2 & 1 & 0 \\ 0 & 2 & 0 \\ 0 & 0 & 2 \end{bmatrix}$ (e) $\begin{bmatrix} 1 & 1 & 0 \\ 0 & 1 & 1 \\ 0 & 0 & 1 \end{bmatrix}$

2. Compute the minimal polynomial of each of the matrices in exercise 1.

3. Suppose the square matrix A has eigenvalues $\lambda_1, \ldots, \lambda_k$ and that the dimensions of the corresponding generalized eigenspaces of A are $n_1, \ldots, n_k$ respectively.

Prove that the characteristic polynomial, $\det(\lambda I - A)$, of the given matrix A, has the form

$$g(\lambda) = (\lambda - \lambda_1)^{n_1} \ldots (\lambda - \lambda_k)^{n_k} \ .$$

Use the following steps:

(a) Show that the given polynomial $g(x)$ is the characteristic polynomial, $\det(\lambda I - J)$, of the Jordan form, J, of A.

(b) Deduce from the theory of "similar matrices" (exercises of section 8.2) that $g(x)$ is also the characteristic polynomial of A.

4. Using the result of the previous exercise, show that the characteristic polynomial, g, of a square matrix A, is divisible by the minimal polynomial, f, of A. That is, for some polynomial, h, the formula $g(\lambda) = f(\lambda)h(\lambda)$ holds.

5. From the previous exercise, deduce the
   Hamilton-Cayley Theorem: Let $g(\lambda)$ be the characteristic polynomial of the square matrix A. Then $g(A) = 0$.

6. (a) Using the Hamilton-Cayley theorem, show that the matrix

$$A = \begin{bmatrix} 2 & a & b \\ 0 & 2 & c \\ 0 & 0 & 1 \end{bmatrix} \quad \text{satisfies the matrix equation}$$

   $A^3 - 5A^2 + 8A - 4I = O$.

   (b) Show that the minimal polynomial of the above matrix A is:

   (i) $f(x) = (x-2)^2(x-1)$ when the parameter $a \neq 0$
   (ii) $f(x) = (x-2)(x-1)$ when the parameter $a = 0$.

7. Show that a square matrix A satisfies an equation of the form $c_q A^q + \cdots + c_1 A = I$ if and only if A is invertible.

8. Show that if an $n \times n$ matrix A is invertible, then there are constants $a_0, \ldots, a_{n-1}$ such that

$$A^{-1} = a_0 I + a_1 A + \cdots + a_{n-1} A^{n-1} \ .$$

9. Suppose the characteristic polynomial of a 4×4 matrix is:

$$g(\lambda) = (\lambda + 1)^2 (\lambda - 1)^2 \ .$$

   Show that $A^{-1} = 2A - A^3$.

10. Show that there is an integer q such that $A^q = 0$ if and only if the sole eigenvalue of A is zero.

11. For each of the following linear systems of differential equations, construct the propagator matrix $P(t)$ and use it to solve the indicated initial value problem.

(a) $\begin{cases} x_1' = 3x_1 + x_2 \\ x_2' = \qquad 3x_2 \end{cases}$
$\mathbf{x}(0) = (4,7)$

(b) $\begin{cases} x_1' = 14x_1 + 16x_2 \\ x_2' = -9x_1 + 10x_2 \end{cases}$
$\mathbf{x}(0) = (4, -3)$

(c) $\begin{cases} x_1' = x_1 + x_2 \\ x_2' = \qquad x_2 + x_3 \\ x_3' = \qquad\qquad x_3 \end{cases}$
$\mathbf{x}(0) = (0,0,1)$

(d) $\begin{cases} x_1' = 6x_1 + x_2 \\ x_2' = \qquad 5x_2 + x_3 \\ x_3' = -x_1 - x_2 + 4x_3 \end{cases}$
$\mathbf{x}(0) = (0,1,0)$

*12. Consider the "initial value problem" for an unknown function $y(t)$ given by the following conditions on $y(t)$:

$$y^{(n)}(t) + c_{n-1}y^{(n-1)}(t) + \cdots + c_0y(t) = 0$$
$$y(0) = a_0, y'(0) = a_1, \ldots, y^{(n-1)}(0) = a_{n-1}' ,$$

where $c_0, \ldots, c_{n-1}, a_0, \ldots, a_{n-1}$ are all constants.

(a) Let $\mathbf{x}_0 = (a_0, \ldots, a_{n-1})$,

$$\mathbf{x}(t) = (y(t), \ldots, y^{(n-1)}(t)) .$$

Show that $y(t)$ satisfies the above *IVP* if and only if $\mathbf{x}(t)$ satisfies the system *IVP*

$$\frac{d\mathbf{x}}{dt} = A\mathbf{x} , \quad A = \begin{bmatrix} 0 & 1 & 0 & . & . & 0 \\ 0 & 0 & 1 & . & . & 0 \\ . & & & . & . & . \\ . & & & . & . & . \\ . & & & & . & 1 \\ -c_0 & -c_1 & . & . & . & -c_{n-1} \end{bmatrix} .$$
$$\mathbf{x}(0) = \mathbf{x}_0$$

(b) Suppose $\det(\lambda I - A) = (\lambda - \lambda_1)^{n_1} \cdots (\lambda - \lambda_k)^{n_k}$.
Using the result of the problem solved at the end of section 9.1, show that $U(t)$ is formed from the blocks $U(t; \lambda; n_j), j = 1, \ldots, k$.

(c) Deduce that there is a solution of the given *IVP* of the form

$$y(t) = q_1(t)e^{\lambda_1 t} + \cdots + q_k(t)e^{\lambda_k t} ,$$

where $q_j(t)$ is a polynomial of degree less than n_j, for $j = 1, \ldots k$.

*13. With A as in the previous exercise, show that

$$\det(\lambda I - A) = \lambda^n + c_{n-1}\lambda^{n-1} + \cdots + c_0 .$$

*14. (a) In the course of constructing the propagator $P(t)$ of a linear system, we showed that *all* solutions of the system $\dfrac{d\mathbf{z}}{dt} = N(p)\mathbf{z}$ are polynomials of degree less than p. Work backwards from this result to show that $\mathbf{x}(t) = P(t)\mathbf{x}_0$ is the *unique* solution of the *IVP*:

$$\begin{cases} \dfrac{d\mathbf{x}}{dt} = A\mathbf{x} \\ x(t_0) = \mathbf{x}_0 \end{cases}.$$

(b) Deduce that the solution $y(t)$ of the *IVP* in the previous exercise is also unique.

*15. Show that $\mathbf{x}(t) = P(t)P(t_0)^{-1}\mathbf{x}_0$ is the (unique) solution of the *IVP*

$$\begin{cases} \dfrac{d\mathbf{x}}{dt} = A\mathbf{x} \\ x(t_0) = \mathbf{x}_0 \end{cases}.$$

*16. Suppose the maximum magnitude of the entries of an $m \times m$ matrix B is denoted $|B|$. That is,

$$|B| = \max_{1 \le i,j \le m} |b_{ij}| .$$

Show that for $n \times n$ matrices B, C,

$$|BC| \le n|B|\,|C| .$$

*17. A sequence $B_1, B_2, B_3, \ldots$ of matrices is said to be bounded if there is a constant M such that

$$|B_j| \le M, \ j = 1, 2, \ldots .$$

Suppose the square matrix A has Jordan form J. Show that the sequence $A, A^2, A^3, \ldots$ of powers of A is bounded if and only if the sequence $J, J^2, J^3, \ldots$ of powers of J is bounded.

*18. Show that the sequence of powers of a Jordan block $J(\lambda;p)$ is bounded if and only if one of the following conditions holds:

(i) $|\lambda| < 1$

or (ii) $|\lambda| = 1$ and $p = 1$.

*19. Show that the sequence of powers of a square matrix A is bounded if and only if every eigenvalue λ of A satisfies one of the following conditions:

(i) $|\lambda| < 1$

or (ii) $|\lambda| = 1$ and $G_\lambda = E_\lambda$.

*20. Let A be a stochastic matrix (see exercises 8.3).

(a) Using the fact that A^r is also a stochastic matrix, show that

$$|A^r| \le 1 \quad \text{for} \quad r = 1, 2, \ldots .$$

(b) Deduce that for a stochastic matrix A (which always has $\lambda = 1$ as an eigenvalue), the condition $G_1 = E_1$ always holds.

(c) Suppose A is strictly positive stochastic matrix, so that (from exercises 8.3), A has a unique (and positive) eigenvector $\mathbf{v}$ corresponding to eigenvalue $\lambda = 1$.

Show that, for any vector $\mathbf{u}$, there is a constant c such that $A^r\mathbf{u}$ approches $c\mathbf{v}$ as r approaches infinity.

Conclude that the stochastic process represented by A approaches a state $\mathbf{v}$ that is independent of its initial state $\mathbf{u}$.

CHAPTER TEN

Numerical Methods

Introduction

Through all the preceding chapters we have tried to illustrate the concepts of matrix algebra with the simplest numerical examples capable of carrying the content of the algebraic theory. Thus most of our numerical matrices were small and started out, at least, with integer entries. We could often complete several steps of a long calculation without encountering any drastic numerical complications. The existence of these transparent, yet algebraically typical, examples greatly facilitated our efforts to grasp the methods of the subject and to see how our mathematical models can solve practical problems.

We are, however, well aware that the real problems to which matrix algebra is applied often involve large numbers of variables and gritty numbers. After all, a matrix entry representing, for instance, the concentration of a mineral in an ore deposit, is usually going to be some featureless number such as $a_{ij} = 1.3714$.

Of course, the numerical calculations for such problems are generally done by computers under the command of prepackaged programs or software. The software that is available covers all the standard computations of matrix theory, including solution of linear systems, construction of eigenvalues and eigenvectors, matrix inversion, and so on. It might seem, then, that except for the small number of specialists who develop and maintain the software, the process of numerical computation need be of little concern.

Nevertheless, it is desirable to have some grasp of the very real problems involved in the design and use of the software. First of all, the reliability of a given software program cannot be assumed. In numerical computation, errors always accumulate as a result of the rounding off of successively computed numbers and can sometimes snowball into a completely inaccurate answer. One therefore wants to know the conditions under which the computation, as dictated by the software, is likely to go wrong. Conversely, if we are familiar with the theory of numerical methods, we can try to design our problems in such a way that smooth sailing is probable.

Second, the cost of computing can be a consideration, even today, if computations with very large matrices must be repeated many times. We will see that the numerical theory is much concerned with limiting the number of steps needed to complete each standard calculation in matrix algebra, and in doing so we will get an idea of where time and money can be saved.

Since our discussion will be in the form of a brief overview only, it should be emphasized that numerical analysis is a very subtle field of study. Indeed, one cannot begin to trace the effects of accumulating error or invent ways of shortcutting long calculations until one has excellent facility in the branch of mathematics to which numerical computation is being applied. The substantial development of numerical analysis is necessarily placed in advanced books devoted solely to that subject.

10.1 Numerical Efficiency

Up to now we have used Gauss-Jordan elimination to solve linear systems. On re-examining this procedure from the point of view of the amount of arithmetic it requires, we find that it saves effort to rearrange the order of the row operations. The revised procedure is called Gaussian elimination. We also study the repeated solution of a system whose constant terms undergo change and find that it can be achieved with relatively little arithmetic.

Further, we present the LDU "decomposition" of a matrix. This concept is a translation into the language of matrix algebra of the new methods just outlined. The theory of elementary matrices serves, as so often, to mediate between concrete equations and pure matrix notation.

A Second Look at Matrix Reduction

Let us take a second look at the Gauss-Jordan method of solving linear systems. For the typical system

Example 1
$$2x_1 + 6x_2 + 4x_3 = 2$$
$$4x_1 + 9x_2 + 3x_3 = 20 \quad ,$$
$$6x_1 + 12x_2 + 4x_3 = 28$$

the first step in the standard reduction sequence is

$$[A|c] = \begin{bmatrix} 2 & 6 & 4 & | & 2 \\ 4 & 9 & 3 & | & 20 \\ 6 & 12 & 4 & | & 28 \end{bmatrix} \xrightarrow{R_2 - 2R_1} \begin{bmatrix} 2 & 6 & 4 & | & 2 \\ 0 & -3 & -5 & | & 16 \\ 6 & 12 & 4 & | & 28 \end{bmatrix} .$$

The arithmetic required for the first row operation, $R_2 - 2R_1$, included a division, multiplications, and subtractions. But divisions and multiplications are much more time-consuming, for people or computers, than subtractions or additions. So we can estimate our investment of time in the row operation by counting the number of divisions and multiplications it required. For convenience, we will call each division or multiplication a "DM."

First, we determined the multiple of row 1 to be subtracted from row 2 by doing the division $\frac{a_{21}}{a_{11}} = \frac{4}{2} = 2$. Then the other entries 6, 4, 2 in row 1 of the augmented matrix were multiplied by 2 and subtracted from the corresponding entries 9, 3, 20 in row 2. This meant three more multiplications, for a total of $1 + 3 = 4$ DM's.

In the same way, we can proceed to determine the total number of DM's required to complete the solution of the linear system. The reasoning isn't especially tied to the 3×3 dimensions of the coefficient matrix; in fact the result is more meaningful if stated, and argued, for the $n \times n$ case.

Lemma 10.1   Suppose A is an $n \times n$ matrix of rank n. Then solution of the system $A\mathbf{x} = \mathbf{c}$ by the Gauss-Jordan method requires at least $\frac{n^3}{2}$ divisions and multiplications.

Note: The exact number of DM's required is $\frac{n(n^2 + 2n - 1)}{2}$.

■**Proof**   When A is $n \times n$, the augmented matrix $[A|\mathbf{c}]$ is $n+1$ entries wide; so the number of DM's needed for the first row operation $R_2 - kR_1$ changes from $3+1$, as obtained in the example with $n = 3$, to $n+1$. Of course row 1 acts in turn on all of the $n-1$ rows below it; so the total number of DM's required for the elimination in column 1 is $(n-1)(n+1)$.

Now elimination in column 2 follows the same pattern as in column 1, and therefore takes almost as many DM's. The only source of relief is that column 1 no longer changes. Thus the width of the augmented matrix is effectively reduced from $n+1$ to n entries and the number of DM's is correspondingly reduced to $(n-1)n$. In fact, at each successive column the effective width of the rows is reduced by 1; consequently the total number of DM's needed for elimination in columns 1 through n is just

$$(n-1)(n+1) + (n-1)n + (n-1)(n-1) + \cdots + (n-1)2 \quad ,$$

or $(n-1\{(n+1) + n + (n-1) + \cdots + 2\})$.

But from elementary algebra, the sum of the first k integers is $\frac{k(k+1)}{2}$, and on using the case $k = n+1$ of this result we see that our formula can be reduced to

$$(n-1)\left\{\frac{(n+1)(n+2)}{2} - 1\right\} \quad \text{or} \quad \frac{n(n^2 + 2n - 3)}{2} \quad .$$

Finally, to change the corners of our nearly reduced matrix to 1's necessitates dividing each constant in the augmented column by the corner entry in its row. This means n further DM's. The total is now $\frac{n(n^2 + 2n - 1)}{2}$, which exceeds $\frac{n^3}{2}$.

Q.E.D.

Thus the solution of a system of, say, 200 equations in 200 unknowns already requires over four million divisions and multiplications. It is clearly worthwhile to find ways of cutting down this number.

Back Substitution

One thing we may have noticed in the course of working through the solution of linear systems is that the elimination of entries above corners can be delayed until after the crossing line has been completed. For the system discussed at the beginning of this section, the first group of row operations would then be as follows:

$$A\mathbf{c} = \begin{bmatrix} 2 & 6 & 4 & | & 2 \\ 4 & 9 & 3 & | & 20 \\ 6 & 12 & 4 & | & 28 \end{bmatrix} \begin{matrix} \\ R_2 - 2R_1 \\ R_3 - 3R_1 \end{matrix} \begin{bmatrix} 2 & 6 & 4 & | & 2 \\ 0 & -3 & -5 & | & 16 \\ 0 & -6 & -8 & | & 22 \end{bmatrix} \cdots$$

$$\cdots \begin{matrix} \\ \\ R_3 - 2R_2 \end{matrix} \begin{bmatrix} 2 & 6 & 4 & | & 2 \\ 0 & -3 & -5 & | & 16 \\ 0 & 0 & 2 & | & -10 \end{bmatrix} \cdot$$

The augmented matrix is now in upper triangular form and corresponds to the list of equations

$$\left\{ \begin{aligned} 2x_1 + 6x_2 + 4x_3 &= 2 \\ -3x_2 - 5x_3 &= 16 \\ 2x_3 &= -10 \end{aligned} \right. .$$

An alternative procedure for completing the solution now suggests itself. From the last equation, $2x_3 = -10$, we see at once that $x_3 = -5$. Then we can substitute this value of x_3 into the second-to-last equation, obtaining

$$-3x_2 + 25 = 16 \quad \text{or} \quad x_2 = 3 \quad .$$

In turn, the values obtained for x_3 and x_2 can be substituted into the preceding equation in the list, yielding

$$2x_1 + 18 - 20 = 2 \quad \text{or} \quad x_1 = 2 \quad .$$

The solution of the system is therefore $\mathbf{x} = (2, 3, -5)$.

The process of working backwards through the list of equations and determining in turn $x_n, x_{n-1}, \ldots, x_1$ is called *back substitution*. The overall strategy of first reducing $[A|\mathbf{c}]$ to upper triangular form and then computing the solution by back substitution is called *Gaussian elimination*.

For the larger systems with which numerical analysis is especially concerned, Gaussian elimination is substantially faster than the Gauss-Jordan method. Here is a precise statement.

Lemma 10.2 Suppose A is an $n \times n$ matrix of rank n. Then the solution of the system $A\mathbf{x} = \mathbf{c}$ by Gaussian elimination requires $\dfrac{(n^3 + 3n^2 - n)}{3}$ divisions and multiplications.

■Proof   Similar to previous lemma; see exercises for an outline.

We conclude that the use of Gaussian elimination reduces the number of DM's roughly from $\frac{n^3}{2}$ to $\frac{n^3}{3}$; a time saving of about one third.

Suppose now that we have to solve the system $A\mathbf{x} = \mathbf{c}$ for many different choices of the vector $\mathbf{c}$. Must we do $\frac{n^3}{3}$ DM's every time? One would think not, because the reduction of the matrix A to upper triangular form is the same every time; only the work in the augmented column is different for each $\mathbf{c}$.

Surely, the software should instruct the computer to record the list $Q_1, \ldots, Q_k$ of type A row operations used in reducing the $n \times n$ matrix A to upper triangular form U, as well as the matrix U itself.

Then, for any constant vector $\mathbf{c}$, the system $A\mathbf{x} = \mathbf{c}$ is equivalent to

$$Q_k \cdots Q_1 A\mathbf{x} = Q_k \cdots Q_1 \mathbf{c} \quad ,$$

or $U\mathbf{x} = \mathbf{c}'$,   where $\mathbf{c}' = Q_k \cdots Q_1 \mathbf{c}$.

Therefore the solution of the system $A\mathbf{x} = \mathbf{c}$ can be obtained by the following steps. First, $\mathbf{c}'$ is computed by applying the row operations $Q_1, \ldots, Q_k$ in turn to $\mathbf{c}$. This requires k multiplications. Then back substitution is applied to the system $U\mathbf{x} = \mathbf{c}'$. It is easily checked (see exercises 10.1) that these two steps require $k = \frac{n(n-1)}{2}$ and $\frac{n(n+1)}{2}$ DM's respectively, for a total of n^2 DM's. Let us summarize.

THEOREM **10.3**

Suppose A is an $n \times n$ matrix of rank n. Suppose also that the following information has been recorded:

 (i) The sequence of row operations used in the Gaussian elimination process to change A to an upper triangular matrix U.

(ii) The upper triangular matrix U itself.

Then for any numerical vector $\mathbf{c}$, the solution of the system $A\mathbf{x} = \mathbf{c}$ requires only n^2 divisions and multiplications.

The economies guaranteed by this theorem are spectacular! For example, suppose A is 300×300. Then to solve a system of the form $A\mathbf{x} = \mathbf{c}$ takes roughly $\frac{n^3}{3} = 9,000,000$ DM's, but to solve it again with a different constant $\mathbf{c}$ on the right side requires only $n^2 = 90,000$ further DM's: a saving of 99%.

We will now show that the accelerated techniques for solving linear systems can easily be distilled into a pure matrix formulation.

The *LDU* Decomposition

The row operations $R_2 - 2R_1$, $R_3 - 3R_1$, $R_3 - 2R_2$, which we used to reduce our illustrative 3×3 matrix A to the upper triangular form,

$$U = \begin{bmatrix} 2 & 6 & 4 \\ 0 & -3 & -5 \\ 0 & 0 & 2 \end{bmatrix} ,$$

are all of type A; and in fact each of them subtracts a multiple of an earlier row from a later row. This means that the type A elementary matrices that implement these row operations are all *lower triangular* matrices. For example, the row operation $R_2 - 2R_1$ is implemented by

$$A(-2;1,2) = \begin{bmatrix} 1 & 0 & 0 \\ -2 & 1 & 0 \\ 0 & 0 & 1 \end{bmatrix} .$$

We denote these matrices L_1, L_2, L_3, so that $L_3 L_2 L_1 A = U$.

The further reduction of U to the identity matrix can be accomplished as follows:

$$U = \begin{bmatrix} 2 & 6 & 4 \\ 0 & -3 & -5 \\ 0 & 0 & 2 \end{bmatrix} \begin{array}{c} (\frac{1}{2})R_1 \\ (-\frac{1}{3})R_2 \\ (\frac{1}{2})R_3 \end{array} \begin{bmatrix} 1 & 3 & 2 \\ 0 & 1 & \frac{5}{3} \\ 0 & 0 & 1 \end{bmatrix} \begin{array}{c} R_1 - 2R_3 \\ R_2 - (\frac{5}{3})R_3 \end{array} \begin{bmatrix} 1 & 3 & 0 \\ 0 & 1 & 0 \\ 0 & 0 & 1 \end{bmatrix}$$

$$R_1 - 3R_2 \begin{bmatrix} 1 & 0 & 0 \\ 0 & 1 & 0 \\ 0 & 0 & 1 \end{bmatrix} = I .$$

The type M row operations used to change the diagonal elements of U to 1's are implemented by *diagonal* matrices, which we denote D_1, D_2, D_3. These are followed by type A row operations, which subtract multiples of later rows from earlier ones, and so are implemented by *upper triangular* elementary matrices. We denote these as U_1, U_2, U_3. Then we have

$$(U_3 U_2 U_1)(D_3 D_2 D_1)(L_3 L_2 L_1)A = I .$$

More generally, consider the reduction of an $n \times n$ matrix of rank n to the identity matrix I. This would require $k = \frac{n(n-1)}{2}$ lower triangular elementary matrices, the same number of upper triangular ones, and also n diagonal elementary matrices.

It is also possible that, during the first stage, the reduction of A to U, row exchanges might be needed to produce nonzero corners. (These, of course, involve no arithmetic and for this reason were not discussed earlier.) But once the reduction of A to U has been worked through, it is easy to see how to rearrange the order of the steps so that all the exchanges are done at the beginning. Let us denote the necessary row exchanges, or type E elementary matrices, by $E_1, \ldots, E_h$. Then all the exchanges are done simultaneously by the product matrix

$$P = E_h \cdots E_1 .$$

The matrix P implements a rearrangement of the rows of A and is called a *permutation matrix*. Permutation matrices are characterized (see exercises 3.3) by having all zero entries except for a single 1 in each row and each column.

The reduction of a general $n \times n$ matrix A of rank n to I can then be represented by a matrix equation,

$$(U_k \cdots U_1)(D_n \cdots D_1)(L_k \cdots L_1)PA = I \quad .$$

The corresponding formula for the solution of the linear system $A\mathbf{x} = \mathbf{c}$ is

$$\mathbf{x} = (U_k \cdots U_1)(D_n \cdots D_1)(L_k \cdots L_1)P\mathbf{c} \quad .$$

In this formula, the application of P to $\mathbf{c}$ simply rearranges the components of $\mathbf{c}$ and requires no arithmetic. Each elementary matrix, when applied in its turn to the vector $\mathbf{c}$, requires one multiplication operation. The total number of DM's needed for the calculation of the solution by this method is therefore

$$k + k + n = \frac{n(n-1)}{2} + \frac{n(n-1)}{2} + n = n^2 \quad ,$$

just as in theorem 10.3.

Example 2 To solve the system
$$\begin{cases} 2x_1 + 6x_2 + 4x_3 = 2 \\ 4x_1 + 9x_2 + 3x_3 = -5 \\ 6x_1 + 12x_2 + 4x_3 = -6 \end{cases},$$

observe that the 3×3 coefficient matrix A is the same as in example 1. Therefore we have only to apply the nine row operations L_1 through U_3 to $\mathbf{c} = (2, -5, -6)$ in turn:

$$\mathbf{c} = \begin{bmatrix} 2 \\ -5 \\ -6 \end{bmatrix} \begin{matrix} \\ R_2 - 2R_1 \\ R_3 - 3R_1 \end{matrix} \begin{bmatrix} 2 \\ -9 \\ -12 \end{bmatrix} \begin{matrix} \\ \\ R_3 - 2R_2 \end{matrix} \begin{bmatrix} 2 \\ -9 \\ 6 \end{bmatrix} \begin{matrix} (\frac{1}{2})R_1 \\ (-\frac{1}{3})R_2 \\ (\frac{1}{2})R_3 \end{matrix} \begin{bmatrix} 1 \\ 3 \\ 3 \end{bmatrix}$$

$$\begin{matrix} R_1 - 2R_3 \\ R_2 - (\frac{5}{3})R_3 \end{matrix} \begin{bmatrix} -5 \\ -2 \\ 3 \end{bmatrix} \begin{matrix} R_1 - 3R_2 \\ \\ \end{matrix} \begin{bmatrix} 1 \\ -2 \\ 3 \end{bmatrix} = \mathbf{x} \quad .$$

That is, $x_1 = 1$, $x_2 = -2$, $x_3 = 3$.

Our reformulation of the Gaussian elimination process in matrix-vector notation clearly contains a statement about the structure of the invertible $n \times n$ matrix A itself. First, we can write

$$PA = (L_1^{-1} \cdots L_k^{-1})(D_1^{-1} \cdots D_n^{-1})(U_1^{-1} \cdots U_k^{-1}) \quad .$$

Next, we recall from section 3.5 that products and inverses of, say, lower triangular matrices are again lower triangular, so that $L = L_1^{-1} \cdots L_k^{-1}$ is lower triangular. Similarly $U = U_1^{-1} \cdots U_k^{-1}$ is upper triangular, and $D = D_1^{-1} \cdots D_n^{-1}$ is diagonal. We therefore have an important theorem.

THEOREM **10.4**

Suppose A is an $n \times n$ matrix of rank n. Then there is a permutation matrix P such that

$$PA = LDU \quad,$$

where L is lower triangular, U is upper triangular, and D is diagonal.

Furthermore, L and U can be chosen so as to have all their diagonal entries equal to 1, and the matrices L, U, and D are then uniquely determined.

■**Proof**   Let us consider further the above construction of L, U, and D.
First, the matrices L_i^{-1}, like the L_i themselves, are actually lower triangular type A elementary matrices. So their diagonal entries are all equal to 1, and it follows easily that the same is true of their product L. The same argument applies to U; therefore L and U both have all their diagonal entries equal to 1.

As for uniqueness, suppose that $PA = L'D'U'$ is a second representation, with all diagonal entries of L' and U' again equal to 1. Then

$$LDU = L'D'U' \quad \text{or} \quad (L')^{-1}L = D'U'U^{-1}D^{-1} \quad.$$

Now since the left and right side of this matrix equation are lower and upper triangular respectively, both sides are actually diagonal. Further, since the lower triangular matrices L' and L have all their diagonal entries equal to 1, it is easily seen that the same is true of the inverse $(L')^{-1}$ and the product $(L')^{-1}L$. Therefore $(L')^{-1}L = I$, or $L' = L$. By cancellation, $DU = D'U'$, and it follows similarly that $D' = D$ and $U' = U$.   Q.E.D.

The LDU decomposition of PA obviously discards a lot of the information in the original, longer, product representation of PA. In different contexts, different amounts of detail about the factorization of PA are needed. Thus, for Gaussian elimination, we need the factors $L_1, \ldots, L_k$, which reduce PA to U; but the consolidated matrix U, rather than the factors $U_1, \ldots, U_k$, can be used in the back substitution process described originally.

The LDU decomposition can be compressed further. In fact, since the factors D and U are both upper triangular, so is their product DU. To put it another way, we say that the factor D can be absorbed into the factor U. That is, we can write

$$PA = LU \quad,$$

where L and U are lower and upper triangular matrices respectively. This is called an LU *decomposition* of PA. It should be noted, however, that the choice of L and U in this decomposition is far from unique.

■ **Problem**   Obtain *LDU* and *LU* decompositions for the matrix
$$A = \begin{bmatrix} 2 & 3 \\ 4 & 7 \end{bmatrix} .$$

■ **Solution**   First we reduce *A* to *I* as follows:

$$A = \begin{bmatrix} 2 & 3 \\ 4 & 7 \end{bmatrix} \underset{R_2 - 2R_1}{} \begin{bmatrix} 2 & 3 \\ 0 & 1 \end{bmatrix} \underset{}{(\frac{1}{2})R_1} \begin{bmatrix} 1 & \frac{3}{2} \\ 0 & 1 \end{bmatrix} \cdots$$

$$R_1 - (\frac{3}{2})R_2 \begin{bmatrix} 1 & 0 \\ 0 & 1 \end{bmatrix} = I .$$

Next we write down the elementary matrices that implement the row operations:

$$L_1 = \begin{bmatrix} 1 & 0 \\ -2 & 1 \end{bmatrix} , \quad D_1 = \begin{bmatrix} \frac{1}{2} & 0 \\ 0 & 1 \end{bmatrix} , \quad U_1 = \begin{bmatrix} 1 & -\frac{3}{2} \\ 0 & 1 \end{bmatrix} .$$

Then we compute

$$L = L_1^{-1} = \begin{bmatrix} 1 & 0 \\ 2 & 1 \end{bmatrix}, D = D_1^{-1} = \begin{bmatrix} 2 & 0 \\ 0 & 1 \end{bmatrix}, U = U_1^{-1} = \begin{bmatrix} 1 & \frac{3}{2} \\ 0 & 1 \end{bmatrix}$$

and conclude that $A = LDU$.

Also, if we collapse the *D* and *U* factors into the product

$$U' = DU = \begin{bmatrix} 2 & 3 \\ 0 & 1 \end{bmatrix} \quad \text{and let } L' = L ,$$

we obtain an *LU* decomposition $A = L'U'$.

■ EXERCISES 10.1

1. Solve the following linear systems by Gaussian elimination.

   (a) $\begin{cases} x_1 + 4x_2 = 4 \\ 3x_1 + 10x_2 = 16 \end{cases}$ (b) $\begin{cases} x_1 + 2x_2 + x_3 = 2 \\ 2x_1 + 6x_2 + 4x_3 = 5 \\ 4x_1 + 6x_2 + 4x_3 = 3 \end{cases}$

   Note: additional practice in Gaussian elimination can be obtained by returning to the exercises of Chapter 1.

2. Let *A* be the coefficient matrix of exercise 1(b), i.e.,
$$A = \begin{bmatrix} 1 & 2 & 1 \\ 2 & 6 & 4 \\ 4 & 6 & 4 \end{bmatrix} .$$

   (a) Find elementary matrices L_i, D_i, U_i such that
$$A = U_k \cdots U_1 D_3 D_2 D_1 L_k \cdots L_1 .$$

(b) Use this formula for A to solve $A\mathbf{x} = \mathbf{c}$ in the case $\mathbf{c} = (1,0,2)$.

3. Find an *LDU* decomposition and an *LU* decomposition for each of the coefficient matrices in exercise 1.

4. Let $A = \begin{bmatrix} 0 & 1 \\ 1 & 0 \end{bmatrix}$.

   (a) Show that A cannot be expressed as the product of an upper and a lower triangular matrix.
   (b) Construct a decomposition of A in the form $PA = LDU$ where P is a permutation matrix, etc.

5. Suppose A and B are $n \times n$ matrices.

   (a) Show that to compute the *ij*th entry of AB by the row-column rule requires n multiplications.
   (b) Deduce that computation of the product matrix AB requires n^3 DM's.

6. Show that if A and B are $m \times k$ and $k \times n$ matrices respectively, then computation of the product matrix AB requires kmn DM's.

7. (Analysis of lemma 10.2) Show by the indicated steps that, if A is an $n \times n$ matrix of rank n, then solution of the system $A\mathbf{x} = \mathbf{c}$ by Gaussian elimination requires $\dfrac{n(n^2+3n-1)}{3}$ multiplications and divisions.

   (a) Show that the number of DM's required for the reduction of $[A|\mathbf{c}]$ to upper triangular form is

   $$(n-1)(n+1) + (n-2)(n) + (n-3)(n-1) + \cdots + (1)(3) = \dfrac{n(2n^2+3n-5)}{6}.$$

   (b) Show that the number of DM's required for back substitution is

   $$1 + (1+1) + (1+2) + \cdots + (1+(n-1)) = \dfrac{n(n+1)}{2}.$$

   (c) Check that the numbers of DM's found in parts (a) and (b) add to the stated total.

8. (Analysis of theorem 10.3) Suppose that in the linear system $A\mathbf{x} = \mathbf{c}$ of the previous exercise, the constant vector $\mathbf{c}$ is changed to, say, $\mathbf{d}$. Show by the steps indicated below that solution of the new system $A\mathbf{x} = \mathbf{d}$ requires only n^2 further DM's.

   (a) Show that the number of type A row operations used in reducing $[A|\mathbf{d}]$ to upper triangular form $[U|\mathbf{d}']$, is $\dfrac{n(n-1)}{2}$.

(b) Show that the number of type M row operations used in reducing $[U|\mathbf{d}']$ to the form $[V|\mathbf{d}'']$, where V is upper triangular with 1's as its diagonal entries, is n.

(c) Show that the number of type A row operations used in reducing $[V|\mathbf{d}'']$ to $[I|\mathbf{x}]$ is $n \dfrac{n(n-1)}{2}$

(d) Deduce that the total number of type A and M *row operations* that must be applied to the constant vector $\mathbf{d}$ to produce the solution vector $\mathbf{x}$ is n^2.

*9. Suppose A is an invertible $n \times n$ matrix. Show that the computation of A^{-1} can be accomplished with roughly n^3 DM's. That is, the number of DM's required is about the same as for computation of A^2 (see exercise 5).

Note: a specific estimate for the number of DM's required is

$$n^3 + \left(\frac{(n^2 - 3n + 2)}{2} \right) .$$

10.2 Error and Approximation

So far we have concentrated on refining our earlier methods so as to reduce the amount of arithmetic needed to solve a system. Now we will consider more drastic changes in computational strategy. First, we discuss some of the sources of serious error and introduce the ''partial pivoting'' method as a defense against it.

Then we address the problem of computing eigenvectors. Our previous methods are extremely inefficient here, and it is necessary to inject entirely new ideas into the subject. In fact the alternative we propose does not aim, even in principle, at computing eigenvectors exactly. Instead it produces a sequence of increasingly accurate approximations to the eigenvectors. This strategy of ''successive approximation'' is widely used in numerical analysis, and even in other branches of mathematics.

Roundoff Error

Suppose a computer is programmed to solve linear systems by Gaussian elimination, working to four-digit accuracy. That is, each constant computed during the solution process is immediately rounded to four significant digits; so that, for example, the sum $12,500 + 32 = 12,532$ would be recorded as 12,530 or 1.253×10^4.

In the case of the linear system

$$(.0001)x_1 + 5x_2 = 8$$
$$2x_1 \quad\ x_2 = 1 \ ,$$

the first step of the computation, as done by the computer, would be

$$\begin{bmatrix} .0001 & 5 & | & 8 \\ 2 & 1 & | & 1 \end{bmatrix}$$

$$R_2 - (20{,}000)R_1 \begin{bmatrix} .0001 & 5 & | & 8 \\ 0 & -99{,}999 & | & -159{,}999 \end{bmatrix} .$$

$$\text{Roundoff} \begin{bmatrix} .0001 & 5 & | & 8 \\ 0 & -100{,}000 & | & -160{,}000 \end{bmatrix}$$

Back substitution would then give first $x_2 = 1.600$, and then

$$(.0001)x_1 + (5.000)(1.600) = 8 \quad \text{or} \quad x_1 = 0 \ .$$

However, this answer, $\mathbf{x} = (0, 1.600)$, is drastically wrong. In fact, when substituted into the left side of the second equation in the given system, it produces the value 1.6 instead of 1.

But because the system is so small, we can easily obtain the exact answer for purposes of comparison. The inversion formula for 2×2 matrices gives

$$\mathbf{x} = A^{-1}\mathbf{c} = \tfrac{1}{(-9.9999)} \begin{bmatrix} 1 & -5 \\ -2 & .0001 \end{bmatrix} \begin{bmatrix} 8 \\ 1 \end{bmatrix}$$

$$= \tfrac{1}{(-9.9999)} \begin{bmatrix} 3 \\ -15.999 \end{bmatrix} \ .$$

Therefore, to four significant digits, the correct answer is

$$x_1 = -0.3000 \ , \quad x_2 = 1.600 \ .$$

The simplicity of the system also helps us to understand why our seemingly cautious roundoff procedure destroys the accuracy of the Gaussian elimination process. The disproportionately small size of the first corner entry, $a_{11} = 10^{-4}$, forces us to multiply row 1 by a large constant, $\frac{a_{21}}{a_{11}} = 2 \times 10^4$, for purposes of elimination in column 1. This introduces the large entries $a_{22} = -99{,}999$ and $c_2 = -159{,}999$ into the augmented matrix. In fact, the original entries $a_{22} = 1$, $c_2 = 1$, are more than four orders of magnitude smaller than these, and so all traces of them are erased in the roundoff to $a_{22} = -100{,}000$, $c_2 = -160{,}000$. But the erasure means that now, after only one row operation, the augmented matrix is exactly what it would have been if the original system had been

$$(.0001)x_1 + 5x_2 = 8$$
$$2x_1 \quad\quad = 0 \quad .$$

From the second equation $2x_1 = 0$ we see why the false result $x_1 = 0$ was obtained.

Fortunately, we can often avoid having to use disproportionately small corner entries simply by doing a row exchange when one occurs. In the above example, we can preface the reduction of the augmented matrix by exchanging rows 1 and 2, thereby changing the course of the Gaussian elimination process to

$$
\begin{array}{c} R_2 \\ R_1 \end{array}
\left[\begin{array}{cc|c}
2 & 1 & 1 \\
.0001 & 5 & 8
\end{array} \right]
$$

$$
R_2 - (.00005)R_1
\left[\begin{array}{cc|c}
2 & 1 & 1 \\
0 & 4.99995 & 7.99995
\end{array} \right]
$$

$$
\text{Roundoff}
\left[\begin{array}{cc|c}
2 & 1 & 1 \\
0 & 5 & 8
\end{array} \right]
$$

and, on back substitution, $x_2 = 1.600$

and $2x_1 + 1(1.600) = 1$ or $x_1 = -0.3000$.

This time our results are correct to four significant digits.

It seems sensible that the safety precaution just described should be used to the fullest possible extent. Thus, before the elimination in each column we select as corner entry the largest entry below the line in the column. A row exchange is used to achieve this when necessary. This tactic is called *partial pivoting*.

Sometimes near-zero corner entries and their pitfalls cannot be avoided, as the outcome of the first row operations on the following augmented matrix shows:

$$
[A|\mathbf{c}] =
\left[\begin{array}{cc|c}
2 & 6 & 2 \\
1 & 3.001 & 9
\end{array} \right]
$$

$$
R_2 - (.5)R_1
\left[\begin{array}{cc|c}
2 & 6 & 2 \\
0 & .0010 & 8
\end{array} \right] .
$$

In this example the coefficient matrix is just barely invertible; in fact its eigenvalues are easily found to be $\lambda_1 = 5.001$, $\lambda_2 = 4.000 \times 10^{-4}$. Generally speaking, whenever a matrix has eigenvalues differing in size by several orders of magnitude, difficulties with numerical computation are to be expected.

Approximation of Eigenvectors

The conventional method for computing the eigenvectors of an $n \times n$ matrix A requires that we first obtain its eigenvalues, which are the roots of the characteristic polynomial

$$
f(\lambda) = \det (\lambda I - A) .
$$

The determinant, however, cannot be evaluated quickly by reduction to upper triangular form, because many of its entries contain the variable λ. Usually repeated Laplace expansion is necessary, and this is extremely slow.

Fortunately, good approximations to the eigenvectors of a matrix can be obtained by far more efficient methods. First we present the underlying idea.

Suppose that the $n \times n$ matrix A has n linearly independent eigenvectors $v_1, \ldots, v_n$, corresponding to eigenvalues $\lambda_1, \ldots, \lambda_n$ respectively. From Chapter 8, we know that if we start with an arbitrary vector $v = c_1 v_1 + \cdots + c_n v_n$, then after applying A to it m times we obtain

$$A^m v = \lambda_1^m c_1 v_1 + \cdots + \lambda_n^m c_n v_n \quad .$$

On factoring out the scalar $c_1 \lambda_1^m$, we see that the vector $A^m v$ is parallel to

$$v_1 + \left(\frac{\lambda_2}{\lambda_1}\right)^m \left(\frac{c_2}{c_1}\right) v_2 + \cdots + \left(\frac{\lambda_n}{\lambda_1}\right)^m \left(\frac{c_n}{c_1}\right) v_n \quad .$$

Suppose now that we have labeled the eigenvalues so that λ_1 is the one with the largest absolute value. Then, as m increases, the factors $\left(\frac{\lambda_j}{\lambda_1}\right)^m$ all converge to zero, for $j = 2, \ldots, n$, and in consequence the direction of the vector $A^m v$ approaches that of the eigenvector v_1. Let us see how this method of approximating an eigenvector works in practice.

Example 1   The matrix $A = \begin{bmatrix} 4 & 3 \\ 1 & 2 \end{bmatrix}$ is so simple that we can compute its eigenvalues and eigenvectors in a minute by the conventional method. These are

$$\lambda_1 = 5 \quad , \quad \lambda_2 = 1$$
$$v_1 = (3,1) \quad , \quad v_2 = (1,-1) \quad .$$

Now let us see how quickly the iterates $A^m v$ converge to v_1. Our choice of v is of course arbitrary; suppose

$$v = (1,0) \quad .$$

Then
$$Av = \begin{bmatrix} 4 & 3 \\ 1 & 2 \end{bmatrix} \begin{bmatrix} 1 \\ 0 \end{bmatrix} = \begin{bmatrix} 4 \\ 1 \end{bmatrix} \quad ,$$
$$A^2 v = \begin{bmatrix} 4 & 3 \\ 1 & 2 \end{bmatrix} \begin{bmatrix} 4 \\ 1 \end{bmatrix} = \begin{bmatrix} 19 \\ 6 \end{bmatrix} \quad .$$

Similarly, $A^3 v = (94,31)$, $A^4 v = (469, 156)$, $A^5 v = (2,344,781)$, etc.

To follow the directions of the successive vectors $A^m v$, it is best to divide each by its largest entry (which doesn't change its direction). We then obtain the "approximating sequence"

$a(0) = (1,0)$
$a(1) = (1,.25)$
$a(2) = (1,.3158)$
$a(3) = (1,.3298)$
$a(4) = (1,.3326)$
$a(5) = (1,.3332)$ etc.

These vectors are approaching the known eigenvector $\mathbf{v}_1 = (1,\tfrac{1}{3}) =$ (1,.3333) quite rapidly. Usually, of course, we do not know the eigenvector in advance, so that when we have what seems to be a good approximation, say, $\mathbf{a}(5) = (1,.3332)$, we check it as follows:

$$A\mathbf{a}(5) = \begin{bmatrix} 4 & 3 \\ 1 & 2 \end{bmatrix} \begin{bmatrix} 1 \\ .3332 \end{bmatrix} = \begin{bmatrix} 4.9996 \\ 1.6664 \end{bmatrix} = (4.9996) \begin{bmatrix} 1 \\ .3333 \end{bmatrix} .$$

This calculation indicates that $\mathbf{a}(5)$ is indeed very nearly an eigenvector, and also that the corresponding eigenvalue is approximately 4.9996; which is, to four-digit accuracy, the correct value, $\lambda_1 = 5$.

An apparent shortcoming of this method is that it computes only the eigenvector corresponding to the largest of the eigenvalues, $\lambda_1, \ldots, \lambda_n$, of A. But the eigenvectors of A are also eigenvectors of A^{-1}, with eigenvalues changed to $\lambda_1^{-1}, \ldots, \lambda_n^{-1}$. Thus if $\mathbf{v}_n$ is the eigenvector of A whose eigenvalue λ_n is the smallest, it is also the eigenvector of A^{-1} whose eigenvalue is the largest. Therefore the iterates $(A^{-1})^m\mathbf{v}$ or $A^{-m}\mathbf{v}$ should lead to the eigenvector $\mathbf{v}_n$.

In the case of example 1, we have

$$A^{-1} = (\tfrac{1}{5}) \begin{bmatrix} 2 & -3 \\ -1 & 4 \end{bmatrix} \quad \text{or} \quad M = 5A^{-1} = \begin{bmatrix} 2 & -3 \\ -1 & 4 \end{bmatrix} .$$

We can delete the factor $\tfrac{1}{5}$ because we are not concerned with scalar factors. Then the iterates $A^{-m}\mathbf{v}$ and approximating vectors $\mathbf{a}(m)$, are

$\mathbf{v} = (1,0)$	$\mathbf{a}(0) = (1,0)$
$A^{-1}\mathbf{v} = (2,-1)$	$\mathbf{a}(1) = (1,-.5)$
$A^{-2}\mathbf{v} = (7,-6)$	$\mathbf{a}(2) = (1,-.8571)$
$A^{-3}\mathbf{v} = (32,-31)$	$\mathbf{a}(3) = (1,-.9688)$
$A^{-4}\mathbf{v} = (157,-156)$	$\mathbf{a}(4) = (1,-.9936)$
$A^{-5}\mathbf{v} = (782,-781)$	$\mathbf{a}(5) = (1,-.9987)$
$A^{-6}\mathbf{v} = (3907,-3906)$	$\mathbf{a}(6) = (1,-.9997)$ etc.

Also, $\quad A\mathbf{a}(6) = (1.001)(1,-.9985)$.

So, to three-digit accuracy, we have recovered the second eigenvector, $\mathbf{v}_2 = (1,-1)$, and its eigenvalue, $\lambda_2 = 1$.

Eigenvectors in Higher Dimensions

When $n \geq 3$, an $n \times n$ matrix will usually have other eigenvalues besides the ones largest and smallest in magnitude. For example, suppose a 3×3 matrix has eigenvalues $\lambda_1, \lambda_2, \lambda_3$ which are close to the values 1, 5, 10 respectively. Then to get at the eigenvalue λ_2 and its corresponding eigenvector, we shift to the matrix $A - 5I$. It has the same eigenvectors as A, but the eigenvalues $\lambda_1, \lambda_2, \lambda_3$ are shifted by -5 to approximately -4, 0, and 5 respectively. So λ_2 becomes the smallest eigenvalue and the iterates $(A - 5I)^{-m}\mathbf{v}$ should lead to the eigenvector $\mathbf{v}_2$.

Shifting should enable us to track down all the eigenvectors eventually, but if we do not know the approximate values of the eigenvalues in advance, we may have to try several shifts before we obtain all the intermediate eigenvalues and eigenvectors.

Clearly our iterative method has its limitations. For example, if two eigenvalues, say λ_1 and λ_2, are very close, many iterations will be needed to reduce the ratio $\left(\frac{\lambda_2}{\lambda_1}\right)^m$ and thereby approximate v_1 well. Also, the cases of multiple eigenvalues and nondiagonalizable matrices require special treatment.

■ **Problem**   Let $A = \begin{bmatrix} 5 & -4 \\ 6 & -5 \end{bmatrix}$, $\mathbf{v} = \begin{bmatrix} 1 \\ 0 \end{bmatrix}$.

Then the successive iterates $A^m\mathbf{v}$ are alternately $(1,0)$ and $(5,6)$, and thus they do not converge to an eigenvector. Is this attributable to a special feature of the matrix A?

■ **Solution**   The eigenvalues of the matrix A are, in fact, $\lambda_1 = 1$, $\lambda_2 = -1$.
So if a vector $\mathbf{v}$ is expressed in terms of the eigenvectors $\mathbf{v}_1$, $\mathbf{v}_2$ as $\mathbf{v} = c_1\mathbf{v}_1 + c_2\mathbf{v}_2$, then

$$A^m\mathbf{v} = c_1\mathbf{v}_1 + (-1)^m c_2\mathbf{v}_2 \quad .$$

Thus the iterates $A^m\mathbf{v}$ alternately equal $\mathbf{v}$ and $c_1\mathbf{v}_1 - c_2\mathbf{v}_2$. Thus there is no convergence unless the starting vector $\mathbf{v}$ happens to be one of the eigenvectors.

This barrier to convergence of the iterates occurs because two of the eigenvalues of A have the same magnitude, and it is overcome simply by shifting to $A - kI$.

■ EXERCISES 10.2

1. Solve, by the method of partial pivoting, each of the following linear systems:

(a) $\left[\begin{array}{cc|c} .01 & 10 & 100 \\ 100 & 1 & -3 \end{array}\right]$ (b) $\left[\begin{array}{ccc|c} .01 & 1 & 2 & 1 \\ 1 & .1 & 1 & 2 \\ 10 & .1 & .01 & 0 \end{array}\right]$

2. (a) Show, by using the 2×2 matrix inversion formula, that the solution obtained for exercise 1(a) was correct to four-digit accuracy.

   (b) Suppose that, in exercise 1(a), Gaussian elimination without partial pivoting had been used, and the result of each arithmetic operation rounded to four-digit accuracy. What result would have been obtained?

3. Using the iterative method of this section, approximate all eigenvectors and eigenvalues of the following matrices.

(a) $\begin{bmatrix} 1 & 5 \\ 2 & 4 \end{bmatrix}$ (b) $\begin{bmatrix} 5 & 7 \\ 8 & 6 \end{bmatrix}$ (c) $\begin{bmatrix} 1 & 2 & 0 \\ 1 & 0 & 0 \\ 0 & 1 & 10 \end{bmatrix}$

In each case, check the results by comparing them with the exact eigenvalues and eigenvectors obtained by the methods of Chapter 8.

4. Let $A = \begin{bmatrix} 11 & -10 \\ 10 & -9 \end{bmatrix}$, and $\mathbf{v} = \begin{bmatrix} 1 \\ 0 \end{bmatrix}$.

(a) Show that the iterates $A^m \mathbf{v}$ and the iterates $A^{-m} \mathbf{v}$ both lead to the same eigenvector $\mathbf{v}_1 = (1,1)$.

(b) Identify a structural feature of the matrix A that accounts for the result obtained in part (a).

CHAPTER ELEVEN

Linear Programming

Introduction

The theory of linear programming addresses a specific category of problems in linear algebra, which usually have a direct or indirect economic interpretation. One attempts to maximize revenues from a gambling game or a manufacturing operation, for example. Someone in the consumer electronics field might have an inventory of components from which several models of pocket calculator can be assembled. Total revenue, if quantities $x_1, \ldots, x_n$ of n different models selling at prices $c_1, \ldots, c_n$ respectively are built, is of course

$$f(x_1, \ldots, x_{n)} = c_1x_1 + \cdots + c_nx_n \quad .$$

But because the supply of components is limited, the more of one model are built, the fewer of the other models can be built. Those constraints can be expressed as linear inequalities of the form

$$a_1x_1 + \cdots + a_nx_n \leq b \quad .$$

In other words, the vector $\mathbf{x} = (x_1, \ldots, x_n)$ representing the manufactured output has to be in the set of vectors satisfying all the linear inequalities. This set is called the ''feasible set'' for the problem. When $n = 2$, it is a subset of the plane and is generally a polygon. When $n = 3$, it is a volume of space bounded by planar surfaces, like a cut diamond.

Maximum revenue generally corresponds to a choice of output vector $\mathbf{x}$ at an extremity of the feasible set, in fact a vertex of the polygon, diamond, or whatever. One would expect, therefore, to solve linear programming problems very simply: just locate the vertices of the feasible set and compare the values of the revenue function at these points. But the method is not practical except for the simplest problems. Indeed, a cut diamond has far more vertices than plane surfaces; and the more substantial linear programming problems have feasible sets with millions of vertices!

Fortunately, a much better approach has been developed. In the *simplex method* of section 11.1, we content ourselves with locating just one vertex, and then we hop from vertex to vertex in such a way as to increase the revenue function at each step. It is possible to select successive vertices by a purely algebraic procedure, so that the potentially daunting geometry of the feasible set need not be visualized. Furthermore, there is a scheme for doing the necessary calculations in the form of row operations on a specially designed matrix called a "tableau."

The proof that the simplex method really produces the best, or "optimal," solution to a programming problem is developed in section 11.2. The reasoning is indirect. Linear programming problems are found to come in pairs, each consisting of a "primal" and a "dual" problem which, in the economic interpretation, correspond to opposing market forces. The optimal solutions of these problems turn out to be linked by the fundamental "Strong Duality Theorem," whose proof falls out easily when elementary matrix methodology is adapted to the tableau format.

11.1 The Simplex Method

As so often in linear algebra, the concepts of linear programming are best introduced through a typical example. We begin with a simple economic problem whose mathematical content is a standard "primal problem." We present a geometric analysis of the problem, which serves as a vehicle for an intuitive understanding of the algebraic simplex method that we subsequently develop. It should be kept in mind, however, that the latter procedure is ultimately shown to give correct results quite independently of the geometric interpretation.

We emphasize the use of the tableau format for numerical calculation in the simplex method. This tidy and reliable method should be used for all concrete examples.

Example 1   A coffee broker has on hand two kinds of coffee beans: 6,000 lb of Java beans and 4,000 lb Colombian beans. Wholesalers buy coffee from him in two blends:

(i) Mild Blend, containing $\frac{3}{4}$ Java and $\frac{1}{4}$ Colombian beans.
(ii) Strong Blend, containing $\frac{1}{2}$ Java and $\frac{1}{2}$ Colombian beans.

The prices paid to the broker are \$3 and \$4 per pound for the Mild and Strong Blends respectively.

Suppose the broker has to raise as much cash as possible, immediately. How many pounds of each blend should he ship to maximize his revenue?

■ **Solution**   (first part) Let x_1, x_2 represent the weights (in thousands of pounds) shipped of the Mild and Strong Blends respectively. Since the total quantities of Java and Colombian beans used cannot exceed the existing inventories, we have some limits on the size of the shipment:

$$\text{Java beans used} = \left(\tfrac{3}{4}\right)x_1 + \left(\tfrac{1}{2}\right)x_2 \leq 6 \text{ thousand pounds.}$$
$$\text{Colombian beans used} = \left(\tfrac{1}{4}\right)x_1 + \left(\tfrac{1}{2}\right)x_2 \leq 4 \text{ thousand pounds.}$$

Total revenue, the quantity to be maximized, is

$f(x_1, x_2) = 3x_1 + 4x_2$ thousands of dollars.

Our economic optimization problem has now been reduced to the following mathematical problem:

Maximize the quantity $f(x_1, x_2) = 3x_1 + 4x_2$,
subject to the conditions

$$\begin{cases} 3x_1 + 2x_2 \leq 24 \\ x_1 + 2x_2 \leq 16 \\ x_1 \geq 0 \\ x_1 \geq 0 \end{cases}$$

The last two inequalities simply affirm that x_1 and x_2, as weights of goods shipped, cannot be negative.

The *feasible points*, or *feasible vectors*, for the problem are those points (x_1, x_2) in R^2 that satisfy all the given linear inequalities. The collection of all feasible points is called the *feasible set*.

To sketch the feasible set for the present problem, note that the points satisfying the inequality $x_2 \geq 0$ are those above the line $x_2 = 0$. Similarly, the points satisfying the inequality $x_1 + 2x_2 \leq 16$ are those below the line $x_1 + 2x_2 = 16$, etc. Evidently the feasible set consists of those points that are on the correct side of each of following lines:

$$\begin{aligned}
\text{below line } l_1 &: \quad 3x_1 + 2x_2 = 24 \\
\text{below line } l_2 &: \quad x_1 + 2x_2 = 16 \\
\text{above } x_1\text{-axis} &: \quad\quad\quad\; x_2 = 0 \\
\text{to right of } x_2\text{-axis} &: \quad\quad\quad\; x_1 = 0 \;\;.
\end{aligned}$$

The feasible set is shaded in the Figure 11.1.

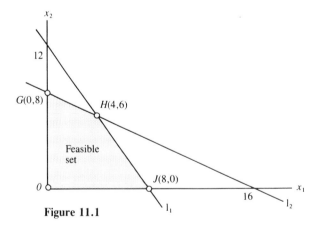

Figure 11.1

The objective of the problem is to maximize the function $f(x_1, x_2)$, subject to the point (x_1, x_2) being in the feasible set. For this reason, $f(x_1, x_2)$ is called the *objective function* of the problem, and a point (x_1, x_2) in the feasible set at which maximization of f occurs is called an *optimal solution*.

Our objective function is $f(x_1,x_2) = 3x_1 + 4x_2$, and it assumes the fixed value k at all points on the line $3x_1 + 4x_2 = k$. These lines are all parallel, having the same slope $-\frac{3}{4}$, and we sketch a few of them in Figure 11.2.

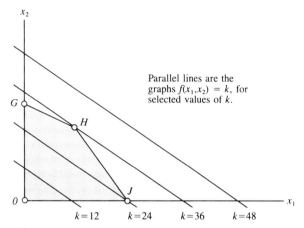

Figure 11.2

We can see that the line $3x_1 + 4x_2 = k$ steadily shifts upward as the parameter k increases from 12 through 24, 36, 48, etc. The line cuts through the feasible set until k reaches the value 36, at which stage there is just one point of intersection of line and set, namely the vertex $H(4,6)$ of the feasible set.

We can conclude, if we accept the geometric reasoning, that the maximum value of the objective function, taken over all points of the feasible set, is $f(4,6) = 36$ and that the optimal solution is the point $(x_1,x_2) = (4,6)$. In other words, the coffee broker can raise a maximum of \$36,000, by selling 4,000 lb of Mild Blend and 6,000 lb of Strong Blend.

Primal Problems

In the theory of linear programming, the foregoing example falls into the important category of *primal problems*. In general, these problems take the following form.

Primal Problem: Maximize the linear function of n variables

$$f(x_1, \ldots, x_n) = c_1x_1 + \cdots + c_nx_n \quad,$$

subject to the $m+n$ conditions

$$\begin{cases} a_{11}x_1 + \cdots + a_{1n}x_n \le b_1 \\ \quad \cdots\cdots\cdots\cdots\cdots\cdots\cdots \\ a_{m1}x_1 + \cdots + a_{mn}x_n \le b_m \end{cases}$$

and $\quad x_1 \ge 0, \ldots, x_n \ge 0 \quad.$

We can reformulate primal problems in matrix-vector notation by introducing the following notational convention. For vectors $\mathbf{x}, \mathbf{y}$ in R^n, $\mathbf{x} \leq \mathbf{y}$ means $x_i \leq y_i$ for $i = 1, \ldots, n$. In other words, vector inequalities are taken componentwise. Then we have the following concise statement.

Primal Problem: Maximize the function

$$f(\mathbf{x}) = \mathbf{c}^t \mathbf{x} \quad,$$

subject to the conditions

$$\begin{cases} A\mathbf{x} \leq \mathbf{b} \\ \quad \mathbf{x} \geq \mathbf{0} \quad, \end{cases}$$

where $\mathbf{c}, \mathbf{x}$ are in R^n, $\mathbf{b}$ is in R^m, and A is an $m \times n$ matrix.

In example 1, the data specifying the primal problem in matrix-vector form would have been

$$\mathbf{c}^t = \begin{bmatrix} 3 & 4 \end{bmatrix} \quad, \quad \mathbf{b} = \begin{bmatrix} 6 \\ 4 \end{bmatrix} \quad, \quad A = \begin{bmatrix} \frac{3}{4} & \frac{1}{2} \\ \frac{1}{4} & \frac{1}{2} \end{bmatrix} \quad.$$

Slack Variables

We want to develop a method of solving primal problems in general. One of our difficulties is that linear inequalities are harder to deal with than linear equations. To ease this problem, we introduce additional variables that "take up the slack" in the given inequalities, sharpening them to equalities. Thus the *slack variables*, z_i, for $i = 1, \ldots, m$ are defined by the equations

$$z_i = b_i - (a_{i1}x_1 + \cdots + a_{in}x_n) \quad,$$

or, in vector notation,

$$\mathbf{z} = \mathbf{b} - A\mathbf{x} \quad.$$

The inequalities $A\mathbf{x} \leq \mathbf{b}$ are then simplified to the inequalities $z_i \geq 0$ for $i = 1, \ldots, m$, or, in short, $\mathbf{z} \geq \mathbf{0}$.

This gives us a further restatement of the primal problem.

Primal Problem: Maximize $\qquad f(\mathbf{x}) = \mathbf{c}^t \mathbf{x} \quad,$

subject to the conditions $\begin{cases} A\mathbf{x} + \mathbf{z} = \mathbf{b} \\ \qquad \mathbf{x} \geq \mathbf{0} \\ \qquad \mathbf{z} \geq \mathbf{0} \quad. \end{cases}$

Thus the points (x_1, x_2) of the feasible set for example 1 are those that satisfy conditions of the form

$$\begin{cases} \left(\frac{3}{4}\right)x_1 + \left(\frac{1}{2}\right)x_2 + z_1 = 6 \\ \left(\frac{1}{4}\right)x_1 + \left(\frac{1}{2}\right)x_2 + z_2 = 4 \\ x_1 \geq 0, \, x_2 \geq 0, \, z_1 \geq 0, \, z_2 \geq 0 \quad. \end{cases}$$

Vertices

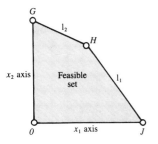

Figure 11.3

At every boundary point of the feasible set for example 1, at least one of the four variables x_1, x_2, z_1, z_2 vanishes. In fact, sides OG and OJ of the feasible set are the x_2 and x_1 axes (see Figure 11.3), on which x_1 and x_2 respectively vanish; while sides JH and GH of the feasible set are lines l_1 and l_2, on which z_1 and z_2 respectively vanish.

Further, since each of the vertices O, G, H, J of the feasible set is the intersection point of two of the four boundary sides, at each vertex two of the four variables must vanish. Thus at J, z_1 and x_2 vanish; at H, z_1 and z_2 vanish, etc.

Note, however, that the converse does not hold. In particular, the point where $x_2 = 0$ and $z_2 = 0$ is not a vertex of the feasible set of example 1. In fact, it is the point of intersection of the x_1-axis and line l_2, which is easily computed to be the point $(16,0)$, well outside the feasible set!

Now, our geometric solution of example 1 showed that, for optimization problems in the plane, i.e., the case when $n=2$ and $\mathbf{x} = (x_1, x_2)$, the feasible set is a polygon and one of its vertices is an optimal solution. In other words, some point of the feasible set where two of the variables $x_1, x_2, z_1,$ $\ldots, z_m$ vanish provides an optimal solution.

Similarly, we might guess that, for a primal problem with $n+m$ variables $x_1, \ldots, x_n, z_1, \ldots, z_m$, some point of the feasible set where n of these variables vanish provides an optimal solution. Let us call the points of the feasible set where n variables vanish, *vertices*.

If our guess is correct, an optimal solution always occurs at a vertex, and we can obtain an optimal solution by comparing the values of the objective function at all the vertices. This method is, however, impractical in general, because when there are more than a few variables in the problem, the number of vertices usually mounts into the thousands or millions.

The Simplex Method

Fortunately, there is a much more sensible procedure, called the *simplex method*. Rather than trying to compute the components of all the vertices, we start by locating just *one* vertex. Then we try to hop from vertex to vertex in such a way as to increase the value of the objective function at each step. Let us see how this can be done.

First we place the $m \times n$ matrix A and the $m \times m$ identity matrix I side by side as the blocks of the $m \times (n+m)$ matrix $[A|I]$. Then the equation $A\mathbf{x} + \mathbf{z} = \mathbf{b}$ can be rewritten as $A\mathbf{x} + I\mathbf{z} = \mathbf{b}$ or

$$[A|I] \begin{bmatrix} \mathbf{x} \\ \hline \mathbf{z} \end{bmatrix} = \mathbf{b} \quad .$$

For example, when the dimensions are $n = m = 2$, this system has the form:

$$\begin{bmatrix} a_{11} & a_{12} & | & 1 & 0 \\ a_{21} & a_{22} & | & 0 & 1 \end{bmatrix} \begin{bmatrix} x_1 \\ x_2 \\ z_1 \\ z_2 \end{bmatrix} = \begin{bmatrix} b_1 \\ b_2 \end{bmatrix} \; .$$

Observe that, because columns 3 and 4 of the matrix are the standard basis vectors $\mathbf{e}_1$ and $\mathbf{e}_2$, we immediately obtain the solution $x_1 = x_2 = 0$, $z_1 = b_1$, $z_2 = b_2$; or $\mathbf{x} = \mathbf{0}$, $\mathbf{z} = \mathbf{b}$. Assuming that $\mathbf{b} \geq \mathbf{0}$, this solution of course satisfies $\mathbf{x} \geq \mathbf{0}$, $\mathbf{z} \geq \mathbf{0}$ and is in the feasible set. We then have a first vertex, $(\mathbf{x},\mathbf{z}) = (\mathbf{0},\mathbf{b})$.

Pivot Operations

Next, suppose we do row operations on the augmented matrix $[A|I|\mathbf{b}]$ of the system, so as to obtain an equivalent system in which $\mathbf{e}_1$ and $\mathbf{e}_2$ still occur as two of the columns. For example, we can turn column 2 into $\mathbf{e}_2$ by changing the entry a_{22} into a 1 and then using it to eliminate in the column. This is just like the use of corners in matrix reduction, except that we are not selecting the corner entry by the old rules. We therefore call the entry a_{22} that is used in the elimination step a *pivot entry*, instead of a corner.

We obtain, by familiar calculations, a new augmented matrix of the form

$$[A'|B'|b'] = \begin{bmatrix} a'_{11} & 0 & | & 1 & b'_{12} & | & b'_1 \\ a'_{12} & 1 & | & 0 & b'_{22} & | & b'_2 \end{bmatrix} \; ,$$

and the corresponding form of the system is

$$\begin{bmatrix} a'_{11} & 0 & | & 1 & b'_{12} \\ a'_{12} & 1 & | & 0 & b'_{22} \end{bmatrix} \begin{bmatrix} x_1 \\ x_2 \\ z_1 \\ z_2 \end{bmatrix} = \begin{bmatrix} b'_1 \\ b'_2 \end{bmatrix} \; .$$

By inspection, a solution is $x_1 = 0$, $x_2 = b'_2$, $z_1 = b'_1$, $z_2 = 0$. Provided $\mathbf{b}' = (b'_1, b'_2)$ also satisfies $\mathbf{b}' \geq \mathbf{0}$, this gives us a new vertex $(0, b'_2, b'_1, 0)$.

Let us illustrate this pivoting procedure by applying it to the system in example 1. There we have

$$[A|I|\mathbf{b}] = \begin{bmatrix} \frac{3}{4} & \frac{1}{2} & | & 1 & 0 & | & 6 \\ \frac{1}{4} & \frac{1}{2} & | & 0 & 1 & | & 4 \end{bmatrix} \; ,$$

and, since $\mathbf{b} = (6,4) \geq \mathbf{0}$, we obtain a first vertex $(0,0,6,4)$.

The pivot operation on $a_{22} = \frac{1}{2}$ requires two row operations:

$$2R_2 \quad \begin{bmatrix} \frac{3}{4} & \frac{1}{2} & | & 1 & 0 & | & 6 \\ \frac{1}{2} & 1 & | & 0 & 2 & | & 8 \end{bmatrix}$$

and $\quad R_1 - \left(\frac{1}{2}\right)R_2 \quad \begin{bmatrix} \frac{1}{2} & 0 & | & 1 & -1 & | & 2 \\ \frac{1}{2} & 1 & | & 0 & 1 & | & 8 \end{bmatrix} \; .$

Since $\mathbf{b}' = (2,8) \geq \mathbf{0}$, we obtain a second vertex $(0,8,2,0)$.

By scanning only the **x**-components of the two vertices just computed, we recognize them as the previously plotted vertices $O(0,0)$ and $G(0,8)$ of the feasible set in the plane. So the pivot operation took us from O to G.

The Theta Ratios of the Pivot Column

In the above example, we were lucky that the passage from $[A|I|\mathbf{b}]$ to $[A'|B'|\mathbf{b}']$ by means of the pivot operation on a_{22} produced a second vertex. In general, this happens only when $\mathbf{b}' \geq \mathbf{0}$.

Clearly, we need a systematic method of selecting the pivot entry a_{ij} that ensures that if $\mathbf{b} \geq \mathbf{0}$, then $\mathbf{b}' \geq \mathbf{0}$ as well. Now, the pivot operation on a_{ij} consists of row operations that we indicate schematically as follows, writing only those entries that are in row i or k and also in column j or the augmented column:

$$[A|I|\mathbf{b}] = \begin{bmatrix} \cdots\cdots\cdots & & \cdots \\ \cdots a_{kj} \cdots & & b_k \\ \cdots\cdots\cdots & & \cdots \\ \cdots a_{ij} \cdots & & b_i \\ \cdots\cdots\cdots & & \cdots \end{bmatrix} \left(\frac{1}{a_{ij}}\right) R_i \begin{bmatrix} \cdots\cdots\cdots & & \cdots \\ \cdots a_{kj} \cdots & & b_k \\ \cdots\cdots\cdots & & \\ \cdots 1 \cdots & & \dfrac{b_i}{a_{ij}} \\ \cdots\cdots\cdots & & \cdots \end{bmatrix}$$

$$R_k - a_{kj} R_i \begin{bmatrix} \cdots\cdots\cdots & & \cdots \\ \cdots 0 \cdots & & \dfrac{b_k - a_{kj} b_i}{a_{ij}} \\ \cdots\cdots\cdots & & \\ \cdots 1 \cdots & & \dfrac{b_i}{a_{ij}} \\ \cdots\cdots\cdots & & \cdots \end{bmatrix} = \begin{bmatrix} \cdots\cdots\cdots & & \cdots \\ \cdots 0 \cdots & & b'_k \\ \cdots\cdots\cdots & & \cdots \\ \cdots 1 \cdots & & b'_i \\ \cdots\cdots\cdots & & \cdots \end{bmatrix}.$$

Thus $\mathbf{b}' \geq \mathbf{0}$ if the following two conditions hold:

(i) $\dfrac{b'_i = b_i}{a_{ij}} \geq 0$

(ii) $b'_k = b_k - \dfrac{a_{kj} b_i}{a_{ij}} \geq 0$, for $k \neq i$.

In order to simplify these somewhat complicated conditions, we define the *θ-ratios of column j* to be the numbers

$$\theta_k = \frac{b_k}{a_{kj}} \quad , \quad \text{for } k = 1, \ldots, m$$

(except that when $a_{kj} = 0$, θ_k is said to be undefined).

The conditions for $\mathbf{b}' \geq \mathbf{0}$ then assume the form:

(i) $\theta_i \geq 0$

(ii) $b_k - a_{kj}\theta_i \geq 0$, for $k \neq i$.

We now state that $\mathbf{b}' \geq \mathbf{0}$ provided that the pivot entry a_{ij} is chosen in such a way that θ_i *is the smallest nonnegative θ-ratio of column j*.

To see that this statement is correct, observe first that, since by assumption $\theta_i \geq 0$ and $b_k \geq 0$, therefore all terms on the left sides of (i) and (ii) are positive, unless $a_{kj} > 0$. But in the remaining case, $a_{kj} > 0$, we have $\theta_k = \dfrac{b_k}{a_{kj}} \geq 0$. Then condition (ii) can be rewritten as

$$a_{kj}\left(\frac{b_k}{a_{kj}} - \theta_i\right) \geq 0 \quad \text{or} \quad \frac{a_{kj}}{(\theta_k - \theta_i)} \geq 0 \quad,$$

which is satisfied since, by definition of θ_i, we have $\theta_k - \theta_i \geq 0$.

To illustrate the use of θ-ratios in selecting a pivot, let us review our earlier example of a pivot operation. We wanted to select a pivot from column $j = 2$. Now the θ-ratios for column 2 are

$$\theta_1 = \frac{b_1}{a_{12}} = \frac{6}{\frac{1}{2}} = 12$$

$$\theta_2 = \frac{b_2}{a_{22}} = \frac{4}{\frac{1}{2}} = 8 \quad.$$

Since the smaller θ-ratio is θ_2, we must choose $i = 2$, and pivot on entry $a_{ij} = a_{22}$. As it happened, we did choose a_{22} as our pivot in the illustrative calculation, and so we got a second vertex.

Selection of the Pivot Column

In a primal problem, the objective function $f(\mathbf{x})$ is expressed in terms of the x-variables alone. By contrast, our representation of vertices in the form $(x_1, \ldots, x_n, z_1, \ldots, z_m)$ puts the introduced slack variables $\mathbf{z}$ on an equal footing with the original $\mathbf{x}$-variables. In order to extend this perspective to the objective function, we must write it as

$$f(\mathbf{x},\mathbf{z}) = c_1 x_1 + \cdots + c_n x_n + 0 z_1 + \cdots + 0 z_m$$

or $\quad f(\mathbf{x},\mathbf{z}) = \mathbf{c}\cdot\mathbf{x} + \mathbf{0}\cdot\mathbf{z} \quad.$

The value of the objective function at the first vertex, $(\mathbf{x},\mathbf{z}) = (\mathbf{0},\mathbf{b})$, can be seen by inspection to be $\mathbf{c}\cdot\mathbf{0} + \mathbf{0}\cdot\mathbf{b} = 0$. The simplicity of this calculation is a consequence of the obvious and important *symmetry between vertex and objective function*:

every variable either   (i) equals zero at the vertex

or   (ii) has zero coefficient in the objective function.

But if we move to a new vertex by pivoting on a_{ij}, then x_j will change from 0 at the old vertex to b_i' at the new vertex, and correspondingly the objective function f will change from 0 to $c_j b_i'$. Since our intention in pivoting is to increase f, it is clear that the pivot column j must be chosen so that $c_j > 0$.

In example 1, the symmetric form of the objective function is initially

$$f(x_1, x_2, z_1, z_2) = 3x_1 + 4x_2 + 0z_1 + 0z_2 \quad.$$

Of the coefficients, $c_1 = 3$ and $c_2 = 4$ are positive, and so either of columns 1 and 2 could serve as pivot column. It is customary to determine the largest of the positive coefficients, say c_j, and then select column j as pivot column. In our example, c_2 is larger than c_1, so column 2 is the correct choice of pivot column. This is the choice we actually made, and as a result we increased the value of f from 0 to

$$c_j b_i' = c_2 b_2' = (4)(8) = 32 \ \ .$$

Restoration of Symmetry

At the new vertex reached by pivoting on a_{ij}, the symmetry between old vertex and objective function is lost. In fact, the variable x_j becomes positive at the new vertex, and yet still has positive coefficient c_j in the objective function. To restore symmetry, we must rewrite the objective function in such a way that the coefficient of x_j becomes zero. This is easy, because the ith row of the augmented matrix $[A|I|\mathbf{b}]$ represents another equation also containing x_j :

$$a_{i1}x_1 + \cdots + a_{ij}x_j + \cdots + a_{in}x_n + z_i - b_i = 0 \ \ .$$

So if we subtract $\dfrac{c_j}{a_{ij}}$ times this zero quantity from the formula

$$f(\mathbf{x},\mathbf{z}) = c_1 x_1 + \cdots + c_j x_j + \cdots + c_n x_n + 0 z_1 + \cdots + 0 z_i + \cdots + 0 z_n \ \ ,$$

we obtain a new formula for the objective function of the form

$$f'(\mathbf{x},\mathbf{z}) = c_1' x_1 + \cdots + 0 x_j + \cdots + c_n' x_n + 0 z_1 + \cdots$$
$$+ s_i' z_i + \cdots + 0 z_m + c_j b_i' \ \ .$$

In our example, the pivot entry was a_{22}. So we rewrite row $i = 2$ of the augmented matrix as the equation

$$\left(\tfrac{1}{4}\right)x_1 + \left(\tfrac{1}{2}\right)x_2 + z_2 - 4 = 0 \ \ ,$$

multiply this equation by $\dfrac{c_2}{a_{22}} = \dfrac{4}{\left(\tfrac{1}{2}\right)} = 8 \ \ ,$

and subtract the result from

$$f(\mathbf{x},\mathbf{z}) = 3x_1 + 4x_2 + 0z_1 + 0z_2 \ \ ,$$

obtaining

$$f'(\mathbf{x},\mathbf{z}) = x_1 + 0x_2 + 0z_1 - 8z_2 + 32 \ \ .$$

By comparing the coefficients of the new objective function f',

which are $\quad (c_1', c_2', s_1', s_2') = (1, 0, 0, -8) \ \ ,$

with the components of the new vertex,

which are $\quad (x_1, x_2, z_1, z_2) = (0, 8, 2, 0) \ \ ,$

we confirm that symmetry has been restored and that the value of f' at the new vertex is therefore given by its constant term, which is 32.

The Tableau

Let us now describe the construction of f' in matrix-vector terms. First of all, f and f' have the forms

$$f = \mathbf{c} \cdot \mathbf{x} + \mathbf{0} \cdot \mathbf{z} + 0$$
$$f' = \mathbf{c}' \cdot \mathbf{x} + \mathbf{s}' \cdot \mathbf{z} + p' \quad,$$

where $\mathbf{c}$ and $\mathbf{c}'$ are in R^n, $\mathbf{s}$ and $\mathbf{s}'$ are in R^m, and the scalars 0 and p' give the values of the objective function at the old and new vertices respectively.

Furthermore, the calculation of f' from f clearly has the character of a row operation involving row i of the augmented matrix $[A|I|\mathbf{b}]$. In fact we can combine the calculation of f' with the pivot operation itself as follows.

Starting with the $m \times (n+m+1)$ matrix $[A|I|\mathbf{b}]$, we add an extra row by placing $\mathbf{c}^t$ below A, $\mathbf{0}^t$ below I, and a further 0 below $\mathbf{b}$. We then have an $(n+1) \times (n+m+1)$ matrix

$$T = \left[\begin{array}{c|c|c} A & I & \mathbf{b} \\ \hline \mathbf{c}^t & \mathbf{0}^t & 0 \end{array} \right] \quad,$$

which is called the *initial tableau* of the primal problem.

In the tableau, c_j is directly below a_{ij}. The pivot a_{ij} can therefore be used to eliminate c_j through the row operation $R_{m+1} - \left(\dfrac{c_j}{a_{ij}} \right) R_i$, exactly matching the subtraction of equations that changed f to f'.

We conclude that the pivot operation on a_{ij}, extended to all $n+1$ rows of the tableau T, produces the tableau

$$T' = \left[\begin{array}{c|c|c} A' & B' & \mathbf{b}' \\ \hline (\mathbf{c}')^t & (\mathbf{s}')^t & -p' \end{array} \right] \quad.$$

Of course the row operation automatically extends to the augmented column, changing the original entry 0 in the lower right corner to

$$0 - \left(\dfrac{c_j}{a_{ij}} \right) b_j \quad \text{or} \quad -c_j b_i' \quad \text{or} \quad -p' \quad.$$

This is very convenient: the value of the objective function appears (with a minus sign attached to it) in the lower right corner of the new tableau!

Let us do the first pivot operation for example 1 in tableau form:

$$T = \left[\begin{array}{cc|cc|c} \frac{3}{4} & \frac{1}{2} & 1 & 0 & 6 \\ \frac{1}{4} & \frac{1}{2} & 0 & 1 & 4 \\ \hline 3 & 4 & 0 & 0 & 0 \end{array} \right] \quad.$$

The pivot on a_{22} produces

$$T' = \left[\begin{array}{cc|cc|c} \frac{1}{2} & 0 & 1 & -1 & 2 \\ \frac{1}{2} & 1 & 0 & 2 & 8 \\ \hline 1 & 0 & 0 & -8 & -32 \end{array} \right] \quad.$$

From the last row of the new tableau we can read off the data

$c_1' = 1, c_2' = 0, s_1' = 0, s_2' = -8, -p' = -32$, which are all in agreement with the previous calculations.

The Pivoting Process

Starting from the data in the tableau T', we can repeat the pivoting procedure to increase the objective function further:

Step 1 *Selection of pivot column.* Find the largest positive coefficient in the list $\mathbf{c'}$, $\mathbf{s'}$ by scanning the extra row of T'. The pivot column is the one containing this coefficient. Call it column j.

In our example, there is only one positive coefficient, $c_1' = 1$, so the pivot column is column $j = 1$.

Step 2 *Selection of pivot row.* Compute the θ-ratios for the pivot column. If $\theta_i = \dfrac{b_i'}{a_{ij}'}$ is the smallest nonnegative one, then row i is the pivot row.

In our example, we compute the θ-ratios for column $j = 1$:

$$\theta_1 = \frac{b_1'}{a_{11}'} = \frac{2}{\left(\frac{1}{2}\right)} = 4$$

$$\theta_2 = \frac{b_2'}{a_{21}'} = \frac{8}{\left(\frac{1}{2}\right)} = 16 \quad .$$

The smallest nonnegative θ-ratio is $\theta_1 = 4$; therefore the pivot row is row $i = 1$.

Step 3 *Implementation of pivot operation.* Pivot on entry a_{ij}'. That is, apply to the tableau T' the row operations that change entry a_{ij}' to 1 and the other entries in column j to 0.

In our example, the pivot entry is $a_{ij}' = a_{11}'$, and the result of the pivot operation is the tableau

$$T'' = \begin{bmatrix} 1 & 0 & 2 & -2 & 4 \\ 0 & 1 & -1 & 3 & 6 \\ 0 & 0 & -2 & -6 & -36 \end{bmatrix} \quad .$$

The new value of the objective function is therefore $p'' = -(-36) = 36$.

Note also that since the standard vectors $\mathbf{e}_1$, $\mathbf{e}_2$ are in columns 1 and 2, respectively, of T'', the third vertex is

$$(x_1'', x_2'', z_1'', z_2'') = (4,6,0,0) \quad .$$

In terms of the feasible set in the plane, the third vertex is $(x_1'', x_2'') = (4,6)$, i.e., vertex $H(4,6)$.

We can continue to increase the value of the objective function by repeated pivoting until, after say k pivots, one of the "features" listed below occurs in the current tableau $T^{(k)}$.

Feature 1 The extra row of $T^{(k)}$ contains no positive entries.

In this case, no further improvement of the objective function by pivoting is possible, and $T^{(k)}$ is called a *final tableau* for the primal problem.

  We will prove later, using the duality theory of section 11.2, that once a final tableau is reached, the vertex associated to it gives an optimal solution of the primal problem.

Feature 2 In the pivot column of $T^{(k)}$, all θ-ratios are negative or undefined.

In this case it can easily be proved (see exercises) that the objective function has no maximum value—it can be arbitrarily large. We give an example of this case below.

Feature 3 In the pivot column of $T^{(k)}$, a θ-ratio $\theta_i = 0$ occurs.

In this case, there is no improvement in the objective function on passage to the next tableau. In fact, the formula for the improvement, derived earlier, is

$$\left(\frac{c_j}{a_{ij}}\right) b_i = c_j \left(\frac{b_i}{a_{ij}}\right) = c_j\theta_i = 0 \quad .$$

This case can only occur when the geometry of the feasible set has atypical features. In particular, there must be vertices having more than n zero components (see exercises). A problem in this category is said to be *degenerate*, and the simplex method may fail to locate an optimal solution for it.

  For example, further pivoting in this situation may only cause us to shuttle endlessly among a set of vertices at which the values of the objective function are all the same. This process is called *cycling*. It seldom occurs in practice but is a theoretical irritant; and there are various methods, beyond the scope of this book, for dealing with it.

In example 1, "feature 1" appears after two pivots, in the tableau T''. In fact, the extra row of T'' gives the list of coefficients

$$(c_1'', c_2'', s_1'', s_2'') = (0, 0, -2, -6) \quad .$$

No coefficient is positive, and T'' is a final tableau. Accordingly, the optimal solution is the vertex associated to T'', namely $(4,6,0,0)$ or $H(4,6)$.

Examples of the Simplex Method

Example 2   Suppose the plight of the coffee broker encountered in example 1 is compounded by an import quota system that allows him to ship only 9,000 lb of coffee at the present time. How much of each blend of coffee should he ship to maximize his revenue in this case?

■ **Solution**   The additional constraint is that $x_1 + x_2$, the sum of the weights of the two blends shipped, cannot exceed 9 (thousands of pounds). So the feasible set is restricted by the further inequality

$$x_1 + x_2 \leq 9 \quad .$$

In particular, the optimal solution to the original problem, $x_1 = 4, x_2 = 6$, is no longer in the feasible set. Indeed the new inequality implies that the broker will not be able to use up his entire inventory of coffee, which totals 10 thousand pounds, and it is not at all obvious what his best strategy is.

Let us construct the initial tableau T for the new problem. The feasible set is now given by the inequalities

$$\left\{\begin{array}{l} \left(\tfrac{3}{4}\right)x_1 + \left(\tfrac{1}{2}\right)x_2 \geq 6 \quad \text{or} \\ \left(\tfrac{1}{4}\right)x_1 + \left(\tfrac{1}{2}\right)x_2 \geq 4 \\ \quad x_1 + \quad x_2 \geq 9 \\ \quad \mathbf{x} \geq \mathbf{0} \end{array}\right. \qquad \left\{\begin{array}{l} \left(\tfrac{3}{4}\right)x_1 + \left(\tfrac{1}{2}\right)x_2 + z_1 = 6 \\ \left(\tfrac{1}{4}\right)x_1 + \left(\tfrac{1}{2}\right)x_2 + z_2 = 4 \\ \quad x_1 + \quad x_2 + z_3 = 9 \\ \quad \mathbf{x} \geq \mathbf{0}, \mathbf{z} \geq \mathbf{0} \quad . \end{array}\right.$$

The objective function, representing total revenue, is, as before,

$$f(x_1, x_2) = 3x_1 + 4x_2 \quad .$$

So the blocks of data to be placed in the tableau are

and $\quad A = \begin{bmatrix} \tfrac{3}{4} & \tfrac{1}{2} \\ \tfrac{1}{4} & \tfrac{1}{2} \\ 1 & 1 \end{bmatrix}$, $\quad \mathbf{b} = \begin{bmatrix} 6 \\ 4 \\ 9 \end{bmatrix}$, $\quad \mathbf{c}^t = [3 \quad 4]$,

and $\quad T = \begin{bmatrix} \tfrac{3}{4} & \tfrac{1}{2} & 1 & 0 & 0 & 6 \\ \tfrac{1}{4} & \tfrac{1}{2} & 0 & 1 & 0 & 4 \\ 1 & 1 & 0 & 0 & 1 & 9 \\ \hline 3 & 4 & 0 & 0 & 0 & 0 \end{bmatrix}$

$\theta_1 = \dfrac{6}{\left(\tfrac{1}{2}\right)} = 12$

$\theta_2 = \dfrac{4}{\left(\tfrac{1}{2}\right)} = 8$

$\theta_3 = \dfrac{9}{1} = 9 \quad .$

The largest positive entry in the extra row is $c_2 = 4$ (circled), so the pivot column is column $j = 2$. The smallest nonnegative θ-ratio for this column is $\theta_2 = 8$, so the pivot row is row $i = 2$. The first pivot is then $a_{22} = \tfrac{1}{2}$ (circled), and leads to the new tableau

$T' = \begin{bmatrix} \tfrac{1}{2} & 0 & 1 & -1 & 0 & 2 \\ \tfrac{1}{2} & 1 & 0 & 2 & 0 & 8 \\ \tfrac{1}{2} & 0 & 0 & -2 & 1 & 1 \\ \hline 1 & 0 & 0 & -8 & 0 & -32 \end{bmatrix}$

$\theta_1 = \dfrac{2}{\left(\tfrac{1}{2}\right)} = 4$

$\theta_2 = \dfrac{8}{\left(\tfrac{1}{2}\right)} = 16$

$\theta_3 = \dfrac{1}{\left(\tfrac{1}{2}\right)} = 2 \quad .$

Since the only positive entry in the extra row this time is $c_1' = 1$, the new pivot column is column $j = 1$. The smallest nonnegative θ-ratio for this column is $\theta_3 = 2$; therefore the new pivot row is row $i = 3$. The second pivot is then $a_{31} = \frac{1}{2}$ and leads to the third tableau

$$T'' = \begin{bmatrix} 0 & 0 & 1 & 1 & -1 & 1 \\ 0 & 1 & 0 & 4 & -1 & 7 \\ 1 & 0 & 0 & -4 & 2 & 2 \\ 0 & 0 & 0 & -4 & -2 & -34 \end{bmatrix} .$$

Now there are no positive entries in the extra row, and thus "feature 1" has occurred, and T'' is a final tableau. According to theory, an optimal solution is therefore given by the associated vertex

$$(x_1, x_2, z_1, z_2, z_3) = (2, 7, 1, 0, 0) \quad ,$$

or, in terms of the x-variables alone, the vertex $K(2,7)$. Furthermore, the maximum value of the objective function is obtained from the lower right entry of the final tableau T'' and is $p'' = -(-34) = 34$.

That is, the maximum revenue the broker can obtain is \$34,000, and he can obtain it by shipping 2,000 lb of the Mild Blend and 7,000 lb of the Strong Blend.

It is easily checked (see exercises) that under the optimal strategy, all the Colombian coffee is used, while 1,000 lb Java is left over. Of course the higher-priced Strong Blend calls for a larger proportion of Colombian coffee than the Mild Blend, and so it is plausible that the optimal strategy does not waste any of the Colombian.

Note, however, that one does not simply pack more and more Strong Blend until the Colombian is all used up. This would result in a shipment of only 8,000 lb of blended coffee and would waste 2,000 lb Java. Instead, the optimal shipment fully uses the 9,000 lb import quota, and it includes some Mild Blend to achieve this.

Example 3 Maximize $f(x_1, x_2) = 3x_1 + 2x_2$,

subject to the conditions
$$x_1 - x_2 \leq 1$$
$$-2x_1 + x_2 \leq 3$$
$$\mathbf{x} \geq \mathbf{0} \quad .$$

■ **Solution** This exercise is stated in the standard form of a primal problem, with $n = 2$ x-variables, $m = 2$ constraint inequalities; and we can write down at a glance the blocks of data needed for the initial tableau T:

$$A = \begin{bmatrix} 1 & -1 \\ -2 & 1 \end{bmatrix} \quad , \quad \mathbf{b} = \begin{bmatrix} 1 \\ 3 \end{bmatrix} \quad , \quad \mathbf{c}^t = \begin{bmatrix} 3 & 2 \end{bmatrix} \quad .$$

Then $T = \begin{bmatrix} A & I_2 & \mathbf{b} \\ \mathbf{c}^t & \mathbf{0} & 0 \end{bmatrix}$

$$= \begin{bmatrix} ① & -1 & 1 & 0 & 1 \\ -2 & 1 & 0 & 1 & 3 \\ ③ & 2 & 0 & 0 & 0 \end{bmatrix} \quad .$$

$\theta_1 = 1/1 \quad\ = \quad 1$
$\theta_2 = 3/(-2) = -3/2$

The largest entry in the extra row is $c_1 = 3$, and so the pivot column is column $j = 1$. Since the only nonnegative θ-ratio is $\theta_1 = 1$, the pivot row is row $i = 1$. Therefore we pivot on $a_{11} = 1$ and obtain

$$T' = \begin{bmatrix} 1 & -1 & 1 & 0 & 1 \\ 0 & -1 & 2 & 1 & 5 \\ \hline 0 & ⑤ & -3 & 0 & -3 \end{bmatrix} \qquad \begin{aligned} \theta_1 &= \left(\frac{1}{-1}\right) = -1 \\ \theta_2 &= \left(\frac{5}{-1}\right) = -5 \end{aligned}$$

Since the only positive entry in the extra row is $c_2' = 5$, column 2 is the new pivot column. But the θ-ratios for column 2 are all negative; and so "feature 2" has occurred.

We conclude that there is no optimal solution; the objective function may be arbitrarily large within the feasible set. Indeed, we can see from Figure 11.4 that the feasible set (shaded) is unbounded and that the objective function will increase without limit as we move away from the origin within the feasible set.

Figure 11.4

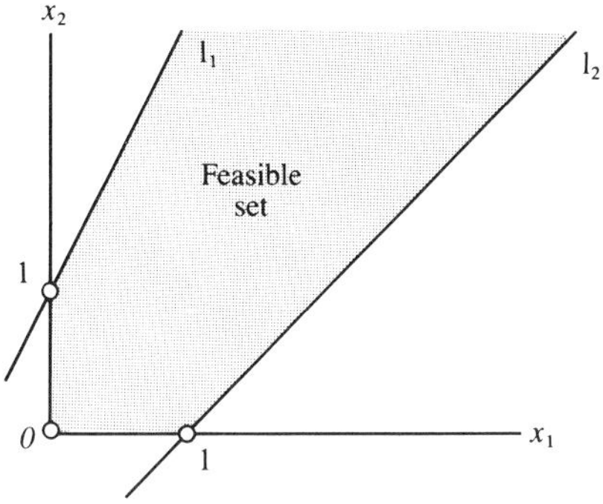

■ **Problem** In the coffee problem of example 1, how much would the price of the Strong Blend have to increase before more revenue could be obtained by packing only Strong Blend than by packing some of both blends?

■ **Solution** Suppose the price of the Strong Blend is increased from 4 to g dollars per pound. Then the objective function changes to

$$\begin{aligned} f(x_1, x_2) &= 3x_1 + gx_2 \\ &= \mathbf{c} \cdot \mathbf{x} \ , \qquad \text{where } \mathbf{c} = (3, g) \quad . \end{aligned}$$

The initial tableau is then modified to

$$T = \begin{bmatrix} \frac{3}{4} & \frac{1}{2} & 1 & 0 & 6 \\ \frac{1}{4} & ⓵ & 0 & 1 & 4 \\ \hline 3 & Ⓖ & 0 & 0 & 0 \end{bmatrix} \qquad \begin{aligned} \theta_1 &= \frac{6}{\left(\frac{1}{2}\right)} = 12 \\ \theta_2 &= \frac{4}{\left(\frac{1}{2}\right)} = 8 \end{aligned} \quad .$$

Since, by assumption, $g \geq 4$, the largest entry in the extra row is $c_2 = g$. The smallest nonnegative θ-ratio for column 2 is $\theta_2 = 8$, and so we pivot on $a_{22} = \frac{1}{2}$, obtaining

$$T' = \begin{bmatrix} \frac{1}{2} & 0 & 1 & -1 & 2 \\ \frac{1}{2} & 1 & 0 & 2 & 8 \\ \hline \frac{6-g}{2} & 0 & 0 & -2g & -8g \end{bmatrix} \qquad \begin{aligned} \theta_1 &= \frac{2}{\left(\frac{1}{2}\right)} = 4 \\ \theta_2 &= \frac{8}{\left(\frac{1}{2}\right)} = 16 \end{aligned} \quad .$$

On inspection of the extra row of T', we see that there are no positive entries when $\frac{(6-g)}{2} < 0$, or $g > 6$.

In this case, T' is a final tableau, and the optimal solution is $(0,8,2,0)$, or $x_1 = 0$, $x_2 = 8$; i.e., vertex $G(0,8)$ is optimal.

That is, only Strong Blend is shipped when its price exceeds \$6 per pound.

In the opposite case, $6 > g > 4$, column 1 is the new pivot column, and the smallest nonnegative θ-ratio is $\theta_1 = 4$. The new pivot is $a_{11} = \frac{1}{2}$, and it produces

$$T'' = \begin{bmatrix} 1 & 0 & 2 & -2 & 4 \\ 0 & 1 & -1 & 3 & 6 \\ \hline 0 & 0 & g-6 & 6-3g & -6g-12 \end{bmatrix} \quad .$$

Now there are no positive entries in the last row, and T'' is a final tableau. The optimal solution is therefore $(4,6,0,0)$, or $x_1 = 4$, $x_2 = 6$.

We conclude that if Strong Blend is priced at between \$4 and \$6 per pound, the optimal solution still involves shipping both Mild and Strong Blends.

■ EXERCISES 11.1

1. For the linear programming problem:

$$\begin{cases} \text{Maximize} & f(x_1, x_2) = 3x_1 + 2x_2 \\ \text{subject to the conditions} & \begin{cases} x_1 + 3x_2 \leq 15 \\ x_1 + x_2 \leq 7 \\ \mathbf{x} \geq \mathbf{0} \end{cases} \end{cases} .$$

(a) Sketch the feasible set.
(b) Compute all vertices of the feasible set.
(c) Compute the value of the objective function at each vertex.
(d) Deduce, on geometric grounds, the optimal solution.

2. Suppose the optimization problem of the previous exercise is modified by the introduction of the additional condition

$$2x_1 + x_2 \leq 12 \quad .$$

Sketch the modified feasible set and determine the new optimal solution by the same geometric reasoning as before.

3. Solve the optimization problem of exercise 1 by the simplex method. Include (for later reference) the initial and final tableaux as part of your answer.

4. Solve the optimization problem of exercise 2 by the simplex method. Include initial and final tableaux as part of your answer.

5. In each of the following cases, maximize the given objective function f, subject to the condition $\mathbf{x} \geq \mathbf{0}$ combined with the additional stated conditions.

(a) $f(x_1, x_2) = 4x_1 + 5x_2$
$$\begin{cases} x_1 + 3x_2 \leq 9 \\ 2x_1 + x_2 \leq 8 \end{cases}$$

(b) $f(x_1, x_2) = 4x_1 + 6x_2$
$$\begin{cases} x_1 + 3x_2 \leq 5 \\ 5x_1 + 3x_2 \leq 13 \end{cases}$$

(c) $f(x_1, x_2, x_3) = 5x_1 + 4x_2 + 14x_3$
$$\begin{cases} x_1 + 2x_2 + 5x_3 \leq 5 \\ 2x_1 + x_2 + 2x_3 \leq 8 \end{cases}$$

(d) $f(x_1, x_2, x_3) = 10x_1 + 9x_2 + 12x_3$
$$\begin{cases} x_1 + 2x_2 + 2x_3 \leq 21 \\ 2x_1 + x_2 + 2x_3 \leq 20 \\ 2x_1 + 2x_2 + x_3 \leq 19 \end{cases}$$

6. (a) Sketch the feasible set for the coffee problem with import quotas (see example 2 of this section).
   (b) Compute the vertices of the feasible set.
   (c) Compute the value of the objective function at each vertex.
   (d) Deduce that the solution obtained earlier by the simplex method is correct.

7. In the coffee problem at the end of this section, the price of Colombian coffee was given the parametric value g. How low would this price have to be before the optimal solution would call for shipping only Mild Blend?

8. In the coffee problem at the end of this section, consider the critical parameter value $g = 6$. Show that the line $f(x_1, x_2) = 48$ coincides with the boundary segment GH of the feasible set. Deduce that the optimal solution is nonunique in this case.

9. Suppose tomatoes, corn, and lettuce are to be planted on a 50-acre farm. Suppose these crops respectively require one, two, and four tons of fertilizer per acre and bring in revenues of 4, 7, and 8 hundred dollars per acre planted. Suppose that 120 tons of fertilizer are available for the whole farm. How many acres should be planted with each crop in order to maximize revenue?

10. Suppose that in the preceding farming problem, the total amount of fertilizer available is only 80 tons. What is the optimal solution?

11. Suppose that in the preceding farming problem, the total amount of fertilizer available is represented as a parameter $g > 0$.

   (a) Show that when the parameter g is in the range $100 > g > 50$, the optimal solution is to plant $100 - g$ acres with tomatoes, $g - 50$ acres with corn, and 0 acres with lettuce.

   (b) Find similar formulas for the optimal solution when g is in the parameter ranges: $g > 200$; $200 > g > 100$; and $50 > g$.

12. Suppose that on the farm of exercise 9, only 160 days' labor are available and that the amounts of labor required for each acre of tomatoes, corn, and lettuce planted are one, four, and four days respectively. What is the optimal solution in this case?

■11.2 Duality Theory

Our computational technique, in the form of pivot operations on tableaux, is running well ahead of our theoretical understanding of the subject of linear programming. We now want to prove some of the fundamental theorems in the subject. This can be best done by an indirect strategy: instead of intensifying our study of the primal problem alone; we will compare it with a closely related problem called the *dual problem*.

Given an $m \times n$ matrix A and vectors **b** in R^m and **c** in R^n, the primal/dual pair of problems are as follows.

Primal problem (P)	Maximize	$f(\mathbf{x}) = \mathbf{c}^t\mathbf{x}$,
	subject to the conditions	$\begin{cases} A\mathbf{x} \leq \mathbf{b} \\ \quad \mathbf{x} \geq \mathbf{0} \end{cases}$.
Dual problem (D)	Minimize	$g(\mathbf{y}) = \mathbf{b}^t\mathbf{y}$,
	subject to the conditions	$\begin{cases} A^t\mathbf{y} \geq \mathbf{c} \\ \quad \mathbf{y} \geq \mathbf{0} \end{cases}$.

At first glance it seems implausible that these problems are closely related; even the dimensions of their vector variables **x** and **y** may be different, since they are in the spaces R^n and R^m respectively. Nevertheless, the objective functions f and g of the paired problems cramp each other in a reciprocal manner.

Lemma 11.1 (Weak duality) Let P and D be paired primal and dual problems. Suppose there exist vectors $\mathbf{x}$ and $\mathbf{y}$ in the feasible sets of problems P and D respectively. Then

(a) $f(\mathbf{x}) \leq g(\mathbf{y})$.

(b) if equality holds, i.e., $f(\mathbf{x}) = g(\mathbf{y})$, the vectors $\mathbf{x}$ and $\mathbf{y}$ are optimal solutions for problems P and D respectively.

■Proof (a) From the statement of the dual problem, we have the inequality of column vectors $\mathbf{c} \leq A^t\mathbf{y}$, or, on transposition, the inequality of row vectors

$$\mathbf{c}^t \leq \mathbf{y}^t A .$$

Now, a vector inequality remains valid if we take its dot product with a positive vector, such as $\mathbf{x}$, and so we obtain

$$\mathbf{c}^t\mathbf{x} \leq \mathbf{y}^t A\mathbf{x} .$$

Similarly, on taking the dot product of the inequality $A\mathbf{x} \leq \mathbf{b}$ with the positive vector $\mathbf{y}^t$, we have

$$\mathbf{y}^t A\mathbf{x} \leq \mathbf{y}^t\mathbf{b} = \mathbf{b}^t\mathbf{y} .$$

Combining these inequalities yields

$$\mathbf{c}^t\mathbf{x} \leq \mathbf{b}^t\mathbf{y} \quad \text{or} \quad f(\mathbf{x}) \leq g(\mathbf{y}) .$$

(b) Suppose $\mathbf{x}'$ is any vector in the feasible set of problem P. Then, by part (a),

$$f(\mathbf{x}') \leq g(\mathbf{y}) .$$

Or, since $g(\mathbf{y}) = f(\mathbf{x})$,

$$f(\mathbf{x}') \leq f(\mathbf{x}) .$$

That is, the objective function f for problem P assumes its maximum value at point $\mathbf{x}$. In other words, $\mathbf{x}$ is an optimal solution of the primal problem. Q.E.D.

COROLLARY 11.2

If the objective function $f(\mathbf{x})$ for a primal problem assumes arbitrarily large values (with $\mathbf{x}$ in the feasible set), then the feasible set for the dual problem is empty.

Similarly, if the objective function $g(\mathbf{y})$ for the dual problem assumes arbitrarily small values (with $\mathbf{y}$ in the feasible set), then the feasible set for the primal problem is empty.

Example 1   To see that a dual problem can have a practical interpretation related to that of the primal problem, let us return to example 1 of section 11.1 and develop the story of the coffee broker further.

Suppose, in fact, that a second buyer hears that the broker has 6,000 lb of Java beans and 4,000 lb of Colombian beans, and wants to get hold of this inventory. The new buyer does not know that the broker is in a hurry to sell; but he knows that if he offers prices for the unblended beans (say y_1 and y_2 dollars per pound for Java and Colombian beans respectively), which make the blended coffees less valuable than the beans they contain, then the broker will sell the beans, unblended, to him.

What prices should the second buyer offer in order to get the inventory at minimal cost to himself?

The purchase cost to be minimized is, of course,

$$g(\mathbf{y}) = 6y_1 + 4y_2 \qquad \text{thousands of dollars} \quad .$$

Now the composition of Mild Blend is $\frac{3}{4}$ Java and $\frac{1}{4}$ Colombian, and so its bean content is worth $\left(\frac{3}{4}\right)y_1 + \left(\frac{1}{4}\right)y_2$ dollars per pound. This must exceed the market value of Mild Blend, which is \$3 per pound. That is, we must have

$$\left(\tfrac{3}{4}\right)y_1 + \left(\tfrac{1}{4}\right)y_2 \geq 3 \quad .$$

Similarly, since Strong Blend is $\frac{1}{2}$ Java and $\frac{1}{2}$ Colombian and has a market value of \$4 per pound, we must have

$$\left(\tfrac{1}{2}\right)y_1 + \left(\tfrac{1}{2}\right)y_2 \geq 4 \quad .$$

The optimization problem faced by the second buyer is therefore to
$$\text{minimize} \qquad g(\mathbf{y}) = \mathbf{c}^t\mathbf{y} \quad , \qquad \text{where } \mathbf{c}^t = (6,4) \quad ,$$

subject to the conditions $\quad \begin{cases} A^t\mathbf{y} \geq \mathbf{b} \quad , \\ \quad \mathbf{y} \geq \mathbf{0} \end{cases}$

where $\quad \mathbf{b} = \begin{bmatrix} 3 \\ 4 \end{bmatrix} \quad ;$ and $\quad A^t = \begin{bmatrix} \frac{3}{4} & \frac{1}{4} \\ \frac{1}{2} & \frac{1}{2} \end{bmatrix} \quad ,$ or $\quad A = \begin{bmatrix} \frac{3}{4} & \frac{1}{2} \\ \frac{1}{4} & \frac{1}{2} \end{bmatrix} \quad .$

The data $\mathbf{b}$, $\mathbf{c}$, A are the same as in the original problem; therefore the present problem is indeed its dual.

From lemma 11.1 we can say at this point that it will cost the second buyer at least \$36,000 dollars to make the deal, since we know $f(\mathbf{x}) = 36$ for the vector $\mathbf{x} = (4,6)$ in the feasible set for the primal problem.

Strong Duality

We now reconsider the concept of the final tableau, which was obtained for a fairly general class of primal problems in section 11.1. We send our workhorse method of elementary matrices into action once again, and it gives us a strong link between initial and final tableaux. From this link, we can deduce with astonishing ease that the final tableau carries within it not only a final vertex $(\mathbf{x}^{(k)}, \mathbf{z}^{(k)})$ for the primal problem, but also a vertex $\mathbf{y}$ for the dual problem; and that

$$f(\mathbf{x}^{(k)}) = g(\mathbf{y}) \quad .$$

In light of the "weak duality" lemma, this means that we have found optimal solutions for the primal and dual problems simultaneously.

THEOREM 11.3

(Strong duality)

Suppose P is a primal problem with

objective function	$f(\mathbf{x}) = \mathbf{c}^t\mathbf{x}$
feasible set	$A\mathbf{x} \leq \mathbf{b}$
	$\mathbf{x} \geq \mathbf{0} \quad .$

Suppose the initial tableau, $T = \left[\begin{array}{c|c|c} A & I & \mathbf{b} \\ \hline \mathbf{c}^t & \mathbf{0}^t & 0 \end{array}\right]^2 \quad ,$

can be transformed into a final tableau,

$$T^{(k)} = \left[\begin{array}{c|c|c} A^{(k)} & B^{(k)} & \mathbf{b}^{(k)} \\ \hline -\mathbf{w}^t & -\mathbf{y}^t & -p^{(k)} \end{array}\right] \quad \cdots$$

$$\cdots \quad , \quad \text{where } \mathbf{w} \geq \mathbf{0}, \mathbf{y} \geq \mathbf{0}, \mathbf{b}^{(k)} \geq \mathbf{0} \quad ,$$

by pivoting k times (on entries not in the extra row or augmented column, of course).

Then $\mathbf{x}^{(k)}$, the vertex of the feasible set associated to $T^{(k)}$, is an optimal solution of the primal problem P, and $\mathbf{y}$ is an optimal solution of the dual problem D.

Note. It is not necessary to assume that $\mathbf{b} \geq \mathbf{0}$ in the primal problem P. The only requirement is that a final tableau in which $\mathbf{w}$, $\mathbf{y}$, and $\mathbf{b}^{(k)}$ are positive can be reached by a finite number of pivots on entries not in the extra row or augmented column of T.

■**Proof**　Since each pivot operation on T consists of row operations, we can say that $T^{(k)}$ is obtained by row operations on T. But row operations are implemented by elementary matrices; thus

$$ST = T^{(k)} \quad ,$$

where　　$S = S_r \cdots S_1 \quad ,$

and each S_q, for $q = 1, \ldots, r$, is an $(m+1) \times (m+1)$ elementary matrix.

Now, it is easily checked (see exercises) that the row operations used in pivoting on an entry in row i are implemented by elementary matrices that differ from the identity matrix only in column i. And since, by hypothesis, row $m+1$ is never the pivot row, column $m+1$ of each elementary matrix S_q is always the same as in the identity matrix. That is,

$$S_q = \left[\begin{array}{c|c} Q_q & \mathbf{0} \\ \hline \mathbf{v}_q^t & 1 \end{array}\right] \quad , \quad \text{for some } m \times m \text{ matrix } Q_q \\ \text{and } 1 \times m \text{ vector } \mathbf{v}_q^t \ .$$

It follows, by inspection, that the product of such matrices has the same form; that is,

$$S = \left[\begin{array}{c|c} Q & \mathbf{0} \\ \hline \mathbf{v}^t & 1 \end{array}\right] \quad , \quad \text{for some } m \times m \text{ matrix } Q \\ \text{and } 1 \times m \text{ vector } \mathbf{v}^t.$$

$$\text{Then} \quad ST = \left[\begin{array}{c|c} Q & \mathbf{0} \\ \hline \mathbf{v}^t & 1 \end{array}\right] \left[\begin{array}{c|c|c} A & I & \mathbf{b} \\ \hline \mathbf{c}^t & \mathbf{0}^t & 0 \end{array}\right]$$

$$= \left[\begin{array}{c|c|c} QA & Q & Q\mathbf{b} \\ \hline \mathbf{v}^t A + \mathbf{c}^t & \mathbf{v}^t & \mathbf{v}^t \mathbf{b} \end{array}\right] = \left[\begin{array}{c|c|c} A^{(k)} & B^{(k)} & \mathbf{b}^{(k)} \\ \hline -\mathbf{w}^t & -\mathbf{y}^t & -p^{(k)} \end{array}\right]$$

$$= T^{(k)} \quad .$$

On comparing some of the corresponding blocks of this matrix equation, we obtain:

(i) $\mathbf{v}^t = -\mathbf{y}^t$ or $\mathbf{v} = -\mathbf{y}$;

(ii) $\mathbf{v}^t \mathbf{b} = -p^{(k)}$ or $\mathbf{y}^t \mathbf{b} = p^{(k)}$

                    or $g(\mathbf{y}) = p^{(k)}$;

(iii) $\mathbf{v}^t A + \mathbf{c}^t = -\mathbf{w}^t$ or $\mathbf{y}^t A - \mathbf{w}^t = \mathbf{c}^t$

                        or $A^t \mathbf{y} - \mathbf{w} = \mathbf{c}$

                        or $A^t \mathbf{y} \geq \mathbf{c}$,   since $\mathbf{w} \geq \mathbf{0}$.

Because $A^t \mathbf{y} \geq \mathbf{c}$ and $\mathbf{y} \geq \mathbf{0}$, the vector $\mathbf{y}$ is in the feasible set for problem D. We also have

$$f(\mathbf{x}^{(k)}) = g(\mathbf{y}) = p^{(k)} \quad ,$$

and so lemma 11.1 implies that $\mathbf{x}^{(k)}$ and $\mathbf{y}$ are optimal solutions for the primal and dual problems respectively.   Q.E.D.

Applications of the Duality Method

Example 1   (concluded) Earlier in this section we introduced the problem of the second coffee buyer who tries to outbid the wholesaler for the broker's inventory of coffee. This problem was of dual type, its primal problem being the original coffee broker problem discussed at the beginning of section 11.1. But we obtained a final tableau for the original problem:

$$T'' = \left[\begin{array}{cc|cc|c} 1 & 0 & 2 & -2 & 4 \\ 0 & 1 & -1 & 3 & 6 \\ \hline 0 & 0 & -2 & -6 & -36 \end{array}\right] \quad .$$

So the Strong Duality Theorem is applicable to this primal/dual pair of problems. In particular, an optimal solution for the dual problem can be read off from the extra row of the final tableau: we have

$$(-w_1, -w_2, -y_1, -y_2) = (0, 0, -2, -6) \quad,$$

and so $y_1 = 2$ and $y_2 = 6$. That is, the second buyer must offer at least \$2 and \$6 per pound for Java and Colombian beans respectively, to be sure of getting the broker's inventory.

Shadow Prices

The data that we were given as a basis for our analysis of the "coffee problems" of this chapter included nothing about the market prices of the unprocessed commodities involved, namely the Java and Colombian coffee beans. For all we know, the beans might have been produced by the coffee broker on his own plantation and might never have had a monetary value assigned to them by market forces. Nevertheless, we have constructed the prices that a rational buyer would have to bid for the commodities, in order to be competitive with their alternative use as raw materials for the manufacture of blended coffees. These prices, called *shadow prices*, or *imputed prices*, are plausible estimates of the *immediate* market value of the commodities under *existing, perhaps momentary, circumstances.*

Example 2 To see that this interpretation of shadow prices is strictly limited in scope, recall that in the problem at the end of section 1, we increased the selling price of Strong Blend to \$g per pound, with $g \geq 6$. The shadow prices y_1, y_2 of Java and Colombian beans in this situation can be read off the extra row of the final tableau that we computed for the problem: referring back to it, we find $y_1 = 0$, $y_2 = 2g$. That is, the shadow price of Java coffee is now zero! But Java coffee would not be grown for very long if it had no market value in the world economy. Rather, the zero shadow price must reflect a momentary, localized trading situation in which Java beans are in surplus. (Remember that only Strong Blend is shipped and that some Java is left over, when $g \geq 6$.)

Example 3 The introduction of import quotas (see example 2 of section 1) means that there are three valuable commodities in the problem: Java coffee, Colombian coffee, and units of the import quota. The shadow prices y_1, y_2, y_3 that a buyer must bid for these are of course read from the extra row of the final tableau computed for the problem:

$$y_1 = 0, \ y_2 = 4, \ y_3 = 2 \quad.$$

So Java coffee again has zero shadow price. But the quota units are valuable; the buyer must offer a bonus of \$2 for every pound of coffee beans that he is allowed to import.

Artificial Variables

Let us now consider a primal problem in which one of the constants b_1 is negative:

Example 4   Maximize $f(x_1,x_2) = x_1 + 2x_2$

subject to the conditions
$$\begin{cases} 3x_1 + 4x_2 \le 12 \\ -2x_1 - x_2 \le -2 \\ \mathbf{x} \ge \mathbf{0} \end{cases}.$$

When we introduce the slack variables z_1, z_2, the conditions defining the feasible set become

$$\begin{cases} 3x_1 + 4x_2 + z_1 = 12 \\ -2x_1 - x_2 + z_2 = -2 \\ \mathbf{x} \ge \mathbf{0}, \quad \mathbf{z} \ge \mathbf{0} \end{cases}.$$

The constant $b_2 = -2$ is negative but becomes positive when we multiply the second equation by -1. Then the feasible set, which we denote F, is given by the conditions

$$\begin{cases} 3x_1 + 4x_2 + z_1 = 12 \\ 2x_1 + x_2 - z_2 = 2 \\ \mathbf{x} \ge \mathbf{0}, \quad \mathbf{z} \ge \mathbf{0} \end{cases}.$$

However, since the coefficient of z_2 is now -1 instead of $+1$, we do not get a first vertex by putting $\mathbf{x} = \mathbf{0}$, $\mathbf{z} = \mathbf{b}$. To overcome this problem, we introduce an *artificial variable* u_1, with coefficient $+1$, on the left side of the second equation. The new feasible set, which we denote F_A, is described as follows:

$$\begin{cases} 3x_1 + 4x_2 + z_1 = 12 \\ 2x_1 + x_2 - z_2 + u_1 = 2 \\ \mathbf{x} \ge \mathbf{0}, \quad \mathbf{z} \ge \mathbf{0}, \quad \mathbf{u} \ge \mathbf{0} \end{cases}.$$

The variables z_1 and u_1 each appear only once, with coefficient $+1$, and so we obtain a first vertex by putting $z_1 = 12$, $u_1 = 2$ and setting all the other variables equal to zero. That is, the first vertex of F_A is

$$(\mathbf{x},\mathbf{z},\mathbf{u}) = (x_1,x_2,z_1,z_2,u_1) = (0,0,12,0,2) \quad.$$

We can now hop to other vertices of the new feasible set F_A by the usual pivoting procedure. And if we reach a vertex $(\mathbf{x},\mathbf{z},\mathbf{u})$ at which $\mathbf{u} = \mathbf{0}$, then on deleting the $\mathbf{u}$-component, we will have a vertex $(\mathbf{x},\mathbf{z})$ of the original feasible set F.

How should we select pivot operations in order to approach the goal $\mathbf{u} = \mathbf{0}$? We do have $\mathbf{u} \ge \mathbf{0}$, and therefore $-u_1 \le 0$, and so we need only increase the function $g(\mathbf{u}) = -u_1$ to its maximum possible value of zero.

That is, we must maximize the *auxiliary objective function* $g(\mathbf{u}) = -u_1$ and obtain for it a maximum value of zero. Therefore, we attach to the usual initial tableau T an extra objective row carrying the coefficients of

$$g = -u_1 = 2x_1 + x_2 - z_2 - 2 \quad .$$

Let us call the enlarged tableau T_A:

$$T_A = \begin{bmatrix} 3 & 4 & 1 & 0 & 0 & 12 \\ ② & 1 & 0 & -1 & 1 & 2 \\ \hline 1 & 2 & 0 & 0 & 0 & 0 \\ \hline ② & 1 & 0 & -1 & 0 & 2 \end{bmatrix} \quad \begin{array}{l} \theta_1 = \frac{12}{3} = 4 \\ \theta_2 = \frac{2}{2} = 1 \\ \text{objective row for } f \\ \text{objective row for } g \quad . \end{array}$$

Note that, in accordance with our standard procedure, the last entry in the objective row for g is

$$-\text{ (present value of } g) = -(-2) = 2 \quad .$$

Our selection of pivot column is governed, at this stage, by the objective row for g. Thus the first pivot column is column 1, and we obtain

$$T'_A = \begin{bmatrix} 0 & \frac{5}{2} & 1 & \frac{3}{2} & -\frac{3}{2} & 9 \\ 1 & \frac{1}{2} & 0 & -\frac{1}{2} & \frac{1}{2} & 1 \\ \hline 0 & \frac{3}{2} & 0 & \frac{1}{2} & -\frac{1}{2} & -1 \\ \hline 0 & 0 & 0 & 0 & -1 & 0 \end{bmatrix} \quad \longleftarrow g = 0 \quad .$$

After just one pivot operation, we have reached a tableau with $g = 0$. This means that $\mathbf{u} = \mathbf{0}$; therefore $(\mathbf{x},\mathbf{z}) = (1,0,9,0)$ is a first vertex for the original feasible set F.

At this point we can delete the $\mathbf{u}$-column and the g-row, obtaining an initial tableau T for the original problem:

$$T = \begin{bmatrix} 0 & \frac{5}{2} & 1 & \frac{3}{2} & 9 \\ 1 & \frac{1}{2} & 0 & -\frac{1}{2} & 1 \\ \hline 0 & \frac{3}{2} & 0 & \frac{1}{2} & -1 \end{bmatrix} \quad \begin{array}{l} \theta_1 = \frac{18}{5} \\ \theta_2 = 2 \quad . \end{array}$$

Now the pivoting proceeds as usual:

$$T' = \begin{bmatrix} -5 & 0 & 1 & ④ & 4 \\ 2 & 1 & 0 & -1 & 2 \\ \hline -3 & 0 & 0 & ② & -4 \end{bmatrix} \quad \begin{array}{l} \theta_1 = 1 \\ \theta_2 = -2 \end{array}$$

$$T'' = \begin{bmatrix} -\frac{5}{4} & 0 & \frac{1}{4} & 1 & 1 \\ \frac{3}{4} & 1 & \frac{1}{4} & 0 & 3 \\ \hline -\frac{1}{2} & 0 & -\frac{1}{2} & 0 & -6 \end{bmatrix} \quad .$$

Since this is a final tableau, the optimal solution is

$$(x_1,x_2,z_1,z_2) = (0,3,0,1), \quad \text{or} \quad x_1 = 0, \, x_2 = 3 \quad ,$$

and the maximum value of f is 6.

Example 5 Maximize $f(x_1, x_2) = x_1 + 3x_2$,

subject to the conditions

$$\begin{cases} -x_1 + x_2 \le 1 \\ x_1 + 2x_2 \le 14 \\ -x_1 + 4x_2 \le 10 \\ \mathbf{x} \le 0 \end{cases}.$$

In the third of the given conditions, the inequality goes the "wrong way" for a primal problem. To produce an equation, we must therefore *subtract* the slack variable, z_3, from the left side. Then we have

$$-x_1 + 4x_2 - z_3 = 10 \quad.$$

Thus we need an artificial variable for this equation, and the new feasible set F_A assumes the form

$$\begin{cases} -x_1 + x_2 + z_1 \qquad\qquad\quad = 1 \\ x_1 + 2x_2 \qquad +z_2 \qquad\quad = 14 \\ -x_1 + 4x_2 \qquad\qquad - z_3 + u_1 = 10 \\ \mathbf{x} \ge 0, \mathbf{z} \ge 0, \mathbf{u} \ge 0 \end{cases}.$$

The auxiliary objective function is:

$$g(\mathbf{u}) = -u_1 = -x_1 + 4x_2 - z_3 - 10 \quad,$$

and the enlarged tableau is

$$T_A = \begin{bmatrix} -1 & ① & 1 & 0 & 0 & 0 & 1 \\ 1 & 2 & 0 & 1 & 0 & 0 & 14 \\ -1 & 4 & 0 & 0 & -1 & 1 & 10 \\ \hline 1 & 3 & 0 & 0 & 0 & 0 & 0 \\ \hline -1 & ④ & 0 & 0 & -1 & 0 & 10 \end{bmatrix} \qquad \begin{matrix} \theta_1 = 1 \\ \theta_2 = 7 \\ \theta_3 = \frac{5}{2} \end{matrix}.$$

Then

$$T'_A = \begin{bmatrix} -1 & 1 & 1 & 0 & 0 & 0 & 1 \\ 3 & 0 & -2 & 1 & 0 & 0 & 12 \\ ③ & 0 & -4 & 0 & -1 & 1 & 6 \\ \hline 4 & 0 & -3 & 0 & 0 & 0 & -3 \\ \hline ③ & 0 & -4 & 0 & -1 & 0 & 6 \end{bmatrix} \qquad \begin{matrix} \theta_1 = -1 \\ \theta_2 = 4 \\ \theta_3 = 2 \end{matrix}$$

$$T''_A = \begin{bmatrix} 0 & 1 & -\frac{1}{3} & 0 & -\frac{1}{3} & \frac{1}{3} & 3 \\ 0 & 0 & 2 & 1 & 1 & -1 & 6 \\ 1 & 0 & -\frac{4}{3} & 0 & -\frac{1}{3} & \frac{1}{3} & 2 \\ \hline 0 & 0 & \frac{7}{3} & 0 & \frac{4}{3} & -\frac{4}{3} & -11 \\ \hline 0 & 0 & 0 & 0 & 0 & -1 & 0 \end{bmatrix} \longleftarrow g = 0.$$

We have now reached a tableau in which $g = 0$, and so we delete the $\mathbf{u}$-column and the g-row. The initial tableau for the original problem is then

$$T = \begin{bmatrix} 0 & 1 & -\frac{1}{3} & 0 & -\frac{1}{3} & 3 \\ 0 & 0 & 2 & 1 & 1 & 6 \\ 1 & 0 & -\frac{4}{3} & 0 & -\frac{1}{3} & 2 \\ \hline 0 & 0 & \frac{7}{3} & 0 & \frac{4}{3} & -11 \end{bmatrix} \qquad \begin{matrix} \theta_1 = -9 \\ \theta_2 = 3 \\ \theta_3 = -\frac{3}{2} \end{matrix}$$

$$T' = \begin{bmatrix} 0 & 1 & 0 & \frac{1}{6} & -\frac{1}{6} & 4 \\ 0 & 0 & 1 & \frac{1}{2} & \frac{1}{2} & 3 \\ 1 & 0 & 0 & \frac{2}{3} & \frac{1}{3} & 6 \\ 0 & 0 & 0 & -\frac{7}{6} & \frac{1}{6} & -18 \end{bmatrix} \quad \begin{matrix} \theta_1 = -24 \\ \theta_2 = 6 \\ \theta_3 = 18 \end{matrix}$$

$$T'' = \begin{bmatrix} 0 & 1 & \frac{1}{3} & \frac{1}{3} & 0 & 5 \\ 0 & 0 & 2 & 1 & 1 & 6 \\ 1 & 0 & -\frac{2}{3} & \frac{1}{3} & 0 & 4 \\ 0 & 0 & -\frac{1}{3} & -\frac{4}{3} & 0 & -19 \end{bmatrix} .$$

From the final tableau T'', we can read off the conclusion that the maximum value of f is 19 and occurs at $(\mathbf{x},\mathbf{z}) = (4,5,0,0,6)$ or simply at $(x_1,x_2) = (4,5)$.

■ **Problem** Suppose factories A and B can produce 600 and 400 engines per week, respectively. Suppose also that assembly plants P and Q require at least 500 and 300 engines per week, respectively. Suppose the cost of shipping an engine from factory to plant is as follows:

unit shipping cost, from factory A to plant P: \$400
unit shipping cost, from factory A to plant Q: \$100
unit shipping cost, from factory B to plant P: \$200
unit shipping cost, from factory B to plant Q: \$300 .

In order to minimize costs, how many engines should be shipped per week from each factory to each plant?

■ **Solution** Let x_1, x_2, x_3, x_4 represent the numbers of engines shipped per week from A to P, A to Q, B to P, and B to Q, respectively. Then total shipping costs are $4x_1 + x_2 + 2x_3 + 3x_4$ hundred dollars per week. In other words, the quantity to be *maximized* is

$$f(\mathbf{x}) = -(\text{cost}) = -4x_1 - x_2 - 2x_3 - 3x_4 .$$

The restrictions on $\mathbf{x}$ are as follows:

weekly shipments from factory A total $x_1 + x_2 \le 6$ hundred engines
weekly shipments from factory B total $x_3 + x_4 \le 4$ hundred engines
weekly deliveries to plant P total $x_1 + x_3 \ge 5$ hundred engines
weekly deliveries to plant Q total $x_2 + x_4 \ge 3$ hundred engines .
Also, $\mathbf{x} \ge \mathbf{0}$.

Since two of the above inequalities go in the wrong direction, two artificial variables, u_1, u_2, are needed. Then the feasible set is given by the conditions

$$\begin{cases} x_1 + x_2 & & + z_1 & & & = 6 \\ & x_3 + x_4 & & + z_2 & & = 4 \\ x_1 & + x_3 & & - z_3 & + u_1 & = 5 \\ & x_2 & + x_4 & & - z_4 & + u_2 = 3 \\ \mathbf{x} \ge \mathbf{0}, & \mathbf{z} \ge \mathbf{0}, & \mathbf{u} \ge \mathbf{0} & . \end{cases}$$

In order to locate a vertex at which $\mathbf{u} = \mathbf{0}$, it suffices to obtain the value zero for the auxiliary objective function

$$g(u_1,u_2) = -u_1 - u_2$$

or $g = x_1 + x_2 + x_3 + x_4 - z_3 - z_4 - 8$.

The enlarged tableau is then

$$T_A = \begin{bmatrix}
1 & 1 & 0 & 0 & 1 & 0 & 0 & 0 & 0 & 0 & 6 \\
0 & 0 & 1 & 1 & 0 & 1 & 0 & 0 & 0 & 0 & 4 \\
① & 0 & 1 & 0 & 0 & 0 & -1 & 0 & 1 & 0 & 5 \\
0 & 1 & 0 & 1 & 0 & 0 & 0 & -1 & 0 & 1 & 3 \\
-4 & -1 & -2 & -3 & 0 & 0 & 0 & 0 & 0 & 0 & 0 \\
① & 1 & 1 & 1 & 0 & 0 & -1 & -1 & 0 & 0 & 8
\end{bmatrix}$$

$$T'_A = \begin{bmatrix}
0 & 1 & -1 & 0 & 1 & 0 & 1 & 0 & -1 & 0 & 1 \\
0 & 0 & 1 & 1 & 0 & 1 & 0 & 0 & 0 & 0 & 4 \\
1 & 0 & 1 & 0 & 0 & 0 & -1 & 0 & 1 & 0 & 5 \\
0 & 1 & 0 & ① & 0 & 0 & 0 & -1 & 0 & 1 & 3 \\
0 & -1 & -2 & -3 & 0 & 0 & -4 & 0 & 4 & 0 & 20 \\
0 & 1 & 0 & ① & 0 & 0 & 0 & -1 & -1 & 0 & 3
\end{bmatrix}$$

$$T''_A = \begin{bmatrix}
0 & 1 & -1 & 0 & 1 & 0 & 1 & 0 & -1 & 0 & 1 \\
0 & -1 & 1 & 0 & 0 & 1 & 0 & 1 & 0 & -1 & 1 \\
1 & 0 & 1 & 0 & 0 & 0 & -1 & 0 & 1 & 0 & 5 \\
0 & 1 & 0 & 1 & 0 & 0 & 0 & -1 & 0 & 1 & 3 \\
0 & 2 & 2 & 0 & 0 & 0 & -4 & -3 & 4 & 3 & 29 \\
0 & 0 & 0 & 0 & 0 & 0 & 0 & 0 & -1 & -1 & 0
\end{bmatrix} \leftarrow g = 0$$

On deletion of the g-row and $\mathbf{u}$-columns, we have the initial tableau

$$T = \begin{bmatrix}
0 & ① & -1 & 0 & 1 & 0 & 1 & 0 & 1 \\
0 & -1 & 1 & 0 & 0 & 1 & 0 & 1 & 1 \\
1 & 0 & 1 & 0 & 0 & 0 & -1 & 0 & 5 \\
0 & 1 & 0 & 1 & 0 & 0 & 0 & -1 & 3 \\
0 & ② & 2 & 0 & 0 & 0 & -4 & -3 & 29
\end{bmatrix}$$

After a few pivots on T, we obtain the optimal solution,

$(\mathbf{x},\mathbf{z}) = (1,3,4,0,2,0,0,0)$,   or $x_1 = 1$, $x_2 = 3$, $x_3 = 4$, $x_4 = 0$.

We conclude that factory A should ship 100 and 300 engines per week to plants P and Q respectively, while factory B should ship 400 engines per week to plant P.

In linear programming theory, problems of the above type are called *transportation problems*. The simplex method is applicable to them but is inefficient when the number of dispatch and delivery points is substantial. Improved and specialized techniques are presented in texts devoted to programming.

■ EXERCISES 11.2

1. For each of the primal problems given in exercises 3, 4, and 5 of section 11.1,

    (a) state the corresponding dual problem;
    (b) obtain an optimal solution for the dual problem by inspection of the final tableau for the primal problem; and
    (c) verify, by substitution, that the solution obtained in part (b) is in the feasible set for the dual problem.

2. In the primal problem of example 3, section 11.1, it was found that the feasible set is unbounded and that the objective function can assume arbitrarily large values.

    (a) Write down the corresponding dual problem.
    (b) Sketch the boundaries of the feasible set for the dual problem and deduce that it is empty, as corollary 11.2 predicts.

3. (a) Write down the shadow prices of fertilizer and farm acreage in the farming problem given in exercise 9 of section 11.1.

    (b) Do the same for the farming problem given in exercise 10 of section 11.1.

    (c) Write down the shadow prices of fertilizer, farm acreage, and labor in the farming problem given in exercise 12 of section 11.1.

4. In the farming problem given in exercise 11 of section 11.1, total fertilizer available is a parameter g. Write down formulas giving the shadow prices of fertilizer and farm acreage as functions of g.

5. Maximize the function $f(x_1, x_2) = -3x_1 + 4x_2$,

    subject to the conditions
    $$\left\{ \begin{array}{rcr} 2x_1 - x_2 & \leq & -1 \\ -x_1 + x_2 & \leq & 4 \\ \mathbf{x} & \geq & \mathbf{0} \end{array} \right. .$$

    (Note that the positivity condition $\mathbf{b} \geq \mathbf{0}$ is not satisfied in this example, and thus the method of artificial variables must be used.)

6. Using the method of artificial variables, solve the following problem:

    maximize $f(x_1, x_2) = x_1 + 2x_2$,

    subject to the conditions
    $$\left\{ \begin{array}{rcr} 2x_1 + 3x_2 & \leq & 12 \\ x_1 - x_2 & \geq & 1 \\ \mathbf{x} & \geq & \mathbf{0} \end{array} \right. .$$

7. Maximize the quantity $f(x_1, x_2, x_3) = 4x_1 + 4x_2 + 2x_3$,
   subject to the conditions
$$\begin{cases} -x_1 - 2x_2 + x_3 \geq \tfrac{1}{2} \\ x_1 - x_2 + x_3 \leq 1 \\ \mathbf{x} \geq \mathbf{0} \end{cases}.$$

8. Maximize the quantity $f(x_1, x_2) = -3x_1 + x_2$,
   subject to the conditions
$$\begin{cases} -x_1 + 2x_2 \leq 2 \\ x_1 + x_2 \leq 6 \\ x_2 \geq 2 \\ \mathbf{x} \geq \mathbf{0} \end{cases}.$$

9. Suppose factories A and B can produce 500 and 800 engines per week, respectively. Suppose also that assembly plants P and Q require at least 400 and 700 engines per week, respectively. Suppose the cost of shipping an engine from factory to plant is as follows:

   unit shipping cost, from factory A to plant P: \$60
   unit shipping cost, from factory A to plant Q: \$50
   unit shipping cost, from factory B to plant P: \$30
   unit shipping cost, from factory B to plant Q: \$40.

   In order to minimize costs, how many engines should be shipped per week from each factory to each plant?

10. In the discussion of θ-ratios of section 1, we saw that the pivot operation on entry a_{ij} of a tableau T is implemented by elementary matrices of the forms

$$\left(\frac{1}{a_{ij}}\right) R_i$$

    and $R_k - a_{kj} R_i$,   for $k \neq i$.

(a) Show that the elementary matrices that implement these row operations differ from the identity matrix only in column i.

   (b) Deduce that the initial and final tableaux T and $T^{(k)}$ of a primal problem are linked by the formula

$$ST = T^{(k)} ,$$

   where S is a square matrix whose last column is the same as the last column of the identity matrix, as specified in the proof of the Strong Duality Theorem.

11. Suppose that, in the gth tableau $T^{(g)}$ of a primal problem, column j is pivot column and all the corresponding θ-ratios are negative or undefined. That is, after g steps of the pivoting process, *feature 2* occurs.
    Show that, starting from vertex $\mathbf{x}^{(g)}$, the jth component of $\mathbf{x}$, and therefore the value of the objective function, can be increased without limit while staying within the feasible set.

12. Suppose that in the gth tableau $T^{(g)}$ of a primal problem, column j is pivot column and one of the corresponding θ-ratios is zero. That is, after g steps of the pivoting process, *feature 3* occurs.

   (a) Show that $b_i^{(g)} = 0$ but $a_{ij}^{(g)} \neq 0$.

   (b) Deduce that, in the new tableau $T^{(g+1)}$ produced by pivoting on entry $a_{ij}^{(g)}$, the associated vertex $\mathbf{x}^{(g+1)}$ has its jth component, as well as n of its other components, equal to zero; i.e., the feasible set of the primal problem has a vertex with $n+1$ zero components.

Complex Numbers

The familiar real number system is usually employed in the initial stages of development of a mathematical theory. However, it often happens that progress beyond a certain point depends on being able to find roots for any given polynomial, and yet even such simple polynomials as $x^2 + 1$ have no real roots. We encounter this problem most urgently in Chapter 8, where the eigenvalues of the rotation matrix $R_{90°}$ are found to be the roots of the polynomial just cited. It is therefore necessary to extend the real numbers to a larger number system that will provide roots for all polynomials. This is the system of "complex numbers" that we discuss in this appendix.

Complex Numbers

A *complex number* z is simply an ordered pair of real numbers, such as (3,4). The first component is called the *real part* of z, and the second component the *imaginary part* of z. Thus in the above example, the real and imaginary parts of z are 3 and 4 respectively.

However, a distinctive notation is provided for complex numbers: instead of writing $z = (a,b)$, we write

$$z = a + bi \quad .$$

That is, the imaginary part, b, of the complex number z is distinguished by having the symbol i attached to it. For example, the complex number w with real part 5 and imaginary part 12 is written $w = 5 + 12i$.

The *sum* of the complex numbers $z = a + bi$ and $w = c + di$ is the complex number

$$z + w = (a + c) + (b + d)i \quad .$$

In particular, if z and w have the concrete values given earlier, then

$$z + w = (3 + 4i) + (5 + 12i) = (3 + 5) + (4 + 12)i$$
$$= 8 + 16i \quad .$$

By identifying the complex number, $z = a + 0i$, which has imaginary part 0, with the real number, a, and writing it simply as $z = a$, we can regard the complex numbers as an extension of the real number system.

For example, we write $0 + 0i = 0$. Clearly every complex number w satisfies $w + 0 = w$, and so we are justified in calling the complex number 0 the "zero element" of the complex number system.

Further, the *additive inverse* of the number $z = a + bi$ is defined to be the number $-z = -a - bi$. Thus $-(3 + 4i)$ is the number $-3 - 4i$.

The *difference $z - w$* of the complex numbers z and w is then defined to be the number $z + (-w)$, or $(a - c) + (b - d)i$. For instance,

$$(3 + 4i) - (5 + 12i) = -2 - 8i \quad .$$

As a special case of this formula, we have $z - z = z + (-z) = 0$, for every complex number z.

Addition and subtraction of complex numbers is, of course, exactly similar to that of 2-tuples. The distinctive properties of complex numbers originate in the definition of complex multiplication, which we now introduce.

Multiplication of Complex Numbers

If we treat the complex numbers $z = a + bi$ and $w = c + di$ as polynomials in i, then we can multiply them in accordance with the laws of elementary algebra:

$$zw = (a + bi)(c + di)$$
$$= ac + (ad + bc)i + bdi^2 \quad .$$

Suppose we now introduce the further rule that the power i^2 is always replaced by -1. Then we have the definition of the *product of complex numbers*:

$$(a + bi)(c + di) = (ac-bd) + (ad+bc)i \quad .$$

For example, $(2 + 3i)(4 + 5i) = \{(2)(4)-(3)(5)\} + \{(2)(5)+(3)(4)\}i$
$$= -7 + 22i \quad .$$

For all complex numbers z, w, u, etc., the following identities are valid:

$$zw = wz \qquad \text{(commutative law of multiplication)}$$
$$(zw)u = z(wu) \qquad \text{(associative law of multiplication)}$$
$$z(w+u) = zw + zu \qquad \text{(distributive law of multiplication)}.$$

These laws can be checked simply by multiplying out their left and right sides. We leave this as an exercise.

As a special case of complex multiplication, note that if one factor happens to be real, say $w = c + 0i$, then the product wz assumes the form

$$wz = c(a + bi) = ca + cbi \quad .$$

Thus $\quad 3(4 + 5i) = 12 + 15i \quad .$

Conversely, we can remove real factors from complex numbers as convenient; for example

$$7 + 14i = 7(1 + 2i) \quad .$$

In particular, the number $w = 1 + 0i = 1$ is a *multiplicative identity*:

$$1z = z \quad , \quad \text{for all complex numbers } z.$$

Complex Conjugate and Multiplicative Inverse

Given a complex number $z = a + bi$, we define the *complex conjugate* of z to be the number $z^* = a - bi$. For example,

$$(3 + 4i)^* = 3 - 4i \quad.$$

It is easily checked that conjugation of complex numbers z, w, etc. has the following properties:

1. $z^* = z$ if and only if z is real.
2. $(z + w)^* = z^* + w^*$
3. $(zw)^* = z^* w^*$
4. $(z^*)^* = z$
5. $z^* z$ is always real and is positive unless $z = 0$.

In fact, to verify property 5, say, we simply compute

$$
\begin{aligned}
z^* z &= (a + bi)^*(a + bi) \\
&= (a - bi)(a + bi) \\
&= (a^2 + b^2) + 0i \\
&= a^2 + b^2 > 0 \quad, \quad \text{unless } a = b = 0 \text{ or } z = 0 \quad.
\end{aligned}
$$

In a functioning number system, every nonzero element z must have an inverse; that is, there must be a number z^{-1} satisfying $z^{-1}z = 1$. But our last calculation shows how the inverse of a complex number $z = a + bi$ can be constructed; for it implies that with $u = (a^2 + b^2)^{-1}$,

$$(uz^*)z = u(z^* z) = (a^2 + b^2)^{-1}(a^2 + b^2) = 1 \quad.$$

So the number uz^* is an inverse for z. We conclude that the number $z = a + bi \neq 0$ has the *multiplicative inverse*:

$$z^{-1} = \frac{a}{a^2 + b^2} - \frac{b}{a^2 + b^2} i \quad.$$

For example, $(3 + 4i)^{-1} = \dfrac{3}{3^2 + 4^2} - \dfrac{4}{3^2 + 4^2} i = \dfrac{3}{25} - \dfrac{4}{25} i \quad.$

Roots of Quadratic Polynomials

The polynomial $x^2 + 1$ in the real variable x has no real roots. However, the corresponding polynomial $z^2 + 1$ in the complex variable z evidently has the complex roots $z = i$ and $z = -i$, and the factorization $z^2 + 1 = (z - i)(z + i)$.

More generally, any quadratic polynomial $az^2 + bz + c$, with real coefficients a, b, c, but, without real roots, has a pair of complex roots z_1, z_2 that are complex conjugates, i.e., $z_2 = z_1^*$. It is easily seen why this is so; consider, for example, the polynomial $z^2 - 6z + 25$. To find its roots, we first "complete the square" by rewriting it as $(z-3)^2 - 9 + 25$ or $(z-3)^2 + 16$. So a root, z, satisfies

$$(z-3)^2 = -16 = (\pm 4i)^2$$

or $\quad z-3 = \pm 4i$

or $\quad z = 3 \pm 4i$.

Therefore the roots are $z_1 = 3 + 4i$, $z_2 = 3 - 4i = z_1^*$.

More generally, the "Fundamental Theorem of Algebra," first proved by Gauss, states that, if $p(z)$ is a polynomial in the complex variable z, with complex coefficients, then $p(z)$ has at least one complex root. Further, a corollary of this result is that $p(z)$ can be expressed as the product of linear factors. The various proofs of this theorem, all difficult, are presented in advanced texts on complex variable theory and algebra.

We can compute the roots of some more polynomials after we have developed the following "polar representation" for complex numbers.

Polar Representation of a Complex Number

To a complex number $z = a + bi$, we can associate the 2-tuple $\mathbf{v} = (a,b)$, or simply the point $P(a,b)$ in the Cartesian plane (see Figure A.1). Of course, the point P has a set of polar coordinates r and θ, linked to its Cartesian coordinates a and b by the formulas

$$a = r \cos \theta \quad \text{and} \quad r = \sqrt{a^2 + b^2} .$$
$$b = r \sin \theta \quad \quad \theta = \arctan \left(\frac{b}{a} \right)$$

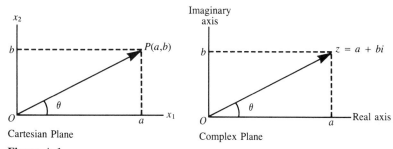

Cartesian Plane   Complex Plane

Figure A.1

The polar coordinate r gives the distance of the point $P(a,b)$, or z, from the origin. This distance is called the *absolute value* of z, and is denoted $|z|$.

Therefore $\quad |z| = r = \sqrt{a^2 + b^2}$.

The polar coordinate θ gives the angle, measured counterclockwise, between the x_1-axis and the position vector $\overrightarrow{OP}$. This angle is called the *argument* of z, and denoted arg z.

Thus $\arg z = \theta = \arctan\left(\frac{b}{a}\right)$.

Then, since $z = a + bi = r \cos \theta + (r \sin \theta)i$, we have the *polar repre-sentation* of the complex number $z = a + bi$:

$z = r(\cos \theta + i \sin \theta)$.

For example, to obtain a polar representation of $z = 1 + \sqrt{3}i$ (see Figure A.2), we compute

$$r = \sqrt{a^2 + b^2} = \sqrt{1 + 3} = 2$$

and $\theta = \arctan\left(\frac{\sqrt{3}}{1}\right) = 60°$ or $\frac{\pi}{3}$ radians,

and obtain $z = 2\ (\cos 60° + i \sin 60°) = 2\ (\cos \frac{\pi}{3} + i \sin \frac{\pi}{3})$.

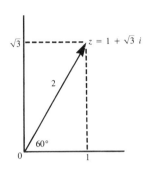

Cartesian and Polar Coordinates of a Complex Number

Figure A.2

Suppose now that, starting from the point z with polar coordinates (r,θ), we advance the argument, θ, by 2π radians (or $360°$), so as to obtain a new point z' with polar coordinates $(r,\theta+2\pi)$. Then we have simply circled once around the origin and returned to the starting point, and in fact $z' = z$. In other words, the sets of polar coordinates (r,θ) and $(r,\theta+2\pi)$ always refer to the same point. Similar reasoning shows that for any choice of the integer n, the set of polar coordinates $(r,\theta+2n\pi)$ always refers to the same point. We conclude, in particular, that the polar coordinates of a point are never unique.

For example, the polar coordinates $(2, \frac{\pi}{3})$ were computed for the point z in the above diagram; but $(2,\frac{7\pi}{3})$ or $(2,-\frac{5\pi}{3})$ would have been just as correct.

De Moivre's Theorem

Observe that the polar representation of a complex number expresses it as the product of two numbers: the real number r, and the complex number $\cos \theta + i \sin \theta$ with absolute value 1.

Given two complex numbers, z and w, with absolute values r and s and arguments θ and ω respectively, their product can be computed as follows:

$$
\begin{aligned}
zw &= (r(\cos \theta + i \sin \theta))(s(\cos \omega + i \sin \omega)) \\
&= (rs)(\cos \theta + i \sin \theta)(\cos \omega + i \sin \omega) \\
&= (rs)\{(\cos \theta \cos \omega - \sin \theta \sin \omega) + i(\cos \theta \sin \omega + \sin \theta \sin \omega)\} \\
&= (rs)(\cos (\theta+\omega) + i \sin (\theta+\omega))   .
\end{aligned}
$$

That is, the product zw has absolute value rs, and argument $\theta + \omega$.

In other words, to *multiply complex numbers* we *multiply their absolute values* and *add their arguments*. In symbols,

$$|zw| = |z||w| ,$$

and $\arg(z+w) = \arg z + \arg w$.

From the special case $z = w$ of this result, we derive the important formulas constituting *De Moivre's Theorem*:

If $z = r(\cos \theta + i \sin \theta)$, then $z^2 = r^2 (\cos 2\theta + i \sin 2\theta)$
and $z^n = r^n (\cos n\theta + i \sin n\theta)$.

For example, suppose we have $z = 1 + i$ and want to compute z^6. First we compute

$$r = \sqrt{1^2 + 1^2} = \sqrt{2}$$

and $\theta = \arctan\left(\frac{1}{1}\right) = \arctan 1 = 45°$ or $\frac{\pi}{4}$ radians.

Then $z = \sqrt{2} \left(\cos \frac{\pi}{4} + i \sin \frac{\pi}{4}\right)$

and $z^6 = (\sqrt{2})^6 \; \{\cos (6(\frac{\pi}{4})) + i \sin (6(\frac{\pi}{4}))\}$

$= 8(\cos \left(\frac{3\pi}{2}\right) + i \sin \left(\frac{3\pi}{2}\right))$

$= 8(0 - i) = -8i$.

Nth Roots of Unity

From De Moivre's Theorem, we know that if the complex number z has polar coordinates (r, θ), then z^n has polar coordinates $(r^n, n\theta)$. Also, the number 1 has polar coordinates $(1, 0)$. Now, z^n and 1 are the same complex number if

(i) they have the same absolute value, or $r^n = 1$,

and (ii) their arguments differ only by an integer multiple of 2π, or

$$n\theta = 0 + 2k\pi \quad .$$

That is, $z^n = 1$ if and only if

(i) $r = 1$

and (ii) $\theta = \frac{2k\pi}{n}$, for some integer k .

We conclude that $z^n = 1$ if and only if

$z = \cos \left(\frac{2k\pi}{n}\right) + i \sin \left(\frac{2k\pi}{n}\right)$, for some integer k.

The complex numbers satisfying $z^n = 1$ are called, naturally enough, the *n*th roots of unity.

Take the specific case $n = 3$. Then from the parameter choices $k = 0, 1, 2$, we obtain the following cube roots of unity:

$z = 1 + 0i = 1$,

$z = \cos \left(\frac{2\pi}{3}\right) + i \sin \left(\frac{2\pi}{3}\right) = -\frac{1}{2} + \frac{3}{2} i$,

and $z = \cos \left(\frac{4\pi}{3}\right) + i \sin \left(\frac{4\pi}{3}\right) = -\frac{1}{2} - \frac{3}{2} i$.

It is easily checked that all other choices of the parameter k lead to one of the three roots just computed. The same holds generally; there are exactly n nth roots of unity, and they can be obtained from the parameter choices $k = 0, 1, \ldots, n-1$.

Since the nth roots of unity satisfy $z^n = 1$ or $z^n - 1 = 0$, we have found the roots of the polynomial $p(z) = z^n - 1$. Handbooks of mathematical tables contain somewhat cumbersome algebraic formulas for the roots of all polynomials of degree less than or equal to 4. When n exceeds 4, the existence of roots is still guaranteed by the Fundamental Theorem of Algebra, but practical computation of these roots can be difficult.

Cramer's Rule

Given an invertible $n \times n$ matrix A, we know that the linear system $A\mathbf{x} = \mathbf{c}$ has the unique solution

$$\mathbf{x} = A^{-1}\mathbf{c} \quad .$$

Further, the inverse of any specific, numerical matrix A can be obtained by row reduction of $[A|I]$ to $[I|A^{-1}]$, as discussed in section 3.4. So the above formula for the solution $\mathbf{x}$ serves perfectly well in numerical problems.

However, it sometimes happens that the entries a_{ij} of A are not known, and we need an algebraic formula for the components x_j of $\mathbf{x}$ in terms of the entries a_{ij} of A and the components c_j of $\mathbf{c}$. Such a formula is developed in this appendix. It is called *Cramer's Rule*.

The 2×2 Case

We can easily derive the desired formula in the 2×2 case, and this in turn will suggest the appropriate construction in the general case. First of all, we have for 2×2 matrices the algebraic inversion formula

$$A = \begin{bmatrix} a_{11} & a_{12} \\ a_{21} & a_{22} \end{bmatrix} \quad , \quad A^{-1} = \frac{1}{(\det A)} \begin{bmatrix} a_{22} & -a_{12} \\ -a_{21} & a_{11} \end{bmatrix} \quad .$$

From this we can derive algebraic formulas for x_1 and x_2 as follows:

$$\begin{bmatrix} x_1 \\ x_2 \end{bmatrix} = \mathbf{x} = A^{-1}\mathbf{c} = \frac{1}{(\det A)} \begin{bmatrix} a_{22} & -a_{12} \\ -a_{21} & a_{11} \end{bmatrix} \begin{bmatrix} c_1 \\ c_2 \end{bmatrix}$$

or
$$\begin{cases} x_1 = \dfrac{(a_{22}c_1 - a_{12}c_2)}{(\det A)} \\ x_2 = \dfrac{(-a_{21}c_1 + a_{11}c_2)}{(\det A)} \end{cases} \quad .$$

Observe that the numerators in these quotients are recognizable as determinants; thus we can write the solution as

$$x_1 = \frac{\det \begin{bmatrix} c_1 & a_{12} \\ c_2 & a_{22} \end{bmatrix}}{\det \begin{bmatrix} a_{11} & a_{12} \\ a_{21} & a_{22} \end{bmatrix}} \quad , \quad x_2 = \frac{\det \begin{bmatrix} a_{11} & c_1 \\ a_{21} & c_2 \end{bmatrix}}{\det \begin{bmatrix} a_{11} & a_{12} \\ a_{21} & a_{22} \end{bmatrix}} \quad .$$

For example, to compute the solution of the system

$$\begin{bmatrix} 1 & 3 \\ 2 & 4 \end{bmatrix} \begin{bmatrix} x_1 \\ x_2 \end{bmatrix} = \begin{bmatrix} 5 \\ 6 \end{bmatrix} \quad ,$$

we substitute as follows:

$$x_1 = \frac{\det \begin{bmatrix} 5 & 3 \\ 6 & 4 \end{bmatrix}}{\det \begin{bmatrix} 1 & 3 \\ 2 & 4 \end{bmatrix}} = \frac{2}{-2} = -1 \quad , \quad x_2 = \frac{\det \begin{bmatrix} 1 & 5 \\ 2 & 6 \end{bmatrix}}{\det \begin{bmatrix} 1 & 3 \\ 2 & 4 \end{bmatrix}} = \frac{-4}{-2} = 2 \quad .$$

The above formulas for x_1 and x_2 can easily be streamlined. Denote the columns of A as $\mathbf{a}_1$ and $\mathbf{a}_2$, so that $A = [\mathbf{a}_1 | \mathbf{a}_2]$. Then we have

$$x_1 = \frac{\det [\mathbf{c} \,| \mathbf{a}_2]}{\det [\mathbf{a}_1 | \mathbf{a}_2]} \quad , \quad x_2 = \frac{\det [\mathbf{a}_1 | \mathbf{c} \,]}{\det [\mathbf{a}_1 | \mathbf{a}_2]} \quad .$$

To summarize this pattern, we note that the formula for x_j would simply be $\frac{(\det A)}{(\det A)}$, except that in the numerator, column j of A is replaced by $\mathbf{c}$. It turns out that the same pattern holds in the $n \times n$ case.

The General Case

> **THEOREM B1**
>
> (Cramer's Rule) Suppose A is an invertible $n \times n$ matrix, with columns $\mathbf{a}_1, \ldots, \mathbf{a}_n$.
>   Then the unique solution of the linear system $A\mathbf{x} = \mathbf{c}$ is
>
> $$x_j = \frac{\det [\mathbf{a}_1| \ldots |\mathbf{a}_{j-1}|\mathbf{c}|\mathbf{a}_{j+1}| \ldots |\mathbf{a}_n]}{\det [\mathbf{a}_1| \ldots |\mathbf{a}_{j-1}|\mathbf{a}_j|\mathbf{a}_{j+1}| \ldots |\mathbf{a}_n]} \quad , \quad j = 1, \ldots, n \quad .$$

Before proving this theorem, let us illustrate its use by applying it to the system (compare exercises 1.3, #3(a))

$$\begin{cases} x_1 \quad\quad\quad + 4x_3 = 1 \\ 2x_1 + 2x_2 + 4x_3 = 2 \\ 2x_1 - 2x_2 + 8x_3 = 6 \end{cases} \quad .$$

Using Laplace expansions across row 1 of our determinants, we obtain

$$x_1 = \frac{\det \begin{bmatrix} 1 & 0 & 4 \\ 2 & 2 & 4 \\ 6 & -2 & 8 \end{bmatrix}}{\det \begin{bmatrix} 1 & 0 & 4 \\ 2 & 2 & 4 \\ 2 & -2 & 8 \end{bmatrix}} = \frac{24 - 0 + 4(-16)}{24 - 0 + 4 \,(-8)} = \frac{-40}{-8} = 5$$

$$x_2 = \frac{\det \begin{bmatrix} 1 & 1 & 4 \\ 2 & 2 & 4 \\ 2 & 6 & 8 \end{bmatrix}}{-8} = \frac{-8 - 8 + 4(8)}{-8} = \frac{16}{-8} = -2$$

$$x_3 = \frac{\det \begin{bmatrix} 1 & 0 & 1 \\ 2 & 2 & 2 \\ 2 & -2 & 6 \end{bmatrix}}{-8} = \frac{16 - 0 - 8}{-8} = \frac{8}{-8} = -1 \quad .$$

So the unique solution is $\mathbf{x} = (5, -2, -1)$.

In this 3×3 example, we had to compute four distinct 3×3 determinants. Similarly, for an $n \times n$ system, there are $n+1$ $n \times n$ determinants to be computed. Clearly Cramer's Rule is an inefficient method of solving large linear systems. It has an important place, however, in theoretical contexts and in algebraic calculation.

■Proof Since the system $A\mathbf{x} = \mathbf{c}$ has a unique solution, we need only show that it is satisfied by the vector $\mathbf{x}$ whose components are as given in the statement of the theorem. But in the formula for x_j, the denominator is simply $\det A$, while the determinant in the numerator can be expanded down column j. So we can rewrite x_j as

$$x_j = \left(\tfrac{1}{\det A}\right) \sum_{i=1}^{n} (-1)^{i+j} c_i \det M_{ij} \quad .$$

Then we can compute the kth component of $A\mathbf{x}$ as

$$(A\mathbf{x})_k = \sum_{i=1}^{n} a_{kj} x_j = \sum_{j=1}^{n} a_{kj} \left(\tfrac{1}{\det A}\right) \sum_{i=1}^{n} (-1)^{i+j} c_i \det M_{ij}$$

$$= \left(\tfrac{1}{\det A}\right) \sum_{i=1}^{n} c_i \left[\sum_{j=1}^{n} (-1)^{i+j} a_{kj} \det M_{ij} \right] \quad .$$

Now the sum in brackets has the form of a Laplace expansion of $\det A$ along row i, except that the coefficient in the jth term is a_{kj} instead of the expected a_{ij}. This means, in effect, that the ith row of A has been replaced by a second copy of the kth row. But the determinant of a matrix with two identical rows is zero, so the bracketed quantity is zero when $i \neq k$. Of course when $i = k$, it is simply $\det A$.

Thus our formula for $(A\mathbf{x})_k$ simplifies to

$$(A\mathbf{x})_k = \left(\tfrac{1}{\det A}\right) c_k \det A = c_k \quad .$$

Since the last formula holds for $k = 1, \ldots, n$, we have $A\mathbf{x} = \mathbf{c}$ as required.

Q.E.D.

An Algebraic Formula for the Matrix Inverse

COROLLARY B2

Suppose A is an invertible $n \times n$ matrix. Then an algebraic formula for the entries of A^{-1} is

$$(A^{-1})_{ij} = (-1)^{i+j} \frac{(\det M_{ji})}{(\det A)}$$

■**Proof**   Recall that for any matrix B, the vector $B\mathbf{e}_j$ is simply column j of B. Suppose now that A is an invertible $n \times n$ matrix. Then the solution, $\mathbf{x}$, of the system $A\mathbf{x} = \mathbf{e}_j$ is the vector

$$\mathbf{x} = A^{-1}\mathbf{e}_j = \text{column } j \text{ of } A^{-1} \quad .$$

In particular, the ith component of $\mathbf{x}$ is

$$x_i = (A^{-1})_{ij} \quad .$$

But, by Cramer's Rule,

$$x_i = \frac{\det \; [\mathbf{a}_1| \ldots |\mathbf{a}_{i-1}|\mathbf{e}_j|\mathbf{a}_{i+1}| \ldots |\mathbf{a}_n]}{\det A} \quad .$$

On taking the Laplace expansion of the numerator down column i, we encounter only one nonzero term and obtain

$$x_i = \frac{((-1)^{i+j} \cdot 1 \cdot \det M_{ji})}{\det A} \quad ,$$

as required.   Q.E.D.

To compute the inverse of an $n \times n$ matrix by this formula would require the evaluation of the determinants of all n^2 minors M_{ji}, as well as $\det A$ itself. Again, the value of the formula is mainly in its theoretical utility.

Solutions to Selected Problems in Text

CHAPTER 1

Section 1.2

1. (a) $R = \begin{bmatrix} 1 & 0 \\ 0 & 1 \end{bmatrix}$ $x = 0$ (b) $R = \begin{bmatrix} 1 & 0 \\ 0 & 1 \end{bmatrix}$ $x = 0$
   $y = 0$ $y = 0$

   (c) $R = \begin{bmatrix} 1 & -2 \\ 0 & 0 \end{bmatrix}$ $x_1 = 2x_2$ (d) $R = \begin{bmatrix} 1 & 1 \\ 0 & 0 \end{bmatrix}$ $x_1 = -x_2$

2. (a) $R = \begin{bmatrix} 1 & 0 & \frac{9}{2} \\ 0 & 1 & -3 \end{bmatrix}$ $x = \left(-\frac{9}{2}\right)z$
   $y = 3z$

   (b) $R = \begin{bmatrix} 1 & 0 & 2 \\ 0 & 1 & 2 \end{bmatrix}$ $x = -2x$
   $y = -2z$

   (c) $R = \begin{bmatrix} 1 & 2 & 0 \\ 0 & 0 & 1 \end{bmatrix}$ $x_1 = -2x_2$
   $x_3 = 0$

   (d) $R = \begin{bmatrix} 1 & -3 & 2 \\ 0 & 0 & 0 \end{bmatrix}$ $x_1 = 3x_2 - 2x_3$

   (e) $R = \begin{bmatrix} 1 & 0 & -\frac{55}{4} & \frac{37}{4} \\ 0 & 1 & 42 & -21 \end{bmatrix}$ $x_1 = \left(\frac{55}{4}\right)x_3 - \left(\frac{37}{4}\right)x_4$
   $x_2 = -42x_3 + 21x_4$

   (f) $R = \begin{bmatrix} 1 & 0 & \frac{4}{3} & \frac{2}{3} & \frac{1}{3} \\ 0 & 1 & 0 & 1 & 0 \end{bmatrix}$ $x_1 = \left(-\frac{4}{3}\right)x_3 - \left(\frac{2}{3}\right)x_4 - \left(\frac{1}{3}\right)x_5$
   $x_2 = \qquad - x_4$

3. (a) $R = \begin{bmatrix} 1 & 0 & 0 \\ 0 & 1 & 0 \\ 0 & 0 & 1 \end{bmatrix}$ $x = 0$ (b) $R = \begin{bmatrix} 1 & 0 & 1 \\ 0 & 1 & -1 \\ 0 & 0 & 0 \end{bmatrix}$ $x = -z$
   $y = 0$ $y = z$
   $z = 0$

   (c) $R = \begin{bmatrix} 1 & 3 & 0 \\ 0 & 0 & 1 \\ 0 & 0 & 0 \end{bmatrix}$ $x_1 = -3x_2$
   $x_3 = 0$

   (d) $R = \begin{bmatrix} 1 & 0 & \frac{2}{3} \\ 0 & 1 & 2 \\ 0 & 0 & 0 \end{bmatrix}$ $x_1 = \left(-\frac{2}{3}\right)x_3$
   $x_2 = -2x_3$

(e) $R = \begin{bmatrix} 1 & -2 & 0 \\ 0 & 0 & 1 \\ 0 & 0 & 0 \end{bmatrix}$ $\begin{array}{l} x_1 = 2x_2 \\ x_3 = 0 \end{array}$

(f) $R = \begin{bmatrix} 1 & 0 & 0 \\ 0 & 1 & 0 \\ 0 & 0 & 1 \end{bmatrix}$ $\begin{array}{l} x = 0 \\ y = 0 \\ z = 0 \end{array}$

4. (a) $R = \begin{bmatrix} 1 & 0 \\ 0 & 1 \\ 0 & 0 \end{bmatrix}$ $\begin{array}{l} x = 0 \\ y = 0 \end{array}$ (b) $R = \begin{bmatrix} 1 & \frac{2}{5} \\ 0 & 0 \\ 0 & 0 \end{bmatrix}$ $x = \left(-\frac{2}{5}\right)y$

(c) $x_1 = \left(-\frac{5}{4}\right)x_3$
$x_2 = \left(-\frac{7}{8}\right)x_3$

5. (a) $\begin{cases} x = 0 \\ y = \left(\frac{3}{2}\right)w \\ z = 2w \end{cases}$ (b) $\begin{cases} x = 4z \\ y = -3z \\ w = 0 \end{cases}$

(c) $\begin{cases} x = y - 5w \\ z = 3w \end{cases}$ (d) $\begin{cases} x_1 = -3x_2 \\ x_3 = 0 \\ x_4 = 0 \end{cases}$

6. (a) $\begin{cases} x = y \\ z = 0 \\ w = 0 \end{cases}$ (b) $\begin{cases} x = 0 \\ y = 0 \\ z = 0 \\ w = 0 \end{cases}$

(c) $\begin{cases} x_1 = \left(-\frac{2}{3}\right)x_2 - \left(\frac{4}{3}\right)x_4 \\ x_3 = \qquad\qquad x_4 \end{cases}$ (d) $\begin{cases} x_1 = -7x_3 - 13x_5 \\ x_2 = -3x_3 - 10x_5 \\ x_4 = 0 \end{cases}$

7. (b)
   (i) $R_1 - 3R_2$ (ii) $\left(\frac{1}{2}\right)R_2$
   (iii) Exchange R_1, R_3 (iv) Exchange R_2, R_3

8. (a) 3 corners   (b) 3 corners
   (c) 4 corners

9. (a) $k = 3$ (b) $\begin{cases} x = -3z \\ y = z \end{cases}$

10. $\begin{cases} x = -13w \\ y = 3w \\ z = w \end{cases}$

Section 1.3

1. (a) $\begin{cases} x = -11 \\ y = 4 \end{cases}$ (b) $x = -4 - 2y$

   (c) No solution

2. (a) $\begin{cases} x_1 = \frac{13}{2} \\ x_2 = -4 \end{cases}$ (b) $\begin{cases} x = -1 - 4z \\ y = 1 \end{cases}$

   (c) No solution (d) $\begin{cases} x_1 = 6 - 9x_3 \\ x_2 = -2 + 3x_3 \end{cases}$

   (e) $\begin{cases} x_1 = -4 - \left(\frac{3}{2}\right)x_2 \\ x_3 = 8 \end{cases}$ (f) $\begin{cases} x_1 = 4 - x_3 \\ x_2 = -1 \end{cases}$

3. (a) $\begin{cases} x = 5 \\ y = -2 \\ z = -1 \end{cases}$ (b) No solution

   (c) $\begin{cases} x = -\frac{1}{3} - \left(\frac{1}{3}\right)z \\ y = \frac{1}{2} - \left(\frac{1}{4}\right)z \end{cases}$ (d) $\begin{cases} x_1 = -6 \\ x_2 = 6 \\ x_3 = 2 \end{cases}$

   (e) No solution

   (f) $\begin{cases} x_1 = -3 \\ x_2 = 2 \\ x_3 = -2 \end{cases}$

4. (a) $\begin{cases} x = -8 + 2z - 5w \\ y = 3 - z + w \end{cases}$ (b) $\begin{cases} x_1 = -3 - 5x_2 - 3x_4 - 5x_5 \\ x_4 = 2 \qquad\qquad + x_5 \end{cases}$

5. (a) $\begin{cases} x = -2 \\ y = 1 - w \\ z = w \end{cases}$ (b) $\begin{cases} x = 6 - 3y - 2w \\ z = -1 \qquad - 5w \end{cases}$

   (c) $\begin{cases} x_1 = -14 - 2x_2 + 15x_4 \\ x_3 = 5 \qquad\qquad - 5x_4 \\ x_5 = 1 \end{cases}$ (d) $\begin{cases} x_1 = 5 - 2x_2 - 3x_4 \\ x_3 = 2 \qquad\qquad + 2x_4 \\ x_5 = 4 \end{cases}$

6. (a) $\begin{cases} x = 4 \\ y = 3 \\ z = 2 \\ w = 1 \end{cases}$ (b) No solution

   (d) $\begin{cases} x_1 = -1 - x_3 \\ x_2 = 2 - x_4 \\ x_5 = -1 \end{cases}$

   (c) No solution

7. $A = \begin{bmatrix} 3 & 5 & 7 & 9 \\ 0 & 2 & 4 & 6 \end{bmatrix}$ rank $(A) = 2$

8. (a) For $q = 6$, system inconsistent

   For $q \neq 6$, system consistent; unique solution

(b) For $q = -3$, system consistent; solution not unique

For $q \neq -3$, system consistent; solution unique

(c) For all q, system consistent; solution not unique

(d) For $q = 4$, system inconsistent

For $q \neq 4$, system consistent; solution unique

9. (a) (i) $p = 13$ *and* $q \neq -1$

(ii) $p \neq 13$, *any* choice of q

(iii) $p = 13$ *and* $q = -1$

(b)
$$\begin{cases} x = -2 + 3w \\ y = \quad\;\; - 2w \\ z = \quad 1 - 3w \end{cases}$$

10. (a) rank $(A) = 1$ (b) rank $(A) = 2$

(c) rank $(A) = 3$

Section 1.4

1. 10 bushels type A, 0 type B, 30 type C

2. 25 bushels type A, 0 type B, 25 type C

3. Yes; 6 different solutions:

($x = 6 - w$, $y = 5$, $z = w$, for $w = 0, 1, 2, 3, 4, 5$; where x, y, z, w denote numbers of containers packed in modes A, B, C, D respectively).

4. 8 mode B, 4 mode C containers

5. 50% from Pine Mine; 10% from Cedar; 40% from Rose

6. Milling level 920 tons/day

84.8% from Pine Mine; 15.2% from Cedar; 0% from Rose

7. Maximum silver production 6,800 oz per day

60% ore from Pine Mine; 40% Cedar; 0% Rose

8. Rent 32 type B, 4 type C flatcars.

9. $3,000 maximum.

10. Least admissible M is 136.

11. Market shares: A: $\frac{17}{35}$, B: $\frac{10}{35}$, C: $\frac{8}{35}$

12. B's rate of gain from C would have to be 70% per year.

CHAPTER 2

Section 2.1

1. (a) (6, 10)   (b) (12, 2)
    (c) (−8, 12)   (d) (−2, 44)
    (e) (3, 2, 9)   (f) (8, 1, 4)
    (g) (3, 6, 9)   (h) (5, 2, −2)
    (i) (7, 21, 14, −7) (j) (0, 6, 8, 16)
    (k) 26   (l) 8
    (m) 4   (n) 11

2. (a) (−1, 1)   (b) (1, 2)
    (c) (−2, −2)   (d) (−1, −2)

3. (a) (3, 9, 6)   (b) (0, 5, 7)
    (c) (8, −1, −19)   (d) $\mathbf{z} = \left(\frac{1}{2}, \frac{1}{4}, -\frac{3}{4}\right)$
    (e) $w = (3, 9, 6)$

4. (a) Yes   (b) Yes
    (c) No

7. (a) 45°   (b) $\mathrm{Cos}^{-1}\left(\frac{4}{5}\right) = 36.9°$

Section 2.2

1. (a) $\begin{bmatrix} 9 \\ -2 \end{bmatrix}$ (b) Not defined

    (c) Not defined   (d) $\begin{bmatrix} 5 \\ 13 \end{bmatrix}$

    (e) $\begin{bmatrix} 5 \\ 19 \end{bmatrix}$ (f) Not defined

    (g) $\begin{bmatrix} a + 2d \\ 3a + 4d \end{bmatrix}$ (h) $\begin{bmatrix} a^2 + bd \\ ac + d^2 \end{bmatrix}$

2. (a) $\begin{bmatrix} 4 \\ 0 \end{bmatrix} + b_1 \begin{bmatrix} 3 \\ 1 \end{bmatrix}$ (b) $\begin{bmatrix} 5 \\ 0 \\ 7 \end{bmatrix} + b_1 \begin{bmatrix} 3 \\ 1 \\ 0 \end{bmatrix}$

    (c) $\begin{bmatrix} 1 \\ 0 \\ 5 \\ 0 \\ 0 \end{bmatrix} + b_1 \begin{bmatrix} 4 \\ 1 \\ 0 \\ 0 \\ 0 \end{bmatrix} + b_2 \begin{bmatrix} 3 \\ 0 \\ -2 \\ 1 \\ 0 \end{bmatrix} + b_3 \begin{bmatrix} 1 \\ 0 \\ 6 \\ 0 \\ 1 \end{bmatrix}$

    (d) $b_1 \begin{bmatrix} -1 \\ 1 \\ 0 \end{bmatrix} + b_2 \begin{bmatrix} 8 \\ 0 \\ 1 \end{bmatrix}$

3. From section 1.2

#1. (c) b_1 $\begin{bmatrix} 2 \\ 1 \end{bmatrix}$

#2. (b) b_1 $\begin{bmatrix} -2 \\ -2 \\ 1 \end{bmatrix}$

#2. (c) b_1 $\begin{bmatrix} -2 \\ 1 \\ 0 \end{bmatrix}$

#2. (d) b_1 $\begin{bmatrix} 3 \\ 1 \\ 0 \end{bmatrix}$ $+$ b_2 $\begin{bmatrix} -2 \\ 0 \\ 1 \end{bmatrix}$

#2. (e) b_1 $\begin{bmatrix} \frac{55}{4} \\ -42 \\ 1 \\ 0 \end{bmatrix}$ $+$ b_2 $\begin{bmatrix} -\frac{37}{4} \\ 21 \\ 0 \\ 1 \end{bmatrix}$

#2. (f) b_1 $\begin{bmatrix} -\frac{4}{3} \\ 0 \\ 1 \\ 0 \\ 0 \end{bmatrix}$ $+$ b_2 $\begin{bmatrix} -\frac{2}{3} \\ -1 \\ 0 \\ 1 \\ 0 \end{bmatrix}$ $+$ b_3 $\begin{bmatrix} -\frac{1}{3} \\ 0 \\ 0 \\ 0 \\ 1 \end{bmatrix}$

3. From section 1.3

#2. (d) $\begin{bmatrix} 6 \\ -2 \\ 0 \end{bmatrix}$ $+$ b_1 $\begin{bmatrix} -9 \\ 3 \\ 1 \end{bmatrix}$ #2. (e) $\begin{bmatrix} -4 \\ 0 \\ 8 \end{bmatrix}$ $+$ b_1 $\begin{bmatrix} -\frac{3}{2} \\ 1 \\ 0 \end{bmatrix}$

#2. (f) $\begin{bmatrix} 4 \\ -1 \\ 0 \end{bmatrix}$ $+$ b_1 $\begin{bmatrix} -1 \\ 0 \\ 1 \end{bmatrix}$

#4. (a) $\begin{bmatrix} -8 \\ 3 \\ 0 \\ 0 \end{bmatrix}$ $+$ b_1 $\begin{bmatrix} 2 \\ -1 \\ 1 \\ 0 \end{bmatrix}$ $+$ b_2 $\begin{bmatrix} -5 \\ 1 \\ 0 \\ 1 \end{bmatrix}$

#4. (b) $\begin{bmatrix} -3 \\ 0 \\ 0 \\ 2 \\ 0 \end{bmatrix}$ $+$ b_1 $\begin{bmatrix} -5 \\ 1 \\ 0 \\ 0 \\ 0 \end{bmatrix}$ $+$ b_2 $\begin{bmatrix} -3 \\ 0 \\ 1 \\ 0 \\ 0 \end{bmatrix}$ $+$ b_3 $\begin{bmatrix} -5 \\ 0 \\ 0 \\ 1 \\ 1 \end{bmatrix}$

4. (a) $\mathbf{x} = \mathbf{v} + b_1\mathbf{u}_1 + b_2\mathbf{u}_2$, where $\mathbf{v} = (0, 3, 0, 4)$
$$\mathbf{u}_1 = (1, 0, 0, 0), \mathbf{u}_2 = (0, -2, 1, 0)$$

(b) $\mathbf{x} = \mathbf{v} + b_1\mathbf{u}_1 + b_2\mathbf{u}_2$, where $\mathbf{v} = (0, 7, 0)$
$$\mathbf{u}_1 = (1, 0, 0), \mathbf{u}_2 = (0, 0, 1)$$

(c) $\mathbf{x} = \mathbf{v}$, where $\mathbf{v} = (5, 8)$

Section 2.3

1. (a) $T((1, 0)) = (1, 0)$ $T((2, 0)) = (4, 0)$
   (b) Let $t = 2$. $T(2(1, 0)) = T((2, 0)) = (4, 0)$
      so $T(2(1, 0)) \neq 2T((1, 0)) = 2(1, 0) = (2, 0)$

3. $A(\mathbf{x} + \mathbf{y}) = \begin{bmatrix} 45 \\ 51 \end{bmatrix}$, $A\mathbf{x} = \begin{bmatrix} 10 \\ 13 \end{bmatrix}$, $A\mathbf{y} = \begin{bmatrix} 35 \\ 38 \end{bmatrix}$
   $A\mathbf{x} + A\mathbf{y} = \begin{bmatrix} 45 \\ 51 \end{bmatrix} = A(\mathbf{x} + \mathbf{y})$, $A(t\mathbf{x}) = \begin{bmatrix} 10t \\ 13t \end{bmatrix} = t(A\mathbf{x})$

4. (a) $(12, 10)$ (b) $(6, 5)$
   (c) $(-4, 4)$ (d) $(-2, 2)$
   (e) $(2, 53)$ (f) $(48, 51)$

5. $(1, -8)$

6. (a) Ker (A) consists of all vectors parallel to $(-7, 2)$.

7. (b) $A\mathbf{x} = \begin{bmatrix} x_1 + 8x_2 \\ 2x_1 + 5x_2 \\ 3x_1 + 7x_2 \end{bmatrix} = x_1 \begin{bmatrix} 1 \\ 2 \\ 3 \end{bmatrix} + x_2 \begin{bmatrix} 8 \\ 5 \\ 7 \end{bmatrix} = x_1 \mathbf{z}_1 + x_2 \mathbf{z}_2$

9. $A = m \begin{bmatrix} 2 & -4 \\ 5 & -10 \end{bmatrix}$, m parameter

10. No

Section 2.4

1. (a) $\mathbf{u} = (4, 2, 1)$, $\mathbf{v} = (4, 1, 1)$, $\mathbf{w} = (8, 3, 2)$

2. (a) $\mathbf{u} = (1, -1, 2)$ (b) $\mathbf{v} = (4, -4, 8)$ (c) $C(6, -1, 13)$

3. (a) $C(2,3,0)$, $B(1,2,-4)$

4. (a) $P(-3,7,-1)$ (b) $Q(-7,9,-3)$ (c) $R(-23,17,-11)$

5. $P(2,3,-2)$

7. (a) $\mathbf{x} = (2,3) + t (3,4)$ (b) $\begin{cases} x_1 = 2 + 3t \\ x_2 = 3 + 4t \end{cases}$
   (c) $4x_1 - 3x_2 + 1 = 0$

8. (a) $\mathbf{x} = (7, 1, 4) + t(-4, 1, 1)$ (b) $\begin{cases} x_1 = 7 - 4t \\ x_2 = 1 + t \\ x_3 = 4 + t \end{cases}$
   (c) $\dfrac{x_1 - 7}{-4} = \dfrac{x_2 - 1}{1} = \dfrac{x_3 - 4}{1}$

9. Point of intersection is $P(1,4,2)$.

10. No point of intersection

11. $2x_1 + x_2 + 3x_3 = 33$

12. (a) $\mathbf{n} = (7, 2, -4)$
    (b) $(7, 2, -4) \cdot (x_1 - 3, x_2, x_3) = 0$

13. $2x_1 + 3x_2 - x_3 = 4$

14. $7x_1 + x_2 + 5x_3 = 0$

16. Point of intersection is $(-5, -6, -4)$.

17. No common point when $p = -4$.

CHAPTER 3

Section 3.1

1. (a) $\begin{bmatrix} 0 & 3 \\ 5 & 29 \end{bmatrix}$ (b) $\begin{bmatrix} 2 & 3 \\ 3 & -1 \end{bmatrix}$

    (c) $\begin{bmatrix} 28 & 6 & 17 \\ 8 & 0 & 4 \\ 0 & 4 & 3 \end{bmatrix}$ (d) $\begin{bmatrix} 1 & 0 & 3 & 5 \\ 2 & 3 & 9 & 16 \\ 5 & 0 & 15 & 25 \end{bmatrix}$

    (e) $[9 \quad 9]$ (f) $[23]$

    (g) $\begin{bmatrix} 7 & 4 & 1 \\ 12 & 8 & 4 \end{bmatrix}$ (h) $\begin{bmatrix} 41 & 10 \\ 6 & 0 \end{bmatrix}$

    (i) $\begin{bmatrix} a + 3c & b + 3d \\ 5a + 4c & 5b + 4d \end{bmatrix}$ (j) $\begin{bmatrix} xa - y & xb + 4y \\ 2a - 3 & 2b + 12 \end{bmatrix}$

2. (a) $2B = \begin{bmatrix} 2 & 0 & 4 \\ -6 & 2 & 8 \end{bmatrix}$ $A + 2B = \begin{bmatrix} 3 & 2 & 7 \\ -2 & 7 & 14 \end{bmatrix}$

        $3(A + 2B) = \begin{bmatrix} 9 & 6 & 21 \\ -6 & 21 & 42 \end{bmatrix}$

    (b) $3A = \begin{bmatrix} 3 & 6 & 9 \\ 12 & 15 & 18 \end{bmatrix}$ $6B = \begin{bmatrix} 6 & 0 & 12 \\ -18 & 6 & 24 \end{bmatrix}$

        $3A + 6B = \begin{bmatrix} 9 & 6 & 21 \\ -6 & 21 & 42 \end{bmatrix}$

3. (a) $AB = \begin{bmatrix} 4 & 4 & 1 \\ 5 & 19 & 3 \end{bmatrix}$ $(AB)C = \begin{bmatrix} 20 & 17 \\ 46 & 51 \end{bmatrix}$

  $BC = \begin{bmatrix} 16 & 17 \\ 2 & 0 \end{bmatrix}$ $A(BC) = \begin{bmatrix} 20 & 17 \\ 46 & 51 \end{bmatrix}$

  (b) No

4. (a) $\begin{bmatrix} -23 & 12 \\ -12 & 19 \end{bmatrix}$ (b) $\begin{bmatrix} -6 & 14 \\ -2 & 2 \end{bmatrix}$

  (c) $\begin{bmatrix} -17 & -2 \\ -10 & 17 \end{bmatrix}$

5. $A^2B + AB^2 - BA^2 - BAB$

6. (d) $(5^6)I$

8. (a) $\begin{bmatrix} 0 & 1 \\ 0 & 0 \end{bmatrix}$ (b) $\begin{bmatrix} 1 & 0 \\ 0 & 0 \end{bmatrix}$

  (c) $\begin{bmatrix} 0 & 1 \\ -1 & 0 \end{bmatrix}$

  (There are many solutions to each of parts a, b, and c.)

9. (b) $\begin{bmatrix} 0 & \frac{1}{4} \\ 4 & 0 \end{bmatrix}$

12. $\pm \begin{bmatrix} 0 & i \\ -i & 0 \end{bmatrix}$

13. (a) 0, $\begin{bmatrix} 0 & b \\ 0 & 0 \end{bmatrix}$, $\begin{bmatrix} 0 & 0 \\ b & 0 \end{bmatrix}$, $\begin{bmatrix} a & b \neq 0 \\ \frac{-a^2}{b} & -a \end{bmatrix}$, a,b parameters

  (b) $0, I$, $\begin{bmatrix} 0 & b \\ 0 & 1 \end{bmatrix}$, $\begin{bmatrix} 0 & 0 \\ b & 1 \end{bmatrix}$, $\begin{bmatrix} 1 & b \\ 0 & 0 \end{bmatrix}$, $\begin{bmatrix} 1 & 0 \\ b & 0 \end{bmatrix}$,

  $\begin{bmatrix} a & b \neq 0 \\ \frac{a(1-a)}{b} & 1-a \end{bmatrix}$, a, b parameters

  (c) $\pm iI$, $\pm \begin{bmatrix} i & b \\ 0 & -i \end{bmatrix}$, $\pm \begin{bmatrix} i & 0 \\ b & -i \end{bmatrix}$, $\pm \begin{bmatrix} a & b \neq 0 \\ \frac{-(1+a^2)}{b} & -a \end{bmatrix}$

Section 3.2

1. (a) $N = [1, -1]$ (c) $[458 \quad 881.8 \quad 494]$

  (d) \$ 3,631,600

2. (a) $\begin{bmatrix} \frac{1}{4} & \frac{1}{2} & 1 \\ \frac{3}{4} & 0 & 0 \\ 0 & \frac{1}{2} & 0 \end{bmatrix}$ (b) $0\%, 0\%, 37.5\%, 9.4\%$

(c) $\mathbf{x} = a(8, 6, 3)$

(d) $\left(\frac{1}{16}\right) \begin{bmatrix} 7 & 10 & 4 \\ 3 & 6 & 12 \\ 6 & 0 & 0 \end{bmatrix}$

(e) No

3. (a)

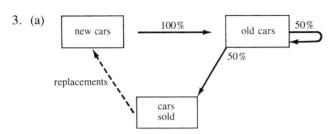

(b) $\begin{bmatrix} 0 & \frac{1}{2} \\ 1 & \frac{1}{2} \end{bmatrix}$

4. (a) $\left(\frac{1}{10}\right) \begin{bmatrix} 7 & 2 & 1 \\ 1 & 6 & 0 \\ 2 & 2 & 9 \end{bmatrix}$ (b) 3 years

6. (a) $G = \left(\frac{1}{10}\begin{bmatrix} 5 & 1 \\ 5 & 9 \end{bmatrix}\right) \left(\frac{1}{10}\begin{bmatrix} 8 & 3 \\ 2 & 7 \end{bmatrix}\right)^3 = \begin{bmatrix} .36 & .31 \\ .64 & .69 \end{bmatrix}$

   (b) $G = \left(\left(\frac{1}{10}\begin{bmatrix} 5 & 1 \\ 5 & 9 \end{bmatrix}\right)\left(\frac{1}{10}\begin{bmatrix} 8 & 3 \\ 2 & 7 \end{bmatrix}\right)\right)^2 = \begin{bmatrix} .304 & .264 \\ .696 & .736 \end{bmatrix}$

Section 3.3

1. (a) $M(8; 2)$ (b) $A(4; 1, 2)$
   (c) Not elementary   (d) Not elementary
   (e) $E(1, 3)$ (f) $A(-5; 3, 2)$
   (g) Not elementary   (h) $M(-1; 2)$

2. $Q_1 = \begin{bmatrix} 1 & 0 \\ -2 & 1 \end{bmatrix}$ $Q_2 = \begin{bmatrix} 1 & 1 \\ 0 & 1 \end{bmatrix}$ $Q_3 = \begin{bmatrix} 1 & 0 \\ 0 & -\frac{1}{3} \end{bmatrix}$

   $Q_1' = \begin{bmatrix} 1 & 0 \\ 2 & 1 \end{bmatrix}$ $Q_2' = \begin{bmatrix} 1 & -1 \\ 0 & 1 \end{bmatrix}$ $Q_3' = \begin{bmatrix} 1 & 0 \\ 0 & -3 \end{bmatrix}$

3. $Q_1 = E(1, 3)$, $Q_2 = A(-3; 2, 3)$, $Q_3 = A(-4; 3, 2)$, $Q_4 = M(\frac{1}{2}; 2)$
   $Q_1' = E(1, 3)$, $Q_2' = A(\ 3; 2, 3)$, $Q_3' = A(\ 4; 3, 2)$, $Q_4' = M(2; 2)$

4. (b) P differs from the identity matrix in all three rows:
       entry a_{ii} in each row i is 0, not 1.
   (c) $P^2 = \begin{bmatrix} 0 & 0 & 1 \\ 1 & 0 & 0 \\ 0 & 1 & 0 \end{bmatrix}$, $P^3 = I$
   (d) $P = E(3, 2) \, E(2, 1)$

5. (a) $E(1, 2) E(3, 4) = E(3, 4) E(1, 2) = \begin{bmatrix} 0 & 1 & 0 & 0 \\ 1 & 0 & 0 & 0 \\ 0 & 0 & 0 & 1 \\ 0 & 0 & 1 & 0 \end{bmatrix}$

   using 4×4 case as example

   (b) $E(1, 2) E(2, 3) = \begin{bmatrix} 0 & 0 & 1 \\ 1 & 0 & 0 \\ 0 & 1 & 0 \end{bmatrix}$,

   $E(2, 3) E(1, 2) = \begin{bmatrix} 0 & 1 & 0 \\ 0 & 0 & 1 \\ 1 & 0 & 0 \end{bmatrix}$ using 3×3 case as example

   (c) $\begin{bmatrix} b_{11} & 0 \\ 0 & b_{22} \end{bmatrix} \begin{bmatrix} c_{11} & 0 \\ 0 & c_{22} \end{bmatrix} = \begin{bmatrix} c_{11} & 0 \\ 0 & c_{22} \end{bmatrix} \begin{bmatrix} b_{11} & 0 \\ 0 & b_{22} \end{bmatrix} = \begin{bmatrix} b_{11}c_{11} & 0 \\ 0 & b_{22}c_{22} \end{bmatrix}$

   (d) Type M elementary matrices are diagonal matrices.

6. $Q_1 = A(-2; 1, 3)$, $Q_2 = A(-4; 2, 1)$

8. $Q_1 = \begin{bmatrix} 1 & 0 \\ 1 & 1 \end{bmatrix}$, $Q_2 = \begin{bmatrix} 1 & -\frac{1}{2} \\ 0 & 1 \end{bmatrix}$

9. $Q_1 = \begin{bmatrix} 0 & 1 \\ 1 & 0 \end{bmatrix}$, $Q_2 = \begin{bmatrix} 1 & 0 \\ 0 & \frac{1}{2} \end{bmatrix}$

Section 3.4

1. (a) $\begin{bmatrix} -7 & 4 \\ 2 & -1 \end{bmatrix}$ (b) $\begin{bmatrix} -2 & \frac{5}{4} \\ 1 & -\frac{1}{2} \end{bmatrix}$

   (c) $\left(\frac{1}{25}\right) \begin{bmatrix} 3 & 4 \\ 4 & -3 \end{bmatrix}$ (d) $\left(\frac{1}{169}\right) \begin{bmatrix} 5 & 12 \\ -12 & 5 \end{bmatrix}$

   (e) $\left(\frac{1}{5}\right) \begin{bmatrix} 36 & -21 & 11 \\ -15 & 10 & -5 \\ 1 & -1 & 1 \end{bmatrix}$ (f) $\begin{bmatrix} -2 & 7 & 0 \\ 1 & -3 & 0 \\ 0 & -2 & \frac{1}{2} \end{bmatrix}$

   (g) $\begin{bmatrix} -2 & 0 & 1 \\ 0 & 3 & -2 \\ 1 & -2 & 1 \end{bmatrix}$ (h) $\begin{bmatrix} -2 & -1 & 12 \\ 1 & 1 & -8 \\ \frac{1}{2} & 0 & -\frac{3}{2} \end{bmatrix}$

   (i) $\begin{bmatrix} 1 & 0 & -4 \\ -6 & 1 & 24 \\ -1 & 0 & 5 \end{bmatrix}$ (j) $\begin{bmatrix} 1 & -2 & 1 \\ 0 & 1 & -2 \\ 0 & 0 & 1 \end{bmatrix}$

   (k) $\begin{bmatrix} 3 & -\frac{5}{2} & \frac{1}{2} \\ -3 & 4 & -1 \\ 1 & -\frac{3}{2} & \frac{1}{2} \end{bmatrix}$ (l) No inverse

   (m) $\begin{bmatrix} 0 & 0 & 0 & \frac{1}{3} \\ 0 & 0 & \frac{1}{2} & 0 \\ 1 & 0 & 0 & 0 \\ 0 & 1 & 0 & 0 \end{bmatrix}$

3. (a) case $p = -5$: rank $(B) = 2$, no inverse.

   case $p \neq -5$: rank $(B) = 3$, inverse exists.

   (b) When $p \neq -5$, $B^{-1} = (p + 5)^{-1} \begin{bmatrix} -2 & 2p & p+1 \\ 1 & -p & 2 \\ 1 & 5 & 2 \end{bmatrix}$

4. $B^{-1} = \begin{bmatrix} 1 & -p & pr-q \\ 0 & 1 & -r \\ 0 & 0 & 1 \end{bmatrix}$

5. $B(3I - B^6) = I$, so $B^{-1} = 3I - B^6$

6. (a) If A has the inverse B, then A^g has an inverse B^g, which is impossible since $A^g = O$.

   (b) Multiply out the product:

   $(I - aA)(I + aA + a^2A^2 + \cdots + a^{g-1}A^{g-1}) = I$.

   Checks.

7. (a) $B^{-1} = \begin{bmatrix} 0 & \frac{5}{4} & 0 \\ 0 & 0 & 1 \\ 1 & -\frac{1}{4} & 0 \end{bmatrix}$

   (b) Last year's fleet vector: $(45, 80, 600)$

   Two years ago: $(100, 600, 25)$

   (c) Three years ago, from the formula $x_{n-3} = B^{-1}x_{n-2}$, the fleet vector would have been $(750, 25, -50)$, which doesn't make sense since the number of third-year cars, computed as -50, cannot be negative.

8. (a) $B^{-1} = (\tfrac{1}{5}) \begin{bmatrix} 7 & -3 \\ -2 & 8 \end{bmatrix}$ (b) 90%

9. (b) When $p + q = 1$, $B = \begin{bmatrix} 1-p & 1-p \\ p & p \end{bmatrix}$; therefore

   rank $(B) = 1$ and B has no inverse. Otherwise, the 2×2 inversion formula gives

   $B^{-1} = (p + q - 1)^{-1} \begin{bmatrix} q-1 & q \\ p & p-1 \end{bmatrix}$.

   (c) No, x_0 cannot be deduced from x_1 when $p + q = 1$, because in this case $x_1 = (1-p, p)$ no matter what x_0 was.

11. (c) $AB = (FS)B = F(SB) = FG = I$

    (d) $S = \begin{bmatrix} 4 & 0 & -1 \\ -1 & 0 & \frac{1}{3} \\ -2 & 1 & 0 \end{bmatrix}$, inverse $A = \begin{bmatrix} 4 & 0 & -1 \\ -1 & 0 & \frac{1}{3} \end{bmatrix}$

    Note: A is nonunique.

    (e) $A'B = (F'S)B = F'(SB) = F'G = I$

Section 3.5

1. $AB = \begin{bmatrix} 3 & 2 \\ 5 & 4 \end{bmatrix} \begin{bmatrix} 2 & 3 & -1 \\ 4 & 0 & 7 \end{bmatrix} = \begin{bmatrix} 14 & 9 & 11 \\ 26 & 15 & 23 \end{bmatrix}$

   $AB^t = \begin{bmatrix} 14 & 26 \\ 9 & 15 \\ 11 & 23 \end{bmatrix}$ $B^tA^t = \begin{bmatrix} 2 & 4 \\ 3 & 0 \\ -1 & 7 \end{bmatrix} \begin{bmatrix} 3 & 5 \\ 2 & 4 \end{bmatrix} = \begin{bmatrix} 14 & 26 \\ 9 & 15 \\ 11 & 23 \end{bmatrix}$

2. $A^t = \begin{bmatrix} 3 & 5 \\ 2 & 4 \end{bmatrix}$ $(A^t)^{-1} = \left(\tfrac{1}{2}\right) \begin{bmatrix} 4 & -5 \\ -2 & 3 \end{bmatrix}$

   $A^{-1} = \left(\tfrac{1}{2}\right) \begin{bmatrix} 4 & -2 \\ -5 & 3 \end{bmatrix}$ $(A^{-1})^t = \left(\tfrac{1}{2}\right) \begin{bmatrix} 4 & -5 \\ -2 & 3 \end{bmatrix}$

3. (a) $AC^t = B$, $C^t = A^{-1}B = \left(\tfrac{1}{2}\right) \begin{bmatrix} 4 & -2 \\ -5 & 3 \end{bmatrix} \begin{bmatrix} 2 & 3 & -1 \\ 4 & 0 & 7 \end{bmatrix}$

   $= \begin{bmatrix} 0 & 6 & -9 \\ 1 & -\tfrac{15}{2} & 13 \end{bmatrix}$, or $C = \begin{bmatrix} 0 & 1 \\ 6 & -\tfrac{15}{2} \\ -9 & 13 \end{bmatrix}$

   (b) $CA = B^t$ or $C = B^tA^{-1} = \begin{bmatrix} 2 & 4 \\ 3 & 0 \\ -1 & 7 \end{bmatrix} \left(\tfrac{1}{2}\right) \begin{bmatrix} 4 & -2 \\ -5 & 3 \end{bmatrix}$

   $= \begin{bmatrix} -6 & 4 \\ 6 & -3 \\ -\tfrac{39}{2} & \tfrac{23}{2} \end{bmatrix}$

4. (a) (i) $A^t = (B + B^t) = B^t + B^{tt} = B^t + B = A$; A symmetric

   (ii) $A^t = (BB^t)^t = B^{tt}B^t = BB^t = A$; A symmetric

   (iii) $A^t = (B - B^t)^t = B^t - B^{tt} = B^t - B = -A$; A antisymmetric

   (iv) $A^t = (B^tB)^t = B^tB^{tt} = B^tB = A$; A symmetric

   (b) $A = \tfrac{1}{2}(A + A^t) + \tfrac{1}{2}(A - A^t)$

5. $U^{-1} = \begin{bmatrix} \tfrac{1}{2} & -\tfrac{3}{2} \\ 0 & \tfrac{1}{3} \end{bmatrix}$.

6. $a_{11} = 1$, $a_{12} = -4$, $a_{32} = 0$, $a_{33} = \tfrac{1}{3}$

*9. $(A^n)_{12} = (A^n)_{23} = na$

   $(A^n)_{13} = nb + \left(\tfrac{1}{2}\right)n(n-1)a^2$

CHAPTER 4

Section 4.1

1. (a) $B = \begin{bmatrix} 1 & 2 \\ -1 & 1 \\ 1 & 1 \end{bmatrix}$ (b) $A = \begin{bmatrix} 1 & -1 & -1 \\ -3 & 2 & 5 \end{bmatrix}$

(c) $S \circ T$ is represented by $AB = \begin{bmatrix} 1 & 0 \\ 0 & 1 \end{bmatrix} = I_2$

(d) $T \circ S$ is represented by $BA = \begin{bmatrix} -5 & 3 & 9 \\ -4 & 3 & 6 \\ -2 & 1 & 4 \end{bmatrix}$, etc.

2. (a) $\begin{bmatrix} 44 & 47 \\ 30 & 32 \end{bmatrix}$ (b) $\begin{bmatrix} 7 & -10 \\ -2 & 3 \end{bmatrix}$

(c) $\begin{bmatrix} 36 & 43 \\ -10 & -12 \end{bmatrix}$

3. (a) $F' = \begin{bmatrix} -1 & 0 \\ 0 & 1 \end{bmatrix}$ $P' = \begin{bmatrix} 0 & 0 \\ 0 & 1 \end{bmatrix}$

4. (a) $T((1, 0)) = (-1, -1)$ (b) $A = \begin{bmatrix} -1 & 3 \\ -1 & 4 \end{bmatrix}$
$T((0, 1)) = (3, 4)$

6. (a) $P = \begin{bmatrix} 1 & 0 & 0 \\ 0 & 1 & 0 \\ 0 & 0 & 0 \end{bmatrix}$ (b) $F = \begin{bmatrix} 1 & 0 & 0 \\ 0 & 1 & 0 \\ 0 & 0 & -1 \end{bmatrix}$

Section 4.2

1. (a) $R_{45°} = \left(\frac{1}{\sqrt{2}}\right) \begin{bmatrix} 1 & -1 \\ 1 & 1 \end{bmatrix}$, $F_{60°} = \left(\frac{1}{2}\right) \begin{bmatrix} -1 & \sqrt{3} \\ \sqrt{3} & 1 \end{bmatrix}$

1. (a) $P_{30°} = \left(\frac{1}{4}\right) \begin{bmatrix} 3 & \sqrt{3} \\ \sqrt{3} & 1 \end{bmatrix}$

(b) $R_{-30°} = \left(\frac{1}{2}\right) \begin{bmatrix} \sqrt{3} & 1 \\ -1 & \sqrt{3} \end{bmatrix}$ $F_{120°} = \left(\frac{1}{2}\right) \begin{bmatrix} -1 & -\sqrt{3} \\ -\sqrt{3} & 1 \end{bmatrix}$

$P_{-90°} = \begin{bmatrix} 0 & 0 \\ 0 & 1 \end{bmatrix}$

3. (a) $R_{-90°}$ or $R_{270°}$ (b) $P_{-45°}$ or $P_{135°}$
(c) $F_{-45°}$ or $F_{135°}$

4. $\alpha = n(90°)$, n an integer

5. (a) $P = P_{90°} = \begin{bmatrix} 0 & 0 \\ 0 & 1 \end{bmatrix}$ (b) $R_{90°} = \begin{bmatrix} 0 & -1 \\ 1 & 0 \end{bmatrix}$

$F_{90°} = \begin{bmatrix} -1 & 0 \\ 0 & 1 \end{bmatrix}$

6. (a) $T = \begin{bmatrix} \frac{3}{10} & -\frac{1}{10} \\ \frac{9}{10} & -\frac{3}{10} \end{bmatrix}$ (b) No (c) $T \circ T = 0$

CHAPTER 5

Section 5.1

1. (a) −2
   (c) 7
   (e) 32
   (g) −6
   (i) 4
   (k) 6

   (b) −1
   (d) 0
   (f) 0
   (h) 3
   (j) 1

2. det $M = 5p - 10$, M invertible unless $p = 2$

4. (a) $C = \begin{bmatrix} 4 & 6 \\ 1 & 3 \end{bmatrix}$, $D = \begin{bmatrix} 3 & 2 \\ 1 & 4 \end{bmatrix}$

   (b) A, B, C, D have determinants 1, 6, 6, 10 respectively.

8. $p = 2$ or 4

9. det $A = 16$ regardless of the unknown entries.

10. (c) $\mathbf{x} = (-3, 6, -3)$

Section 5.2

1. (a) 31
   (c) −8
   (e) −540

   (b) −3
   (d) −306
   (f) 72

3. (b) 25

   (c) 1,600

8. −16

Section 5.3

1. (a) 3
   (c) $\frac{1}{3}$
   (e) 24

   (b) 9
   (d) 3

2. (a) 1
   (c) 0

   (b) 1

3. $A = \begin{bmatrix} 2 & 0 \\ 0 & \frac{1}{2} \end{bmatrix}$

CHAPTER 6

Section 6.1

1. (a) l.i.   (b) l.d.; $-a\mathbf{v}_1 + 2a\mathbf{v}_2 - 3a\mathbf{v}_3 = \mathbf{0}$
   (c) l.i.   (d) l.d.; $-a\mathbf{v}_1 - a\mathbf{v}_2 + a\mathbf{v}_3 = \mathbf{0}$
   (e) l.i.   (f) l.d.; $-2a\mathbf{v}_1 + a\mathbf{v}_2 = \mathbf{0}$

3. The set is l.d. only for $p = 5$; then $\mathbf{v}_3 = 3\mathbf{v}_1 - 4\mathbf{v}_2$.

Section 6.2

1. (a) Basis; $(-19, 9)$ (b) Not a basis

2. (a) Basis; $(1, 4, -2)$ (b) Not a basis
   (c) Basis; $(-1, -1, 3)$

3. (b) $(2, 3)$ (c) $(4, 7, 8)$ is not in Ker(A)

4. (b) $(4, -2)$ (c) $(3, 0, 6)$ is not in Range(A)

7. (b) A basis for S is $\mathbf{u}_1 = (-2, 1, 0)$, $\mathbf{u}_2 = (-4, 0, 1)$; dim $(S) = 2$

8. (b) A basis for S is $\mathbf{u}_1 = (-3, 5, 1, 0, 0)$, $\mathbf{u}_2 = (0, -3, 0, 1, 0)$,
   $\mathbf{u}_3 = (-4, 3, 0, 0, 1)$; dim $(S) = 3$

Section 6.3

1. *Note:* the bases for Range(A) and Ker(A) are not unique.

   (a) dim(Range(A)) $= 1$; a basis is $(2, 4)$
       dim(Ker(A)) $= 1$; a basis is $(1, 0)$
       $\mathbf{x}$ is in Ker(A), $\mathbf{y}$ is not in Range(A).
   (b) dim(Range(A)) $= 2$; a basis is $(1, 2)$, $(1, 3)$.
       dim(Ker(A)) $= 1$; a basis is $(1, -1, 0)$.
       $\mathbf{x}$ is not in Ker(A), $\mathbf{y}$ is in Range(A).
   (c) dim(Range(A)) $= 2$; a basis is $(1, 1, 1)$, $(1, 2, 3)$.
       dim(Ker(A)) $= 0$; no basis
       $\mathbf{x}$ is not in Ker(A), $\mathbf{y}$ is not in Range(A).
   (d) dim(Range(A)) $= 2$; a basis is $(1, 2, 3)$, $(0, 0, 1)$.
       dim(Ker(A)) $= 1$; a basis is $(-3, 0, 1)$.
       $\mathbf{x}$ is in Ker(A), $\mathbf{y}$ is in Range(A).
   (e) dim(Range(A)) $= 1$; a basis is $(1, 2, 8)$.
       dim(Ker(A)) $= 2$; a basis is $(-1, 0, 1)$, $(-1, 1, 0)$.
       $\mathbf{x}$ is not in Ker(A), $\mathbf{y}$ is not in Range(A).
   (f) dim(Range(A)) $= 2$; a basis is $(1, 3, 4)$, $(1, 0, 5)$.
       dim(Ker(A)) $= 2$; a basis is $(-2, 1, 0, 0)$, $(-1, 0, -1, 1)$.
       $\mathbf{x}$ is in Ker(A), $\mathbf{y}$ is not in Range(A).

2. (a) $0 \le \dim(\text{Range}(A)) \le 14$ (b) $3 \le \dim(\text{Ker}(A)) \le 17$
   (c) $\dim(\text{Ker}(A)) = 6$

3. (a) $0 \le \dim(\text{Range}(A)) \le 15$ (b) $0 \le \dim(\text{Ker}(A)) \le 15$
   (c) $\dim(\text{Ker}(A)) = 4$

4. (a) $\dim(\text{Range}(A)) = 2$ or 3; $\dim(\text{Ker}(A)) = 2$ or 1
   (b) $\mathbf{x} = (1, 1, -1, 0)$ (not unique)

5. (a) Basis $\mathbf{v}_1, \mathbf{v}_2, \mathbf{v}_4$; $\mathbf{w} = 3\mathbf{v}_1 + 2\mathbf{v}_2 - \mathbf{v}_4$
   (b) Basis $\mathbf{v}_1, \mathbf{v}_2, \mathbf{v}_4$; $\mathbf{w} = 3\mathbf{v}_1 - 9\mathbf{v}_2 + 6\mathbf{v}_4$
   (c) Basis $\mathbf{v}_1, \mathbf{v}_2, \mathbf{v}_4$; $\mathbf{w} = -2\mathbf{v}_1 - \mathbf{v}_2 - 2\mathbf{v}_4$
   (d) Basis $\mathbf{v}_1, \mathbf{v}_2$; $\mathbf{w}$ not in S
   (e) Basis $\mathbf{v}_1, \mathbf{v}_2$; $\mathbf{w} = 3\mathbf{v}_1 + 2\mathbf{v}_2$
   (f) Basis $\mathbf{v}_1, \mathbf{v}_2, \mathbf{v}_3$; $\mathbf{w} = \mathbf{v}_1 - 2\mathbf{v}_2 + \mathbf{v}_3$

6. (a) All vectors parallel to $(3, -1, 1, -3)$
   (b) All vectors parallel to $(1, 0, 0, 1)$
   (c) All vectors parallel to $(1, 2, 8)$

9. (a) $P = \begin{bmatrix} 8 & 5 \\ -3 & -2 \end{bmatrix}$ (b) $P = \begin{bmatrix} 2 & 3 \\ 1 & 1 \end{bmatrix}$

10. (c) $P = \begin{bmatrix} 1 & 1 & 2 \\ 1 & 1 & 0 \\ 1 & 0 & 0 \end{bmatrix}$

Section 6.4

1. (a) $A' = -\frac{1}{7} \begin{bmatrix} 19 & 41 \\ -10 & -19 \end{bmatrix}$ (b) $A'' = \frac{1}{7} \begin{bmatrix} 196 & 245 \\ -157 & -196 \end{bmatrix}$

2. $A' = \begin{bmatrix} 24 & 12 \\ -38 & -19 \end{bmatrix}$

4. $\mathbf{v}_1 = (1, 0, 1)$ $\mathbf{v}_2 = (0, 1, 1)$ $A = \begin{bmatrix} 0 & 1 \\ 1 & -1 \end{bmatrix}$
   $\mathbf{w}_1 = (1, 2, 0)$ $\mathbf{w}_2 = (0, 0, 1)$

5. $\mathbf{v}_1 = (1, 4, 4)$ $\mathbf{v}_2 = (-1, -3, 0)$ $A = \begin{bmatrix} 5 & 8 \\ -1 & -3 \end{bmatrix}$

CHAPTER 7

Section 7.1

1. (a) $\mathbf{u}_1 = \frac{1}{\sqrt{10}}(3, 1)$ $\mathbf{u}_2 = \frac{1}{\sqrt{10}}(1, -3)$
   $c_1 = \frac{7}{\sqrt{10}}$; $c_2 = -\frac{11}{\sqrt{10}}$

(b) $\mathbf{u}_1 = \frac{1}{13}(5, 12)$ $\quad\quad\quad\quad\quad$ $\mathbf{u}_2 = \frac{1}{13}(-12, 5)$

$\quad\quad$ $c_1 = \frac{99}{13}; c_2 = -\frac{1}{13}$

(c) $\mathbf{u}_1 = \frac{1}{9}(1, 8, -4)$ $\quad\quad\quad\quad$ $\mathbf{u}_2 = \frac{1}{9}(8, 1, 4)$

$\quad\quad$ $\mathbf{u}_3 = \frac{1}{9}(4, -4, -7)$ $\quad\quad\quad$ $c_1 = \frac{187}{9}; c_2 = \frac{218}{9}; c_3 = -\frac{17}{9}$

(d) $\mathbf{u}_1 = \frac{1}{3}(1, -2, 2)$ $\quad\quad\quad\quad$ $\mathbf{u}_2 = \frac{1}{15}(2, 11, 10)$

$\quad\quad$ $\mathbf{u}_3 = \frac{1}{15}(-14, -2, 5)$ $\quad\quad$ $c_1 = 0; c_2 = 1; c_3 = -2$

(e) $\mathbf{u}_1 = \frac{1}{\sqrt{2}}(1, 1, 0, 0)$ $\quad\quad\quad$ $\mathbf{u}_2 = \frac{1}{\sqrt{2}}(0, 0, 1, 1)$

$\quad\quad$ $\mathbf{u}_3 = \frac{1}{2}(-1, 1, 1, -1)$ $\quad\quad$ $\mathbf{u}_4 = \frac{1}{2}(-1, 1, -1, 1)$

$\quad\quad$ $c_1 = 0; c_2 = 0; c_3 = 2; c_4 = 4.$

2. (a) $\mathbf{u}_1 = \frac{1}{\sqrt{2}}(0, 1, 1)$ $\quad\quad\quad\quad$ $\mathbf{u}_2 = \frac{1}{\sqrt{6}}(2, -1, 1)$

$\quad\quad$ $\mathbf{w}$ is in S: coordinates are $c_1 = \frac{1}{\sqrt{2}}$ $c_2 = \frac{9}{\sqrt{6}}$

(b) $\mathbf{u}_1 = \frac{1}{7}(2, 3, 6)$ $\quad\quad\quad\quad\quad$ $\mathbf{u}_2 = \frac{1}{7}(-3, 6, -2)$

$\quad\quad$ $\mathbf{w}$ is not in S

(c) $\mathbf{u}_1 = \frac{1}{11}(2, 6, 9)$ $\quad\quad\quad\quad$ $\mathbf{u}_2 = \frac{1}{11}(6, 7, -6)$

$\quad\quad$ $\mathbf{w}$ is not in S

(d) $\mathbf{u}_1 = \frac{1}{2}(1, 1, 1, 1)$ $\quad\quad\quad\quad$ $\mathbf{u}_2 = \frac{1}{2}(1, 1, -1, -1)$

$\quad\quad$ $\mathbf{u}_3 = \frac{1}{2}(1, -1, 1, -1)$

$\quad\quad$ $\mathbf{w}$ is in S; coordinates are $c_1 = 7; c_2 = 3; c_3 = 2$

3. (a) $\mathbf{u}_1 = \left(\frac{1}{\sqrt{5}}\right)(-2, 1, 0); \mathbf{u}_2 = \left(\frac{1}{\sqrt{30}}\right)(-1, -2, 5)$ is one solution; many solutions are possible.

(b) $\mathbf{u}_1 = \left(\frac{1}{\sqrt{6}}\right)(1, 2, 1, 0); \mathbf{u}_2 = \left(\frac{1}{\sqrt{21}}\right)(2, -2, 2, 3)$ is one solution; many solutions are possible.

5. (a) $\mathbf{u}_1 = \left(\frac{1}{9}\right)(4, 4, 7); \mathbf{u}_2 = \left(\pm \frac{1}{\sqrt{2}}\right)(1, -1, 0); \mathbf{u}_3 = \left(\pm \frac{1}{9\sqrt{2}}\right)$
$\quad$ $(7, 7, -8)$

(b) Exactly four such bases exist; only the choice of plus or minus signs for $\mathbf{u}_2$ and $\mathbf{u}_3$ is at our disposal.

Section 7.2

1. $T = \frac{1}{13}\begin{bmatrix} 5 & -12 \\ 12 & 5 \end{bmatrix}$ $\quad\quad$ or $\frac{1}{13}\begin{bmatrix} 5 & 12 \\ 12 & -5 \end{bmatrix}$

2. $T = \frac{1}{13}\begin{bmatrix} 3 & 12 & 4 \\ 4 & 3 & -12 \\ 12 & -4 & 3 \end{bmatrix}$ $\quad\quad$ or $\frac{1}{11}\begin{bmatrix} 3 & -12 & 4 \\ 4 & -3 & -12 \\ 12 & 4 & 3 \end{bmatrix}$

3. $T = \frac{1}{2}\begin{bmatrix} 1 & \sqrt{3} \\ \sqrt{3} & -1 \end{bmatrix}$ $\quad\quad$ or $\frac{1}{2}\begin{bmatrix} 1 & -\sqrt{3} \\ -\sqrt{3} & -1 \end{bmatrix}$

$\quad$ or $\frac{1}{2}\begin{bmatrix} 1 & -\sqrt{3} \\ \sqrt{3} & 1 \end{bmatrix}$ $\quad\quad$ or $\frac{1}{2}\begin{bmatrix} 1 & \sqrt{3} \\ -\sqrt{3} & 1 \end{bmatrix}$

4. (a) $T = \begin{bmatrix} 0 & 0 & 1 \\ 0 & \pm 1 & 0 \\ 1 & 0 & 0 \end{bmatrix}$

(b) $T = \begin{bmatrix} 0 & 1 & 0 & 0 \\ \frac{4}{5} & 0 & 0 & \frac{3}{5} \\ 0 & 0 & 1 & 0 \\ \frac{3}{5} & 0 & 0 & -\frac{4}{5} \end{bmatrix}$ or $\begin{bmatrix} 0 & 1 & 0 & 0 \\ -\frac{4}{5} & 0 & 0 & \frac{3}{5} \\ 0 & 0 & 1 & 0 \\ \frac{3}{5} & 0 & 0 & \frac{4}{5} \end{bmatrix}$

or $\begin{bmatrix} 0 & 1 & 0 & 0 \\ \frac{4}{5} & 0 & 0 & \frac{3}{5} \\ 0 & 0 & 1 & 0 \\ -\frac{3}{5} & 0 & 0 & \frac{4}{5} \end{bmatrix}$ or $\begin{bmatrix} 0 & 1 & 0 & 0 \\ -\frac{4}{5} & 0 & 0 & \frac{3}{5} \\ 0 & 0 & 1 & 0 \\ -\frac{3}{5} & 0 & 0 & -\frac{4}{5} \end{bmatrix}$

6. (a) $T = \begin{bmatrix} 1 & 0 \\ 0 & -1 \end{bmatrix}$ or $\begin{bmatrix} 0 & -1 \\ 1 & 0 \end{bmatrix}$

(b) None exist

15. Basis for V: $\mathbf{u}_1$, $\mathbf{u}_2$; dim $V = 2$
Basis for W: $(3, -1, 1, 0)$, $(-11, 3, 0, 1)$; dim $W = 2$

Section 7.3

1. (b) $c_1 = i$, $c_2 = 5$

2. (b) $\mathbf{u}_1 = \left(\frac{1}{3}\right)(2+i, 2)$; $\mathbf{u}_2 = \left(\frac{1}{3}\right)(2i, -1-2i)$
(c) $c_1 = 2 + 5i$, $c_2 = -5i - 6$

3. (a) $\mathbf{u}_1 = \left(\frac{1}{3}\right)(1, 2, 2i)$; $\mathbf{u}_2 = \left(\frac{1}{3}\right)(-2, 2, -i)$

4. $U = \begin{bmatrix} \frac{3}{5} & -\frac{4i}{5} \\ \frac{4}{5} & \frac{3i}{5} \end{bmatrix}$, or $U = \begin{bmatrix} \frac{3}{5} & -ib \\ b^* & \frac{3i}{5} \end{bmatrix}$, where $|b| = \frac{4}{5}$

CHAPTER 8

Section 8.1

1. (a) $A\mathbf{v}_1 = (1, -1)$; $A\mathbf{v}_2 = (33, 77)$

(b) $\lambda_1 = 1$; $\lambda_2 = 11$ (c) $A^{100}\mathbf{v}_1 = \mathbf{v}_1$

(d) $A^m\mathbf{v}_2 = 11^m\mathbf{v}_2$

2. (a) $\lambda_1 = 1$; $\lambda_2 = -1$ (b) $c_1 = 3$; $c_2 = -4$
(c) $\begin{bmatrix} 15 - 8\,(-1)^m \\ -9 + 4\,(-1)^m \end{bmatrix}$ (d) $(23, -13)$

3. (a) $\lambda^2 - 4\lambda - 5 = 0$
(b) $\lambda_1 = 5$; $\lambda_2 = -1$; $\mathbf{v}_1 = (1, 2)$; $\mathbf{v}_2 = (1, -1)$
(c) $5^m\mathbf{v}_1 + 3(-1)^m\mathbf{v}_2$

4. $\mathbf{v}_1 = (1, 0, 0)$; $\mathbf{v}_2 = (3, 1, 0)$; $\mathbf{v}_3 = (10, 4, 1)$

5. $\theta = 0°$, $180°$, or any integer multiple of $180°$.

6. (a) $\lambda_1 = 3$; $\lambda_2 = -2$; $\mathbf{v}_1 = (2, 1)$; $\mathbf{v}_2 = (1, -2)$
   (b) $\lambda_1 = 2$; $\mathbf{v}_1 = (4, -3)$
   (c) No real eigenvalues; no eigenvectors
   (d) $\lambda_1 = 19$; $\lambda_2 = -2$; $\mathbf{v}_1 = (1, 2)$; $\mathbf{v}_2 = (-8, 5)$
   (e) $\lambda_1 = 8$; $\lambda_2 = -4$; $\mathbf{v}_1 = (1, 2)$; $\mathbf{v}_2 = (-5, 2)$
   (f) $\lambda_1 = 3$; $\mathbf{v}_1 = (-2, 1)$

7. (a) $\lambda_1 = -1$; $\lambda_2 = 1$; $\lambda_3 = 4$; $\mathbf{v}_1 = (-1, 3, 0)$; $\mathbf{v}_2 = (1, -1, 6)$;
   $\mathbf{v}_3 = (1, 2, 0)$
   (b) $\lambda_1 = 3$; no other real eigenvalues; $\mathbf{v}_1 = (1, 0, 0)$
   (c) $\lambda_1 = 1$; $\lambda_2 = 1$; $\lambda_3 = 2$; $\mathbf{v}_1 = (1, 0, -1)$; $\mathbf{v}_2 = (1, 1, -1)$;
   $\mathbf{v}_3 = (-1, 1, 2)$
   (d) $\lambda_1 = 2$; $\lambda_2 = 3$; $\mathbf{v}_1 = (1, 0, 0)$; $\mathbf{v}_2 = (0, 1, 0)$
   (e) $\lambda_1 = 0$; $\mathbf{v}_1 = (1, 2, -2)$
   (f) $\lambda_1 = 1$; $\lambda_2 = 4$; $\lambda_3 = -4$; $\mathbf{v}_1 = (1, -1, 1)$; $\mathbf{v}_2 = (1, 2, 1)$;
   $\mathbf{v}_3 = (1, 0, -1)$
   (g) $\lambda_1 = 1$; $\lambda_2 = 3$; $\lambda_3 = 7$; $\mathbf{v}_1 = (3, 3, -2)$; $\mathbf{v}_2 = (0, 1, 1)$;
   $\mathbf{v}_3 = (0, 0, 1)$
   (h) $\lambda_1 = -1$; $\lambda_2 = -1$; $\lambda_3 = -2$; $\mathbf{v}_1 = (2, 0, 1)$; $\mathbf{v}_2 = (2, 1, 0)$;
   $\mathbf{v}_3 = (3, -1, 3)$
   (i) $\lambda_1 = 3$; $\lambda_2 = 4$; $\lambda_3 = 5$; $\mathbf{v}_1 = (0, 2, 3)$; $\mathbf{v}_2 = (1, 2, 2)$;
   $\mathbf{v}_3 = (1, 1, 0)$

8. (a) $\lambda_1 = 0$; $\lambda_2 = 0$; $\lambda_3 = -2$; $\lambda_4 = 2$; $\mathbf{v}_1 = (1, 0, 0, -1)$;
   $\mathbf{v}_2 = (0, 1, -1, 0)$; $\mathbf{v}_3 = (1, 1, 1, 1)$; $\mathbf{v}_4 = (-1, 1, 1, -1)$
   (b) $\lambda_1 = -1$; $\lambda_2 = 1$; $\lambda_3 = -3$; $\lambda_4 = 3$; $\mathbf{v}_1 = (1, -1, 1, -1)$;
   $\mathbf{v}_2 = (1, -1, -1, 1)$; $\mathbf{v}_3 = (-1, -1, 1, 1)$; $\mathbf{v}_4 = (1, 1, 1, 1)$

9. (b) There is one new l.i. eigenvector, $\mathbf{y} = (-2, 1, 2)$.

13. Let $A = \begin{bmatrix} 0 & -1 \\ 1 & 0 \end{bmatrix}$. Then $A^2 = -I$; therefore $\mathbf{e}_1$ is an
eigenvector of A^2, but not of A.

Section 8.2

1. (a) $D = \begin{bmatrix} 1 & 0 \\ 0 & -2 \end{bmatrix}$ $\qquad$ $P = \begin{bmatrix} 1 & 1 \\ 4 & 1 \end{bmatrix}$
   (c) $A^{10} = \begin{bmatrix} 1,365 & -341 \\ 1,364 & -340 \end{bmatrix}$

2. (a) $A^m = \frac{1}{10} \begin{bmatrix} 7 & -3 \\ -7 & 3 \end{bmatrix} + \frac{(11^m)}{10} \begin{bmatrix} 3 & 3 \\ 7 & 7 \end{bmatrix}$

(b) $A^m = \begin{bmatrix} -5 & -10 \\ 3 & 6 \end{bmatrix} + (-1)^m \begin{bmatrix} 6 & 10 \\ -3 & -5 \end{bmatrix}$

(c) $A^m = \frac{(5^m)}{3} \begin{bmatrix} 1 & 1 \\ 2 & 2 \end{bmatrix} + \left(\frac{(-1)^m}{3}\right) \begin{bmatrix} 2 & -1 \\ -2 & 1 \end{bmatrix}$

3. (a) c_1-axis parallel to eigenvector $\mathbf{v}_1 = (1, 1)$
   c_2-axis parallel to eigenvector $\mathbf{v}_2 = (-1, 1)$
   Transformation represents reflection across c_1-axis.

   (b) c_1-axis parallel to eigenvector $\mathbf{v}_1 = (1, 2)$
   c_2-axis parallel to eigenvector $\mathbf{v}_2 = (-2, 1)$
   Transformation stretches c_2-coordinate by a factor of 6.

7. (a) $A = PDP^{-1}$ and $B = PGP^{-1}$, where
$$P = \begin{bmatrix} 2 & 1 \\ 1 & -2 \end{bmatrix} \quad, \quad D = \begin{bmatrix} 5 & 0 \\ 0 & -5 \end{bmatrix} \quad, \quad G = \begin{bmatrix} 5 & 0 \\ 0 & 0 \end{bmatrix}$$
   (b) None exists.

Section 8.3

1. $\mathbf{x}_m = \left(\frac{1}{5}\right)^m \begin{bmatrix} 3 \\ 3 \end{bmatrix} + \left(\frac{7}{10}\right)^m \begin{bmatrix} -2 \\ 3 \end{bmatrix}$

2. $\mathbf{x}_m = \left(\frac{11}{10}\right)^m \begin{bmatrix} 2 \\ -3 \end{bmatrix}$

4. The market shares approach $\frac{4}{15}$ for brand A, $\frac{1}{15}$ for brand B, and $\frac{2}{3}$ for brand C.

5. Ultimate age distribution: $\frac{5}{13}$, $\frac{4}{13}$, and $\frac{4}{13}$ of the fleet being first, second, and third year cars, respectively.

8. (a) $\begin{bmatrix} 1 & 1 \\ 0 & 0 \end{bmatrix}$ is a stochastic matrix that is not invertible.

Section 8.4

1. (a) $\mathbf{x}(t) = e^{3t}(3, 1) + 2e^{4t}(2, 1)$
   (b) $\mathbf{x}(t) = 3e^{8t}(1, 2) + e^{-4t}(-5, 2)$
   (c) $\mathbf{x}(t) = (1, 0, 1) - e^{-t}(1, -1, -1)$
   (d) $\mathbf{x}(t) = 2e^{t}(1, 1, -1)$

2. $\mathbf{x}(t) = 50e^{-5t}(1, -2, 1) + 100e^{-4t}(0, 1, -1) + 50(0, 0, 1)$

3. Market shares of X, Y, Z are $\frac{(1-e^{-1})}{2}$, $\frac{(1-e^{-1})}{4}$, $\frac{(1+3e^{-1})}{4}$ respectively.

7. (a) $A = \begin{bmatrix} 0 & 1 \\ 4 & -3 \end{bmatrix}$ (c) $x(t) = 2e^{-4t} + 5e^t$

Section 8.5

1. (a) $P = \dfrac{1}{\sqrt{2}} \begin{bmatrix} 1 & 1 \\ -1 & 1 \end{bmatrix}$ $D = \begin{bmatrix} 2 & 0 \\ 0 & 4 \end{bmatrix}$

   (b) $P = \dfrac{1}{\sqrt{5}} \begin{bmatrix} 1 & -2 \\ 2 & 1 \end{bmatrix}$ $D = \begin{bmatrix} 5 & 0 \\ 0 & -5 \end{bmatrix}$

   (c) $P = \dfrac{1}{\sqrt{5}} \begin{bmatrix} 2 & 1 \\ 1 & -2 \end{bmatrix}$ $D = \begin{bmatrix} 5 & 0 \\ 0 & 10 \end{bmatrix}$

   (d) $P = \dfrac{1}{3} \begin{bmatrix} 2 & 1 & 2 \\ -2 & 2 & 1 \\ -1 & -2 & 2 \end{bmatrix}$ $D = \begin{bmatrix} 1 & 0 & 0 \\ 0 & 4 & 0 \\ 0 & 0 & 7 \end{bmatrix}$

   (e) $P = \dfrac{1}{\sqrt{6}} \begin{bmatrix} \sqrt{3} & 1 & \sqrt{2} \\ 0 & -2 & \sqrt{2} \\ -\sqrt{3} & 1 & \sqrt{2} \end{bmatrix}$ $D = \begin{bmatrix} -1 & 0 & 0 \\ 0 & -1 & 0 \\ 0 & 0 & 2 \end{bmatrix}$

   (f) $P = \begin{bmatrix} \sqrt{3} & \sqrt{2} & 1 \\ 0 & -\sqrt{2} & 2 \\ \sqrt{3} & -\sqrt{2} & -1 \end{bmatrix}$ $D = \begin{bmatrix} 1 & 0 & 0 \\ 0 & -1 & 0 \\ 0 & 0 & 5 \end{bmatrix}$

9. (a) $U = \left(\dfrac{1}{\sqrt{2}}\right) \begin{bmatrix} 1 & 1 \\ -i & i \end{bmatrix}$ (b) $U = \left(\dfrac{1}{\sqrt{26}}\right) \begin{bmatrix} 3+4i & -1 \\ 1 & 3-4i \end{bmatrix}$

   $D = \begin{bmatrix} 1 & 0 \\ 0 & -1 \end{bmatrix}$ $D = \begin{bmatrix} 13 & 0 \\ 0 & -13 \end{bmatrix}$

   (c) $U = \left(\dfrac{1}{\sqrt{6}}\right) \begin{bmatrix} 1+2i & -1 \\ 1 & 1-2i \end{bmatrix}$ (d) $U = \begin{bmatrix} \dfrac{1}{\sqrt{3}} & 0 & \dfrac{2i}{\sqrt{6}} \\ \dfrac{i}{\sqrt{3}} & \dfrac{1}{\sqrt{2}} & \dfrac{1}{\sqrt{6}} \\ \dfrac{i}{\sqrt{3}} & \dfrac{-1}{\sqrt{2}} & \dfrac{1}{\sqrt{6}} \end{bmatrix}$

   $D = \begin{bmatrix} 3 & 0 \\ 0 & -3 \end{bmatrix}$ $D = \begin{bmatrix} -2 & 0 & 0 \\ 0 & 1 & 0 \\ 0 & 0 & 1 \end{bmatrix}$

Section 8.6

1. (a) Ellipse   (b) Hyperbola
   (c) Ellipse   (d) Hyperbola
   (e) Hyperbola   (f) Ellipsoid
   (g) Hyperboloid of two sheets   (h) Hyperboloid of one sheet

2. (b) Hyperbola   (g) Hyperboloid of one sheet
   (h) Hyperboloid of two sheets

5. Ellipse for $k > 7$; hyperbola for $7 > k > -3$.

6. Ellipsoid for $k > 3$

   Hyperboloid of one sheet for $3 > k > 0$

   Hyperboloid of two sheets for $0 > k > -3$

CHAPTER 9

Section 9.1

1. (a) The only eigenvalue is $\lambda = 0$.

   (b) dim $G_0 = 4$, dim $E_0 = 3$.

   (c) 2 strings of length 1; 1 string of length 2.

2. (a) $\lambda = 2$;   dim $G_2 = 2$, dim $E_2 = 1$;
      1 string of length 2

   (b) $\lambda = 2 + 3i$; dim $G_{2+3i} = $ dim $E_{2+3i} = 1$;
      1 string of length 1

      $\lambda = 2 - 3i$; dim $G_{2-3i} = $ dim $E_{2-3i} = 1$;
      1 string of length 1

   (c) $\lambda = 3$;   dim $G_3 = 2$, dim $E_3 = 1$;
      1 string of length 2

   (d) $\lambda = 4$;   dim $G_4 = 3$, dim $E_4 = 2$;
      1 string of length 1, 1 string of length 2

   (e) $\lambda = 2$;   dim $G_2 = $ dim $E_2 = 1$;
      1 string of length 1

      $\lambda = 3$;   dim $G_3 = 2$, dim $E_3 = 1$;
      1 string of length 2

   (f) $\lambda = 0$;   dim $G_0 = 3$, dim $E_0 = 1$;
      1 string of length 3

6. The sole eigenvalue is $\lambda = 0$.

   Compute: $d_1 = 2$, $d_2 = 4$, $d_p = 4$ for $p > 2$;
      $h_1 = 0$, $h_2 = 2$

7. (a) Sole eigenvalue is $\lambda = 0$; $d_1 = 1$, $d_2 = 3$, $d_3 = 4$;
      $d_p = 4$ for $p > 4$; $h_1 = 1$, $h_2 = 0$,
      $h_3 = 1$

   String pattern is $\begin{cases} \mathbf{a}_1 \to \mathbf{b}_1 \to \mathbf{c}_1 \to \mathbf{0} \\ \qquad\qquad \mathbf{c}_2 \to \mathbf{0} \end{cases}$

   (b) For eigenvalue $\lambda = 2$; $d_1 = 1$, $d_2 = 2$, $d_p = 2$ for $p > 2$;
      $h_1 = 0$, $h_2 = 1$

   String pattern is: $\mathbf{a}_1 \to \mathbf{b}_1 \to \mathbf{0}$

   For eigenvalue $\lambda = 0$; $d_1 = 1$, $d_2 = 2$, $d_p = 2$ for $p > 2$;
      $h_1 = 0$, $h_2 = 1$

   String pattern is: $\mathbf{a}_2 \to \mathbf{b}_2 \to \mathbf{0}$

Section 9.2

1. (Jordan forms for matrices from section 9.1)

   exercise 1 $J = \begin{bmatrix} \boxed{\begin{matrix} 0 & 1 \\ 0 & 0 \end{matrix}} & & \\ & \boxed{0} & \\ & & \boxed{0} \end{bmatrix}$

   exercise 2 (a) $J = \begin{bmatrix} 2 & 1 \\ 0 & 2 \end{bmatrix}$ (b) $J = \left[\begin{array}{c|c} 2+3i & 0 \\ \hline 0 & 2-3i \end{array}\right]$

                    (c) $J = \begin{bmatrix} 3 & 1 \\ 0 & 3 \end{bmatrix}$ (d) $J = \left[\begin{array}{cc|c} 4 & 1 & \\ 0 & 4 & \\ \hline & & 4 \end{array}\right]$

                    (e) $J = \left[\begin{array}{c|cc} 2 & & \\ \hline & 3 & 1 \\ & 0 & 3 \end{array}\right]$ (f) $J = \begin{bmatrix} 0 & 1 & 0 \\ 0 & 0 & 1 \\ 0 & 0 & 0 \end{bmatrix}$

   exercise 6 $J = \left[\begin{array}{cc|cc} 0 & 1 & & \\ 0 & 0 & & \\ \hline & & 0 & 1 \\ & & 0 & 0 \end{array}\right]$

   exercise 7 (a) $J = \left[\begin{array}{ccc|c} 0 & 1 & 0 & \\ 0 & 0 & 1 & \\ 0 & 0 & 0 & \\ \hline & & & 0 \end{array}\right]$ (b) $J = \left[\begin{array}{cc|cc} 2 & 1 & & \\ 0 & 2 & & \\ \hline & & 0 & 1 \\ & & 0 & 0 \end{array}\right]$

2. (a) $J = \begin{bmatrix} \boxed{\begin{matrix} 0 & 1 \\ 0 & 0 \end{matrix}} & & \\ & \boxed{0} & \\ & & \boxed{1} \end{bmatrix}$ (b) $J = \left[\begin{array}{cc|cc} 3 & 1 & & \\ 0 & 3 & & \\ \hline & & 4 & 1 \\ & & 0 & 4 \end{array}\right]$

  (c) $J = \begin{bmatrix} \boxed{3} & & \\ & \boxed{3} & \\ & & \boxed{\begin{matrix} 4 & 1 \\ 0 & 4 \end{matrix}} \end{bmatrix}$

3. $J = \left[\begin{array}{ccc|cc|cc|c} 4 & 1 & 0 & & & & & \\ 0 & 4 & 1 & & & & & \\ 0 & 0 & 4 & & & & & \\ \hline & & & 4 & 1 & & & \\ & & & 0 & 4 & & & \\ \hline & & & & & 7 & 1 & \\ & & & & & 0 & 7 & \\ \hline & & & & & & & 7 \end{array}\right]$

4. $J =$
$$\begin{bmatrix} 2 & 1 & 0 & & & & & \\ 0 & 2 & 1 & & & & & \\ 0 & 0 & 2 & & & & & \\ & & & 2 & 1 & & & \\ & & & 0 & 2 & & & \\ & & & & & 5 & 1 & \\ & & & & & 0 & 5 \end{bmatrix}$$

8. $K = \left[\begin{array}{cc|cc} 1 & 2 & 1 & 0 \\ -2 & 1 & 0 & 1 \\ \hline 0 & 0 & 1 & 2 \\ 0 & 0 & -2 & 1 \end{array}\right.$ $\left.\begin{array}{cc} & \\ & \\ 0 & 1 \\ -1 & 0 \end{array}\right]$ $= \left[\begin{array}{c|c} Y_{1+2i} & I_2 \\ \hline O & Y_{1+2i} \end{array}\right.$ $\left.\begin{array}{c} \\ Y_i \end{array}\right]$

Section 9.3

1. (a) $A^2 = \begin{bmatrix} 1 & 2 \\ 0 & 1 \end{bmatrix}$, $A^3 = \begin{bmatrix} 1 & 3 \\ 0 & 1 \end{bmatrix}$, $A^r = \begin{bmatrix} 1 & r \\ 0 & 1 \end{bmatrix}$

   (b) $A^2 = \begin{bmatrix} 9 & 6 \\ 0 & 9 \end{bmatrix}$, $A^3 = \begin{bmatrix} 27 & 27 \\ 0 & 27 \end{bmatrix}$, $A^r = \begin{bmatrix} 3^r & r3^{r-1} \\ 0 & 3^r \end{bmatrix}$

   (c) $A^2 = \begin{bmatrix} 52 & 64 \\ -36 & -44 \end{bmatrix}$, $A^3 = \begin{bmatrix} 152 & 192 \\ -108 & -136 \end{bmatrix}$,

   $A^r = 2^{r-1} \begin{bmatrix} 2+12r & 16r \\ -9r & 2-12r \end{bmatrix}$

   (d) $A^2 = \begin{bmatrix} 4 & 4 & 0 \\ 0 & 4 & 0 \\ 0 & 0 & 4 \end{bmatrix}$, $A^3 = \begin{bmatrix} 8 & 12 & 0 \\ 0 & 8 & 0 \\ 0 & 0 & 8 \end{bmatrix}$, $A^r = \begin{bmatrix} 2^r & r2^{r-1} & 0 \\ 0 & 2^r & 0 \\ 0 & 0 & 2^r \end{bmatrix}$

   (e) $A^2 = \begin{bmatrix} 1 & 2 & 1 \\ 0 & 1 & 2 \\ 0 & 0 & 1 \end{bmatrix}$, $A^3 = \begin{bmatrix} 1 & 3 & 3 \\ 0 & 1 & 3 \\ 0 & 0 & 1 \end{bmatrix}$, $A^r = \begin{bmatrix} 1 & r & \frac{1}{2}r(r-1) \\ 0 & 1 & r \\ 0 & 0 & 1 \end{bmatrix}$

2. (a) $(\lambda - 1)^2$ (b) $(\lambda - 3)^2$ (c) $(\lambda - 2)^2$
   (d) $(\lambda - 2)^2$ (e) $(\lambda - 1)^3$

11. (a) $P(t) = e^{3t} \begin{bmatrix} 1 & t \\ 0 & 1 \end{bmatrix}$, $\mathbf{x}(t) = e^{3t} \begin{bmatrix} 4 + 7t \\ 7 \end{bmatrix}$

    (b) $P(t) = e^{2t} \begin{bmatrix} 1 + 12t & 16t \\ -9t & 1 - 12t \end{bmatrix}$, $\mathbf{x}(t) = e^{2t} \begin{bmatrix} 4 \\ -3 \end{bmatrix}$

    (c) $P(t) = e^t \begin{bmatrix} 1 & t & \frac{t^2}{2} \\ 0 & 1 & t \\ 0 & 0 & 1 \end{bmatrix}$, $\mathbf{x}(t) = e^t \left(\frac{t^2}{2}, t, 1\right)$

(d) $P(t) = e^{5t} \begin{bmatrix} 1 + t + \frac{t^2}{2} & t + \frac{t^2}{2} & \frac{t^2}{2} \\ -t^2 & 1 - t^2 & t - t^2 \\ -t & -t & 1 - t \end{bmatrix}$

$\mathbf{x}(t) = e^{5t}(t + \frac{t^2}{2}, 1 - t^2, -t)$

CHAPTER 10

Section 10.1

1. (a) $[A|\mathbf{c}] \rightarrow [U|\mathbf{c}'] = \begin{bmatrix} 1 & 4 & | & 4 \\ 0 & -2 & | & 4 \end{bmatrix}$;

   back substitution gives $x_2 = 12$, $x_1 = -2$.

   (b) $[A|\mathbf{c}] \rightarrow [U|\mathbf{c}'] = \begin{bmatrix} 1 & 2 & 1 & | & 2 \\ 0 & 2 & 2 & | & 1 \\ 0 & 0 & 2 & | & -4 \end{bmatrix}$;

   back substitution gives $x_3 = -2, x_2 = \frac{5}{2}, x_1 = -1$.

2. (a) $L_1 = \begin{bmatrix} 1 & 0 & 0 \\ -2 & 1 & 0 \\ 0 & 0 & 1 \end{bmatrix}$, $L_2 = \begin{bmatrix} 1 & 0 & 0 \\ 0 & 1 & 0 \\ -4 & 0 & 1 \end{bmatrix}$, $L_3 = \begin{bmatrix} 1 & 0 & 0 \\ 0 & 1 & 0 \\ 0 & 1 & 1 \end{bmatrix}$

   $D_1 = I,$ $D_2 = \begin{bmatrix} 1 & 0 & 0 \\ 0 & \frac{1}{2} & 0 \\ 0 & 0 & 1 \end{bmatrix}$, $D_3 = \begin{bmatrix} 1 & 0 & 0 \\ 0 & 1 & 0 \\ 0 & 0 & \frac{1}{2} \end{bmatrix}$

   $U_1 = \begin{bmatrix} 1 & 0 & -1 \\ 0 & 1 & 0 \\ 0 & 0 & 1 \end{bmatrix}$, $U_2 = \begin{bmatrix} 1 & 0 & 0 \\ 0 & 1 & -1 \\ 0 & 0 & 1 \end{bmatrix}$, $U_3 = \begin{bmatrix} 1 & -2 & 0 \\ 0 & 1 & 0 \\ 0 & 0 & 1 \end{bmatrix}$

   (b) $\mathbf{x} = U_3 U_2 U_1 D_3 D_2 D_1 L_3 L_2 L_1 \mathbf{c} = (1, 1, -2)$

3. (a) *LDU* decomposition: $L = \begin{bmatrix} 1 & 0 \\ 3 & 1 \end{bmatrix}$, $D = \begin{bmatrix} 1 & 0 \\ 0 & -2 \end{bmatrix}$, $U = \begin{bmatrix} 1 & 4 \\ 0 & 1 \end{bmatrix}$

   LU decomposition: $L = \begin{bmatrix} 1 & 0 \\ 3 & 1 \end{bmatrix}$, $U = \begin{bmatrix} 1 & 4 \\ 0 & -2 \end{bmatrix}$

   (b) *LDU* decomposition:

   $L = \begin{bmatrix} 1 & 0 & 0 \\ 2 & 1 & 0 \\ 4 & -1 & 1 \end{bmatrix}$, $D = \begin{bmatrix} 1 & 0 & 0 \\ 0 & 2 & 0 \\ 0 & 0 & 2 \end{bmatrix}$, $U = \begin{bmatrix} 1 & 2 & 1 \\ 0 & 1 & 1 \\ 0 & 0 & 1 \end{bmatrix}$

   LU decomposition: $L = \begin{bmatrix} 1 & 0 & 0 \\ 2 & 1 & 0 \\ 4 & -1 & 1 \end{bmatrix}$, $U = \begin{bmatrix} 1 & 2 & 1 \\ 0 & 2 & 2 \\ 0 & 0 & 2 \end{bmatrix}$

4. (b) $P = \begin{bmatrix} 0 & 1 \\ 1 & 0 \end{bmatrix}$, $L = I, D = I, U = I$

1. (a) $\mathbf{x} = (.1300, 10.00)$

   (b) $\mathbf{x} = (.03222, -3.444, 2.222)$

2. (b) $\mathbf{x} = (0, 10.00)$ (false solution to 1(a))

3. (a) $\lambda_1 = 6, \mathbf{v}_1 = (1, 1)$

      $\lambda_2 = -1, \mathbf{v}_2 = (5, -2)$

   (b) $\lambda_1 = 13, \mathbf{v}_1 = (7, 8)$

      $\lambda_2 = -2, \mathbf{v}_2 = (1, -1)$

   (c) $\lambda_1 = 10, \mathbf{v}_1 = (0, 0, 1)$

      $\lambda_2 = 2, \mathbf{v}_2 = (16, 8, -1)$

      $\lambda_3 = -1, \mathbf{v}_3 = (11, -11, 1)$

4. (b) The matrix A is nondiagonalizable; has only 1 eigenvalue $\lambda = 1$
   and 1 l.i. eigenvector. Its Jordan form is

$$J = \begin{bmatrix} 1 & 1 \\ 0 & 1 \end{bmatrix} \quad , \quad \text{and} \quad J^m = \begin{bmatrix} 1 & m \\ 0 & 1 \end{bmatrix}$$

   The iterates $J^m\mathbf{v}$ lead to a scalar times $\mathbf{v}_1 = (1,0)$ regardless of
   the choice of $\mathbf{v}$.

CHAPTER 11

Section 11.1

1. (a & b)

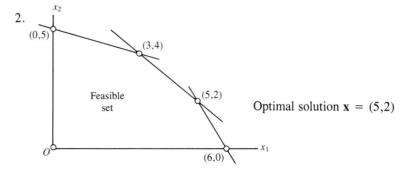

   (c) $f(0,0) = 0, f(0,5) = 10, f(3,4) = 17, f(7,0) = 21$

   (d) Optimal solution $\mathbf{x} = (7,0)$

2.

Optimal solution $\mathbf{x} = (5,2)$

3. Initial tableau $T = \begin{bmatrix} 1 & 3 & | & 1 & 0 & | & 15 \\ 1 & 1 & | & 0 & 1 & | & 7 \\ \hline 3 & 2 & | & 0 & 0 & | & 0 \end{bmatrix}$ $\theta_1 = 15$
$\theta_2 = 7$

$T' = \begin{bmatrix} 0 & 2 & | & 1 & -1 & | & 8 \\ 1 & 1 & | & 0 & 1 & | & 7 \\ \hline 0 & -1 & | & 0 & -3 & | & -21 \end{bmatrix}$ = final tableau

Optimal solution $(x_1,x_2) = (7,0)$

4. Initial Tableau $T = \begin{bmatrix} 1 & 3 & | & 1 & 0 & 0 & | & 15 \\ 1 & 1 & | & 0 & 1 & 0 & | & 7 \\ 2 & 1 & | & 0 & 0 & 1 & | & 12 \\ \hline 3 & 2 & | & 0 & 0 & 0 & | & 0 \end{bmatrix}$

Final tableau $T'' = \begin{bmatrix} 0 & 0 & | & 1 & -5 & 2 & | & 4 \\ 0 & 1 & | & 0 & 2 & -1 & | & 2 \\ 1 & 0 & | & 0 & -1 & 1 & | & 5 \\ \hline 0 & 0 & | & 0 & -1 & -1 & | & -19 \end{bmatrix}$

Optimal solution $(x_1, x_2) = (5,2)$

5. (a) $(x_1,x_2) = (3,2)$ (c) $(x_1,x_2,x_3) = \left(\frac{15}{4}, 0, \frac{1}{4}\right)$
   (b) $(x_1,x_2) = (2,1)$ (d) $(x_1,x_2,x_3) = (3,4,5)$

6. (a & b)

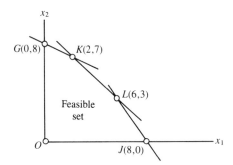

(c) $f(0,0) = 0$, $f(0,8) = 32$, $f(2,7) = 34$, $f(6,3) = 30$,
    $f(8,0) = 24$.
(d) Optimal solution is the point $K(2,7)$.

7. Price of Strong Blend would have to fall below $2 a pound.

9. Plant 0 acres tomatoes, 40 acres corn, 10 acres lettuce.

10. Plant 20 acres tomatoes, 30 acres corn, 0 acres lettuce.

11.

		Value of Parameter g (Total Fertilizer Supply)			
		$g > 200$	$200 > g > 100$	$100 > g > 50$	$50 > g$
Acres	Tomatoes	0	0	$100 - g$	g
planted	Corn	0	$100 - \frac{g}{2}$	$g - 50$	0
	Lettuce	50	$\frac{g}{2} - 50$	0	0

12. Plant $\frac{40}{3}$ acres tomatoes, 20 acres corn, $\frac{50}{3}$ acres lettuce.

Section 11.2

1. Solutions for the dual problems to primal problems from exercises 11.1:

For #3, $(y_1, y_2) = (0, 3)$

For #4, $(y_1, y_2, y_3) = (0, 1, 1)$

For #5(a), $(y_1, y_2) = \left(\frac{6}{5}, \frac{7}{5}\right)$

For #5(b), $(y_1, y_2) = \left(\frac{3}{2}, \frac{1}{2}\right)$

For #5(c), $(y_1, y_2) = \left(\frac{9}{4}, \frac{11}{8}\right)$

For #5(d), $(y_1, y_2, y_3) = \left(\frac{12}{5}, \frac{17}{5}, \frac{2}{5}\right)$

2. (a) The dual problem to example 3 of section 11.1 is the following:

minimize $g(\mathbf{y}) = y_1 + 3y_2$,

subject to the conditions
$$\begin{cases} y_1 - y_2 \geq 3 \\ -2y_1 + y_2 \geq 2 \\ \mathbf{y} \geq \mathbf{0} \end{cases}.$$

(b)

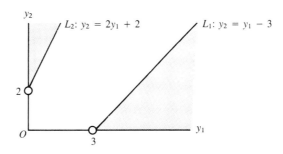

To be in the feasible set for the dual problem, the point $\mathbf{y}$ must be above L_2 and below L_1, which is impossible. So the feasible set for the dual problem is empty.

3. (a) Shadow prices: fertilizer, $50 per ton
              acreage, $600 per acre

(b) Shadow prices: fertilizer, $300 per ton
              acreage, $100 per acre

4.

| | | \multicolumn{4}{c}{**Value of Parameter g (Total Fertilizer Supply)**} |
		$g > 200$	$200 > g > 100$	$100 > g > 50$	$50 > g$
Shadow	Fertilizer	0	50	300	400
prices	Acreage	800	600	100	0

5. Maximum value of f is 19, optimal solution $(x_1, x_2) = (3,7)$.

6. Maximum value of f is 7, optimal solution $(x_1, x_2) = (3,2)$.

7. Maximum value of f is $\frac{5}{2}$, optimal solution $(x_1, x_2, x_3) = \left(\frac{1}{4}, 0, \frac{3}{4}\right)$.

8. Maximum value of f is -4, optimal solution $(x_1, x_2) = (2,2)$.

9. Factory A should ship 300 engines per week to plant Q, while factory B should ship 400 and 400 engines per week to plants P and Q respectively.

Index